Fifty Years of Cytochrome P450 Research

Hiroshi Yamazaki
Editor

Fifty Years of Cytochrome P450 Research

Springer

Editor
Hiroshi Yamazaki
Showa Pharmaceutical University
Tokyo, Japan

ISBN 978-4-431-54991-8 ISBN 978-4-431-54992-5 (eBook)
DOI 10.1007/978-4-431-54992-5
Springer Tokyo Heidelberg New York Dordrecht London

Library of Congress Control Number: 2014939417

Printed on acid-free paper

Springer is part of Springer Science+Business Media (www.springer.com)

Preface

The year 2012 celebrates the 50th anniversary of the first publication on cytochrome P450 (P450) by Tsuneo Omura and Ryo Sato, "A New Cytochrome in Liver Microsomes," in *The Journal of Biological Chemistry* (237:1375–1376) in 1962. Fifty years of research on P450s have revealed that many forms of P450 exist in animals, plants, and microorganisms. Research on P450s has expanded into many different fields ranging from molecules to human bodies by attracting biochemists, structural biologists, and pharmaceutical/medical researchers. P450s have drawn the attention of industry for bioengineering applications such as drug development as well as the creation of the "blue rose" as a new product. The basic catalytic mechanism of P450 is the activation of oxygen to a reactive form in a complex cycle (see Fig. 2.1). The research on nuclear receptors, which has extended outward from the research on the regulatory mechanisms of the P450 genes, has become a very important area in biology, medical science, pharmacology, and/or clinical medicines. P450 research has developed tremendously, from the early studies with rat livers to personalized medicines for individual patients in the twenty-first century (for a timeline of major discoveries, see Fig. 3.1).

One of the authors of that first article in *The Journal of Biological Chemistry*, Dr. Tsuneo Omura, now Professor Emeritus, Kyushu University, is actively reviewing a variety of research areas that have flourished with P450s to date. The readers of this book will find in the Epilogue a special letter from the editors of *The Journal of Biological Chemistry* to Dr. Omura with remarks on the 50th anniversary of this seminal work. It should also be noted that the contribution of researchers in this field worldwide has been significant in the last 50 years. Taking the unique opportunity provided by this 50th anniversary, a scientific symposium was conducted with leading scientists around the world in Kyushu, Japan, in December 2012 (Fig. 1).

The success of the research has had tremendous implications in fields such as medicine, agriculture, and biotechnology. The planning for this book was in three parts: to collect a comprehensive coverage of major progress, to discuss future

Fig. 1 Participants at the 50th Anniversary Symposium on P450 in Fukuoka, Japan, held December 2–3, 2012

directions of the research on P450s, and to invite young researchers to join this important and exciting world of P450. I hope that this book will encourage many young students and postdocs in P450 research to try new approaches, and as an editor I will be greatly pleased if our book can support their research work.

Tokyo, Japan Hiroshi Yamazaki

Contents

Part I
Cytochrome P450 History

Chapter 1
Pioneers in the Early Years of Cytochrome P450 Research

Tsuneo Omura

Abstract Since the discovery of "cytochrome P450" in 1962, many biochemists have intensively studied various problems of this unique hemoprotein and contributed to the elucidation of its molecular properties and physiological functions. This review chapter describes the important contributions by the pioneer scientists in the early years of research on P450. A few important events that preceded and were pertinent to the discovery of P450 are also described.

Keywords Coordination state of P450 heme • Cytochrome P450 • Discovery of P450 • P450 • P450 enzyme systems • Physiological function of P450 • Pioneers in P450 research • Primary structure of P450 • Purification of P450s

1.1 Microsomes and Oxygenases

Two important discoveries, one in cell biology and another in enzymology, preceded the discovery of cytochrome P450 (P450). A. Claude and his collaborators of the Rockefeller Institute for Medical Research, New York, established an efficient method for cell fractionation of animal tissues in the 1940s and discovered a novel subcellular fraction, which was named "microsomes" by Claude in 1943 (Claude 1943). G.E. Palade and P. Siekevitz investigated the morphological and biochemical properties of microsomes isolated from rat liver and identified them as the fragmented vesicles of the endoplasmic reticulum of the hepatocytes (Palade and Siekevitz 1956).

The discovery of the novel subcellular fraction, microsomes, attracted the attention of many biochemists, and they found NADPH-cytochrome *c* reductase (Horecker 1950), cytochrome b_5 (Strittmatter and Ball 1951), and NADH-cytochrome b_5 reductase

T. Omura (✉)
Kyushu University, Fukuoka, Japan
e-mail: omurat@mxs.mesh.ne.jp

H. Yamazaki (ed.), *Fifty Years of Cytochrome P450 Research*,
DOI 10.1007/978-4-431-54992-5_1, © Springer Japan 2014

(Strittmatter and Velick 1956a) in liver microsomes. The possible contribution of these microsomal NAD(P)H-linked electron-transfer components to cellular respiration was speculated and examined, but no solid evidence for their physiological functions was obtained in the 1950s. Another NADH-reducible component, which could bind carbon monoxide in the reduced form, was spectrally detected in rat liver microsomes (Klingenberg 1958), but its molecular nature was not elucidated.

The discovery of "oxygenase" by H.S. Mason, W.L. Folks, and E. Peterson of the University of Oregon Medical School, Portland, and by O. Hayaishi, M. Katagiri, and S. Rothberg of the National Institute of Health, Bethesda, in 1955 was another important prelude to the discovery of P450. Although the oxygenase reactions they independently discovered were catalyzed by a mushroom enzyme, phenolase (Mason et al. 1955), and by a bacterial enzyme, pyrocatecase (Hayaishi et al. 1955), respectively, the contribution of oxygenase-type reactions to several important metabolic activities in animal tissues was soon found. Steroid hormone biosynthesis in the adrenal gland (Hayano et al. 1955) and the oxidative metabolism of drugs in the liver (Axelrod 1955) were found to be oxygenase-type reactions, and intensive studies were carried out to isolate and purify the oxygenases responsible for these reactions. However, the oxygenases involved in these metabolic activities of animal tissues were apparently membrane bound, and were found to be highly labile to various solubilization treatments. The nature of these oxygenases remained unknown until the discovery of P450.

1.2 Discovery of "Cytochrome P450"

M. Klingenberg was studying the spectral properties of rat liver microsomes in the laboratory of B. Chance at the Johnson Research Foundation of the University of Pennsylvania, Philadelphia, when he found a novel carbon monoxide-binding pigment in the microsomes in 1955. The microsomes showed a prominent optical absorption peak at 450 nm when carbon monoxide was added to the microsomal suspension reduced with dithionite or NADH. The carbon monoxide-difference spectrum of the novel microsomal pigment was not similar to any of the known colored proteins, and the pigment seemed to be highly labile to various solubilization treatments. He found that the content of protoheme in the microsomes was much more in excess over the amount of cytochrome b_5, which was the only hemoprotein known to be present in liver microsomes at that time, but the molecular nature of the novel pigment was not elucidated. His spectral study on "microsomal carbon monoxide-binding pigment" was published in 1958 (Klingenberg 1958).

When Klingenberg was working in Chance's laboratory, he had two good friends, a visitor from Japan, R. Sato and an American postdoc, R.W. Estabrook, who happened to be in the same laboratory at that time. Klingenberg left the Johnson Research Foundation in 1956 and went back to Germany to work on a different research subject.

Sato came back to Japan in 1956 and was promoted to professor at the Institute for Protein Research of the Osaka University, Osaka, in 1958. He invited T. Omura, who was in the Shizuoka University at that time, to join his new laboratory. Sato was interested in the physiological function of microsomal electron-transfer components, and asked Omura, who came to his laboratory in 1960, to investigate the molecular nature of "microsomal carbon monoxide-binding pigment" found by Klingenberg, who described in his paper (Klingenberg 1958) that the pigment was highly labile to solubilization treatments and its characteristic carbon monoxide-binding spectrum, which showed a prominent peak at 450 nm, disappeared completely upon the addition of detergents to microsomes.

Omura repeated the experiments described in Klingenberg's paper using rabbit liver microsomes and unexpectedly found that the disappearance of the 450-nm peak of the carbon monoxide-binding spectrum of the microsomal pigment upon treatment with detergents was accompanied by the parallel appearance of a new spectral peak at 420 nm. The carbon monoxide-difference spectrum of the detergent-solubilized microsomes looked similar to that of hemoglobin, but contamination of the original microsomes with hemoglobin was excluded. To distinguish the original microsome-bound pigment from its solubilized form, the former was named P-450 and the latter P-420 ("P" is the abbreviation of "pigment") (Omura and Sato 1962).

P-420 was soluble. It could be separated from cytochrome b_5 by conventional purification procedures, and a partially purified P-420 preparation, which was free from cytochrome b_5, was prepared from detergent-solubilized rabbit liver microsomes. The optical absorption spectra of oxidized and reduced P-420 resembled those of b-type cytochromes. The extinction coefficient of the carbon monoxide-difference spectrum of P-420 per mole of heme, together with the quantitative data of the spectral conversion of P-450 to P-420 upon solubilization of the microsomes with detergents, enabled the calculation of the molar extinction coefficient of the carbon monoxide-difference spectrum of P-450. The amount of P-450 in microsomes was thus calculated from the carbon monoxide-difference spectrum, and the content of protoheme in microsomes was found to be explained by the sum of P-450 and cytochrome b_5. These observations confirmed the hemoprotein nature of "microsomal carbon monoxide-binding pigment" (Omura and Sato 1962, 1963). Omura and Sato examined various properties of P-450 and P-420, and full accounts of their study were published in 1964 (Omura and Sato 1964a, b).

When Omura and Sato were studying "microsomal carbon monoxide-binding pigment," H.S. Mason of the University of Oregon Medical School, Portland, and his collaborators, Y. Hashimoto and T. Yamano, who were examining the electron paramagnetic resonance (EPR) spectra of the subcellular fractions from animal tissues, detected a novel EPR signal in rabbit liver microsomes. The novel EPR spectrum suggested the involvement of iron. The signal decreased when the microsomal preparation was reduced with NADH or NADPH and disappeared completely when carbon monoxide was added to NADH-reduced microsomes. They named the putative iron compound responsible for the novel EPR signal "microsomal Fex," and reported their findings in 1962 (Hashimoto et al. 1962). Mason and his collaborators continued their study on "microsomal Fex" (Murakami and Mason 1967), which was soon found to be identical with P-450.

1.3 Physiological Function of P450

The oxygenase-type enzymes responsible for the biosynthesis of steroid hormones in the adrenal gland and the gonads were highly labile to solubilization treatments, and their molecular nature had remained unelucidated until the beginning of the 1960s. However, inhibition experiments of the steroid hydroxylase activities of adrenal cortex microsomes suggested the presence of a heavy metal, possibly iron, in the steroid hydroxylases because the hydroxylase activities were strongly inhibited by carbon monoxide and the inhibition was reversed by white-light irradiation (Ryan and Engel 1957).

D.Y. Cooper in the laboratory of O. Rosenthal at the Harrison Department of Surgical Research of the University of Pennsylvania, Philadelphia, was studying the regulation of the steroid hormone synthesis in the adrenal gland in collaboration with R.W. Estabrook in the same university. Stimulated by the novel information that the carbon monoxide-binding pigment in liver microsomes is a hemoprotein (Omura and Sato 1962), they examined the carbon monoxide-difference spectrum of bovine adrenal cortex microsomes and found P450 in the microsomes. They then examined the photo-reversal of the carbon monoxide inhibition of C-21 hydroxylation of 17-hydroxyprogesterone catalyzed by adrenal cortex microsomes. The action spectrum of the photo-reversal showed a peak at 450 nm, which was identical with the peak of the carbon monoxide-difference spectrum of P450, indicating the participation of P450 in the hydroxylase activity (Estabrook et al. 1963). They analyzed the effect of the carbon monoxide to oxygen ratio on steroid hydroxylase activity, concluding that P450 is the oxygenase activating the oxygen molecule for steroid hydroxylation, and that carbon monoxide competes with molecular oxygen on the active site of P450 (Cooper et al. 1963). This was the first evidence for the physiological function of P450.

Because the photochemical action spectrum method used by Estabrook, Cooper, and Rosenthal was highly successful in the elucidation of the participation of P450 in the steroid hormone synthesis in the adrenal gland, the method was soon applied widely to the elucidation of the involvement of P450 in various other metabolic activities in animal tissues. The oxidative metabolism of drugs by liver microsomes was found to be catalyzed by P450 (Cooper et al. 1965a, b). Steroid hormone synthesis in the adrenal gland and drug metabolism in the liver were two major research subjects in the early years of P450 research.

1.4 P450 Enzyme Systems

The monooxygenase reactions catalyzed by P450 require molecular oxygen and NAD(P)H to introduce the oxygen atom to the substrate molecules. Elucidation of the mechanism of supply of reducing equivalents from NAD(P)H to P450 was an important subject in the early years of P450 research. Intensive studies revealed the presence of three different types of the P450 enzyme system (Omura 2010).

1.4.1 Mitochondrial P450 Enzyme System

It was known that both microsomes and mitochondria contribute to the synthesis of steroid hormones from cholesterol in the adrenal gland. The presence of P450 in the mitochondria isolated from rat adrenal cortex was first found by B.W. Harding, S.H. Wong, and D.H. Nelson in 1964 (Harding et al. 1964), which suggested the participation of P450 in the steroid hydroxylation reactions catalyzed by mitochondria in the adrenal cortex.

D.Y. Cooper and R.W. Estabrook tried to confirm the participation of P450 in steroid hydroxylation reactions catalyzed by mitochondria isolated from bovine adrenal cortex using the photochemical action spectrum method. Although the method gave decisive evidence for the role of P450 in the microsomal steroid hydroxylation system, it did not work well with the mitochondrial system. Cooper then tried solubilization of the steroid hydroxylation system from mitochondria and prepared a non-sedimentable hydroxylase preparation by sonicating mitochondria.

Estabrook invited T. Omura from Japan in 1964 to work in his laboratory for 1 year to study the steroid hydroxylase system of adrenal cortex mitochondria in collaboration with Cooper. Omura examined Cooper's sonicated mitochondrial preparation and found the P450 in the preparation was associated with small membrane fragments that could be sedimented by extensive ultracentrifugation. The P450 in the original preparation was reducible with NADPH, but the sedimented P450 was no longer reduced by NADPH. Addition of the supernatant to the sedimented P450 restored the reducibility of P450 by NADPH, indicating the presence of NADPH-P450 reductase activity in the supernatant. Column chromatography of the supernatant revealed the presence of two colored components, a yellow flavoprotein and a red ferredoxin-type non-heme iron protein, both of which were needed for the reduction of the membrane-associated P450 by NADPH. The NADPH-dependent steroid hydroxylase activity could be reconstituted from the three components: membrane-bound P450, soluble flavoprotein, and soluble non-heme iron protein. It was concluded that the flow of electrons from NADPH to mitochondrial P450 was mediated by the NADPH-linked flavoprotein and the non-heme iron protein (Omura et al. 1965, 1966). This was the first elucidation of the constitution of a P450 enzyme system.

When Omura, Estabrook, and Cooper were studying the components involved in the steroid hydroxylase system of adrenal cortex mitochondria, K. Suzuki and T. Kimura of the College of Science, St. Paul's University, Tokyo, were also studying the steroid hydroxylation system of bovine adrenal cortex mitochondria. They isolated a soluble non-heme iron protein from the sonic extract of the mitochondria, named it "adrenodoxin," and confirmed the reduction of adrenodoxin by NADPH, which was catalyzed by a soluble flavoprotein in the mitochondrial extract (Suzuki and Kimura 1965; Kimura and Suzuki 1965).

1.4.2 Bacterial P450 Enzyme System

The presence of P450 in bacteria was first reported by C.A. Appleby of the Division of Plant Industry, CSIRO, Canberra, in 1967. He discovered the presence of P450 in the homogenates of *Rhizobium* bacteroids. In contrast to the P450 of animal tissues, the *Rhizobium* P450 was soluble. It was partially purified, but its function was not elucidated (Appleby 1967).

In the next year, another bacterial P450 was reported by M. Katagiri, B.N. Ganguli, and I.C. Gunsalus of the Chemistry Department, University of Illinois, Urbana. They were studying the NADH-dependent camphor-hydroxylating system of camphor-grown *Pseudomonas putida*, and obtained a soluble NADH-dependent camphor hydroxylase preparation from the bacterium. They could separate the hydroxylase preparation into three components: putidaredoxin reductase, putidaredoxin, and P450, which was named P450cam (Katagiri et al. 1968). Putidaredoxin reductase and putidaredoxin mediated the electron transfer from NADH to P450cam. Similar soluble P450 enzyme systems were found in a few other bacterial species in the following years.

Soluble reductase-P450 fusion enzymes were also found later in some bacterial species. The first of such unique self-sufficient fusion P450 enzymes was reported by L.O. Narhi and A.J. Fulco in 1986 (Narhi and Fulco 1986).

1.4.3 Microsomal P450 Enzyme System

Elucidation of the constitution of microsomal NADPH-dependent P450 oxygenase system was difficult. In contrast to the mitochondrial P450 system, all components of the microsomal P450 system were apparently bound to the membrane, and solubilization of microsomes with detergents converted P450 to P420, a denatured inactive form of P450, resulting in the complete loss of oxygenase activity. Because NADPH-cytochrome *c* reductase was the only NADPH-linked electron-transfer enzyme found in liver microsomes, its contribution to the supply of electrons from NADPH to P450 was likely, but no solid evidence for that was available.

The breakthrough in this problem was made by Y. Ichikawa and T. Yamano of the Osaka University Medical School, Osaka, in 1967. They showed that P420 formed by the treatment of liver microsomes with detergents or sulfhydryl reagents could be converted back to P450 by treatment with glycerol or reduced glutathione (Ichikawa and Yamano 1967). The presence of high concentrations, 20–30 %, of glycerol in the medium was found to prevent the conversion of P450 to P420 when microsomes were solubilized with detergents.

The next year, A.Y.H. Lu and M.J. Coon of the University of Michigan, Ann Arbor, reported successful solubilization of the fatty acid ω-hydroxylation system of rabbit liver microsomes with deoxycholate in the presence of glycerol and dithiothreitol. The solubilized preparation could be separated into three fractions: a P450-containing fraction, an NADPH-cytochrome *c* reductase-containing fraction, and a heat-stable fraction. The three fractions were needed to reconstitute the NADPH-dependent fatty acid hydroxylase activity (Lu and Coon 1968).

Solubilization and purification of NADPH-cytochrome *c* reductase from liver microsomes had already been reported in 1962. C.H. Williams and H. Kamin solubilized the reductase from microsomes by pancreatic lipase treatment. A.H. Phillips and R.G. Langdon solubilized the reductase by trypsin digestion of microsomes. The solubilized and purified reductase contained FAD and retained NADPH-cytochrome *c* reductase activity (Williams and Kamin 1962; Phillips and Langdon 1962). However, Y. Miyake, J.L. Gaylor, and H.S. Mason reported in 1968 that the reductase purified from liver microsomes by the procedure of Williams and Kamin was unable to reduce P450 in a particle fraction prepared from microsomes by Lubrol WX treatment that removed NADPH-cytochrome *c* reductase (Miyake et al. 1968). A.Y.H. Lu, K.W. Junk, and M.J. Coon were also unable to reduce the P450 in the P450 fraction of the solubilized fatty acid ω-hydroxylase with the NADPH-cytochrome *c* reductase solubilized and purified from liver microsomes by the procedure of Williams and Kamin (Lu et al. 1969).

An important clue to solve this problem was provided by A. Ito and R. Sato of Osaka University, Osaka. When they solubilized cytochrome b_5 from rabbit liver microsomes with a detergent and purified it, they found that the molecular weight of the detergent-solubilized cytochrome b_5, which was called d-b_5, was significantly larger than that of the previously reported preparation (t-b_5) solubilized with trypsin from microsomes (Strittmatter and Velick 1956b). Ito and Sato examined the properties of d-b_5 and concluded that the molecule of native cytochrome b_5 is amphipathic; it binds to the microsomal membrane by the hydrophobic portion of the molecule. The heme resides in the hydrophilic portion, which may be detached from the membrane by trypsin treatment as t-b_5, leaving the hydrophobic portion in the membrane (Ito and Sato 1968). This finding suggested a similar topology for microsomal NADPH-cytochrome *c* reductase in microsomal membrane.

Solubilization of NADPH-cytochrome *c* reductase from microsomes without protease digestion was soon attempted by several groups. K. Ichihara, E. Kusunose, and M. Kusunose solubilized NADPH-cytochrome *c* reductase from porcine liver microsomes with Triton X-100, purified it to homogeneity, and confirmed the reconstitution of fatty acid ω-hydroxylase activity by adding the detergent-solubilized NADPH-cytochrome *c* reductase to the P450 fraction prepared from porcine kidney cortex microsomes. Treatment of the detergent-solubilized reductase with trypsin killed its ability of reconstituting the hydroxylase activity with the P450 fraction (Ichihara et al. 1973). The microsomal P450 system was a two-component system consisting of P450 and NADPH-cytochrome *c* reductase, which was soon renamed NADPH-P450 reductase.

Another important property of NADPH-P450 reductase was reported by T. Iyanagi and H.S. Mason in 1973. They found that the reductase solubilized and purified from liver microsomes contained both FAD and FMN (Iyanagi and Mason 1973). Kinetic analysis of the oxidation–reduction cycle of the reductase in the reduction of P450 by NADPH confirmed the essential roles of the two flavins in the transfer of electrons from NADPH to P450 (Iyanagi and Mason 1974).

1.5 Mechanism of P450-Catalyzed Oxygenase Reactions

The P450-catalyzed mono-oxygenation reaction consumes one mole each of NADPH and O_2 to introduce one atom of oxygen to the substrate molecule, which means the supply of two electrons to P450 in one cycle of the reaction. The mechanism of P450-catalyzed oxygenase reaction was a large subject of research.

The first important information on the mechanism of P450-catalyzed oxygenation reaction was obtained by S. Narasimhulu in the laboratory of D.Y. Cooper and O. Rosenthal in 1965. She found a small shift of the Soret peak of P450 in adrenal cortex microsomes when a steroid substrate, 17-hydroxyprogesterone, was added to the suspension of microsomes. The substrate-induced difference spectrum showed a broad maximum at 390 nm and a minimum at 420 nm and disappeared when the substrate was hydroxylated by the addition of NADPH. Nonsubstrate steroids did not induce such a spectral change (Narasimhulu et al. 1965). A similar spectral change was also observed with adrenal cortex mitochondria upon the addition of a substrate steroid of mitochondrial P450 (Cooper et al. 1965b). These observations suggested specific binding of substrate steroids to the oxidized forms of microsomal and mitochondrial P450s.

A similar substrate-induced spectral change of P450 was soon observed with liver microsomes upon the addition of various drugs, the substrates of liver microsomal P450 (Remmer et al. 1966; Imai and Sato 1966). J.B. Schenkman, H. Remmer, and R.W. Estabrook analyzed the drug-induced spectral changes of liver microsomal P450 and concluded that the binding of substrates to the oxidized form of P450 is an essential step of the P450-catalyzed oxygenation reaction because the concentration of a substrate necessary for the half-maximal reaction was similar to the concentration of the same substrate that induced the half-maximal spectral change (Schenkman et al. 1967).

The second important acquisition of information on the mechanism of P450-catalyzed oxygenation reaction was the detection of the "oxygenated form" of P450 by R.W. Estabrook and his collaborators. They detected a novel spectral intermediate, which was presumed to be an oxygenated form of reduced P450, in the oxygenation reactions catalyzed by liver microsomal P450 (Estabrook et al. 1971) and bacterial P450cam (Ishimura et al. 1971). By analyzing the appearance and disappearance of the "oxygenated form" in the oxygenation reaction, they concluded that the oxygenated form of reduced P450 is an intermediate in the overall oxygenase reaction. They also noticed a change in the reduction level of microsomal cytochrome b_5 during the oxygenation reaction and concluded that cytochrome b_5, in addition to NADPH-P450 reductase, serves as a donor of electrons to the oxygenated form of reduced P450. Based on these observations, a cyclic reaction mechanism of P450-catalyzed oxygenation reactions was proposed (Estabrook et al. 1971). The mechanism was widely accepted. The substrate-induced spectral change of P450 was utilized for the detection of the substrates of various P450s in the following years.

1.6 Coordination State of the Heme of P450

The unique optical absorption spectrum of the carbon monoxide compound of P450 and its conversion to the "normal" carbonyl hemoglobin-like spectrum of P420 upon solubilization from membranes attracted much attention from the biochemists studying hemoproteins. Because the heme of P450 was protoheme, the abnormal spectral properties of P450 had to be explained by some novel ligand coordinated to the heme.

It was known that P450 was converted to P420 by treating microsomes with sulfhydryl reagents (Murakami and Mason 1967), which suggested involvement of a sulfhydryl group in the unique spectral characteristics of P450. Stabilization of P450 to solubilization treatments by thiol compounds (Ichikawa and Yamano 1967) also indicated the role of the sulfhydryl group in the spectral properties of P450.

The first strong evidence for the coordination state of P450 heme was reported by J.O. Stern and J. Peisach of the Albert Einstein College of Medicine, Yeshiva University, New York, in 1974. They prepared a model compound of P450 by dissolving hemin chloride in dimethylsulfoxide-ethanol solvent containing tetramethylammonium hydroxide and reducing the hemin with 2-mercaptoethanol in the presence of carbon monoxide. The model compound showed an optical absorption spectrum with a prominent peak at 450 nm similar to that of P450. Because the presence of a thiol compound and a strong base in the hemin solution were essential for the model compound showing a P450-like absorption spectrum, they concluded that an ionized sulfhydryl group, thiolate anion, is a ligand of the heme of P450 (Stern and Peisach 1974). This conclusion was confirmed later by the elucidation of the molecular structure of a P450, P450cam, by X-ray crystallographic analysis (Poulos et al. 1985).

P450 was the first hemoprotein found with a thiolate anion as the ligand of protoheme. Two years later, chloroperoxidase of *Caldaryomyces fumago* was found to be another heme-thiolate protein with P450-like spectral properties (Dawson et al. 1976). More heme-thiolate proteins with diverse physiological functions were found in the following years (Omura 2005).

1.7 Purification of P450s

The soluble P450 of *Pseudomonas putida*, P450cam, was purified in 1968 (Katagiri et al. 1968), but the purification of membrane-bound microsomal and mitochondrial P450s of animals was difficult, mainly because of the denaturation of P450 to P420 upon solubilization from the membranes. The discovery of the stabilization of membrane-bound P450s by glycerol and thiol compounds to solubilization treatments reported in 1967 (Ichikawa and Yamano 1967) paved the way to the purification of microsomal and mitochondrial P450s.

Another important discovery contributing to the purification of microsomal P450 was the induction of liver microsomal drug-metabolizing P450 by various chemical compounds, including drugs. A.H. Conney, E.C. Miller, and J.A. Miller of the University of Wisconsin Medical School, Madison, reported in 1956 a marked increase of the metabolism of 3-methyl-4-monomethylaminoazobenzene by rat liver microsomes when the animals were pretreated with 3-methylcholanthrene (Conney et al. 1956). Similar increase of the microsomal drug-metabolizing activities in the livers of the animals treated with various drugs was reported in the following years (Remmer 1959; Conney et al. 1960; Kato 1960), and the induction of the drug-metabolizing activities was explained by the increase of the content of P450 in liver microsomes (Remmer and Merker 1963; Orrenius, et al. 1965). Remarkable increase of P450 in the liver microsomes of drug-treated animals was highly useful for the purification of P450. Most of the purification studies in the following years used rats or rabbits treated with various chemical compounds, including phenobarbital and 3-methylcholanthrene.

In the beginning of the 1970s, several research groups reported partial purification of liver microsomal P450 solubilized with detergents in the presence of glycerol. Purification of homogeneous preparations of several forms of microsomal P450 from the livers of drug-treated rats, mice, and rabbits was then achieved by the middle of the 1970s (Imai and Sato 1974; Van der Hoeven et al. 1974; Ryan et al 1975; Haugen and Coon 1976; Huang et al. 1976). Mitochondrial P450 was purified from bovine adrenal cortex mitochondria (Takemori et al. 1975). Purification of P450 from the microsomes of the yeast *Saccharomyces cerevisiae* was reported in 1977 (Yoshida et al. 1977). Purity of the final P450 preparations was usually assessed by sodium dodecyl sulfate-polyacrylamide gel electrophoresis (SDS-PAGE). Determination of the partial amino acid sequence at the amino-terminus of the purified P450 preparations by chemical sequencing was also utilized to confirm the distinction between various forms of P450. The immunochemical Ouchterlony double-diffusion test was utilized in some studies to confirm the absence of contaminating proteins.

Isolation of several forms of P450 showing different mobilities on SDS-PAGE and different substrate specificities in reconstituted systems established the existence of multiple forms of P450 in animal liver microsomes. Selective induction of different P450s by the administration of various chemical inducers was also confirmed. Purification studies on microsomal and mitochondrial P450s were extended from liver to nonhepatic tissues, and many P450s with different substrate specificities were purified and characterized in the following years.

1.8 Primary Structures of P450s

Elucidation of the primary structure of P450 protein was a big target of biochemical study on P450. Purification of microsomal and mitochondrial P450s enabled biochemists to start working on this problem in the latter half of the 1970s.

The first successful complete sequencing of two P450s, a bacterial soluble P450 and a liver microsomal P450, was reported in 1982. The former, P450cam of *Pseudomonas putida*, was sequenced by chemical sequencing by M. Haniu,

L.G. Armes, M. Tanaka, K.T. Yasunobu, B.S. Shastry, G.C. Wagner, and I.C. Gunsalus of the University of Hawaii, Manoa (Haniu et al. 1982). The amino acid sequence of the latter, a phenobarbital-inducible P450 of rat liver microsomes, was deduced from the nucleotide sequence of a cloned cDNA by Y. Fujii-Kuriyama, Y. Mizukami, K. Kawajiri, K. Sogawa, and M. Muramatsu of the Cancer Institute, Tokyo (Fujii-Kuriyama et al. 1982).

The use of cDNA cloning technique in sequencing a P450 protein was the first application of molecular biological methods to P450 research. Determination of the amino acid sequences of P450 proteins by cloning their cDNAs proved to be more convenient than the traditional chemical sequencing, and the complete amino acid sequences of many microsomal and mitochondrial P450s were elucidated in the following years by cloning their cDNAs.

Elucidation of the primary structures of P450s was an epoch-making achievement in P450 research. It enabled biochemists to study and discuss many unsolved problems about P450, such as the cysteine residue donating the thiolate anion ligand to the heme, conserved amino acid sequences that characterize the basic molecular structure of P450s, the evolutional relationship of various P450s, and the molecular mechanism of subcellular distribution of microsomal and mitochondrial P450s. It also marked the end of the first stage of P450 research where traditional biochemists were main players. The research on P450 proceeded further to the next stage, where molecular biological study on P450 genes and the regulation of their expression was one of the major targets of research. Elucidation of the tertiary structures of various P450s by X-ray crystallographic analysis opened up new ways in the analysis of the molecular properties of P450s and the mechanism of P450-catalyzed reactions. The research on P450 expanded and proceeded further in the years following the early times described in this review.

1.9 Epilogue

This review describes the contributions of the pioneers who made highly important contributions to the discovery of P450 and the elucidation of its molecular and functional properties in the early years, for about two decades from the beginning of the research on this unique hemoprotein, focusing on biochemical aspects of P450 research. It is my great pleasure to record their contributions. I regret and apologize, however, that I was unable to describe many important contributions by other scientists because of the space limitations of this chapter.

I was fortunate to be in the field of P450 research from the beginning and to have been acquainted with most of the pioneers described here. They have already retired from active research, and, to my deep regret, many of them have passed away. This review is my tribute to these pioneers who devoted themselves to the elucidation of various difficult problems on this remarkable hemoprotein, cytochrome P450, which plays a highly important diversified role in the biological world. I recently wrote another review paper describing the early years of P450 research with my personal recollections (Omura 2011).

References

Appleby CA (1967) A soluble hemoprotein P-450 from nitrogen-fixing *Rhizobium* bacteroids. Biochim Biophys Acta 147:399–402

Axelrod J (1955) The enzymatic deamination of amphetamine (Benzederine). J Biol Chem 214:753–763

Claude A (1943) The constitution of protoplasm. Science 97:451–456

Conney AH, Miller EC, Miller JA (1956) The metabolism of methylated aminoazo dyes. V. Evidence for induction of enzyme synthesis in the rat by 3-methylcholanthrene. Cancer Res 16:450–459

Conney AH, Davidson C, Gastel R, Burns JJ (1960) Adaptive increases in drug-metabolizing enzymes induced by phenobarbital and other drugs. J Pharmacol Exp Ther 130:1–8

Cooper DY, Estabrook RW, Rosenthal O (1963) The stoichiometry of C21 hydroxylation of steroids by adrenocortical microsomes. J Biol Chem 238:1320–1323

Cooper DY, Levin SS, Narasimhulu S, Rosenthal O, Estabrook RW (1965a) Photochemical action spectrum of the terminal oxidase of mixed function oxidase systems. Science 147:400–402

Cooper DY, Narasimhulu S, Slade A, Raich W, Foroff O, Rosenthal O (1965b) Hemoprotein content and activity of solubilized steroid 11b-hydroxylase preparation from adrenocortical mitochondria. Life Sci 4:2109–2114

Dawson JH, Trudell JR, Barth G, Linder RE, Bunnenberg E, Djerassi C, Chiang R, Hager LP (1976) Chloroperoxidase. Evidence for P-450 type heme environment from magnetic circular dichroism spectroscopy. J Am Chem Soc 98:3709–3710

Estabrook RW, Cooper DY, Rosenthal O (1963) The light reversible carbon monoxide inhibition of the steroid C21-hydroxylase system of the adrenal cortex. Biochem Z 338:741–755

Estabrook RW, Hildebrandt AG, Baron J, Netter KJ, Leibman K (1971) A new spectral intermediate associated with cytochrome P-450 function in liver microsomes. Biochem Biophys Res Commun 42:132–139

Fujii-Kuriyama Y, Mizukami Y, Kawajiri K, Sogawa K, Muramatsu M (1982) Primary structure of a cytochrome P-450: coding nucleotide sequence of phenobarbital-inducible cytochrome P-450 cDNA from rat liver. Proc Natl Acad Sci USA 79:2793–2797

Haniu M, Armes LG, Tanaka M, Yasunobu KT, Shastry BS, Wagner GC, Gunsalus IC (1982) The primary structure of the monooxygenase cytochrome P450CAM. Biochem Biophys Res Commun 105:889–894

Harding BW, Wong SH, Nelson DH (1964) Carbon monoxide-binding substances in rat adrenal. Biochim Biophys Acta 92:415–417

Hashimoto Y, Yamano T, Mason HS (1962) An electron spin resonance study of microsomal Fex. J Biol Chem 237:PC3843–PC3844

Haugen DA, Coon MJ (1976) Properties of electrophoretically homogeneous phenobarbital-inducible and β-naphthoflavone-inducible forms of liver microsomal cytochrome P-450. J Biol Chem 251:7929–7939

Hayaishi O, Katagiri M, Rothberg S (1955) Mechanism of the pyrocatecase reaction. J Am Chem Soc 77:5450–5451

Hayano M, Lindberg MC, Dorfman RI, Hancock JEH, Doering WE (1955) On the mechanism of the C-11 β-hydroxylation of steroids. J Am Chem Soc 77:5450–5451

Horecker BL (1950) Triphosphopyridine nucleotide-cytochrome *c* reductase in liver. J Biol Chem 183:593–605

Huang MT, West SB, Lu AYH (1976) Separation, purification, and properties of multiple forms of cytochrome P-450 from the liver microsomes of phenobarbital-treated mice. J Biol Chem 251:4659–4665

Ichihara K, Kusunose E, Kusunose M (1973) Some properties of NADPH-cytochrome *c* reductase reconstitutively active in fatty acid ω-hydroxylation. Eur J Biochem 38:463–472

Ichikawa Y, Yamano T (1967) Reconversion of detergent- and sulfhydryl reagent-produced P-420 to P-450 by polyols and glutathione. Biochem Biophys Acta 131:490–497

Imai Y, Sato R (1966) Substrate interaction with hydroxylase system in liver microsomes. Biochem Biophys Res Commun 22:620–626
Imai Y, Sato R (1974) A gel-electrophoretically homogeneous preparation of cytochrome P-450 from liver microsomes of phenobarbital-treated rabbits. Biochem Biophys Res Commun 60:8–14
Ishimura Y, Ullrich V, Peterson JA (1971) Oxygenated cytochrome P-450 and its possible role in enzymatic hydroxylation. Biochem Biophys Res Commun 42:140–146
Ito A, Sato R (1968) Purification by means of detergents and properties of cytochrome b_5 from liver microsomes. J Biol Chem 243:4922–4930
Iyanagi T, Mason HS (1973) Some properties of hepatic reduced nicotinamide adenine dinucleotide phosphate-cytochrome *c* reductase. Biochemistry 12:2297–2308
Iyanagi T, Mason HS (1974) Redox properties of the reduced nicotinamide adenine dinucleotide phosphate-cytochrome P-450 and reduced nicotinamide dinucleotide-cytochrome b_5 reductase. Biochemistry 13:1701–1710
Katagiri M, Ganguli BN, Gunsalus IC (1968) A soluble cytochrome P-450 functional in methylene hydroxylation. J Biol Chem 243:3543–3546
Kato R (1960) Induced increase of meprobamate metabolism in rats treated with phenobarbital orphenaglycodol. Med Exp 3:95–100
Kimura T, Suzuki K (1965) Enzymatic reduction of non-heme iron protein (adrenodoxin) by reduced nicotinamide adenine dinucleotide phosphate. Biochem Biophys Res Commun 20:373–379
Klingenberg M (1958) Pigments of rat liver microsomes. Arch Biochem Biophys 75:376–386
Lu AYH, Coon MJ (1968) Role of hemoprotein P-450 in fatty acid ω-hydroxylation in a soluble enzyme system from liver microsomes. J Biol Chem 243:1331–1332
Lu AYH, Junk KW, Coon MJ (1969) Resolution of the cytochrome P-450-containing ω-hydroxylation system of liver microsomes into three components. J Biol Chem 244:3714–3721
Mason HS, Fowlks WL, Peterson E (1955) Oxygen transfer and electron transport by the phenolase complex. J Am Chem Soc 77:2914–2915
Miyake Y, Gaylor JL, Mason HS (1968) Properties of a submicrosomal particle containing P-450 and flavoprotein. J Biol Chem 243:5788–5797
Murakami K, Mason HS (1967) An electron resonance study of microsomal Fex. J Biol Chem 242:1102–1110
Narasimhulu S, Cooper DY, Rosenthal O (1965) Spectrophotometric properties of a triton-clarified steroid 21-hydroxylase system of adrenocortical microsomes. Life Sci 4:2102–2107
Narhi LO, Fulco AL (1986) Characterization of a catalytically self-sufficient 119,000-dalton cytochrome P-450 monooxygenase induced by barbiturates in *Bacillus megaterium*. J Biol Chem 261:7160–7169
Omura T (2005) Heme-thiolate proteins. Biochem Biophys Res Commun 338:404–409
Omura T (2010) Structural diversity of cytochrome P450 enzyme system. J Biochem 147:297–306
Omura T (2011) Recollection of the early years of the research on cytochrome P450. Proc Jpn Acad Series B 87:617–640
Omura T, Sato R (1962) A new cytochrome in liver microsomes. J Biol Chem 237:PC1375–PC1376
Omura T, Sato R (1963) Fractional solubilization of hemoproteins and partial purification of carbon monoxide-binding cytochrome from liver microsomes. Biochem Biophys Acta 71:224–226
Omura T, Sato R (1964a) The carbon monoxide-binding pigment of liver microsomes. I. Evidence for its hemoprotein nature. J Biol Chem 239:2370–2378
Omura T, Sato R (1964b) The carbon monoxide-binding pigment of liver microsomes. II. Solubilization, purification, and properties. J Biol Chem 239:2379–2385
Omura T, Sato R, Cooper DY, Rosenthal O, Estabrook RW (1965) Function of cytochrome P-450 of microsomes. Fed Proc 24:1181–1189

Omura T, Sanders E, Estabrook RW, Cooper DY, Rosenthal O (1966) Isolation from adrenal cortex of a non-heme iron protein and a flavoprotein functional as a reduced triphosphopyridine nucleotide-cytochrome P-450 reductase. Arch Biochem Biophys 117:660–673
Orrenius S, Erisson JLE, Ernster L (1965) Phenobarbital-induced synthesis of the microsomal drug-metabolizing enzyme system and its relationship to the proliferation of endoplasmic reticulum: a morphometric and biochemical study. J Cell Biol 25:627–639
Palade GE, Siekevitz P (1956) Liver microsomes, an integrated morphological and biochemical study. J Biophys Biochem Cytol 2:171–198
Phillips AH, Langdon RG (1962) Hepatic triphosphopyridine nucleotide-cytochrome *c* reductase: isolation, characterization, and kinetic studies. J Biol Chem 237:2652–2660
Poulos TL, Finzel BC, Gunsalus IC, Wagner GC, Kraut J (1985) The 2.6-Å crystal structure of *Pseudomonas putida* cytochrome P450. J Biol Chem 260:16122–16130
Remmer H (1959) Die Beschleunigung der Evipan Oxydation und der Methylierung von Methylaminoantipyrin durch Barbitrate. Arch Exp Pathol Pharmakol 237:296–307
Remmer H, Merker HJ (1963) Enzyminduktion und Vermehrung von endoplasmatischen Reticulum in der Leberzelle wahrend der Behandlung mit Phenobaribital (Luminal). Klin Wochenschr 41:276–283
Remmer H, Schenkman J, Estabrook RW, Sasame H, Gillette J, Narasimhulu S, Cooper DY, Rosenthal O (1966) Drug interaction with hepatic microsomal cytochrome. Mol Pharmacol 2:187–190
Ryan KJ, Engel LL (1957) Hydroxylation of steroids at carbon 21. J Biol Chem 225:103–114
Ryan D, Lu AYH, Kawalek J, West SB, Levin W (1975) Highly purified cytochrome P-448 and P-450 from rat liver microsomes. Biochem Biophys Res Commun 64:1134–1141
Schenkman JB, Remmer H, Estabrook RW (1967) Spectral studies of drug interaction with hepatic microsomal cytochrome. Mol Pharmacol 3:113–123
Stern JO, Peisach J (1974) A model compound study of the CO-adduct of cytochrome P-450. J Biol Chem 249:7495–7498
Strittmatter CF, Ball EG (1951) A hemochromogen component of liver microsomes. Proc Natl Acad Sci USA 38:19–25
Strittmatter P, Velick SF (1956a) A microsomal cytochrome reductase specific for diphosphopyridine nucleotide. J Biol Chem 221:277–286
Strittmatter P, Velick SF (1956b) The isolation and properties of microsomal cytochrome. J Biol Chem 221:253–264
Suzuki K, Kimura T (1965) An iron protein as a component of steroid 11β-hydroxylase complex. Biochem Biophys Res Commun 19:340–345
Takemori S, Suhara K, Hashimoto S, Hashimoto M, Sato H, Gomi T, Katagiri M (1975) Purification of cytochrome P-450 from bovine adrenocortical mitochondria by an "aniline-Sepharose" and the properties. Biochem Biophys Res Commun 63:588–593
Van der Hoeven TA, Haugen DA, Coon MJ (1974) Cytochrome P-450 purified to apparent homogeneity from phenobarbital-induced rabbit liver microsomes: catalytic activity and other properties. Biochem Biophys Res Commun 60:569–575
Williams CH, Kamin H (1962) Microsomal triphosphopyridine nucleotide-cytochrome *c* reductase of liver. J Biol Chem 237:587–595
Yoshida Y, Aoyama Y, Kumaoka H, Kubota S (1977) A highly purified preparation of cytochrome P-450 from microsomes of anaerobically grown yeast. Biochem Biophys Res Commun 70:723–728

Chapter 2
Fifty Years of Progress in Drug Metabolism and Toxicology: What Do We Still Need to Know About Cytochrome P450 Enzymes?

F. Peter Guengerich

Abstract The 50 years following the discovery of the cytochrome P450 system have been years of remarkable progress in the basic science and in its application to important problems, particularly in medicine. This chapter reviews what has been done, at both basic and applied levels. My own views on the still unresolved basic issues are presented, along with some thoughts about opportunities for future development in practical applications.

Keywords Agricultural applications • Basis of pharmacokinetics • Catalytic mechanisms • Clinical applications • Cytochrome P450 • Drug metabolism • Endocrinology • Metabolic diseases • Metabolism of carcinogens • Processivity of P450 reactions • Toxicology

2.1 Introduction

I wish to begin this chapter with two personal notes, in celebrating the 50th anniversary of the first real report of cytochrome P450 (P450). One is a round of thanks to and acknowledgment of Professor Tsuneo Omura, who co-authored the original paper in *The Journal of Biological Chemistry* (Omura and Sato 1962) 50 years ago as a graduate student with his mentor Professor Ryo Sato. He is still a brilliant but humble man who follows the field and attends the meetings. The other note is that I entered the P450 field as a postdoctoral fellow with

F.P. Guengerich (✉)
Department of Biochemistry and Center in Molecular Toxicology, Vanderbilt University School of Medicine, 638 Robinson Research Building, 2200 Pierce Avenue, Nashville, TN 37232-0146, USA
e-mail: f.guengerich@vanderbilt.edu

H. Yamazaki (ed.), *Fifty Years of Cytochrome P450 Research*,
DOI 10.1007/978-4-431-54992-5_2,

Professor Minor J. Coon in 1973, four decades ago at the time I finished this article. I had no idea that I would continue to work on this same enzyme system for 40 years. I work on other enzyme and nucleic acid systems too, but there have just been too many interesting things to do with P450 and I do not think I can ever stop.

The field of P450 research began, in part, with individuals who were interested in a liver protein with unusual spectral properties (Garfinkel 1958; Klingenberg 1958; Omura and Sato 1962, 1964a, b). However, what has driven research in the P450 field has been its relevance to a number of areas. Reactions that we now know are catalyzed by P450s were already described in chemical carcinogenesis (Mueller and Miller 1948, 1953), drug metabolism (Gillette et al. 1957), and steroid metabolism (Ryan 1958) when the pigment with the unusual spectrum was identified. Key studies along the way were the carbon monoxide inhibition studies establishing the newly found P450 as the terminal oxidase in microsomal electron transport (Estabrook et al. 1963), several lines of investigation suggesting the existence of multiple P450s (Alvares et al. 1967; Hildebrandt et al. 1968; Sladek and Mannering 1969), and the separation and reconstitution of the components of the microsomal P450 system (Lu and Coon 1968) (see also the chapter by Professor Tsuneo Omura in this monograph). Another notable contribution was the work of Professor Irwin Gunsalus and his associates (Katagiri et al. 1968) with the bacterial model $P450_{cam}$ (CYP101A1), which provided interesting biophysical insights into the structures (Poulos et al. 1985) and catalytic mechanisms (Tyson et al. 1972) for both bacterial and mammalian P450s (Mueller et al. 1995; McLean et al. 2005).

The field of P450 research is still fueled by its relevance in the fields of metabolism in chemical carcinogenesis, drug metabolism, and endocrinology. However, there are also many practical applications in medicine (Nebert and Russell 2002), nutrition (Plum and DeLuca 2010), agriculture (Mizutani and Sato 2011), and biotechnology, including the use of P450s as designed catalysts (Guengerich 2002; Coelho et al. 2013). The number of academic research papers on P450 continue to increase each year, and more than 33,000 papers related to P450 research have been published (http://webtools.mf.uni-lj.si/public/medsum.html) with about 2,000 per year still being added.

2.2 What We Know About P450

The following list is not intended to be comprehensive but is a summary of what I consider the most important. Even with these there are some missing pieces of information.

First of all, the number of P450 genes is now known in many organisms, including humans (57) (Table 2.1), Defining these numbers was not trivial, and the final answers really came with the completion of genomic sequences. It is of note that some microorganisms have a fairly high number of P450s (e.g., 32 in *Streptomyces avermitelus*) and plants have hundreds.

Table 2.1 Classification of human P450s based on major substrate class

Sterols	Xenobiotics	Fatty acids	Eicosanoids	Vitamins	Unknown
1B1[a]	1A1[a]	2J2	4F2	2R1[a]	2A7
7A1[a]	1A2[a]	4A11	4F3	24A1[a]	2S1
7B1	2A6[a]	4B1	4F8	26A1	2U1
8B1	2A13[a]	4F12	5A1	26B1	2W1
11A1[a]	2B6[a]		8A1[a]	26C1	3A43
11B1	2C8[a]			27B1	4A22
11B2	2C9[a]				4F11
17A1[a]	2C18				4F22
19A1[a]	2C19[a]				4V2
21A2[b]	2D6[a]				4X1
27A1	2E1[a]				4Z1
39A1	2F1				20A1
46A1[a]	3A4[a]				27C1
51A1[a]	3A5				
	3A7				

[a]X-ray crystal structure(s) reported (for human enzymes)
[b]Bovine X-ray crystal structure reported (Zhao et al. 2012)
Source: Guengerich (2005)

Another success, which may not be appreciated today, is definition of the number of different proteins involved in P450 function. Bacteria have a myriad of different redox systems (Guengerich and Munro 2013). Mammalian microsomal P450s all accept electrons from NADPH-P450 reductase (POR), with cytochrome b_5 being involved in some but not all P450 reactions. Mammalian mitochondrial P450s use an electron-transfer system related to some bacteria, involving the iron–sulfur protein adrenodoxin and the flavoprotein NADPH-adrenodoxin reductase. Although the literature contains early (Nelson et al. 1973) and more recent (Hughes et al. 2007) reports of other proteins being involved, the current literature has not confirmed the roles of any of these.

2.2.1 *Electron Transfer*

The source of reducing equivalent for P450s, with only a few exceptions (Guengerich and Munro 2013), is NAD(P)H. Pyridine nucleotides are two-electron donors and, with one exception (Shiro et al. 1995), do not reduce P450, a one-electron acceptor. Flavoproteins are biological transformers and can be involved in both one- and two-electron transfers. With microsomal P450s, electrons come from POR, a two-flavin protein, in a system where electrons flow from NADPH to FAD to FMN then to P450s. In some cases cytochrome b_5 donates the "second" electron (to the FeO_2^{2+} complex). For the seven mitochondrial P450s [11A1, 11B1, 11B2, 24A1, 27A1, 27B1, 27C1 (note: 27C1 is tentative)], electrons flow from NADPH to the flavoprotein NADPH-adrenodoxin reductase to adrenodoxin to the P450.

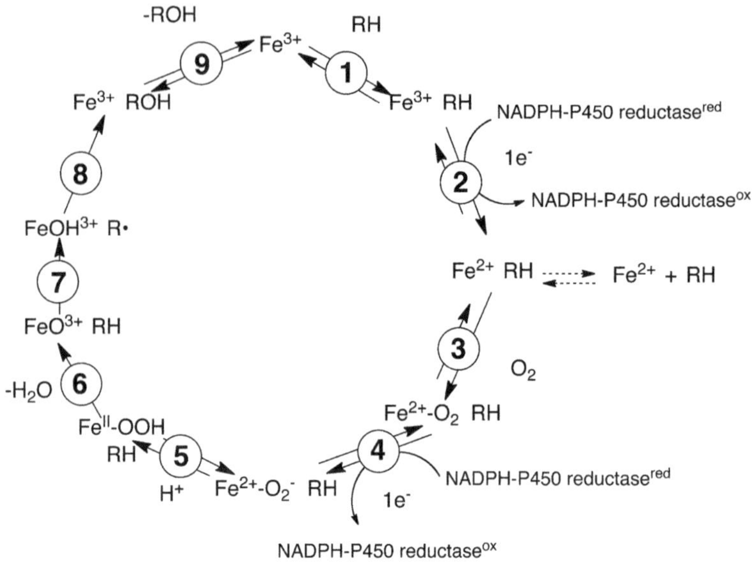

Fig. 2.1 General catalytic cycle for P450 reactions (Guengerich 2001)

Rates of electron transfer to many of the human P450s have been measured (Guengerich and Johnson 1997). Some are dependent upon the binding of a substrate to the P450, but this is not always the case (Guengerich and Johnson 1997). However, the concentration of total P450 is about 20 fold higher than POR in the liver (Estabrook et al. 1971). Reduction rates are usually biphasic in microsomes, with the P450 in closest proximity being reduced first (Peterson et al. 1976).

Cytochrome b_5 can provide the "first" electron (to ferric P450), but the rate is slow, probably because of the unfavorable redox potential (West et al. 1974; Yamazaki et al. 1996). Such a functional system can apparently occur in vivo as well (Henderson et al. 2013). Measurement of the rate of cytochrome b_5 electron transfer to the P450 FeO_2^{2+} complex is technically difficult and has only been done in a few settings (Yun et al. 2005; Zhang et al. 2007).

2.2.2 *Basic Catalytic Mechanism*

The activation of oxygen to a reactive form is complex and involves unstable high-valent iron intermediates that have been difficult to study (Fig. 2.1). The subject has been discussed at length elsewhere (Ortiz de Montellano and De Voss 2002; Ortiz de Montellano and De Voss 2005).

The key species is the perferryl oxygen complex, FeO^{3+}. This complex is formed in the indicated pathway (Fig. 2.1) and then abstracts a hydrogen atom to leave a carbon-centered radical. "Rebound" of oxygen from the resulting $FeOH^{3+}$ species forms an alcohol (or an equivalent product). One variation on this theme involves the abstraction of a non-bonded electron from a low redox potential substrate (e.g., nitrogen), followed by possible rearrangements and then an oxygen rebound (Guengerich 2001; Ortiz de Montellano and De Voss 2005).

The only other viable oxidant at present is the FeO_2^- species, a precursor of FeO^{3+}. This nucleophilic species can explain some P450 reactions with aldehydes (Akhtar et al. 1982). The possibility exists that even some of those reactions might also have at least a partial contribution from a FeO^{3+}-based mechanism (Hackett et al. 2005).

Using these basic mechanisms, it has been possible to rationalize almost all the reported P450 reactions, with the inclusion of rearrangements of enzyme intermediates or reaction products (Guengerich 2001; Ortiz de Montellano and De Voss 2005; Isin and Guengerich 2007; Guengerich and Munro 2013; Guengerich and Isin 2014).

2.2.3 Multiple Rate-Limiting Steps

In early P450 research, there was a quest to find the "rate-limiting step" in the P450 reaction cycle (Diehl et al. 1970; Gigon et al. 1969). The subject has been addressed many times, utilizing pre-steady-state kinetics and kinetic isotope effect studies. There is evidence that steps **2**, **4**, **7**, and **9** in Fig. 2.1 can all contribute to rate determination (Guengerich 2013).

Of course, some of the reactions in Fig. 2.1 are difficult to measure (e.g., **3**, **5**, **6**, **8**), and rates are not known but are assumed to be fast. Two other points can be made. One is that the overall rate of a P450 reaction can be a function of the frequency of uncoupling (i.e., diversion of intermediate species to reduction of O_2 to O_2^-, H_2O_2, and H_2O) (Gorsky et al. 1984). The other point is that even substrate binding (step **1** in Fig. 2.1) can be a complex, multi-step pathway moving the substrate from the periphery of the P450 to the active site (Isin and Guengerich 2006; Sevrioukova and Poulos 2012).

2.2.4 Regulation of Expression

In retrospect, early work in this area was extremely difficult because of the lack of appropriate technology and reagents. Today we recognize that most of the regulation of the P450 genes is at the transcriptional level. The general model follows that developed for steroid nuclear receptors: a cytosolic receptor binds a ligand, heterodimerizes with a partner protein, and moves to the nucleus. The loaded heterodimer binds to a specific ("consensus") site in the 5′-regulatory sequence ("enhancer") and alters the gene/chromatic structure to open the promoter region for RNA polymerase to copy the P450 gene faster. Indeed, several of the major receptors involved in P450 gene regulation were in the steroid nuclear receptor "orphan" group (e.g., PXR, CAR, PPARα). The well-known AhR/ARNT pathway for induction by polycyclic aromatic hydrocarbons, etc. follows a similar model (with unrelated proteins) (Williams et al. 2005).

Some regulation is more complex. The CAR pathway involves kinases and intersects with the epidermal growth factor receptor-signaling pathway (Mutoh et al. 2013). There are elements of posttranscriptional regulation in some systems, and microRNA regulation has been implicated in the regulation of P450s (Gomez and Ingleman-Sundberg 2009). Rodents show considerable gender-linked P450 regulation, which is the result of steroid, growth hormone, and STAT pathway regulation (Waxman and Holloway 2009) (this is not seen, at least at this level, in humans; Yang et al. 2010).

2.2.5 Polymorphisms

P450 differences among individuals have been suspected since early discoveries of individuals with inherited errors of metabolism in endocrine issues. These differences began to be characterized with the availability of recombinant DNA technology (White et al. 1984). The incidence of these variations is relatively low.

Polymorphisms are generally defined as variations in the population at a level greater than 1 % (Kalow 1962). One impetus for work in this area was the stratification of smokers into three groups based on the inducibility of aryl hydrocarbon hydroxylase (now associated with P450s 1A1 and 1B1) (Kellerman et al. 1973a, b). This area was driven by work by Professor Robert Smith and others, who found people (including Professor Smith himself) who could not effectively oxidize certain drugs (i.e., debrisoquine, sparteine, metoprolol) (Mahgoub et al. 1977). This "poor metabolizer" phenotype was consistent within individuals and showed Mendelian inheritance. Ultimately this phenomenon was understood in the context of P450 2D6 (Distlerath et al. 1985; Gut et al. 1986; Gonzalez et al. 1988). Today we know that there are not only two genotypes of P450 2D6 but more than 100. This situation is not atypical for the P450s, and today the collected genotypes are collected and available online (www.cypalleles.ke.se). Today this variability of P450s and its effects on drug metabolism are a major component of "personalized medicine" (Evans and McLeod 2003).

2.2.6 Cellular Localization

Classically, mammalian P450s have been considered to be either microsomal (i.e., in the endoplasmic reticulum) or mitochondrial. Human P450s 11A1, 11B1, 11B2, 24A1, 27A1, 27B1, and (probably) 27C1 are considered to be mitochondrial. However, work by Prof. Narayan Avadhani has shown that fractions of some of the microsomal P450s can be localized in the mitochondria (Niranjan and Avadhani 1980). In at least some cases this localization is the result of cryptic import signals, which can be manifested by proteolytic cleavage (Addya et al. 1997). Some are sensitive to phosphorylation (e.g., P450 2E1) (Bansal et al. 2010), and polymorphisms in human P450s can determine the

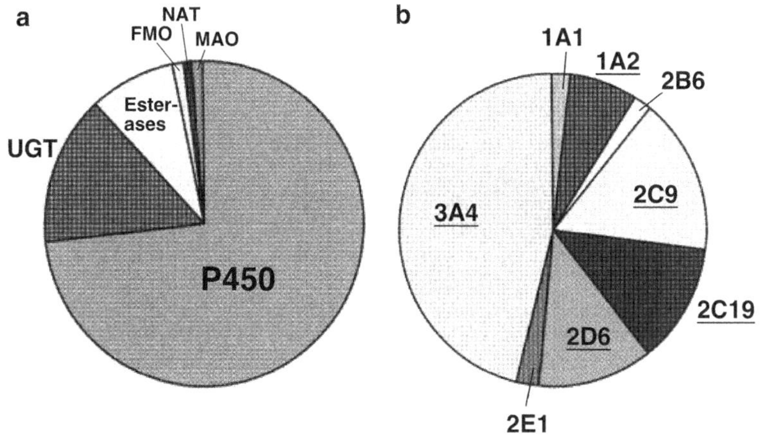

Fig. 2.2 Drug metabolism reactions. (**a**) Contributions of different enzymes. *UGT* UDG glucuronosyl transferase, *FMO* flavin-containing monooxygenase, *NAT* *N*-acetyltransferase, *MAO* monoamine oxidase. (**b**) Contributions of individual human P450 enzymes to (P450) drug metabolism (Williams et al. 2004)

partitioning between the endoplasmic reticulum and mitochondria, for example, P450 2D6 (Bajpai et al. 2013). Interestingly, the P450s that locate in the mitochondria can efficiently utilize the adrenodoxin electron-transport pathway in their function.

2.2.7 Roles of P450s in Individual Reactions

Determination of which P450s are involved in reactions is now a relatively straightforward in vitro process. Selective inhibitors, antibodies, comparisons with established markers, and purified P450s render the procedures very direct, given appropriate consideration of levels of expression (Beaune et al. 1986; Guengerich and Shimada 1991). Accordingly, there is extensive information about the human P450 regarding drug substrates, inhibitors, and inducers (http://medicine.iupui.edu/clinpharm/ddis/main-table/) (Fig. 2.2) and also carcinogen substrates (Rendic and Guengerich 2012) (Fig. 2.3).

2.2.8 Importance of P450s in Medical Practice

The significance of P450 research in medicine is considerable, as seen in several areas.

Endocrinology Steroid metabolism is complex, and many inborn errors of metabolism can now be explained in terms of deficiencies of individual P450s or NADPH-P450 reductase (Miller and Auchus 2011). Included among these are the

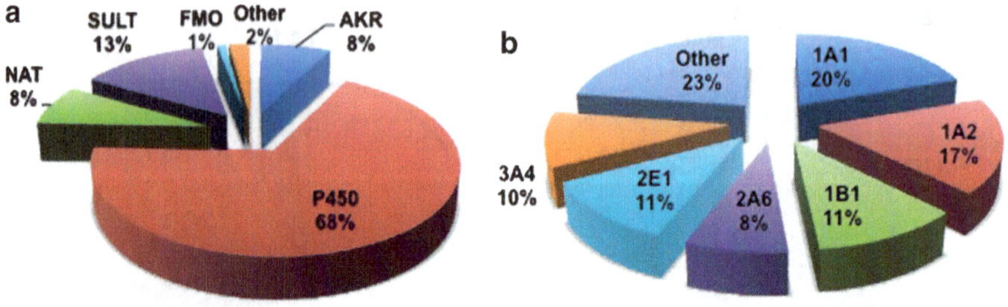

Fig. 2.3 Carcinogen activation by human enzymes. (**a**) Contributions of different (human) enzyme systems. *FMO* flavin-containing monooxygenase, *NAT N*-acetyltransferase, *SULT* sulfotransferase, *AKR* aldo-keto reductase, *COX* cyclooxygenase/prostaglandin synthase. (**b**) Contributions of individual human P450s to the P450 sector of carcinogen activation (Rendic and Guengerich 2012)

more than 100 different genotypes associated with P450 21A2 deficiency, including salt-wasting syndrome (Wedell 2011; Zhao et al. 2012). In addition, deficiencies in the P450s involved in the metabolism of vitamins A and D are the basis of diseases (Nebert and Russell 2002).

Drug Metabolism Before understanding of human P450s developed, prediction of human pharmacokinetic behavior of a drug candidate was very difficult, and human pharmacokinetic problems were a major reason for failure of drug candidates in clinical trials (Kola and Landis 2004).

P450s are involved in about 75 % of the enzymatic reactions involved in drug metabolism (Fig. 2.2) (Wienkers and Heath 2005; Williams et al. 2004), and today it is possible to determine which P450s are involved in the metabolism of a new drug candidate using in vitro approaches (Guengerich and Shimada 1991). There are reasonably good approaches to extrapolating to in vivo situations and predicting variability in human populations (Ito et al. 1998; Guest et al. 2011). In addition, both induction and inhibition can be studied in vitro. Other knowledge of the inducers and inhibitors of individual P450s (Guengerich 2005) enables prediction of potential drug–drug interactions (Andersson et al. 2005).

Knowledge about individual P450s involved in reactions and genetic polymorphisms (vide supra) is also used to guide drug prescriptions and use, and this is a major element in "personalized medicine." As an example, the maintenance levels of the anticoagulant warfarin are related to polymorphism in the *CYP2C9* gene (P450 2C9) (Daly et al. 2002; Garcia and Hylek 2009). Several side effects of drugs (e.g., debrisoquine, perhexiline) are related to P450 2D6 (Idle et al. 1978; Idle and Smith 1979; Oates et al. 1981; Shah et al. 1982). Serious adverse reactions of terfenadine are related to the inhibition of its metabolism by P450 3A4 (and the resulting increased plasma and tissue levels) (Yun et al. 1993; Thompson and Oster 1996; Guengerich 2013, 2014). Induction of P450 3A4 by rifampicin, barbiturates, and herbal medicines containing hyperforin increases the metabolism of the oral contraceptive 17α-ethynylestradiol, which can lead to unexpected breakthrough bleeding and pregnancy (Bolt et al. 1975; Guengerich 1988a).

Knowledge of P450 oxidation of drug candidates, induction, and inhibition is widely used today in the overall process of drug development (Humphreys 2008).

P450s as Drug Targets Another aspect of P450s in medical practice is their undesired effects and targeting by drugs. One example is the steroid aromatase, P450 19A1, which converts androgens to estrogens (Brodie 1985); this is an issue in estrogen-stimulated tumors. Another target is P450 17A1, which forms androgens and is a target in prostate cancer (DeVore and Scott 2012). Several fungal conditions are treated with antimycotic inhibitors of fungal and yeast P450 51A1 (Aoyama et al. 1998). In this regard, some P450s of *Mycobacterium tuberculosis* have been shown to be required for viability or virulence, and efforts to develop drugs are in progress (Seward et al. 2006; Johnston et al. 2010).

2.3 What Have We Left to Learn About P450s: Basic Questions

Another author might provide a different list, but the following is my own opinion. More practical questions about P450s follow later. This selection is biased in part on my own research interests, although we are not working on all aspects.

2.3.1 What Are the Functions of the Orphan P450s?

P450s can be classified on the basis of their substrates (Table 2.1). Almost one-fourth of the human P450s are grouped as “orphans” (Guengerich 2005), a term adopted from the orphan steroid nuclear receptor family (Mangelsdorf and Evans 1995). Our laboratory has been involved in systematic searches for functions of these, and some of the progress has been reviewed (Guengerich and Cheng 2011). Recent (and unexpected) results include the oxidations of lysphophospholipids by P450 2W1 (Xiao and Guengerich 2012) and of *N*-arachidonoylserotonin by P450 2U1 (Siller et al. 2014). Some drug substrates have been identified for several of the orphans (Nishida et al. 2010; Xiao et al. 2011; Wang and Guengerich 2012; Edson et al. 2013), but there is little current information about the overall contribution of these. Only one of the orphans (2W1; Table 2.1) has been found to activate carcinogens (Wu et al. 2006).

At the present time only speculation is possible as to whether any of these orphan P450s will be shown to have important physiological roles. P450 4F11 hydroxylates vitamin K (Edson et al. 2013). *Cyp2s1*(−/−) mice are phenotypically normal (X. Ding, personal communication).

When can a P450 be considered to be "deorphanized?" P450 2R1 was, when an important role in vitamin A metabolism was defined (Cheng et al. 2003). However, in another sense, all the P450s under the "Xenobiotics" heading in Table 2.1 can be considered orphans in the sense that no critical physiological reactions have been identified. Elimination of the apparent orthologues in mice has, in most cases, no major observable phenotype in mice, and humans missing some of these are known but are normal unless exposed to certain drugs.

The deorphanization of the myriad of bacterial, insect, and plant P450s is difficult but has important implications in agriculture, pest control, and other practical issues. In many cases there are advantages (compared to mammals and humans) in that gene knockouts can be done and phenotypes can be observed; for example, two P450 genes in *Streptomyces coelicolor* are functional in sporulation (Cheng et al. 2010; Tian et al. 2013), although the reactions underlying these phenomena have not been defined. Exactly how defects in sterol metabolism relate to *M. tuberculosis* is not yet clear (Seward et al. 2006; Johnston et al. 2010). Elucidation of the functions of the function of P450 genes in crop plants, weeds, insects, and fungi has great potential in agriculture (Kinney 2006).

2.3.2 Is There More to Learn About the Nature of Oxidizing Species of P450s?

As mentioned earlier, all the early proposals about the nature of oxidizing species have now largely culminated with two entities, FeO^{3+} ("Compound I") and FeO_2^-, its precursor (Ortiz de Montellano and De Voss 2005). The latter has been used to rationalize some unusual reactions, mainly those with aldehyde substrates (Akhtar et al. 1982). Strong evidence for the role of FeO^{3+} has come from the detailed characterization of this entity by Prof. Michael Green and his associates (Rittle and Green 2010). The major evidence for the FeO_2^- mechanism comes from (1) site-directed mutagenesis work, mainly with active site Thr mutants (Vaz et al. 1996), and (2) ^{18}O labeling experiments with some steroids (Akhtar et al. 1982, 1994).

With this as a background, what are the remaining issues? One need is the extension of the studies on the characterized Compound I (FeO^{3+}) to more P450s, thus testing hypotheses about alternate forms. Further, proposals about the spin-state duality of FeO^{3+} have been made by Prof. Sasson Shaik on the basis of theoretical considerations (Harris et al. 2000; Shaik et al. 2005), but these have not been experimentally tested. Access to defined Compound I (FeO^{3+}) P450 species should allow these hypotheses to be addressed.

The role of FeO_2^- species has been considered, but the duality of alternate reactions (FeO^{3+} and FeO_2^-) acting together has not. In some cases both mechanisms have been proposed, such as with P450 19A1 (Hackett et al. 2005; Sen and Hackett 2012). Site-directed mutagenesis approaches cannot be used in a quantitative approach, but ^{18}O (and other labeling approaches?) can be. These experiments are now possible, and our preliminary $^{18}O_2$ work with P450 19A1 indicates that the FeO^{3+} species is involved in the third oxygenation step in the reaction sequence (Yoshimoto and Guengerich, in preparation).

2.3.3 Allosteric Systems and Two Ligands

Nonhyperbolic kinetics (homotropic cooperativity) and direct P450 stimulation (heterotropic cooperativity) were first reported 20 years ago (Guengerich et al. 1994; Shou et al. 1994). Since then, these phenomena have been studied extensively, and several indirect lines of evidence led to the proposal that these phenomena could be explained by a model with two (or more) ligands (Shou et al. 1994; Hosea et al. 2000; Davydov et al. 2005; Sligar and Denisov 2007). For many years, however, there was no direct physical proof of this, such as isothermal calorimetry. However, a number of X-ray crystal structures of P450s have now been reported with two ligands in the active site (Zhao et al. 2005, 2012; Ekroos and Sjögren 2006; Schoch et al. 2008). In no case have two *different* ligands both been reported together, and the likelihood of success in such an endeavor is probably very low.

It should be emphasized that not all reports of homotropic cooperativity are necessarily valid. The data for low substrate concentrations, which drive Hill n values, etc., are especially sensitive to substrate depletion in reactions. Low n values and excessive deconvolution are suspect.

One of the outstanding issues is explanation and prediction of cooperativity. In particular, cooperativity has not been reported for some of the P450s for which dual occupancy has been observed (Schoch et al. 2008; Zhao et al. 2012). This lack leaves several questions: can we invoke multiple occupancy as a basis for cooperativity if it is not observed? If multiple occupancy is the basis of cooperativity, should it not be observed when the X-ray structures clearly indicate multiple occupancy? Can we predict when we should expect cooperativity? Can the available crystal structures help in this regard?

2.3.4 Do We Really Understand Cytochrome b_5 Effects?

What is clear is that cytochrome b_5 can stimulate a number of P450 catalytic activities. Although multiple explanations have been proposed (Schenkman and Jansson 2003), most of the explanations fit into two hypotheses: (1) providing the "second" electron (step 4) in the general catalytic cycle (Fig. 2.1) and (2) a general allosteric effect that promotes catalytic efficiency. Either effect may reduce the extent of abortive oxygen loss in the cycle. The distinction between the two proposed mechanisms is usually provided by experiments in which the effects of

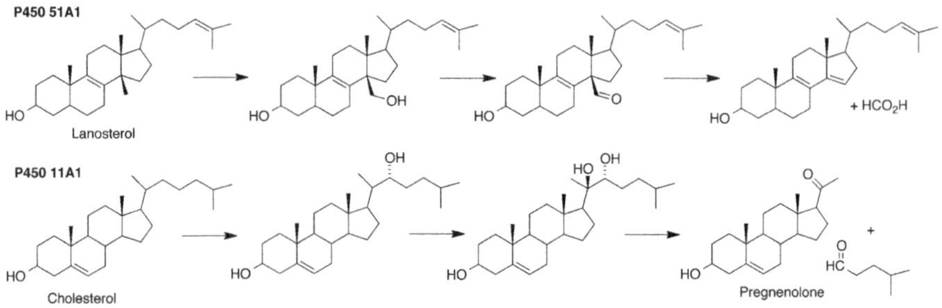

Fig. 2.4 Two multi-step human P450 reactions

holo-cytochrome b_5 and apo-cytochrome b_5 (devoid of heme) are compared. Among the human P450 enzymes that are stimulated by cytochrome b_5, P450s 2A6 (Yun et al. 2005), 2E1, and 4A11 appear to involve electron transfer [i.e., apo-cytochrome is not effective, but P450s 2B6, 2C8, 2C9, 2C19, 3A4, 3A5, and 17A1 (Auchus et al. 1998) do not involve electron transport (Yamazaki et al. 2002)]. [The proposal that the apo-cytochrome b_5 results can be explained by heme transfer from P450s (Guryev et al. 2001) has been dismissed (Yamazaki et al. 2001).]

Biophysical studies with several P450s have provided evidence that the anionic cytochrome b_5 protein is bound to a positively charged region of P450s (Bridges et al. 1998; Estrada et al. 2013). One issue is that this is the same region of P450s proposed to bind NADPH-P450 reductase. Thus, the reductase and cytochrome b_5 would have to switch positions in each cycle of catalysis (Fig. 2.1). This proposal is problematic in a P450 such as 17A1, where cytochrome b_5 only stimulates the second ("lyase") reaction of a two-step sequence, thus influencing the product distribution (Estrada et al. 2013). Further, the reaction is effectively stimulated by apo-cytochrome b_5, arguing against a requirement for electron transfer (Auchus et al. 1998). Thus, if cytochrome b_5 occupies the same site as NADPH-P450 reductase (Estrada et al. 2013), then the reductase must be attached to P450 17A1 from steps **2–4** of the catalytic cycle (Fig. 2.1) and then be released for cytochrome b_5 to bind in steps **5–8**. Given the instability of the FeO_2^- species, this scenario does not seem intuitive but would be consistent with the observation (Estrada et al. 2013). Further investigation of this phenomenon is still in order, even more than 40 years since the first reports of the cytochrome b_5 effects (Correia and Mannering 1973; Hildebrandt and Estabrook 1971).

2.3.5 Why Are Some Multi-Step P450 Reactions Processive?

At least five physiological P450 reactions involve multiple steps (Fig. 2.4) (Guengerich et al. 2011). Moreover, multiple oxidations of drugs are commonly observed (Ortiz de Montellano and De Voss 2005; Isin and Guengerich 2007). Although one might view processive reactions (i.e., no equilibration of the intermediate products with the medium) as being more efficient, such a phenomenon would make specifically inhibiting one step of a multi-step reaction impossible, for example, the so-called lyase reaction of P450 17A1 (DeVore and Scott 2012). Indeed, drugs have been developed to inhibit only this lyase reaction of P450 17A1 (Hara et al. 2013). We have shown the P450 19A1-catalyzed three-step conversion of the androgen androstenedione to estrone is very distributive (the opposite of processive), with each of the intermediates being released from the enzyme and then bound again. Studies on the processivity of the other multi-step P450 steroid oxidations are in progress.

However, some multi-step P450 reactions are processive, such as oxidations of nitrosamines to carboxylic acids by P450s 2A6 and 2E1 (Chowdhury et al. 2010, 2012). The P450 2E1 oxidation of ethanol to acetic acid is also processive (Bell-Parikh and Guengerich 1999), and Prof. Kurt Kunze and his associates have shown that some amine oxidations are processive (Hanson et al. 2010). These are not physiological substrates, and the question is why these are processive. Many of these processive P450 reactions involve carbonyl intermediates, although there does not appear to be any obvious P450 affinity for them (Guengerich et al. 2011). More information is needed in this area.

2.4 What Have We Left to Learn About the Practical Issues of P450s?

This list is also not intended to be comprehensive, and it may well be that totally new fields will develop that utilize P450s. This view results from the ability of P450s to catalyze oxidation of so many substrates. In a recent conversation I had with Professor Minor J. Coon, he quipped that if organic chemicals are ever found on another planet, they will probably be substrates for at least one of the (terrestrial) P450s.

2.4.1 Can We Use our Knowledge of P450 Structures Productively?

The answer is already "yes," but the real question is how well we can do this in a prospective manner. We now have structures of at least 20 different human P450s (Table 2.1), including all the major "drug-metabolizing" P450s. Most P450s have shown extensive changes upon binding ligands, and therefore ligand-free P450 crystal structures are limited in their usefulness. The malleability of P450 structures is a problem in that different ligands can yield different protein structures for a single P450, for example, P450 3A4 (Ekroos and Sjögren 2006) (rabbit) and P450 2B4 (Shah et al. 2013). Thus, even having a structure of a P450–substrate complex does not necessarily predict the structure of that P450 with a new substrate or ligand.

In practice, drug discovery and development operate on what is sometimes called a “2-week” scale, when new leads need to be reevaluated in a short time. The time needed to obtain crystals and solve a structure is still longer than two weeks and therefore not compatible with industry needs. The hope is that as scientists collect more P450 structures, each P450 will have a finite but limited number of major conformations and that there can be used practically with new substrates. Obviously, this is a very important area for practical research. My own opinion is that actual experiments will still need to be done after virtual screening. The prediction of rates of oxidation is much more difficult than (qualitative) prediction of sites of oxidation in a molecule.

2.4.2 Can We Relate P450 Polymorphisms to Chronic Diseases?

As mentioned earlier, knowledge about human P450s has been very useful in advancing drug development and even clinical practice. Polymorphisms can have dramatic influences on drug metabolism. As also pointed out earlier, deficiencies in the steroid-metabolizing P450s have major consequences.

However, what is not yet very clear is how variations in P450s affect chronic diseases. There are several issues here. One is cancer. P450s were studied extensively in part because of chemical carcinogenesis (Conney 1982; Guengerich 1988a, b), and two thirds of the bioactivation reactions with carcinogens are catalyzed by P450s (one half of which are family 1 P450s) (Fig. 2.3) (Rendic and Guengerich 2012). Changes in P450s in animal models can have dramatic effects on cancer incidence (Guengerich 1988b; Gonzalez 2004), but to date relationships in humans have not been so clear. The early relationship of inducibility of aryl hydrocarbon hydroxylation activation with lung cancer (Kellerman et al. 1973a, b) is probably best understood in the context of P450 1B1 (Toide et al. 2003), although this field does not seem to have been addressed again recently. Despite early excitement (Ayesh et al. 1984), the association of decreased lung cancer with the P450 2D6 poor metabolizer phenotype could never be validated (Rostami-Hodjegan et al. 1998). One of the issues in this field is that epidemiology proceeds in the absence of basic science. There is a weak association of P450 1A2 with colon cancer, but only if *N*-acetyltransferase status and consumption of well-done meat are factored in (Lang et al. 1994). Another possibility is P450 2A6 and tobacco-related cancer, which is complicated in that individuals with a deficiency may smoke less because they lack the ability to clear nicotine (Swan et al. 2005). Other epidemiological association of cancers with P450 polymorphisms have either been weak or not confirmed in more extensive studies (d’Errico et al. 1996; Tamaki et al. 2011).

Other efforts have shown weak associations between P450 2D6 status and Parkinson’s disease (Halling et al. 2008).

An interesting relationship has been reported for a polymorphism in P450 4A11 (rs1126742) and hypertension (Gainer et al. 2005). The basis is still unclear. P450 4A11 converts arachidonic acid to its 20-hydroxy product (20-HETE), and the polymorphic variant (coding for a P434S mutation) has been reported to have 60 % of the catalytic efficiency of the wild-type protein (Gainer et al. 2005). However, *Cyp4a10*(−/−) mice show hypertension but do not produce 20-HETE (Nakagawa et al. 2006), and a transgenic mouse that expresses the human 20-hydroxylase, P450 4A11 (Savas et al. 2009), is also hypertensive (E.F. Johnson, personal communication). In mouse models, deletion of *Cyp2c* subfamily genes can also make animals hypersensitive (Sun et al. 2012). Other work with animals has led to postulates of roles of P450 4F and 2J enzymes in hypertension and other cardiovascular effects (Deng et al. 2011; Yang et al. 2001).

These examples show the complexity of the association. Clearly, more questions remain to be addressed.

2.4.3 Application of P450s in Toxicology and Other Assays

Another challenge regarding P450s is their effective application in toxicology. Many successes have already been realized with transgenic mouse models, for example, roles of individual P450s in the toxicity of certain chemicals (Lee et al. 1996). These approaches will continue to develop, particularly as key enzyme and receptor systems are "humanized," that is, mouse systems are replaced (in the mice) by the human counterparts. The success of this approach has already been demonstrated in mice expressing the human AhR (aryl hydrocarbon receptor) and PPARα (peroxisomal proliferating activator receptor-α) receptors (Moriguchi et al. 2003; Yang et al. 2008). Another developing area with transgenic mice may be in response to the need to characterize "human-specific" metabolites (i.e., those constituting ≥10 % of metabolites not found in animals) for safety testing (Guengerich 2006).

Another important need is in high-throughput toxicology assays. The importance of metabolic capability is generally appreciated in the pharmaceutical industry, given the history (Kola and Landis 2004), and the emphasis on adsorption/distribution/metabolism/excretion work in drug development. However, this is not necessarily the case with other chemicals. A battery of in vitro toxicity tests is being applied to a very large number of compounds in the Toxcast testing program supported by several agencies of the United States government (Sipes et al. 2013), but in the absence of metabolism systems. Although this approach may yield information about the intrinsic toxicities of individual chemicals, the predictive value of the exercise may be limited, just as was the relationship between mutagens and carcinogens before the incorporation of liver extracts with the Ames test (Ames et al. 1973). Aflatoxin B_1 would not be considered very toxic without bioactivation, and the mutagenicity of dinitropyrenes would be overestimated. Exactly which systems should be used needs to be decided for large-scale, low-volume screens. Individual P450 systems are probably not the most appropriate for broad screening, and a multi-P450 system (including P450s known to oxidize most of the known drugs and carcinogens) (Rendic and Guengerich 2012; Wienkers and Heath 2005; Williams et al. 2004) should be considered, coexpressed with NADPH-P450 reductase. This is one possibility, and others might be better, but this topic deserves attention.

2.4.4 Applications with Plants and Agriculture

The final area for consideration is agriculture. As mentioned earlier, plant genomes contain hundreds of P450s. We know little about which of these might be significant in the production of important crops, such as rice, corn (maize), and soybeans. State-of-the-art genomic approaches can now be applied, such as genome-wide association studies (GWAS) for traits and analysis of phenotypic function with gene knockout technology. Using such approaches, we may well find that certain P450s are beneficial (and can be overexpressed) and that others are detrimental to desired properties (and could be targets for development of inhibitory chemicals).

The approaches are designed to increase crop production. Another important aspect is control of pests, including weeds, insects, and fungi. These genomes are being developed rapidly, as are similar approaches (GWAS, knockouts) to find targets for inhibition of function. As an example, one could develop a herbicide from an inhibitor of a key P450 in a weed. Another approach is to overexpress a herbicide-metabolizing P450 gene into a cereal crop (e.g., rice or corn). These approaches are very feasible and can be used without the ethical issues inherent in human medicine. I realize that there is political resistance to biotechnology in Europe, but the need to feed the 7 billion people in the world should trump such concerns, given the very safe record of genomic agriculture today.

The last point under agricultural application of P450s involves animals and veterinary issues. P450 applications in human medicine have developed rapidly, but the same concerns about drug–drug interactions, polymorphisms, etc. certainly apply in veterinary pharmacology. I recently reviewed a grant application (from another country) and learned that the P450 repertoire in horses is largely unknown, beyond the genomic level, and the prescription of drugs cannot be guided by such information. I am sure that we also need to know more about P450s in important domestic animals such as cattle, swine, and sheep (as well as sport and companion animals, e.g., horses, dogs, cats) to treat them more effectively.

2.5 Epilogue

The P450 world has been exciting for many of us. We have already seen tremendous advancement at the basic level, and experiments we could only dream of 40 years ago are routine today. The number of papers published on P450s continues to grow (http://webtools.mf.uni-lj.si/public/medsum.html). As the field has matured, the bar for publishing P450 papers continues to be set higher. The success of P450 research in medicine has been remarkable, more than first imagined. The application of P450 research continues to grow in many fields, and we will undoubtedly see more of this progress.

Finally, it remains for me to thank two of our senior members of the P450 field, Professors Tsuneo Omura and Minor J. Coon, for their discoveries and contributions to the field, which are very much appreciated.

Acknowledgments I thank Kathleen Trisler for her assistance in preparation of the manuscript. P450 research in this laboratory is currently supported by National Institutes of Health grants R37 CA090426, P01 DK038226, and R01 GM103937.

Finally, this chapter is dedicated to the memory of two of the pioneers in this field, Professors Ronald W. Estabrook and Allan H. Conney. Both died in 2013 (August and September, respectively). Professor Estabrook was involved in critical experiments that established P450 as the terminal oxidase in the microsomal electron transport chain. Professor Conney was involved in the discovery of P450 induction as a graduate student with Professors James and Elizabeth Miller, made important contributions regarding the multiplicity of P450s, and, together with Dr. Donald Jerina, established the bay-region diol epoxide pathway for activation of polycyclic aromatic hydrocarbons and its significance in chemical carcinogenesis.

References

Addya S, Anandatheerthavarada HK, Biswas G, Bhagwat SV, Mullick J, Avadhani NG (1997) Targeting of NH_2-terminal-processed microsomal protein to mitochondria: a novel pathway for the biogenesis of hepatic mitochondrial $P450_{mt2}$. J Cell Biol 139:589–599

Akhtar M, Calder MR, Corina DL, Wright JN (1982) Mechanistic studies on C-19 demethylation in oestrogen biosynthesis. Biochem J 201:569–580

Akhtar M, Corina D, Miller S, Shyadehi AZ, Wright JN (1994) Mechanism of the acyl-carbon cleavage and related reactions catalyzed by multifunctional P-450s: studies on cytochrome $P450_{17\alpha}$. Biochemistry 33:4410–4418

Alvares AP, Schilling G, Levin W, Kuntzman R (1967) Studies on the induction of CO-binding pigments in liver microsomes by phenobarbital and 3-methylcholanthrene. Biochem Biophys Res Commun 29:521–526

Ames BN, Durston WE, Yamasaki E, Lee FD (1973) Carcinogens are mutagens: a simple test system combining liver homogenates for activation and bacteria for detection. Proc Natl Acad Sci USA 70:2281–2285

Andersson T, Flockhart DA, Goldstein DB, Huang SM, Kroetz DL, Milos PM, Ratain MJ, Thummel K (2005) Drug-metabolizing enzymes: evidence for clinical utility of pharmacogenomic tests. Clin Pharmacol Ther 78:559–581

Aoyama Y, Horiuchi T, Gotoh O, Noshiro M, Yoshida Y (1998) *Cyp51*-like gene of *Mycobacterium tuberculosis* actually encodes a P450 similar to eukaryotic CYP51. J Biochem (Tokyo) 124:694–696

Auchus RJ, Lee TC, Miller WL (1998) Cytochrome b_5 augments the 17,20-lyase activity of human P450c17 without direct electron transfer. J Biol Chem 273:3158–3165

Ayesh R, Idle JR, Ritchie JC, Crothers MJ, Hetzel MR (1984) Metabolic oxidation phenotypes as markers for susceptibility to lung cancer. Nature (Lond) 312:169–170

Bajpai P, Sangar MC, Tang W, Chowdhury G, Cheng Q, Fang J-K, Martin MV, Guengerich FP, Avadhani NG (2013) Metabolism of 1-methyl-4-phenyl-1,2,3,6-tetrahydropyridine by

mitochondria-targeted cytochrome P450 2D6: implications for Parkinson's disease. J Biol Chem 288:4436–4451
Bansal S, Liu C-P, Sepuri NBV, Anandatheerthavarada HK, Guengerich FP, Avadhani NG (2010) Mitochondria-targeted cytochrome P450 2E1 preferentially induces oxidative damage and augments alcohol mediated mitochondrial dysfunction in cultured cells. J Biol Chem 285:24609–24619
Beaune P, Kremers PG, Kaminsky LS, de Graeve J, Guengerich FP (1986) Comparison of monooxygenase activities and cytochrome P-450 isozyme concentrations in human liver microsomes. Drug Metab Dispos 14:437–442
Bell-Parikh LC, Guengerich FP (1999) Kinetics of cytochrome P450 2E1-catalyzed oxidation of ethanol to acetic acid via acetaldehyde. J Biol Chem 274:23833–23840
Bolt HM, Kappus H, Bolt M (1975) Effect of rifampicin treatment on the metabolism of oestradiol and 17α-ethinyloestradiol by human liver microsomes. Eur J Clin Pharmacol 8:301–307
Bridges A, Gruenke L, Chang Y-T, Vakser IA, Loew G, Waskell L (1998) Identification of the binding site on cytochrome P450 2B4 for cytochrome b_5 and cytochrome P450 reductase. J Biol Chem 273:17036–17049
Brodie AMH (1985) Aromatase inhibition and its pharmacologic implications. Biochem Pharmacol 34:3213–3219
Cheng JB, Motola DL, Mangelsdorf DJ, Russell DW (2003) De-orphanization of cytochrome P450 2R1: a microsomal vitamin D 25-hydroxylase. J Biol Chem 278:38084–38093
Cheng Q, Lamb DC, Kelly SL, Li L, Guengerich FP (2010) Cyclization of a cellular dipentaenone by *Streptomyces coelicolor* cytochrome P450 154A1 without oxidation reduction. J Am Chem Soc 132:15173–15175
Chowdhury G, Calcutt MW, Guengerich FP (2010) Oxidation of *N*-nitrosodimethylamine and *N*-nitrosodiethylamine by human cytochrome P450 2A6: sequential oxidation to carboxylic acids and analysis of reaction steps. J Biol Chem 285:8031–8044
Chowdhury G, Calcutt MW, Nagy LD, Guengerich FP (2012) Oxidation of methyl and ethyl nitrosamines by cytochromes P450 2E1 and 2B1. Biochemistry 51:9995–10007
Coelho PS, Brustad EM, Kannan A, Arnold FH (2013) Olefin cyclopropanation via carbene transfer catalyzed by engineered cytochrome P450 enzymes. Science 339:307–310
Conney AH (1982) Induction of microsomal enzymes by foreign chemicals and carcinogenesis by polycyclic aromatic hydrocarbons: G. H. A. Clowes Memorial Lecture. Cancer Res 42:4875–4917
Correia MA, Mannering GJ (1973) Reduced diphosphopyridine nucleotide synergism of the reduced triphosphopyridine nucleotide-dependent mixed-function oxidase system of hepatic microsomes. II. Role of the type I drug-binding site of cytochrome P-450. Mol Pharmacol 9:470–485
d'Errico A, Taioli E, Chen X, Vineis P (1996) Genetic metabolic polymorphisms and the risk of cancer: a review of the literature. Biomarkers 1:149–173
Daly AK, Day CP, Aithal GP (2002) CYP2C9 polymorphism and warfarin dose requirements. Br J Clin Pharmacol 53:408–409
Davydov DR, Botchkareva AE, Davydova NE, Halpert JR (2005) Resolution of two substrate-binding sites in an engineered cytochrome $P450_{eryf}$ bearing a fluorescent probe. Biophys J 89:418–432
Deng Y, Edin ML, Theken KN, Schuck RN, Flake GP, Kannon MA, DeGraff LM, Lih FB, Foley J, Bradbury JA, Graves JP, Tomer KB, Falck JR, Zeldin DC, Lee CR (2011) Endothelial CYP epoxygenase overexpression and soluble epoxide hydrolase disruption attenuate acute vascular inflammatory responses in mice. FASEB J 25:703–713
DeVore NM, Scott EE (2012) Structures of cytochrome P450 17A1 with prostate cancer drugs abiraterone and TOK-001. Nature (Lond) 482:116–119
Diehl H, Schädelin J, Ullrich V (1970) Studies on the kinetics of cytochrome P-450 reduction in rat liver microsomes. Hoppe Seylers Z Physiol Chem 351:1359–1371

Distlerath LM, Reilly PEB, Martin MV, Davis GG, Wilkinson GR, Guengerich FP (1985) Purification and characterization of the human liver cytochromes P-450 involved in debrisoquine 4-hydroxylation and phenacetin *O*-deethylation, two prototypes for genetic polymorphism in oxidative drug metabolism. J Biol Chem 260:9057–9067

Edson K, Prasad B, Unadkat JD, Suhara Y, Okano T, Guengerich FP, Rettie AE (2013) Cytochrome P450 dependent catabolism of vitamin K: initiation of ω-hydroxylation of human CYP4F2 and CYP4F11. Biochemistry 52:8276–8285

Ekroos M, Sjögren T (2006) Structural basis for ligand promiscuity in cytochrome P450 3A4. Proc Natl Acad Sci USA 103:13862–13867

Estabrook RW, Cooper DY, Rosenthal O (1963) The light reversible carbon monoxide inhibition of the steroid C21-hydroxylase system of the adrenal cortex. Biochem Z 338:741–755

Estabrook RW, Franklin MR, Cohen B, Shigamatzu A, Hildebrandt AG (1971) Biochemical and genetic factors influencing drug metabolism. Influence of hepatic microsomal mixed function oxidation reactions on cellular metabolic control. Metabolism 20:187–199

Estrada DF, Laurence JS, Scott EE (2013) Substrate-modulated cytochrome P450 17A1 and cytochrome b_5 interactions revealed by NMR. J Biol Chem 288:17008–17018

Evans WE, McLeod HL (2003) Pharmacogenomics–drug disposition, drug targets, and side effects. N Engl J Med 348:538–549

Gainer JV, Bellamine A, Dawson EP, Womble KE, Grant SW, Wang Y, Cupples LA, Guo CY, Demissie S, O'Donnell CJ, Brown NJ, Waterman MR, Capdevila JH (2005) Functional variant of CYP4A11 20-hydroxyeicosatetraenoic acid synthase is associated with essential hypertension. Circulation 111:63–69

Garcia DA, Hylek E (2009) Warfarin pharmacogenetics. N Engl J Med 360:2474, author reply 2475

Garfinkel D (1958) Studies on pig liver microsomes. I. Enzymic and pigment composition of different microsomal fractions. Arch Biochem Biophys 77:493–509

Gigon PL, Gram TE, Gillette JR (1969) Studies on the rate of reduction of hepatic microsomal cytochrome P-450 by reduced nicotinamide adenine dinucleotide phosphate: effect of drug substrates. Mol Pharmacol 5:109–122

Gillette JR, Brodie BB, La Du BN (1957) The oxidation of drugs by liver microsomes: on the role of TPNH and oxygen. J Pharmacol Exp Ther 119:532–540

Gomez A, Ingleman-Sundberg M (2009) Epigenetic and microRNA-dependent control of cytochrome P450 expression: a gap between DNA and protein. Pharmacogenomics 10:1067–1076

Gonzalez FJ (2004) Cytochrome P450 humanised mice. Hum Genomics 1:300–306

Gonzalez FJ, Skoda RC, Kimura S, Umeno M, Zanger UM, Nebert DW, Gelboin HV, Hardwick JP, Meyer UA (1988) Characterization of the common genetic defect in humans deficient in debrisoquine metabolism. Nature (Lond) 331:442–446

Gorsky LD, Koop DR, Coon MJ (1984) On the stoichiometry of the oxidase and monooxygenase reactions catalyzed by liver microsomal cytochrome P-450: products of oxygen reduction. J Biol Chem 259:6812–6817

Guengerich FP (1988a) Oxidation of 17α-ethynylestradiol by human liver cytochrome P-450. Mol Pharmacol 33:500–508

Guengerich FP (1988b) Roles of cytochrome P-450 enzymes in chemical carcinogenesis and cancer chemotherapy. Cancer Res 48:2946–2954

Guengerich FP (2001) Common and uncommon cytochrome P450 reactions related to metabolism and chemical toxicity. Chem Res Toxicol 14:611–650

Guengerich FP (2002) Cytochrome P450 enzymes in the generation of commercial products. Nat Rev Drug Discov 1:359–366

Guengerich FP (2005) Human cytochrome P450 enzymes. In: Ortiz de Montellano PR (ed) Cytochrome P450: structure, mechanism, and biochemistry, 3rd edn. Kluwer Academic/Plenum Press, New York, pp 377–530

Guengerich FP (2006) Safety assessment of stable drug metabolites. Chem Res Toxicol 19:1559–1560

Guengerich FP (2013) Kinetic deuterium isotope effects in cytochrome P450 reactions. J Labelled Comp Radiopharm 56:428–431

Guengerich FP (2014) Cytochrome P450-mediated drug interactions and cardiovascular toxicity: the Seldane to Allegra transformation. In: Wang J, Urban L (eds) Predictive ADMET: integrated approaches in drug discovery and development. Wiley, New York, Chap. 23, pp 523–534

Guengerich FP, Cheng Q (2011) Orphans in the human cytochrome P450 family: approaches to discovering function and relevance to pharmacology. Pharmacol Rev 63:684–699

Guengerich FP, Isin EM (2014) Unusual metabolic reactions and pathways. In: Lee P, Aizawa H, Gau L, Prakash C, Zhong D (eds) The handbook of metabolic pathways of xenobiotics. Wiley, Chichester, UK, pp 147–197.

Guengerich FP, Johnson WW (1997) Kinetics of ferric cytochrome P450 reduction by NADPH-cytochrome P450 reductase: rapid reduction in absence of substrate and variations among cytochrome P450 systems. Biochemistry 36:14741–14750

Guengerich FP, Munro AW (2013) Unusual cytochromes P450: enzymes and reactions. J Biol Chem 288:17063–17069

Guengerich FP, Shimada T (1991) Oxidation of toxic and carcinogenic chemicals by human cytochrome P-450 enzymes. Chem Res Toxicol 4:391–407

Guengerich FP, Kim B-R, Gillam EMJ, Shimada T (1994) Mechanisms of enhancement and inhibition of cytochrome P450 catalytic activity. In: Lechner MC (ed) Proceedings of the 8th international conference on cytochrome P450: biochemistry, biophysics, and molecular biology. John Libbey Eurotext, Paris, pp 97–101

Guengerich FP, Sohl CD, Chowdhury G (2011) Multi-step oxidations catalyzed by cytochrome P450 enzymes: processive vs. distributive kinetics and the issue of carbonyl oxidation. Arch Biochem Biophys 507:126–134

Guest EJ, Rowland-Yeo K, Rostami-Hodjegan A, Tucker GT, Houston JB, Galetin A (2011) Assessment of algorithms for predicting drug–drug interactions via inhibition mechanisms: comparison of dynamic and static models. Br J Clin Pharmacol 71:72–87

Guryev OL, Gilep AA, Usanov SA, Estabrook RW (2001) Interaction of apo-cytochrome b_5 with cytochromes P4503A4 and P45017A: relevance of heme transfer reactions. Biochemistry 40:5018–5031

Gut J, Catin T, Dayer P, Kronbach T, Zanger U, Meyer UA (1986) Debrisoquine/sparteine-type polymorphism of drug oxidation: purification and characterization of two functionally different human liver cytochrome P-450 isozymes involved in impaired hydroxylation of the prototype substrate bufuralol. J Biol Chem 261:11734–11743

Hackett JC, Brueggemeier RW, Hadad CM (2005) The final catalytic step of cytochrome P450 aromatase: a density functional theory study. J Am Chem Soc 127:5224–5237

Halling J, Petersen MS, Grandjean P, Weihe P, Brosen K (2008) Genetic predisposition to Parkinson's disease: CYP2D6 and HFE in the Faroe Islands. Pharmacogenet Genomics 18:209–212

Hanson KL, VandenBrink BM, Babu KN, Allen KE, Nelson WL, Kunze KL (2010) Sequential metabolism of secondary alkyl amines to metabolic-intermediate complexes: opposing roles for the secondary hydroxylamine and primary amine metabolites of desipramine, (*S*)-fluoxetine, and *N*-desmethyldiltiazem. Drug Metab Dispos 38:963–972

Hara T, Kouno J, Kaku T, Takeuchi T, Kusaka M, Tasaka A, Yamaoka M (2013) Effect of a novel 17,20-lyase inhibitor, orteronel (TAK-700), on androgen synthesis in male rats. J Steroid Biochem Mol Biol 134:80–91

Harris N, Cohen S, Filatov M, Ogliaro F, Shaik S (2000) Two-state reactivity in the rebound step of alkane hydroxylation by cytochrome P-450: origins of free radicals with finite lifetimes. Angew Chem Int Ed 39:2003–2007

Henderson CJ, McLaughlin LA, Wolf CR (2013) Evidence that cytochrome b_5 and cytochrome b_5 reductase can act as sole electron donors to the hepatic cytochrome P450 systems. Mol Pharmacol 83:1209–1217

Hildebrandt A, Estabrook RW (1971) Evidence for the participation of cytochrome b_5 in hepatic microsomal mixed-function oxidation reactions. Arch Biochem Biophys 143:66–79

Hildebrandt A, Remmer H, Estabrook RW (1968) Cytochrome P-450 of liver microsomes: one pigment or many. Biochem Biophys Res Commun 30:607–612

Hosea NA, Miller GP, Guengerich FP (2000) Elucidation of distinct binding sites for cytochrome P450 3A4. Biochemistry 39:5929–5939

Hughes AL, Powell DW, Bard M, Eckstein J, Barbuch R, Link AJ, Espenshade PJ (2007) DAP1/PGRMC1 binds and regulates cytochrome P450 enzymes. Cell Metab 5:143–149

Humphreys WG (2008) Drug metabolism research as an integral part of the drug discovery process. In: Zhang D, Zhu M, Humphreys WG (eds) Drug metabolism in drug design and development. Wiley, Hoboken, Chap. 8, pp 239–260

Idle JR, Smith RL (1979) Polymorphisms of oxidation at carbon centers of drugs and their clinical significance. Drug Metab Rev 9:301–317

Idle JR, Mahgoub A, Lancaster R, Smith RL (1978) Hypotensive response to debrisoquine and hydroxylation phenotype. Life Sci 22:979–984

Isin EM, Guengerich FP (2006) Kinetics and thermodynamics of ligand binding by cytochrome P450 3A4. J Biol Chem 281:9127–9136

Isin EM, Guengerich FP (2007) Complex reactions catalyzed by cytochrome P450 enzymes. Biochim Biophys Acta 1770:314–329

Ito K, Iwatsubo T, Kanamitsu S, Ueda K, Suzuki H, Sugiyama Y (1998) Prediction of pharmacokinetic alterations caused by drug–drug interactions: metabolic interaction in the liver. Pharmacol Rev 50:387–411

Johnston JB, Ouellet H, Ortiz de Montellano PR (2010) Functional redundancy of steroid C26-monooxygenase activity in *Mycobacterium tuberculosis* revealed by biochemical and genetic analyses. J Biol Chem 285:36352–36360

Kalow W (1962) Pharmacogenetics. Saunders, Philadelphia

Katagiri M, Ganguli BN, Gunsalus IC (1968) A soluble cytochrome P450 functional in methylene hydroxylation. J Biol Chem 243:3543–3546

Kellerman G, Luyten-Kellerman M, Shaw CR (1973a) Genetic variation of aryl hydrocarbon hydroxylase in human lymphocytes. Am J Hum Genet 25:327–331

Kellerman G, Shaw CR, Luyten-Kellerman M (1973b) Aryl hydrocarbon hydroxylase inducibility and bronchogenic carcinoma. N Engl J Med 298:934–937

Kinney AJ (2006) Metabolic engineering in plants for human health and nutrition. Curr Opin Biotechnol 17:130–138

Klingenberg M (1958) Pigments of rat liver microsomes. Arch Biochem Biophys 75:376–386

Kola I, Landis J (2004) Can the pharmaceutical industry reduce attrition rates? Nat Rev Drug Discov 3:711–715

Lang NP, Butler MA, Massengill J, Lawson M, Stotts RC, Maurer-Jensen M, Kadlubar FF (1994) Rapid metabolic phenotypes for acetyltransferase and cytochrome P4501A2 and putative exposure to food-borne heterocyclic amines increase the risk for colorectal cancer or polyps. Cancer Epidemiol Biom 3:675–682

Lee SST, Buters JTM, Pineau T, Fernandez-Salguero P, Gonzalez FJ (1996) Role of Cyp2e1 in the hepatotoxicity of acetaminophen. J Biol Chem 271:12063–12067

Lu AYH, Coon MJ (1968) Role of hemoprotein P-450 in fatty acid ω-hydroxylation in a soluble enzyme system from liver microsomes. J Biol Chem 243:1331–1332

Mahgoub A, Idle JR, Dring LG, Lancaster R, Smith RL (1977) Polymorphic hydroxylation of debrisoquine in man. Lancet 2:584–586

Mangelsdorf DJ, Evans RM (1995) The RXR heterodimers and orphan receptors. Cell 83:841–850

McLean KJ, Sabri M, Marshall KR, Lawson RJ, Lewis DG, Clift D, Balding PR, Dunford AJ, Warman AJ, McVey JP, Quinn AM, Sutcliffe MJ, Scrutton NS, Munro AW (2005) Biodiversity of cytochrome P450 redox systems. Biochem Soc Trans 33:796–801

Miller WL, Auchus RJ (2011) The molecular biology, biochemistry, and physiology of human steroidogenesis and its disorders. Endocr Rev 32:81–151

Mizutani M, Sato F (2011) Unusual P450 reactions in plant secondary metabolism. Arch Biochem Biophys 507:194–203

Moriguchi T, Motohashi H, Hosoya T, Nakajima O, Takahashi S, Ohsako S, Aoki Y, Nishimura N, Tohyama C, Fujii-Kuriyama Y, Yamamoto M (2003) Distinct response to dioxin in an arylhydrocarbon receptor (*Ahr*)-humanized mouse. Proc Natl Acad Sci USA 100:5652–5657

Mueller GC, Miller JA (1948) The metabolism of 4-dimethylaminoazobenzene by rat liver homogenates. J Biol Chem 176:535–544

Mueller GC, Miller JA (1953) The metabolism of methylated aminoazo dyes. II. Oxidative demethylation by rat liver homogenates. J Biol Chem 202:579–587

Mueller EJ, Loida PJ, Sligar SG (1995) Twenty-five years of $P450_{cam}$ research: mechanistic insights into oxygenase catalysis. In: Ortiz de Montellano PR (ed) Cytochrome P450: structure, mechanism, and biochemistry, 2nd edn. Plenum, New York, Chap. 3, pp 83–124

Mutoh S, Sobhany M, Moore R, Perera L, Pedersen L, Sueyoshi T, Negishi M (2013) Phenobarbital indirectly activates the constitutive active androstane receptor (CAR) by inhibition of epidermal growth factor receptor signaling. Sci Signal 6:ra31

Nakagawa K, Holla VR, Wei Y, Wang WH, Gatica A, Wei S, Mei S, Miller CM, Cha DR, Price E Jr, Zent R, Pozzi A, Breyer MD, Guan Y, Falck JR, Waterman MR, Capdevila JH (2006) Salt-sensitive hypertension is associated with dysfunctional *Cyp4a10* gene and kidney epithelial sodium channel. J Clin Invest 116:1696–1702

Nebert DW, Russell DW (2002) Clinical importance of the cytochromes P450. Lancet 360:1155–1162

Nelson DO, Lorusso DJ, Mannering GJ (1973) Requirement of a soluble protein for maximal activity of the monooxidase system of hepatic microsomes. Biochem Biophys Res Commun 53:995–1001

Niranjan BG, Avadhani NG (1980) Activation of aflatoxin B_1 by a monooxygenase system localized in rat liver mitochondria. J Biol Chem 255:6575–6578

Nishida CR, Lee M, Ortiz de Montellano PR (2010) Efficient hypoxic activation of the anticancer agent AQ4N by CYP2S1 and CYP2W1. Mol Pharmacol 78:497–502

Oates NS, Shah RR, Idle JR, Smith RL (1981) Phenformin-induced lactic acidosis associated with impaired debrisoquine hydroxylation. Lancet 1:837–838

Omura T, Sato R (1962) A new cytochrome in liver microsomes. J Biol Chem 237:1375–1376

Omura T, Sato R (1964a) The carbon monoxide-binding pigment of liver microsomes. I. Evidence for its hemoprotein nature. J Biol Chem 239:2370–2378

Omura T, Sato R (1964b) The carbon monoxide-binding pigment of liver microsomes. II. Solubilization, purification, and properties. J Biol Chem 239:2379–2385

Ortiz de Montellano PR, De Voss JJ (2005) Substrate oxidation by cytochrome P450 enzymes. In: Ortiz de Montellano PR (ed) Cytochrome P450: structure, mechanism, and biochemistry, 3rd edn. Kluwer Academic/Plenum Press, New York, pp 183–245

Ortiz de Montellano PR, DeVoss JJ (2002) Oxidizing species in the mechanism of cytochrome P450. Nat Prod Rep 19:477–493

Peterson JA, Ebel RE, O'Keeffe DH, Matsubara T, Estabrook RW (1976) Temperature dependence of cytochrome P-450 reduction. A model for NADPH-cytochrome P-450 reductase: cytochrome P-450 interaction. J Biol Chem 251:4010–4016

Plum LA, DeLuca HF (2010) Vitamin D, disease and therapeutic opportunities. Nat Rev Drug Discov 9:941–955

Poulos TL, Finzel BC, Gunsalus IC, Wagner GC, Kraut J (1985) The 2.6-Å crystal structure of *Pseudomonas putida* cytochrome P-450. J Biol Chem 260:16122–16130

Rendic S, Guengerich FP (2012) Contributions of human enzymes in carcinogen metabolism. Chem Res Toxicol 25:1316–1383

Rittle J, Green MT (2010) Cytochrome P450 compound I: capture, characterization, and C–H bond activation kinetics. Science 330:933–937

Rostami-Hodjegan A, Lennard MS, Woods HF, Tucker GT (1998) Meta-analysis of studies of the CYP2D6 polymorphism in relation to lung cancer and Parkinson's disease. Pharmacogenetics 8:227–238

Ryan KJ (1958) Conversion of androstenedione to estrone by placental microsomes. Biochim Biophys Acta 27:658–662

Savas U, Machemer DE, Hsu MH, Gaynor P, Lasker JM, Tukey RH, Johnson EF (2009) Opposing roles of peroxisome proliferator-activated receptor alpha and growth hormone in the regulation of *CYP4A11* expression in a transgenic mouse model. J Biol Chem 284:16541–16552

Schenkman JB, Jansson I (2003) The many roles of cytochrome b_5. Pharmacol Ther 97:139–152

Schoch GA, Yano JK, Sansen S, Dansette PM, Stout CD, Johnson EF (2008) Determinants of cytochrome P450 2C8 substrate binding: structures of complexes with montelukast, troglitazone, felodipine, and 9-*cis*-retinoic acid. J Biol Chem 283:17227–17237

Sen K, Hackett JC (2012) Coupled electron transfer and proton hopping in the final step of CYP19-catalyzed androgen aromatization. Biochemistry 51:3039–3049

Sevrioukova IF, Poulos TL (2012) Structural and mechanistic insights into the interaction of cytochrome P450 3A4 with bromoergocryptine, a type I ligand. J Biol Chem 287:3510–3517

Seward HE, Roujeinikova A, McLean KJ, Munro AW, Leys D (2006) Crystal structure of the *Mycobacterium tuberculosis* P450 CYP121-fluconazole complex reveals new azole drug-P450 binding mode. J Biol Chem 281:39437–39443

Shah RR, Oates NS, Idle JR, Smith RL, Lockhart JDF (1982) Impaired oxidation of debrisoquine in patients with perhexiline neuropathy. Br Med J 284:295–299

Shah MB, Kufareva I, Pascual J, Zhang QH, Stout CD, Halpert JR (2013) A structural snapshot of CYP2B4 in complex with paroxetine provides insights into ligand binding and clusters of conformational states. J Pharmacol Exp Ther 346:113–120

Shaik S, Kumar D, de Visser SP, Altun A, Thiel W (2005) Theoretical perspective on the structure and mechanism of cytochrome P450 enzymes. Chem Rev 105:2279–2328

Shiro Y, Fujii M, Iizuka T, Adachi S, Tsukamoto K, Nakahara K, Shoun H (1995) Spectroscopic and kinetic studies on reaction of cytochrome $P450_{nor}$ with nitric oxide: implication for its nitric oxide reduction mechanism. J Biol Chem 270:1617–1623

Shou M, Grogan J, Mancewicz JA, Krausz KW, Gonzalez FJ, Gelboin HV, Korzekwa KR (1994) Activation of CYP3A4: evidence for the simultaneous binding of two substrates in a cytochrome P450 active site. Biochemistry 33:6450–6455

Siller M, Goyal S, Yoshimoto FK, Xiao Y, Wei S, Guengerich FP (2014) Oxidation of endogenous N-arachidonoylserotonin by human cytochrome P450 2U1. J Biol Chem 289: 10476–10487

Sipes NS, Martin MT, Kothiya P, Reif DM, Judson RS, Richard AM, Houck KA, Dix DJ, Kavlock RJ, Knudsen TB (2013) Profiling 976 toxcast chemicals across 331 enzymatic and receptor signaling assays. Chem Res Toxicol 26:878–895

Sladek NE, Mannering GJ (1969) Induction of drug metabolism. II. Qualitative differences in the microsomal N-demethylating systems stimulated by polycyclic hydrocarbons and by phenobarbital. Mol Pharmacol 5:186–199

Sligar SG, Denisov IG (2007) Understanding cooperativity in human P450 mediated drug–drug interactions. Drug Metab Rev 39:567–579

Sun P, Antoun J, Lin DH, Yue P, Gotlinger KH, Capdevila J, Wang WH (2012) Cyp2c44 epoxygenase is essential for preventing the renal sodium absorption during increasing dietary potassium intake. Hypertension 59:339–347

Swan GE, Benowitz NL, Lessov CN, Jacob P 3rd, Tyndale RF, Wilhelmsen K (2005) Nicotine metabolism: the impact of CYP2A6 on estimates of additive genetic influence. Pharmacogenet Genomics 15:115–125

Tamaki Y, Arai T, Sugimura H, Sasaki T, Honda M, Muroi Y, Matsubara Y, Kanno S, Ishikawa M, Hirasawa N, Hiratsuka M (2011) Association between cancer risk and drug-metabolizing enzyme gene (CYP2A6, CYP2A13, CYP4B1, SULT1A1, GSTM1, AND GSTT1) polymorphisms in cases of lung cancer in Japan. Drug Metab Pharmacokinet 26:516–522

Thompson D, Oster G (1996) Use of terfenadine and contraindicated drugs. JAMA 275:1339–1341

Tian Z, Cheng Q, Yoshimoto FK, Lei L, Lamb DC, Guengerich FP (2013) Cytochrome P450 107U1 is required for sporulation and antibiotic production in *Streptomyces coelicolor*. Arch Biochem Biophys 530:101–107

Toide K, Yamazaki H, Nagashima R, Itoh K, Iwano S, Takahashi Y, Watanabe S, Kamataki T (2003) Aryl hydrocarbon hydroxylase represents *CYP1B1*, and not *CYP1A1*, in human freshly isolated white cells: trimodal distribution of Japanese population according to induction of CYP1B1 mRNA by environmental dioxins. Cancer Epidemiol Biomarkers Prev 12:219–222

Tyson CA, Lipscomb JD, Gunsalus IC (1972) The roles of putidaredoxin and $P450_{cam}$ in methylene hydroxylation. J Biol Chem 247:5777–5784

Vaz ADN, Pernecky SJ, Raner GM, Coon MJ (1996) Peroxo-iron and oxenoid-iron species as alternative oxygenating agents in cytochrome P450-catalyzed reactions: switching by threonine-302 to alanine mutagenesis of cytochrome P450 2B4. Proc Natl Acad Sci USA 93:4644–4648

Wang K, Guengerich FP (2012) Oxidation of fluorinated 2-aryl-benzothiazole antitumor molecules by human cytochromes P450 1A1 and 2W1. Deactivation by cytochrome P450 2S1. Chem Res Toxicol 25:1740–1751

Waxman DJ, Holloway MG (2009) Sex differences in the expression of hepatic drug metabolizing enzymes. Mol Pharmacol 76:215–228

Wedell A (2011) Molecular genetics of 21-hydroxylase deficiency. Endocr Dev 20:80–87

West SB, Levin W, Ryan D, Vore M, Lu AYH (1974) Liver microsomal electron transport systems. II. The involvement of cytochrome b_5 in the NADH-dependent hydroxylation of 3,4-benzpyrene by a reconstituted cytochrome P-448-containing system. Biochem Biophys Res Commun 58:516–522

White PC, New MI, Dupont B (1984) HLA-linked congenital adrenal hyperplasia results from a defective gene encoding a cytochrome P-450 specific for steroid 21-hydroxylation. Proc Natl Acad Sci USA 81:7505–7509

Wienkers LC, Heath TG (2005) Predicting *in vivo* drug interactions from in vitro drug discovery data. Nat Rev Drug Discov 4:825–833

Williams SN, Dunham E, Bradfield CA (2005) Induction of cytochrome P450 enzymes. In: Ortiz de Montellano PR (ed) Cytochrome P450: structure, mechanism, and biochemistry, 3rd edn. Kluwer Academic/Plenum, New York, pp 323–346

Williams JA, Hyland R, Jones BC, Smith DA, Hurst S, Goosen TC, Peterkin V, Koup JR, Ball SE (2004) Drug–drug interactions for UDP-glucuronosyltransferase substrates: a pharmacokinetic explanation for typically observed low exposure (AUC_I/AUC) ratios. Drug Metab Dispos 32:1201–1208

Wu Z-L, Sohl CD, Shimada T, Guengerich FP (2006) Recombinant enzymes over-expressed in bacteria show broad catalytic specificity of human cytochrome P450 2W1 and limited activity of human cytochrome P450 2S1. Mol Pharmacol 69:2007–2014

Xiao Y, Guengerich FP (2012) Metabolomic analysis and identification of a role for the orphan human cytochrome P450 2W1 in selective oxidation of lysophospholipids. J Lipid Res 53:1610–1617

Xiao Y, Shinkyo R, Guengerich FP (2011) Cytochrome P450 2S1 is reduced by NADPH-cytochrome P450 reductase. Drug Metab Dispos 39:944–946

Yamazaki H, Johnson WW, Ueng Y-F, Shimada T, Guengerich FP (1996) Lack of electron transfer from cytochrome b_5 in stimulation of catalytic activities of cytochrome P450 3A4: characterization of a reconstituted cytochrome P450 3A4/NADPH-cytochrome P450 reductase system and studies with apo-cytochrome b_5. J Biol Chem 271:27438–27444

Yamazaki H, Shimada T, Martin MV, Guengerich FP (2001) Stimulation of cytochrome P450 reactions by apo-cytochrome b_5. Evidence against transfer of heme from cytochrome P450 3A4 to apo-cytochrome b_5 or heme oxygenase. J Biol Chem 276:30885–30891

Yamazaki H, Komatsu T, Ohyama K, Nakamura M, Asahi S, Shimada N, Guengerich FP, Nakajima A, Yokoi T (2002) Roles of NADPH-P450 reductase and apo- and holo-cytochrome b_5 on xenobiotic oxidations catalyzed by 12 recombinant human cytochrome P450s expressed in membranes of *Escherichia coli*. Protein Express Purif 24:329–337

Yang B, Graham L, Dikalov S, Mason RP, Falck JR, Liao JK, Zeldin DC (2001) Overexpression of cytochrome P450 CYP2J2 protects against hypoxia-reoxygenation injury in cultured bovine aortic endothelial cells. Mol Pharmacol 60:310–320

Yang Q, Nagano T, Shah Y, Cheung C, Ito S, Gonzalez FJ (2008) The PPARα-humanized mouse: a model to investigate species differences in liver toxicity mediated by PPARα. Toxicol Sci 101:132–139

Yang X, Zhang B, Molony C, Chudin E, Hao K, Zhu J, Gaedigk A, Suver C, Zhong H, Leeder JS, Guengerich FP, Strom SC, Schuetz E, Rushmore TH, Ulrich RG, Slatter JG, Schadt EE, Kasarskis A, Lum PY (2010) Genetic and genomic analysis of cytochrome P450 enzyme activities in human liver. Genome Res 20:1020–1036

Yun C-H, Okerholm RA, Guengerich FP (1993) Oxidation of the antihistaminic drug terfenadine in human liver microsomes: role of cytochrome P450 3A4 in N-dealkylation and C-hydroxylation. Drug Metab Dispos 21:403–409

Yun C-H, Kim K-H, Calcutt MW, Guengerich FP (2005) Kinetic analysis of oxidation of coumarins by human cytochrome P450 2A6. J Biol Chem 280:12279–12291

Zhang H, Im S-C, Waskell L (2007) Cytochrome b_5 increases the rate of product formation by cytochrome P450 2B4 and competes with cytochrome P450 reductase for a binding site on cytochrome P450 2B4. J Biol Chem 282:29766–29776

Zhao B, Guengerich FP, Bellamine A, Lamb DC, Izumikawa M, Funa N, Lei L, Podust LM, Sundamoorthy M, Reddy LM, Kelly SL, Stec D, Voehler M, Falck JR, Moore BS, Shimada T, Waterman MR (2005) Binding of two flaviolin substrate molecules, oxidative coupling, and crystal structure of *Streptomyces coelicolor* A3(2) cytochrome P450 158A2. J Biol Chem 280:11599–11607

Zhao B, Kagawa N, Sundaramoorthy M, Banerjee S, Nagy LD, Guengerich FP, Waterman MR (2012) A three-dimensional structure of steroid 21-hydroxylase (cytochrome P450 21A2) with binary substrate occupancy reveals locations of disease-associated variants. J Biol Chem 287:10613–10622

Chapter 3
Fifty Years of Cytochrome P450 Research: Examples of What We Know and Do Not Know

David C. Lamb and Michael R. Waterman

Abstract The discovery of a new hemoprotein from rabbit liver in 1962 is one of the key moments in biology. Today, cytochrome P450 enzymes constitute a protein superfamily found in all domains of life and have even been described in some viruses. Herein, we describe some of the early experimental groundwork including P450 multiplicity and activity, electron transfer, and cellular localization that led to our early understanding of the molecular properties of P450 enzymes. Subsequently, following the advent of recombinant DNA technology and the development of heterologous expression systems and genomic sequencing, the traditional understanding of what actually defined a P450 was challenged: these included P450s with unusual molecular properties such as differences in electron partner proteins, fusion proteins, posttranslational modification(s) and unique catalytic activities. Selected examples from our own research findings are highlighted.

Keywords Biodiversity • Cellular localization • Cytochrome P450 • Heterologous expression • Posttranslational modification • Purification • Redox partners • Structure/function • Unusual properties

3.1 Introduction

In 1962, Tsuneo Omura and Ryo Sato published a short communication in *The Journal of Biological Chemistry* entitled "A New Cytochrome in Liver Microsomes" (Omura and Sato 1962). Therein, the authors presented conclusive spectral

---

D.C. Lamb
Institute of Life Science, Medical School, Swansea University, Singleton Park, Swansea SA2 8PP, UK

M.R. Waterman (✉)
Department of Biochemistry, Vanderbilt University School of Medicine, Nashville, TN 37232-0146, USA
e-mail: michael.waterman@vanderbilt.edu

H. Yamazaki (ed.), *Fifty Years of Cytochrome P450 Research*,
DOI 10.1007/978-4-431-54992-5_3,

evidence of the hemoprotein nature of this newly observed rabbit liver microsomal pigment and proposed for it the provisional name P450, which meant "a pigment with an absorption at 450 nm when it is in the reduced state and complexed with carbon monoxide." Simultaneously, Howard Mason and colleagues, using electron spin resonance, proved that the rabbit pigment, which they had termed Fex, was a ferric hemoprotein (Hashimoto et al. 1962). Both papers were the culmination of the research exploits of many distinguished scientists including R.T. Williams, Jim and Elizabeth Miller, Bernard B. Brodie, Jim Gillette, Julius Axelrod, Britton Chance, G. Ron Williams, David Garfinkel, and Martin Klingenberg, and the reader is directed to the following reviews regarding the early history of the discovery of cytochrome P450 (Estabrook 2003; Omura 2011). Later, Ron Estabrook and colleagues demonstrated the first function of a P450, that of the bovine steroid 21-hydroxylase in the adrenal cortex, as a monooxygenase being able to activate atmospheric dioxygen (O_2) and insert one atom of O_2 into the steroid substrate molecule while the second oxygen atom is reduced to a water molecule (Estabrook et al. 1963). This report was quickly followed by the confirmation that P450s can hydroxylate drugs and xenobiotics (Cooper et al. 1965). During the subsequent decades, scientists have demonstrated that the principal role of the cytochrome P450 enzymes is to hydroxylate thousands of structurally diverse xeno- and endobiotics, according to the following general scheme:

$$\mathrm{NADPH} + \mathrm{H}^+ + \mathrm{RH} + \mathrm{O}_2 \rightarrow \mathrm{NADP}^+ + \mathrm{H}_2\mathrm{O} + \mathrm{ROH} \quad (\mathrm{RH} = \text{substrate})$$

Many, but not all, of the key scientific discoveries in cytochrome P450 research during the past 50 years are highlighted in Fig. 3.1. Today, cytochrome P450 (CYP; P450) hemoprotein enzymes constitute one of the largest protein superfamilies found in nature. P450 enzymes play essential roles in steroid hormone and sterol biosynthesis, in vitamin biosynthesis, in the detoxification and activation of many drugs and environmental pollutants, in the biosynthesis of vast arrays of secondary metabolites in bacteria, fungi, plants, and insects, and in the enzymatic activation of carcinogens (Ortiz de Montellano 2005). Remarkably, more than 18,000 P450s are known to biology, with examples being found in all domains of life, and many more will be uncovered by genomic analyses in the years to come.

3.2 Initial Reports on the Distribution of Cytochromes P450 in Nature

Although the early experimental investigations on the biology of P450 enzymes were dominated by mammalian microsomal studies such as those using rabbit, rat, mouse, and human microsomes from various tissues, its presence in mitochondria was soon discovered by Boyd Harding and coworkers in 1964 (Harding et al. 1964). In this work, study of the biosynthesis of steroid hormones in the rat adrenal gland

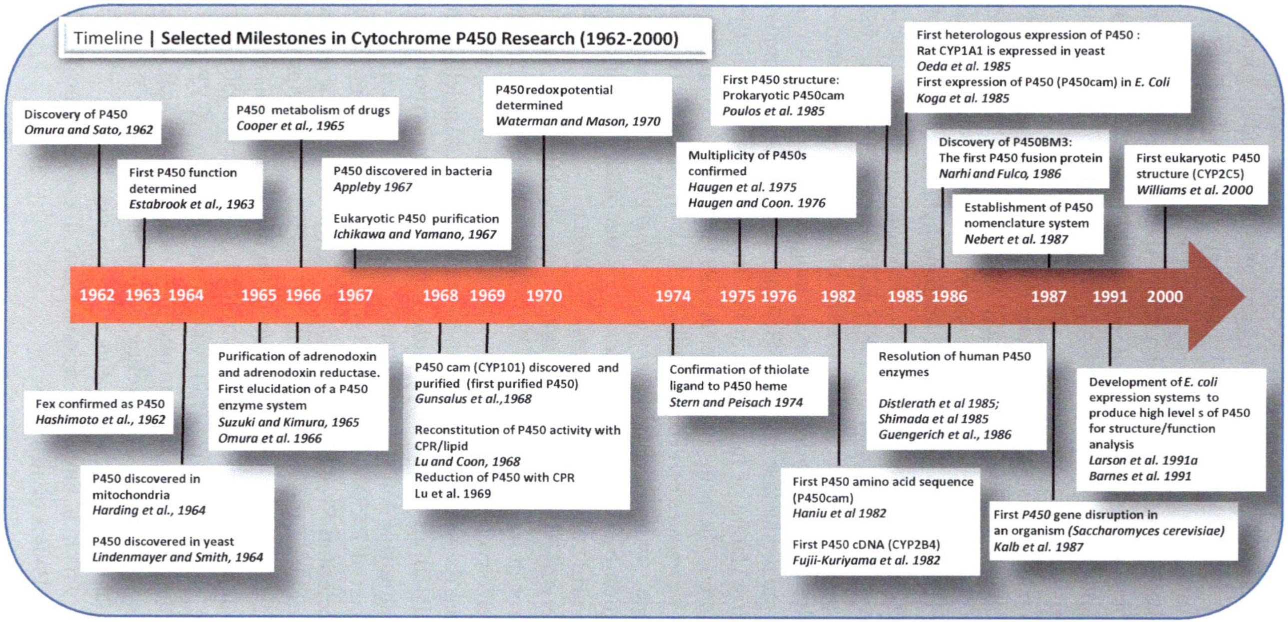

Fig. 3.1 A timeline highlighting the major scientific discoveries in cytochrome P450 research from 1962 to 2000

was being undertaken and the presence of P450 was detected in the mitochondrial fraction prepared from adrenal glands. The first discovery of P450 in a nonmammalian organism was also reported in 1964 when the presence of P450 in the unicellular yeast *Saccharomyces cerevisiae* was described by Lindenmayer and Smith (Lindenmayer and Smith 1964). Less than 10 years later, the presence of P450 in filamentous fungi was proposed when it was observed that the mycelia of the fungus *Cunninghamella bainieri* showed NADPH-dependent monooxygenase activities to compounds such as aniline and naphthalene and that such activities could be inhibited by carbon monoxide (Ferris et al. 1973). Regarding fungal P450s, up until the advent of genomic sequencing in 1998, only approximately 40 *P450* genes had been isolated from more than 20 fungal species, and only eight enzymatic reactions could be associated and attributable to P450 (Vanden Bossche and Koymans 1998): the conversion of oleic acid into 17-hydroxyoleic acid in *Torulopsis* sp. (Heinz et al. 1970); the 14α-demethylation of lanosterol (Aoyama and Yoshida 1978); Δ^{22}-desaturation of 22, 23-dihydroergosterol in the final step of ergosterol biosynthesis (Hata et al. 1981; Kelly et al. 1997); the formation of dityrosine, a sporulation-specific component of the yeast ascospore wall (Briza et al. 1994); the terminal hydroxylation of long-chain *n*-alkanes for use as sole carbon source and production of energy (Schunck et al. 1989; Sanglard and Fiechter 1989); the hydroxylation of fatty acids in the production of industrially important long-chain dicarboxylic acids (Müller et al. 1991); the hydroxylation of benzo[*a*] pyrene in *Saccharomyces cerevisiae* (Kelly et al. 1993); and the 11α-hydroxylation of progesterone by *Rhizopus niger* and *Aspergillus ochraceous* and the 11β-hydroxylation by *Cochliobolous lunatus* in the production of corticosteroids (for review, see Hudnik-Plevnik and Breskvar 1991).

The discovery of P450 in bacteria was first described by Cyril Appleby in 1967 from the nitrogen-fixing bacterium *Rhizobium japonicum* (Appleby 1967). Although the eukaryotic microsomal P450s investigated thus far were membrane bound, requiring the use of detergents in their preparation, Appleby noted that the *R. japonicum* P450 was apparently soluble and could be purified from cytosolic extracts. Of historical P450 scientific importance was the report of the occurrence of another bacterial P450 reported by Gunsalus and coworkers in 1968, which allowed the bacterium *Pseudomonas putida* to grow on camphor as a sole carbon source (Katagiri et al. 1968). On obtaining an extract of the cytosolic fraction of *P. putida* that catalyzed the initial hydroxylation of camphor, this extract could be resolved into three fractions, one of which contained the P450, the first P450 enzyme to be purified. This P450, known as $P450_{cam}$ (CYP101) was soon established as the model P450 protein of choice in the biochemical and biophysical investigations and allowed the unraveling of many aspects of the P450 oxygenation reaction cycle as well as structure–function analysis. In 1985 CYP101 was the first P450 for which a three-dimensional structure was determined by X-ray crystallography (Poulos et al. 1985).

3.3 Key Events Leading to the Establishment of the P450 Superfamily: Multiplicity of P450s and *P450* Gene Cloning

Following the initial observations that P450 preparations from mammalian origin could hydroxylate numerous lipophilic compounds, it was a striking observation that the P450 monooxygenase system lacked substrate specificity. Allan Conney and others described that treatment of animals with different microsomal inducers, such as phenobarbital and 3-methylcolanthrene, resulted in different profiles of monooxygenase activity for the metabolism of xenobiotics and steroid hormones (Conney 2003). Other observations in different monooxygenase catalytic profiles were also reported around this time when either animal species utilized, diet, sex, age, or exposure to chemical agents were taken into account (Gillette 1967; el-Masry et al. 1974; Kato 1977). Such experimental data strongly suggested the occurrence of multiple P450 enzymes with different substrate specificities in singularly prepared microsomal extracts. In contrast, another school of thought at that time suggested that such variable P450 catalytic profiles observed in the induced microsomes could be attributed to alterations in membrane environment, P450 posttranslational modification(s), or proteolysis or changes in redox partner protein levels. As early as 1966 indirect evidence for the presence of multiple forms of P450s based on anomalies in P450 spectra were presented by the Sato and Manning laboratories (Imai and Sato 1966; Sladek and Mannering 1966). The major hindrance to resolving this conundrum at this time was the difficulty in purifying P450 in an active form rather than the denatured P420 form. However, the discovery that membrane-bound P450s could be stabilized following detergent solubilization by the addition of glycerol allowed for the purification of membrane-bound P450 for the first time (Ichikawa and Yamano 1967) and resolution of the issue of P450 multiplicity.

In two papers published by the laboratory of Jud Coon in 1975 and 1976, four isoforms of P450 were purified from microsomes of rabbit liver (Haugen et al. 1975; Haugen and Coon 1976). The latter paper described the purification of the principal phenobarbital-inducible (CYP2B4) and β-naphthoflavone-inducible (CYP1A2) P450 forms, and showed beyond doubt that the two P450s were different isozymes as shown by differences in molecular masses, differences in amino acid composition, COOH- and NH_2-terminal amino acid residue sequencing, spectral properties, and differences in catalytic activities. Research in several laboratories then addressed the question of the numbers of P450s in different species with multiple P450 isoforms described in rat, mouse, and rabbit (Coon 2005). Ultimately, in the 1980s the laboratory of Fred Guengerich published three papers in the *Journal of Biological Chemistry* that described the purification and characterization of four P450s that metabolized specific drugs in the human liver (Distlerath et al. 1985; Shimada et al. 1986; Guengerich et al. 1986). For the first time these findings allowed for the pharmaceutical testing of drugs for human toxicity via individual human P450 metabolism analysis rather than inferring toxicity predictions from animal studies. With the advent of genomics, we now know that there are 57 human cytochromes P450 (Guengerich 2005).

Following on from the purification of bacterial, yeast, mammalian, and mitochondrial P450s, the next research goal was the determination of the complete amino acid sequence of a P450. This objective was achieved in 1982 when the primary sequence of *Pseudomonas putida* P450cam was determined by chemical sequencing (Haniu et al. 1982). Finally, the amino acid sequence of the phenobarbital-induced P450 of rat liver was deduced from the nucleotide sequence of a cloned cDNA (Fujii-Kuriyama et al. 1982). Following this vital event in the P450 research field, the application of such molecular biological techniques to the field of P450 research resulted in the elucidation of the complete amino acid sequences of numerous P450s. Such generated sequence data allowed the establishment of a systematic classification system for the naming of P450s. Hence, based on the similarity of the primary amino acid sequences, P450s were classified into families and subfamilies and the symbol "CYP," for cytochrome P450, was proposed (Nebert et al. 1987).

3.4 The Cytochrome P450 Superfamily

The current system adopted for the naming of P450s is maintained by David Nelson (Nelson 2009; http://drnelson.uthsc.edu/CytochromeP450.html). The most highly related P450 proteins are grouped into families whose members share greater than 40 % amino acid identity and are designated numbers (CYP1, CYP2, etc.). Families are divided into subfamilies whose members share greater than 55 % amino acid identity and are given alphabetical characters (CYP1A, CYP1B, CYP2A, CYP2B, etc.). Finally, the subfamilies are divided into individual loci designated with a specific number (CYP1A1, CYP1A2, etc.). Subsequently, P450 sequences that share more than 97 % amino acid identity are described as being allelic variants and designated with additional nomenclature (v1, v2, etc.). This highly structured nomenclature system allows for the comparison of P450 sequences found within an individual organism as well as establishing whether P450 families, and potentially associated function(s), are conserved in different species and specific kingdoms.

3.5 Diversity of P450 Expression Systems

The production of P450 enzymes in an active form has been critical to the advancement of several areas of P450 molecular biology and biochemistry, including enzyme structure and function, gene regulation, and biotechnological applications. Following the isolation of *P450* gene coding sequences and cDNAs, successful heterologous expression systems led to an unlimited source of protein for research applications. Catalytically functional cytochrome P450 enzymes have been successfully expressed in bacterial, yeast, insect, and mammalian cells.

Generally, in the early days of developing P450 heterologous expression systems, the methods employed were developed for the expression of animal P450s. However, the vectors and strains employed today are now used to express *P450* genes isolated from nearly all organisms.

The first heterologous P450 express system used to successfully express a P450 was yeast. Rat CYP1A1 was produced in *S. cerevisiae* following expression via an alcohol dehydrogenase constitutive promoter using the vector pAAH5, with 1 % of the total yeast protein accounted for by the P450 enzyme (Oeda et al. 1985). Subsequently, a variety of vectors containing auxotrophic markers have been employed for the expression of P450s in yeast with variable levels of expression reported. The most commonly used strain of *S. cerevisiae* for P450 expression is AH22 and, although *S. cerevisiae* contains three P450s, during the minimal growth conditions used for heterologous P450 expression, endogenous P450 levels are usually undetectable so that P450 detected can be attributable to the expressed enzyme. Additionally, P450s expressed in yeast are catalytically self-sufficient as the endogenous CPR can support the activity of the heterologously expressed P450. The use of an inducible promoter can be advantageous in generating large amounts of P450 protein. For example, the pYEP-based vectors can be utilized for high-level P450 expression via a galactose inducible promoter, as shown for the expression of *Candida albicans* CYP51 in AH22: >2.5 μmol P450/l yeast culture was obtained (Shyadehi et al. 1996). Yeast was also the first system used to produce a functional P450-CPR fusion protein as demonstrated with rat CYP1A1 and CPR (Murakami et al. 1987).

By far the most commonly used system today for the expression of P450 is bacteria, with *Escherichia coli* the bacterium of choice. Traditionally, prokaryotic P450s were isolated and purified from their original bacterial cell source. However, the advent of gene cloning accelerated focus on the development of *E. coli* heterologous expression systems to study P450 structure–function relationships. One of the earliest P450 expression systems was for CYP101, when the *camC* gene, which encodes CYP101, was cloned into the shuttle vector pKT240 and recovered as the recombinant pKG201. CYP101 was then expressed constitutively in *P. putida* and in *E. coli* (Koga et al. 1985). Thus, the use of bacterial expression systems allowed for site-directed mutagenesis studies of P450 to be initiated, as originally performed with CYP101 (Atkins and Sligar 1988), allowing for key insights into the mechanics of P450 function to be unraveled at the molecular level. Another example includes the expression of the native protein form of CYP102A1 in *E. coli* (Narhi et al. 1988) as well as its individual domains (Oster et al. 1991; Li et al. 1991). Subsequently, different expression vectors, for example, pET and pCWori$^+$ systems, were utilized to obtain high-level prokaryotic and eukaryotic P450 production in *E. coli*. These systems also aided in the biochemical characterization of individual P450s identified by genomic sequencing in organisms with multiple *P450* genes and where P450 expression could not be detected in vivo. Additionally, such expression systems provided a platform for the structural characterization of P450 enzymes, thus providing important links between primary sequence, activity, and tertiary structure.

The high-level expression of eukaryotic P450s, originally mammalian in origin, was difficult to achieve without modifications of the amino-terminal region of the enzyme. *E. coli* expression of mammalian P450 was first achieved with rabbit CYP2E1 where low levels of holoprotein of the native enzyme were produced (Larson et al. 1991a). Subsequently, removal of amino acid residues 3 through 29 did not allow for increased levels of CYP2E1 expression but allowed for the production of an active enzyme tightly bound to the bacterial lipid bilayers, providing the first evidence that the hydrophobic amino-terminus membrane anchor is not absolutely required for anchoring the eukaryotic P450 to the membrane (Larson et al. 1991b). A next step forward in developing eukaryotic P450 expression in *E. coli* was achieved through the expression of cholesterol 7α-hydroxylase (CYP7A1) when an alanine codon was placed at codon 2 that had previously been shown to be the preferred codon for high-level expression of T7 phage proteins in *E. coli* (Looman et al. 1987). Interestingly, CYP2E1 naturally encodes alanine at position 2, which may explain the original successful expression of this protein (Larson et al. 1991a).

Finally, a genetically engineered form of bovine CYP17A1 was produced in *E. coli* in which the second codon was changed to a GCT-encoding alanine and codons 4 and 5 were modified to TTA to make an AT-rich 5′-mRNA (Barnes et al. 1991). Additionally, the last nucleotides of codons 6 and 7 were changed to A and T, respectively, to minimize secondary structure effects. This modified *CYP17A1* gene when expressed resulted in levels of expression up to 16 mg P450 produced per liter of *E. coli* culture. In the absence of these modifications, no detectable P450 was observed. The importance of the NH_2-terminal modifications resulting in the functional expression of CYP17A1 can be seen in the application of these modifications for successful expression of (1) other mammalian P450s and (2) to the production of recombinant P450s from diverse eukaryotic sources including those from plants, fungi, insects, and protozoa. In most instances the P450s produced in this manner were functionally active and their associated activities can be reconstituted with redox partner proteins. Importantly, this approach allowed for the generation of large amounts of eukaryotic P450 protein, which was ultimately a key step in allowing for the elucidation of the first eukaryotic P450 three-dimensional structure, rabbit CYP2C5, by X-ray crystallography (Williams et al. 2000) and subsequently the elucidation of many others (Johnson and Stout 2013).

As well as yeast and bacteria, P450s have also been successfully expressed in alternative systems including mammalian cells. The COS cell system is among the most widely used for the heterologous expression of P450s in mammalian cells. This system was originally developed for the expression of bovine CYP17A1 (Zuber et al. 1986) using the vector pCD, and using this system it was demonstrated that CYP17A1 had dual catalytic activities of 17α-hydroxylation of progesterone followed by 17, 20-lyase activity to produce dehydroepiandrosterone. The advantages of this system are that the *P450* cDNAs do not require modification, enzyme activities can be determined after a few days following cloning, and that the COS1 cells have adequate CPR and cytochrome b_5 levels to support the expressed P450 activities. Baculovirus expression in insect cells has also been employed for the expression of P450s where high levels of expression can be obtained, although the baculovirus host cells are deficient in electron-transfer proteins required for P450 activity, thus necessitating the purification of the P450 following expression to test activity (Asseffa et al. 1989). Additional P450 expression systems that have been employed primarily for the expression of mammalian P450s include vaccinia virus (Aoyama et al. 1990), B lymphoblastoid cells (Crespi et al. 1990), retroviruses (Battula 1989), and V79 Chinese hamster cells (Doehmer and Oesch 1991).

3.6 Diversity of P450 Systems and Redox Partners

Generally, P450 enzyme reactions require the supply of electrons from NADPH or NADH for catalysis to occur. Thus, P450s are normally found associated with a NAD(P)H-linked reductase or reducing system in the cell. In animal cells P450s are associated with both the endoplasmic reticulum (ER) and mitochondria whereas in plants and lower eukaryotes P450 has so far only been found associated with the endoplasmic reticulum except in one instance, fungal CYP55A1. Conversely, in prokaryotic organisms, P450s are soluble enzymes being located in the cytosol of the cell. Hence the reductase component of microsomal P450 systems is membrane bound, whereas in the mitochondria the P450 is membrane bound while the reductase components are soluble. Finally, in prokaryotes, the reducing system is soluble. The advent of genome sequencing and the uncovering of the diversity of P450 enzymes from organisms across biological kingdoms established the biodiversity of P450-redox systems and highlighted unusual forms. To date, 13 different classes of P450 systems can be described (**I–XIII**, described in following text and in Fig. 3.2) based upon the unique nature and arrangement (or lack of) associated redox partner protein(s).

In the early days of P450 research, P450 redox systems were divided into three broad categories (I–III, herein). (**I**) Class I (for prokaryotic P450s), consisting of the cytosolic and soluble three-component P450-ferredoxin (Fdx)-ferredoxin reductase (FDR) electron chain. Originally, the class I system was elucidated for the *P. putida* CYP101 enzyme system, and the three components were individually purified and activity reconstituted by Katagiri et al. in 1968. This three-component constitution is similar in organization to the mitochondrial P450 reductase systems found in animals, with a major difference being that in the bacterial system the P450 is soluble and generally utilizes NADH as the electron source whereas the mitochondrial P450 enzymes are membrane bound and prefer NADPH. (**II**) Class II (for eukaryotic P450s) consisting of the ER membrane-bound P450-CPR electron chain. In two seminal papers by Anthony Lu and Jud Coon in 1968 and 1969, results were described identifying P450 as a fatty acid hydroxylase in rat microsomes and that the microsomal P450 system could be resolved into three fractions: P450, cytochrome P450 reductase (CPR), and lipid (Lu and Coon 1968; Lu et al. 1969). This work conclusively proved that CPR can transfer reducing equivalent to microsomal P450, and the procedures described therein for the reconstitution of the microsomal

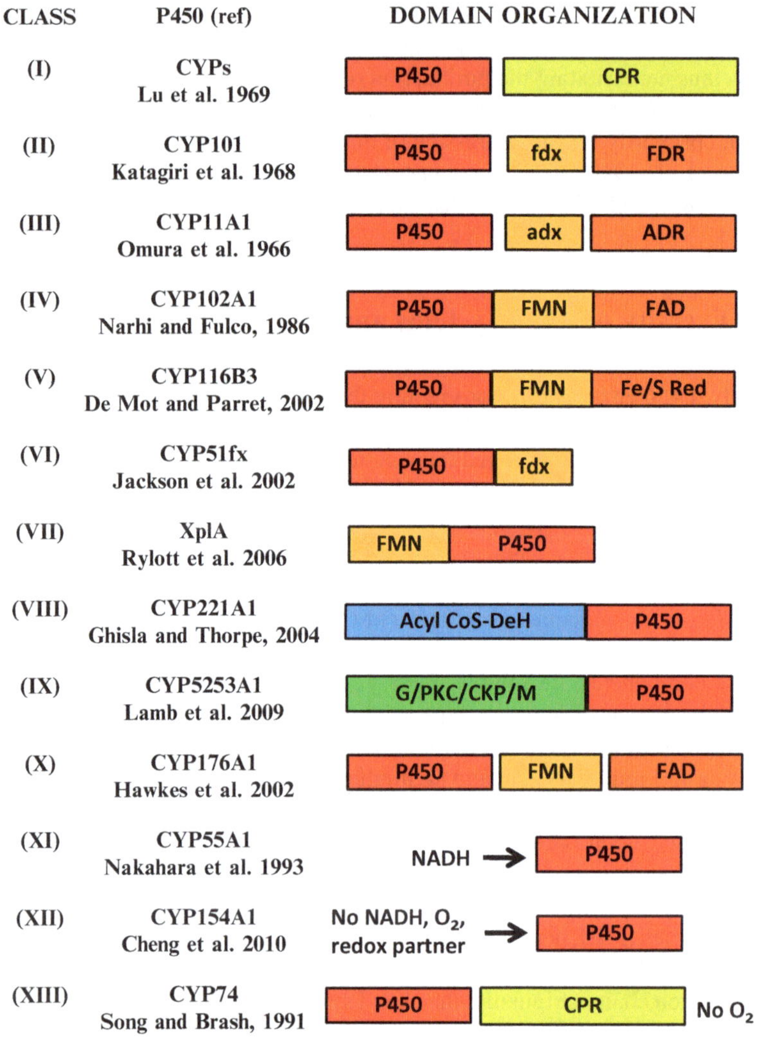

Fig. 3.2 Different arrangements of 13 distinct classes of P450 redox partners and associated fusion proteins. Color scheme: P450 (*red*); P450 reductase, CPR (*yellow*); ferredoxin (fdx), adrenodoxin (adx), flavodoxin (fld), and FMN (*light orange*); ferredoxin reductase (FDR), adrenodoxin reductase (ADR), Fe/S reductase domain of phthalate dioxygenase (Fe/S *red*), and FAD (*dark orange* as labeled); Acyl CoA dehydrogenase (Acyl CoS-DeH) domain (*blue*); protein domain fused to Mimivirus P450 (labeled G, glycosylation site; PKC, protein kinase C phosphorylation site; C, four casein kinase II phosphorylation sites; M, three myristoylation sites) (*green*). Each class is described in the text

P450 systems became the established approach for subsequent work on eukaryotic P450s internationally. (**III**) Class III, for mitochondrial P450s, consisting of the inner mitochondrial membrane-bound P450 and a soluble 2Fe-2S ferredoxin, such as protein (adrenodoxin), and a soluble FAD-containing flavoprotein (NADPH adrenodoxin reductase). Adrenodoxin was first isolated and purified in 1965 (Suzuki and Kimura 1965) and andrenodoxin reductase the following year (Omura et al. 1966). This work allowed for the elucidation of the mechanics of mitochondrial P450 steroid hormone biosynthesis, specifically for the P450s CYP11A1 (which converts cholesterol to pregnenolone, the first committed step in steroid hormone biosynthesis), CYP11B1 (which converts deoxycorticosterone to corticosterone and 11-deoxycortisol to cortisol), and CYP11B2 (which converts corticosterone to aldosterone).

These initial discoveries gave clues to unique redox partner associations for individual P450 enzymes. The first deviation from the class I/II standard just described was discovered by Linda Narhi and Armand Fulco in 1986, who isolated (**IV**), a naturally occurring P450 fusion protein (CYP102A1; P450BM3) consisting of an N-terminal P450 heme domain linked to a NADPH-dependent P450 reductase-like domain containing FMN and FAD in the bacterium *Bacillus megaterium* (Narhi and Fulco 1986). CYP102A1 is located in the cytosol and catalyzes ω-1-, ω-2-, and ω-3-hydroxylations of long-chain fatty acids and alcohols with the highest turnover rate reported for any P450 monooxygenase (17,000 min^{-1}). This high catalytic rate is probably reflected in the rapid intra-transfer of electrons between the fused protein domains in contrast with the separate domains of most P450s and redox partner proteins (Miles et al. 1992). Genome sequencing also revealed similar CYP102A1-like enzymes with the same domain organization in other bacteria. Two CYP102A1 homologues (CYP102A2 and CYP102A3) are present in *Bacillus subtilis*, and CYP102 fusion protein homologues have been found in the bacteria *Ralstonia metallidurans*, *Bradyrhizobium japonicum*, and various bacilli and streptomycetes (Gustafsson et al. 2004). A eukaryotic counterpart of bacterial CYP102A1 was initially discovered in the fungus *Fusarium oxysporum* in 1996 (Nakayama et al. 1996) and named CYP505A1, with homologues detectable in various *Neurospora* and *Aspergillus* species. However, bacterial CYP102As are all soluble enzymes whereas CYP505s are membrane-bound enzymes in the ER. CYP505As have fatty acid hydroxylase activity, although one member of the subfamily CYP505B1 is a hydroxylase in the production of the mycotoxin fumonisin (Seo et al. 2001). In contrast to the CYP102A-reductase fusion proteins just described, a CYP102 subfamily (CYP102B) was described that exists solely as a single P450 heme domain (Lamb et al. 2010). CYP102B1 from *Streptomyces coelicolor* can be reconstituted with spinach ferredoxin and ferredoxin reductase to catalyze the turnover of arachidonic acid at levels 1,000 fold less than CYP102A1.

Another arrangement of a P450 linked to a redox partner was found in the bacterium *Rhodococcus* sp. NCIMB 9784 (Roberts et al. 2003). In this protein (**V**), a soluble P450 heme domain is fused at the C-terminus to a FMN- and a 2Fe2S-containing reductase, a redox domain that resembles phthalate dioxygenase reductase, and the complete P450 protein was named CYP116B2. Further homologues of this P450-reductase fusion arrangement had been found in *Ralstonia metallidurans*

(CYP116B1) and *Rhodococcus ruber* (CYP116B3) (De Mot and Parret 2002). Although the endogenous function of these P450s is not known, it has been shown that the phthalate reductase domain can accept reducing equivalents and drive CYP116B2 metabolism of 7-ethoxycoumarin and CYP116B3 metabolism of polycyclic aromatic hydrocarbons (Hunter et al. 2005).

A sixth class of P450–redox partner association was found in the bacterium *Methylococcus capsulatus*, which is one of the few bacteria that can synthesize sterol de novo. Encoded in the genome of *M. capsulatus* was (**VI**) a soluble CYP51 (sterol 14α-demethylase) P450 heme domain fused at the C-terminus to a ferredoxin domain (Jackson et al. 2002). The physiological role for this P450 fusion protein was established as a true CYP51 in that it catalyzed the three-step oxidation of the sterol lanosterol to 4,4,-dimethyl-5α-cholesta-8,14,24-diene-3β-ol, utilizing an exogenous ferredoxin reductase as the third electron-transfer protein component. Another novel P450 fusion arrangement was found in the bacterium *Rhodococcus rhodocrous* consisting of (**VII**) a fusion between a soluble C-terminal P450 domain (XplA) and an FMN-containing N-terminal flavodoxin domain FMN (XplB) (Rylott et al. 2006). This P450 was shown to catalyze the degradation of the nitramine explosive and pollutant hexahydro-1,3,5-trinitro-1,3,5-triazine by reductive denitration (Jackson et al. 2007). Further, when this P450 was expressed in transgenic plants, hexahydro-1,3,5-trinitro-1,3,5-triazine is removed from contaminated soil, thus having application in bioremediation of contaminated military sites (Rylott et al. 2011). A unique mechanistic observation for this P450 enzyme was that oxygen is not required for the degradation of hexahydro-1,3,5-trinitro-1,3,5-triazine, but its presence and binding are absolutely required to determine the final degradation products (Sabbadin et al. 2009). Finally, more examples of a P450 heme domain fused to peptides of known and unknown function (and which have no homology to redox proteins) have emerged and will continue to do so. For example, in the bacterium *Pseudomonas fluorescens* (**VIII**), a novel but uncharacterized P450 was found consisting of a P450 domain fused to a acyl CoA dehydrogenase domain and named CYP221A1 (Ghisla and Thorpe 2004) In another example, a P450 was found in the Mimivirus genome (CYP5253A1, the first viral P450) and shown to be (**IX**) a N-terminal P450 domain fused at the C-terminus to a protein containing several putative posttranslational modification sites including one N-glycosylation site, a protein kinase C phosphorylation site, four casein kinase II phosphorylation sites, and three myristoylation sites (Lamb et al. 2009).

As well as novel P450 fusion protein arrangements, unique genomic arrangements of *P450* and putative redox partner genes were observed and published that deviate from the class I and II systems originally proposed. For example, the gene *CYP176A1* from the bacterium *Citrobacter braakii* is arranged in an operon consisting of genes encoding for (**X**) a flavodoxin (cindoxin) and a FAD-containing flavodoxin reductase (cidoxin reductase) (Hawkes et al. 2002). *E. coli* flavodoxin reductase was shown to reconstitute CYP176A1, and cindoxin mediated hydroxylation of the monoterpene cineole, allowing the bacterium to grow on cineole (Hawkes et al. 2002). Of particular interest is the finding that *E. coli*, which has no *P450* genes encoded in its genome, encodes genes for flavodoxin and flavodoxin reductase (Jenkins and Waterman 1994). For many P450s these *E. coli* redox proteins have been able to drive their specific catalytic activities, leading to the hypothesis that the ancestral eukaryotic CPR was encoded and arose from gene fusion of the genes encoding flavodoxin and flavodoxin reductase.

Two examples of P450 that function without any redox partner have been described. The first P450, a eukaryotic example, is (**XI**) CYP55A1 (P450nor), discovered in the fungus *Fusarium oxysporum* (Nakahara et al. 1993). Biochemically, CYP55A1 catalyzes the reduction of nitric oxide to nitrous oxide. CYP55A1 is a soluble P450 and, although encoded by the same gene, is located in both the cytoplasm and mitochondria in the fungal cell. The P450 heme prosthetic group receives electrons for nitric oxide reduction directly from NADH without participation of any NADH-linked reductase or reducing system. CYP55A1-like soluble homologues have also been found in the yeast *Trichosporon cutaneum* (Zhang et al. 2001) and in the fungi *Cylindrocarpon tonkinense*, *Histoplasma capsulatum*, and *Aspergillus oryzae* (Omura 2010).

The second P450, a prokaryotic example, is (**XII**) CYP154A1 from *S. coelicolor*, which was shown to catalyze the intramolecular cyclization of a novel dipentaenone with a high degree of conjugation to a Paternò–Büchi-like product (Cheng et al. 2010) In an unprecedented biochemical reaction described for any P450 studied to date, CYP154A1 requires no redox partner, reducing equivalent (NADPH), or oxygen to function, although CYP154A1 retains key residues in its primary sequence seen in other P450s and produces a Soret maximum at 450 nm in the reduced CO-difference spectrum.

Hydrogen peroxide (H_2O_2) can also support P450-catalyzed oxygenation reactions, and some P450s specifically require H_2O_2 for their catalytic activity. Two bacterial P450s, CYP152A1 of *Bacillus subtilis* and CYP152B1 of *Sphingomonas paucimobilis*, catalyze H_2O_2-dependent α- and β-hydroxylation of fatty acids (Matsunaga et al. 2002). However, both P450s were named peroxygenases rather than monooxygenases. Specific P450s can also catalyze the rearrangement of the oxygen atoms in the substrate molecule itself. For example, (**XIII**) synthesis of prostacycline from prostaglandin H2 is catalyzed by the microsomal P450, CYP8A1 (DeWitt and Smith 1983). Also, thromboxane A2 is synthesized from prostaglandin H2 by the activity of the microsomal P450, CYP5A1 (Haurand and Ullrich 1985). In both instances, these enzyme reactions involve molecular rearrangements of the substrate molecule itself and do not require the external supply of electrons to the P450. Another example of this class of P450 enzyme is CYP74s from plants. *CYP74* encode allene oxide synthases, and they catalyze the conversion of fatty acid hydroperoxides to allene epoxides, also without the supply of reducing equivalents (Song and Brash 1991).

3.7 Diversity in P450 Cellular Localization and Posttranslational Modification

In the early work on P450 enzymes, it generally became assumed that all prokaryotic P450s were soluble, being located in the cytosol of the cell, and that all eukaryotic P450s were membrane bound, being located in the ER or mitochondria of animal cells or the ER of all other eukaryotes. Following 50 years of P450 research, it is still true that all the prokaryotic P450s biochemically studied to date are soluble, and hence cytosolically located, enzymes. Generally, analysis of prokaryotic P450 primary sequences does not reveal the presence of N-terminal hydrophobic membrane-spanning regions, and a prokaryotic membrane-bound P450 has yet to be described experimentally. However, it should be noted that with the vast amount of prokaryotic P450 gene sequences deposited in various databases and yet to be experimentally characterized, the existence of a membrane-bound prokaryotic P450 cannot be ruled out.

Microsomal P450s are integral membrane proteins with each P450 containing a single N-terminal transmembrane-spanning segment. These P450s are directed and retained in the ER by a signal sequence, which also prevents translocation through the ER membrane (Sakaguchi et al. 1984). Although P450 signal sequence domains have been identified in many P450s, the majority of eukaryotic P450s do not have a specific and identifiable signal sequence, and it is thought that overall hydrophobicity at the NH_2-terminus of a P450 peptide directs targeting to the ER membrane (Avadhani et al. 2011). In contrast, mitochondrial P450s involved in steroid biosynthesis and vitamin D metabolism have canonical and cleavable mitochondrial targeting signals at their N-termini, which are distinct from the signal sequences of ER-bound P450s. Thus, mitochondrial P450s are synthesized as pre-proteins containing N-terminal cleavable pre-sequences in the cytosol and posttranslationally targeted to the mitochondria (Fig. 3.3). After translocation into the matrix, the targeting sequence is cleaved, and the mature P450 enzyme associates with the inner membrane (Avadhani et al. 2011). To date, CYP55A1 is the only eukaryotic P450 found that is located in both the cytoplasm and the mitochondria in the fungal cell. The biological significance of this occurrence is still not known.

A further unusual and unexplained finding of some mammalian P450 members belonging to the one, two, and three families is they are bimodally targeted proteins, being located in the mitochondria and plasma membranes in addition to the ER (Anandatheerthavarada et al. 1999). The traditional thought was that all animal P450s were exclusively located in the ER of the animal cell. However, for the following P450s it was experimentally shown that they are bimodally targeted to the mitochondria: CYP1A1, CYP1A2, CYP1B1, CYP2B1, CYP2E1, CYP2D6, CYP2C11, CYP2C6, CYP3A2, and CYP4A1 (Avadhani et al. 2011). Within the NH_2-termini of these P450s are chimeric signals that target each P450 to both locations. Such chimeric targeting signals are difficult to identify but are thought to consist of an ER-targeting signal (part of the transmembrane anchor) flanked by a cryptic mitochondria-targeting signal located at amino acid residues 20–36 in each

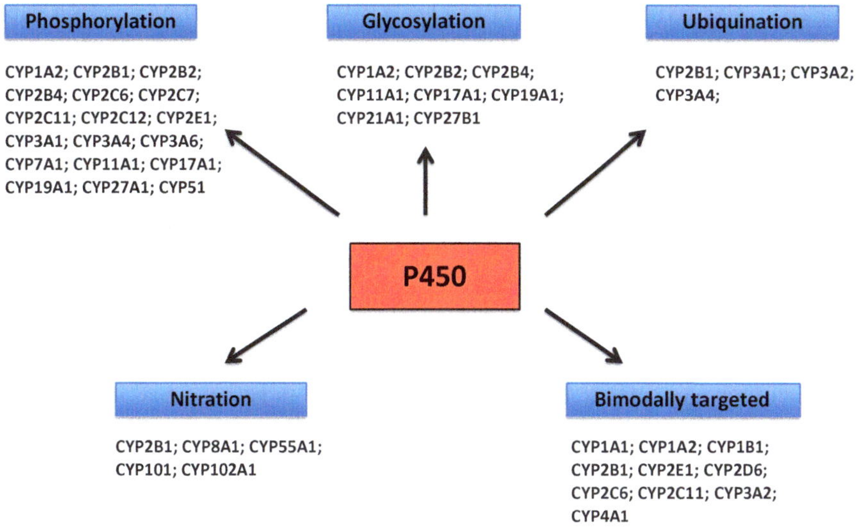

Fig. 3.3 Schematic representation of P450 posttranslational modifications and bimodal targeting. Examples of P450s, which undergo posttranslational modifications (phosphorylation, glycosylation, ubiquination, and nitration), are given below each subheading. Those P450s that are targeted to both the endoplasmic reticulum (ER) and mitochondria are also named under the subheading bimodally targeted

of the different P450s named above. Within the mitochondrial-targeting signal are two to five positively charged amino acid residues that are essential for mitochondrial targeting (Robin et al. 2002). One further feature of the mitochondrial targeted P450s is that their catalytic activities are supported by adrenodoxin and adrenodoxin reductase, redox partners used to drive mitochondrial P450 steroid activity. Although the true physiological role(s) for these targeted P450s are unknown, toxicological implications regarding their presence in the mitochondria are now emerging. For example, increased protein levels of mitochondrial-targeted CYP2E1 results in depletion of glutathione, a key indicator of increased oxidative stress, resulting in augmented alcohol-mediated cellular and tissue injury (Bansal et al. 2010, 2013). In another study, mitochondrial-targeted CYP2D6 was shown to activate the pro-neurotoxin 1-methyl-4-phenyl-1,2,3,6-tetrahydropyridine (MPTP) to the toxic 1-methyl-4-phenylpyridinium ion (MPP(+)), in dopaminenergic neurons and thus induce neuronal damage (Bajpai et al. 2013). Classically, MPTP is a known inducing agent of Parkinson's disease.

As well as P450 diversity in the cellular localization, individual P450s themselves can show diversity through posttranslational modification of specific forms. Again, in a deviation from the traditional idea it was thought P450s were not modified further once the final folded peptide had been synthesized. However, for certain eukaryotic P450 enzymes posttranslational modification occurs: included are phosphorylation (e.g., CYP2B1, CYP2B4, CYP2C6, CYP2D6, CYP2E1, CYP3A4, CYP11A1, CYP17A1, CYP19A1), ubiquitination (e.g., CYP3A4,

CYP2B1), glycosylation (e.g., CYP11A1, CYP19A1), and nitration (CYP4A subfamily) (Lamb and Waterman 2013). Molecular requirements for P450 phosphorylation include the presence of a cytosolically exposed protein kinase A (PKA) recognition sequence (RRXS) with the Ser residue as the kinase target (Oesch-Bartlomowicz and Oesch 2003). For the P450s that do not contain the PKA motif, other cryptic amino acid target sequences are phosphorylated by additional protein kinases, such as protein kinase C. The impact of P450 phosphorylation remains uncertain. It can serve as a marker for P450 loss of function or degradation or both. For example, phosphorylation of CYP2E1 Ser129 and CYP2B1 Ser128 results in immediate inactivation of the proteins, more rapidly than transcriptional downregulation can reduce activity through reduced protein levels (Oesch-Bartlomowicz and Oesch 2003). Conversely, PKA-mediated P450 phosphorylation has been shown to activate the cryptic mitochondrial-targeting signal in CYP2B1, CYP2E1, and CYP2D6 (Avadhani et al. 2011) by disrupting the ability of these P450s to be targeted to the ER via disruption of signal recognition particle binding.

P450s involved in animal steroid hormone biosynthesis in particular show high degrees of posttranslational modifications. Differences in these modifications can be observed in the individual P450 enzyme belonging to the same P450 family and with P450 families between species. For example, human, bovine, and equine CYP19As are glycosylated but the porcine ovary and placenta isoforms are not (Jo Corbin et al. 2003). Additionally, analysis of some species shows that CYP19As undergo glycosylation and phosphorylation. The human CYP19A1 sequence shows potential N-glycosylation sites at Asn12 and Asn180, and it has been predicted that the glycosylation site is localized at the N-terminal membrane-spanning region (Shimozawa et al. 1993). Most mammalian CYP19s, including human, bovine, and equine, have the sequence NXT/S (Asn12-Ile13-Thr14) motif at the N-terminal region, but porcine ovary and placenta sequences lack this motif. The NXT/S motif is present only in mammalian and amphibian species but not in birds, reptiles, and both isoforms of fishes, even though a few exceptions are found. Site-directed mutagenesis of mouse CYP19 has demonstrated that Ser118 is a potential phosphorylation site, and mutation of this residue resulted in decreased the specific activity (Miller et al. 2008). In contrast to the less conserved N-terminal glycosylation site, the predicted phosphorylation site Ser118 is conserved in all CYP19A sequences analyzed in the present study, except in the ovarian isoform of zebrafish. Further, the differences in the levels of conservation of glycosylation and phosphorylation correlate with their effect on the function of the P450 enzyme. Although the poorly conserved N-glycosylation site does not show any effect on the activity of CYP19A1, the nearly absolutely conserved phosphorylation site has a significant effect on decreasing the enzyme activity once phosphorylated (Miller et al. 2008). In direct contrast, phosphorylation of serines and threonines of CYP11A1 by PKC and in CYP17A1 by PKA results in an increase of both P450 enzyme activities (Vilgrain et al. 1984; Zhang et al. 1995).

3.8 Diversity in P450 Biochemical Properties and Implications for the Superfamily

As we have already described, the early experimental findings regarding the biochemical properties of P450 enzymes suggested a conservation of molecular properties that were predicted to be attributable to all the subsequent P450 enzymes isolated and characterized. Critical in this analysis was the presence of a heme prosthetic group, the cysteine residue that forms the thiolate ligand to the heme iron as first described by Stern and Peisach (1974), and the reduced CO-difference spectrum at 450 nm. Indeed, P450 diversity was originally predicted only to be observed in the substrate recognized and the oxidative reaction catalyzed by each specific P450 enzyme. However, with the advent of the postgenomic area, many unexpected P450 properties have been discovered challenging the traditional view of what defines a P450 enzyme. Some are variations of the original properties, whereas others are difficult to explain because of their unique nature relative to the biochemical properties of the rest of the P450 superfamily. In this final section, we describe highlights from our own experimental findings that we believe have impact on the true definition of a P450 enzyme.

3.8.1 Diversity in P450 Amino Acid Residue Conservation

Until 2006 it was universally thought that three amino acid residues were absolutely conserved in all P450 primary sequences: the conserved cysteine, which forms the thiolate ligand to the heme Fe atom, and the EXXR motif, forming a charge pair in the P450 K helix and which is involved in heme binding and overall P450 fold topology and stability. However, the first inkling of a deviation from the three conserved residue hypothesis was suggested when the completed genome sequence of *S. coelicolor* was published. Of the 18 *P450*s identified in this genome, 3—*CYP156B1*, *CYP157A1*, and *CYP157C1*—were predicted to encode P450s that do not contain the EXXR motif and thus challenged this dogma (Fig. 3.4a) (Lamb et al. 2002).

Examination of the CYP157C1 amino acid sequence revealed the presence of EQSLW in place of EXXR (Fig. 3.4a). However, following expression and isolation, the ferrous–CO complex of the purified enzyme gave a Soret maximum at 448 nm, typical of a "normal" P450 (Rupasinghe et al. 2006). Site-directed mutagenesis, to create mutant EXXR forms of CYP157C1, was undertaken but all the mutants produced did not result in correctly folded P450, only the incorrectly folded P420 form (Rupasinghe et al. 2006). This work proved that the EXXR motif is not required in all P450s, and the only residue absolutely conserved in all P450 sequences was the cysteine that coordinates with the heme Fe. However,

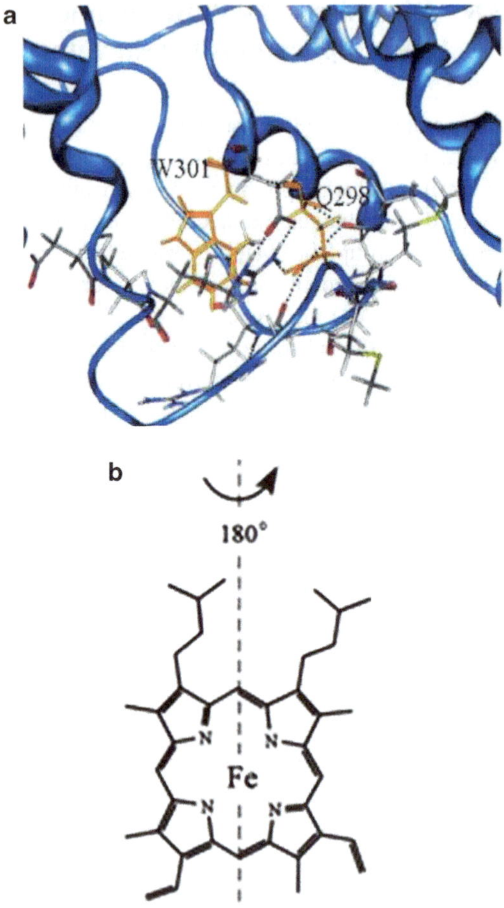

Fig. 3.4 Unusual structural features of streptomycete P450s. (**a**) Molecular model of CYP157C1 depicting the replacement of the conserved P450 glutamic acid and arginine residues of the EXXR motif with glutamine and tryptophan residues. The absence of EXXR in CYP157C1 still results in correctly folded P450. (**b**) Heme orientation in CYP154A1 is 100 % in the opposite position compared with that seen in most other P450s

continuing genome sequencing has revealed that members of the CYP408 family do not contain the canonical cysteine residue but do contain the EXXR motif and have sufficient sequence identity to be classified as P450s (Nelson 2009). Thus, it may well be that no single residue is essential for P450 function, although the latter observation requires experimental verification.

3.8.2 Diversity in Heme Topology

Until 2003, all P450 X-ray crystal structures revealed that following heme incorporation into the P450 molecule, the heme prosthetic group adopted a unique and defined orientation based upon the positions of the two protoheme vinyl groups. However, the crystal structure of CYP121A1 from *Mycobacterium tuberculosis* revealed that the heme group could assume two distinct molecular orientations, the normal orientation and a small fraction in the opposite orientation (Leys et al. 2003). CYP121A1 catalyzes the formation of an intramolecular C–C bond between two tyrosyl carbon atoms of cyclodipeptide cyclo (*L*-Tyr–*L*-Tyr) (cYY), but it is not known whether a particular orientation of the prosthetic heme is favored (Belin et al. 2009; Fonvielle et al. 2013).

The crystal structure of *S. coelicolor* CYP154A1 revealed that the heme orientation is 100 % opposite to that of all reported P450 structures (Fig. 3.4a) (Podust et al. 2004). This difference is not a unique characteristic of the CYP154 family because the resolved structure of the another family member, CYP154C1, revealed all its heme in the normal P450 orientation (Podust et al. 2003). It is possible that the heme orientation may influence CYP154A1 enzymatic activities. CYP154A1 has been shown to catalyze the unique cyclo-addition of a dipentaenone in an unprecedented P450-catalyzed reaction that requires no redox partner, NADPH or O_2. It may be that the unusual orientation of the heme in CYP154A1 contributes a significant role to this novel reaction undertaken (Cheng et al. 2010).

3.8.3 Diversity in P450 Substrate Binding and Protonation

Streptomyces coelicolor contains a three-gene operon encoding a type III polyketide synthase, CYP158A2, and a quinione-forming monooxygenase (momA). The polyketide synthase sequentially converts five molecules of malonyl CoA into 1,3,6,8-tetahydroxynapthalene, which is oxidized to flaviolin by momA. Finally, CYP158A2 catalyzes C–C bond formation to polymerize flaviolin into the di- and trimer forms (Fig. 3.5a) (Zhao et al. 2005a). However, the *S. coelicolor* genome contains a second *CYP158A* gene, *CYP158A1*, and the protein shares 61 % amino acid sequence identity with CYP158A2 (Lamb et al. 2002). It was shown CYP158A1 can catalyze the dimerization of flaviolin as for CYP158A2, but with differing regiospecificity (Fig. 3.5a) (Zhao et al. 2007). In the flaviolin-bound CYP158A1 crystal structure, one flaviolin molecule is positioned over the CYP158A1 heme similar to that found in CYP158A2 but in CYP158A1 the second flaviolin molecule is bound at the entrance to the substrate access channel (Zhao et al. 2007). This unexpected finding prevents understanding in how CYP158A1 catalyzes dimerization compared to CYP158A2. Thus, the different modes of substrate binding for two different P450 enzymes, catalyzing the same biochemical reaction, is a unique observation for the P450 superfamily.

Furthermore, the crystal structure of CYP158A2 revealed that the 2-OH group of flaviolin anchors the substrate in the active site whereas the 5-OH and 7-OH groups stabilize water molecules important for enzyme catalysis (Zhao et al. 2005b). The use of a substrate analogue, α-hydroxynapthalene (which is missing both the 5-OH and 7-OH groups), revealed 70 fold less activity in dimerization, suggesting the

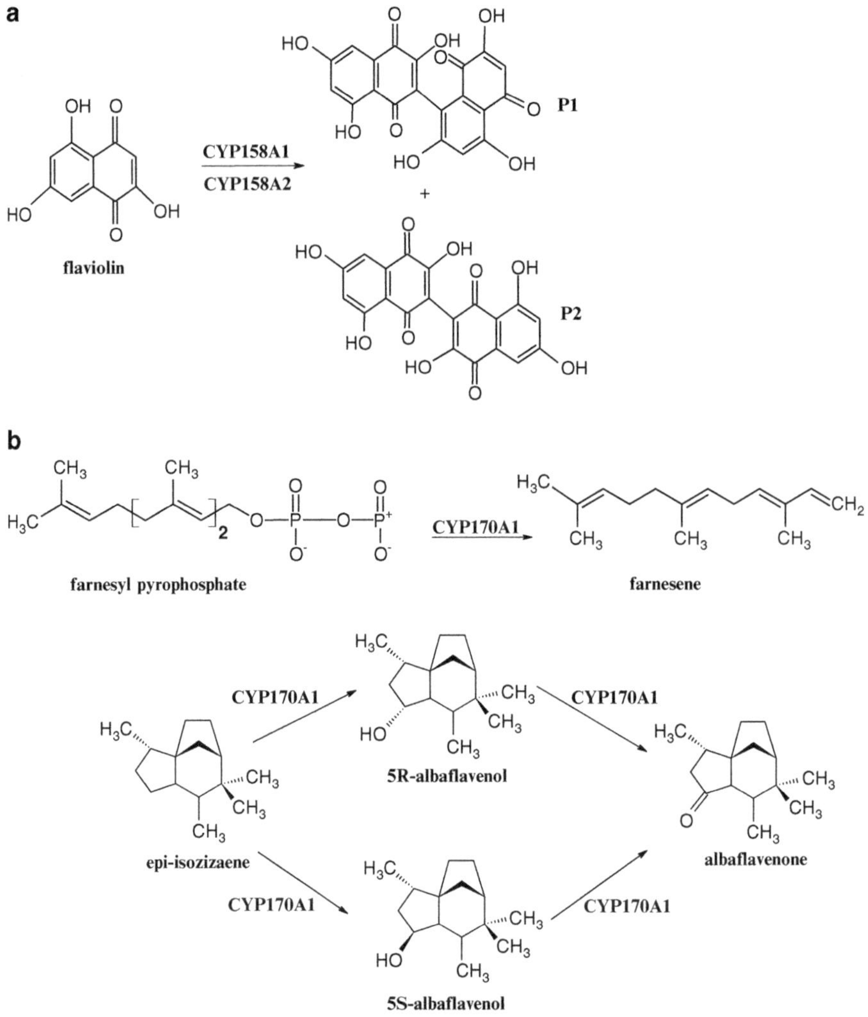

Fig. 3.5 Unique biochemical reactions catalyzed by *Streptomyces coelicolor* P450s. (**a**) CYP158A1 and CYP158A2 bind the substrate flaviolin in different orientations but catalyze the dimerization of this molecule to produce two distinct products, **P1** and **P2**. CYP158A1 produces more of the P2 product compared to CYP158A1, which produces more P1. (**b**) CYP170A1 catalyzes the conversion of farnesyl diphosphate to farnesene, which occurs in the moonlighting active site. Within the heme-containing active site, the oxidation of epi-isozizaene to albaflavenone occurs

water molecules may form a proton relay pathway to the bulk solvent (Zhao et al. 2005b). During the general P450 catalytic cycle, dioxygen binding, protonation, and splitting of the oxygen–oxygen bond are critical steps for product formation. Two unique mechanisms have previously been described for the protonation step in catalysis. In CYP101A1, a hydroxyl group of a threonine residue in the P450 I helix hydrogen bonds to the dioxygen molecule in conjunction with a water molecule. This water molecule provides the proton for P450 catalysis (Raag et al. 1991). This conserved threonine has been found in the majority of P450 sequences to date. In CYP107A1, the conserved threonine is absent and the crystal structure suggests that a hydroxyl group of the P450 substrate itself, 6-deoxyerythronolide B, directly donates a proton to the Fe-linked dioxygen for the proton transfer (Nagano et al. 2005).

3.8.4 Diversity in P450 Bifunctionality

A P450 whose normal protein structure is responsible for more than one biochemical function is referred to as a moonlighting P450. To date, three P450s can be classified as being moonlighting: eukaryotic CYP17A1 and CYP7B1 and prokaryotic CYP170A1 (Zhao and Waterman 2011). CYP17A1 and CYP7B1 both have one enzymatic active site but are considered moonlighting as their catalytic reaction is changed dependent upon (1) protein interaction for CYP17A1 (Katagiri et al. 1995) or (2) tissue location for CYP7B1 (Stiles et al. 2009). However, CYP170A1 is a clear bifunctional P450 enzyme containing two enzyme active sites in a single P450 protein molecule (Zhao et al. 2009).

First, CYP170A1 carries out two sequential allylic oxidations to convert epi-isozizaene to an epimeric mixture of albaflavenols and ultimately to the single ketone sesquiterpene, the antibiotic albaflavenone (Fig. 3.5b; Zhao et al. 2008). Although numerous examples have been described of three-step oxidations of methyl groups to carboxylic acids that are catalyzed by a single P450, the two-step oxidation of an allylic methylene to a conjugated ketone by a single P450 is unusual. Second, during studies of the CYP170A1 activity, an unexpected additional catalytic activity was observed. It was shown that purified CYP170A1 alone can convert farnesyl diphosphate to a mixture of farnesene isomers without the need for redox partners and NADPH but with Mg^{2+} being essential for this reaction (Fig. 3.5b). The crystal structure of CYP170A1 revealed the presence of a novel terpene synthase active site moonlighting on the normal P450 structure (Zhao et al. 2009); this included signature sequences for Mg^{2+} binding. Structurally, the CYP170A1 terpene synthase active site is unusual because it consists of only four helices rather than six as found in all other terpene synthases. The presence of two such distinct active sites and unrelated catalytic activities in a single P450 molecule is unparalleled within the superfamily. Finally, in another unexplained deviation from the original belief that all P450s should have a reduced CO Soret peak at 450 nm, CYP170A1 has its Soret peak at 440 nm but is clearly a P450 monooxygenase. Hence, even the most fundamental defining characteristic of identifying and naming a P450 can be challenged.

3.9 Conclusions and Outlook

Since the discovery of cytochrome P450 in 1962, much information regarding the biology of this unique class of hemoprotein has been revealed, but there is much more to be learned. New properties of P450s at the molecular level, including structural differences and unique physiological functions, are being found. These properties influence the structure and function of a small number of specific P450s but can teach us new roles that P450s can play beyond the traditional ones first discovered. In our own research endeavors we have described (in Section 8) a small number of examples of streptomycete P450 properties, which we cannot yet explain in a biological context. Researchers can expect to find additional novel properties of P450s as the number of different gene families are studied in greater experimental detail. In the opening paragraph of the seminal paper written by Tsuneo Omura and Ryo Sato in 1962, they include the following sentence regarding the newly discovered hemoprotein: "Little information is, however, available concerning its nature and functions." Fifty years later, tens of thousands of P450 sequences are known, but the lack of knowledge of P450 activities is still the primary weakness in understanding the superfamily. Thus, without the ability to accurately predict function(s) of the many members of this superfamily, the expectations of Omura and Sato in 1962 remain unfulfilled.

References

Anandatheerthavarada HK, Biswas G, Mullick J, Sepuri NB, Otvos L, Pain D, Avadhani NG (1999) Dual targeting of cytochrome P4502B1 to endoplasmic reticulum and mitochondria involves a novel signal activation by cyclic AMP-dependent phosphorylation at ser128. EMBO J 18:5494–5504

Aoyama Y, Yoshida Y (1978) The 14-alpha-demethylation of lanosterol by a reconstituted cytochrome P-450 system from yeast microsomes. Biochem Biophys Res Commun 85:28–34

Aoyama T, Yamano S, Guzelian PS, Gelboin HV, Gonzalez FJ (1990) Five of 12 forms of vaccinia virus-expressed human hepatic cytochrome P450 metabolically activate aflatoxin B1. Proc Natl Acad Sci USA 87:4790–4793

Appleby CA (1967) A soluble hemoprotein P450 from nitrogen-fixing *Rhizobium* bacteroids. Biochim Biophys Acta 147:399–402

Asseffa A, Smith SJ, Nagata K, Gillette J, Gelboin HV, Gonzalez FJ (1989) Novel exogenous heme-dependent expression of mammalian cytochrome P450 using baculovirus. Arch Biochem Biophys 274:481–490

Atkins WM, Sligar SG (1988) The roles of active site hydrogen bonding in cytochrome P-450cam as revealed by site-directed mutagenesis. J Biol Chem 263:18842–18849

Avadhani NG, Sangar MC, Bansal S, Bajpai P (2011) Bimodal targeting of cytochrome P450s to endoplasmic reticulum and mitochondria: the concept of chimeric signals. FEBS J 278:4218–4229

Bajpai P, Sangar MC, Singh S, Tang W, Bansal S, Chowdhury G, Cheng Q, Fang JK, Martin MV, Guengerich FP, Avadhani NG (2013) Metabolism of 1-methyl-4-phenyl-1,2,3,6-tetrahydropyridine by mitochondrion-targeted cytochrome P450 2D6: implications in Parkinson disease. J Biol Chem 288:4436–4451

Bansal S, Liu CP, Sepuri NB, Anandatheerthavarada HK, Selvaraj V, Hoek J, Milne GL, Guengerich FP, Avadhani NG (2010) Mitochondria-targeted cytochrome P450 2E1 induces oxidative damage and augments alcohol-mediated oxidative stress. J Biol Chem 285:24609–24619

Bansal S, Anandatheerthavarada HK, Prabu GK, Milne GL, Martin MV, Guengerich FP, Avadhani NG (2013) Human cytochrome P450 2E1 mutations that alter mitochondrial targeting efficiency and susceptibility to ethanol-induced toxicity in cellular models. J Biol Chem 288:12627–12644

Barnes HJ, Arlotto MP, Waterman MR (1991) Expression and enzymatic activity of recombinant cytochrome P450 17 alpha-hydroxylase in *Escherichia coli*. Proc Natl Acad Sci USA 88:5597–5601

Battula N (1989) Transduction of cytochrome P3-450 by retroviruses: constitutive expression of enzymatically active microsomal hemoprotein in animal cells. J Biol Chem 264:2991–2996

Belin P, Le Du MH, Fielding A, Lequin O, Jacquet M, Charbonnier JB, Lecoq A, Thai R, Courçon M, Masson C, Dugave C, Genet R, Pernodet JL, Gondry M (2009) Identification and structural basis of the reaction catalyzed by CYP121, an essential cytochrome P450 in *Mycobacterium tuberculosis*. Proc Natl Acad Sci USA 106:7426–7431

Briza P, Eckerstorfer M, Breitenbach M (1994) The sporulation-specific enzymes encoded by the DIT1 and DIT2 genes catalyze a two-step reaction leading to a soluble LL-dityrosine-containing precursor of the yeast spore wall. Proc Natl Acad Sci USA 91:4524–4528

Cheng Q, Lamb DC, Kelly SL, Lei L, Guengerich FP (2010) Cyclization of a cellular dipentaenone by *Streptomyces coelicolor* cytochrome P450 154A1 without oxidation/reduction. J Am Chem Soc 132:15173–15175

Conney AH (2003) Induction of drug-metabolizing enzymes: a path to the discovery of multiple cytochromes P450. Annu Rev Pharmacol Toxicol 43:1–30

Coon MJ (2005) Cytochrome P450: nature's most versatile biological catalyst. Annu Rev Pharmacol Toxicol 45:1–25

Cooper DY, Levin SS, Narasimhulu S, Rosenthal O, Estabrook RW (1965) Photochemical action spectrum of the terminal oxidase of mixed function oxidase systems. Science 147:400–402

Crespi CL, Steimel DT, Aoyama T, Gelboin HV, Gonzalez FJ (1990) Stable expression of human cytochrome P450IA2 cDNA in a human lymphoblastoid cell line: role of the enzyme in the metabolic activation of aflatoxin B1. Mol Carcinog 3:5–8

De Mot R, Parret AHA (2002) A novel class of self-sufficient cytochrome P450 monooxygenases in prokaryotes. Trends Microbiol 10:502–508

DeWitt DJ, Smith WL (1983) Purification of prostacyclin synthase from bovine aorta by immunoaffinity chromatography. Evidence that the enzyme is hemoprotein. J Biol Chem 258:3285–3293

Distlerath LM, Reilly PE, Martin MV, Davis GG, Wilkinson GR, Guengerich FP (1985) Purification and characterization of the human liver cytochromes P-450 involved in debrisoquine 4-hydroxylation and phenacetin O-deethylation, two prototypes for genetic polymorphism in oxidative drug metabolism. J Biol Chem 260:9057–9067

Doehmer J, Oesch F (1991) V79 Chinese hamster cells genetically engineered for stable expression of cytochromes P450. Methods Enzymol 206:117–123

el-Masry S e-D, Cohen GM, Mannering GJ (1974) Sex-dependent differences in drug metabolism in the rat. I. Temporal changes in microsomal drug-metabolizing system of the liver during sexual maturation. Drug Metab Dispos 2:267–278

Estabrook RW (2003) A passion for P450s (remembrances of the early history of research on cytochrome P450). Drug Metab Dispos 31:1461–1473

Estabrook RW, Cooper DY, Rosenthal O (1963) The light reversible carbon monoxide inhibition of the steroid C21-hydroxylase system of the adrenal cortex. Biochem Z 338:741–755

Ferris JP, Fasco MJ, Stylianopoulou FL, Jerina DM, Daly JW, Jeffrey AM (1973) Monooxygenase activity in *Cunninghamella bainieri*: evidence for a fungal system similar to liver microsomes. Arch Biochem Biophys 156:97–103

Fonvielle M, Le Du MH, Lequin O, Lecoq A, Jacquet M, Thai R, Dubois S, Grach G, Gondry M, Belin P (2013) Substrate and reaction specificity of *Mycobacterium tuberculosis* cytochrome P450 CYP121: insights from biochemical studies and crystal structures. J Biol Chem 288:17347–17359

Fujii-Kuriyama Y, Mizukami Y, Kawajiri K, Sogawa K, Muramatsu M (1982) Primary structure of a cytochrome P-450: coding nucleotide sequence of phenobarbital-inducible cytochrome P-450 cDNA from rat liver. Proc Natl Acad Sci USA 79:2793–2797

Ghisla S, Thorpe C (2004) Acyl-CoA dehydrogenases. A mechanistic overview. Eur J Biochem 271:494–508

Gillette JR (1967) Comments on comparative patterns of drug metabolism. Fed Proc 26:1040–1043

Guengerich FP (2005) Human cytochrome P450 enzymes. In: Ortiz de Montellano PR (ed) Cytochrome P450: structure, mechanism and biochemistry, 3rd edn. Kluwer Academic/Plenum Press, New York, pp 377–530

Guengerich FP, Martin MV, Beaune PH, Kremers P, Wolff T, Waxman DJ (1986) Characterization of rat and human liver microsomal cytochrome P-450 forms involved in nifedipine oxidation, a prototype for genetic polymorphism in oxidative drug metabolism. J Biol Chem 261:5051–5060

Gustafsson MC, Roitel O, Marshall KR, Noble MA, Chapman SK, Pessegueiro A, Fulco AJ, Cheesman MR, von Wachenfeldt C, Munro AW (2004) Expression, purification, and characterization of *Bacillus subtilis* cytochromes P450 CYP102A2 and CYP102A3: flavocytochrome homologues of P450 BM3 from *Bacillus megaterium*. Biochemistry 43:5474–5487

Haniu M, Armes LG, Tanaka M, Yasunobu KT, Shastry BS, Wagner GC, Gunsalus IC (1982) The primary structure of the monooxygenase cytochrome P450CAM. Biochem Biophys Res Commun 105:889–894

Harding BW, Wong SH, Nelson DH (1964) Carbon monoxide-binding substances in rat adrenal. Biochim Biophys Acta 92:415–417

Hashimoto Y, Yamano T, Mason HS (1962) An electron spin resonance study of microsomal Fex. J Biol Chem 237:3843–3844

Hata S, Nishino T, Komori M (1981) Involvement of cytochrome P450 in Δ^{22}-desaturation in ergosterol biosynthesis of yeast. Biochem Biophys Res Commun 103:272–277

Haugen DA, Coon MJ (1976) Properties of electrophoretically homogeneous phenobarbital-inducible and β-naphthoflavone-inducible forms of liver microsomal cytochrome P-450. J Biol Chem 251:7929–7939

Haugen DA, van der Hoeven TA, Coon MJ (1975) Purified liver microsomal cytochrome P-450. Separation and characterization of multiple forms. J Biol Chem 250:3567–3570

Haurand M, Ullrich V (1985) Isolation and characterization of thromboxane synthase from human platelets as a cytochrome P-450 enzyme. J Biol Chem 260:15059–15067

Hawkes DB, Adams GW, Burlingame AL, Ortiz de Montellano PR, De Voss JJ (2002) Cytochrome P450(cin) (CYP176A), isolation, expression, and characterization. J Biol Chem 277:27725–27732

Heinz E, Tulloch AP, Spencer JFT (1970) Hydroxylation of oleic acid by cell free extracts of a species of *Torulopsis*. Biochim Biophys Acta 202:49–55

Hudnik-Plevnik T, Breskvar K (1991) Cytochrome P450 from *Rhizopus niger*. In: Ruckpaul K, Rein H (eds) Microbial and plant cytochromes P450: biochemical characteristics, genetic engineering and practical implications. Akadenic Verlag, Berlin, pp 149–168

Hunter DJ, Roberts GA, Ost TW, White JH, Muller S, Turner NJ, Flitsch SL, Chapman SK (2005) Analysis of the domain properties of the novel cytochrome P450 RhF. FEBS Lett 579:2215–2220

Ichikawa Y, Yamano T (1967) Reconversion of detergent- and sulfhydryl reagent-produced P-420 to P-450 by polyols and glutathione. Biochim Biophys Acta 131:490–497

Imai Y, Sato R (1966) Evidence for two forms of P-450 hemoprotein in microsomal membranes. Biochem Biophys Res Commun 23:5–11

Jackson CJ, Lamb DC, Marczylo TH, Warrilow AG, Manning NJ, Lowe DJ, Kelly DE, Kelly SL (2002) A novel sterol 14alpha-demethylase/ferredoxin fusion protein (MCCYP51FX) from *Methylococcus capsulatus* represents a new class of the cytochrome P450 superfamily. J Biol Chem 277:46959–46965

Jackson RG, Rylott EL, Fournier D, Hawari J, Bruce NC (2007) Exploring the biochemical properties and remediation applications of the unusual explosive-degrading P450 system XplA/B. Proc Natl Acad Sci USA 104:16822–16827

Jenkins CM, Waterman MR (1994) Flavodoxin and NADPH-flavodoxin reductase from *Escherichia coli* support bovine cytochrome P450c17 hydroxylase activities. J Biol Chem 269:27401–27408

Jo Corbin C, Mapes SM, Lee YM, Conley AJ (2003) Structural and functional differences among purified recombinant mammalian aromatases: glycosylation, N-terminal sequence and kinetic analysis of human, bovine and the porcine placental and gonadal isozymes. Mol Cell Endocrinol 206:147–157

Johnson EF, Stout CD (2013) Structural diversity of eukaryotic membrane cytochrome p450s. J Biol Chem 288:17082–17090

Kalb VF, Woods CW, Turi TG, Dey CR, Sutter TR, Loper JC (1987) Primary structure of the P450 lanosterol demethylase gene from *Saccharomyces cerevisiae*. DNA 6:529–537

Katagiri M, Ganguli BN, Gunsalus IC (1968) A soluble cytochrome P-450 functional in methylene hydroxylation. J Biol Chem 243:3543–3546

Katagiri M, Kagawa N, Waterman MR (1995) The role of cytochrome b(5) in the biosynthesis of androgens by human P450c 17. Arch Biochem Biophys 317:343–347

Kato R (1977) Drug metabolism under pathological and abnormal physiological states in animals and man. Xenobiotica 7:25–92

Kelly SL, Lamb DC, Baldwin BC, Kelly DE (1993) Benzo(a)pyrene hydroxylase activity in yeast is mediated by P450 other than sterol 14 alpha-demethylase. Biochem Biophys Res Commun 197:428–432

Kelly SL, Lamb DC, Baldwin BC, Corran AJ, Kelly DE (1997) Characterization of *Saccharomyces cerevisiae* CYP61, sterol delta22-desaturase, and inhibition by azole antifungal agents. J Biol Chem 272:9986–9988

Koga H, Rauchfuss B, Gunsalus IC (1985) P450cam gene cloning and expression in *Pseudomonas putida* and *Escherichia coli*. Biochem Biophys Res Commun 130:412–417

Lamb DC, Waterman MR (2013) Unusual properties of the cytochrome P450 superfamily. Philos Trans R Soc Lond B Biol Sci 368:20120434

Lamb DC, Skaug T, Song HL, Jackson CJ, Podust LM, Waterman MR, Kell DB, Kelly DE, Kelly SL (2002) The cytochrome P450 complement (CYPome) of *Streptomyces coelicolor* A3(2). J Biol Chem 277:24000–24005

Lamb DC, Lei L, Warrilow AG, Lepesheva GI, Mullins JG, Waterman MR, Kelly SL (2009) The first virally encoded cytochrome P450. J Virol 83:8266–8269

Lamb DC, Lei L, Zhao B, Yuan H, Jackson CJ, Warrilow AGS, Skaug T, Dyson PJ, Dawson ES, Kelly SL, Hachey DL, Waterman MR (2010) *Streptomyces coelicolor* A3(2) CYP102 protein, a novel fatty acid hydroxylase encoded as a heme domain without an N-terminal redox partner. Appl Environ Microbiol 76:1975–1980

Larson JR, Coon MJ, Porter TD (1991a) Alcohol-inducible cytochrome P-450IIE1 lacking the hydrophobic NH_2-terminal segment retains catalytic activity and is membrane-bound when expressed in *Escherichia coli*. J Biol Chem 266:7321–7324

Larson JR, Coon MJ, Porter TD (1991b) Purification and properties of a shortened form of cytochrome P-450 2E1: deletion of the NH_2-terminal membrane-insertion signal peptide does not alter the catalytic activities. Proc Natl Acad Sci USA 88:9141–9145

Leys D, Mowat CG, McLean KJ, Richmond A, Chapman SK, Walkinshaw MD, Munro AW (2003) Atomic structure of *Mycobacterium tuberculosis* CYP121 to 1.06 Å reveals novel features of cytochrome P450. J Biol Chem 278:5141–5147

Li HY, Darwish K, Poulos TL (1991) Characterization of recombinant *Bacillus megaterium* cytochrome P-450 BM-3 and its two functional domains. J Biol Chem 266:11909–11914
Lindenmayer A, Smith L (1964) Cytochromes and other pigments of baker's yeast grown aerobically and anaerobically. Biochim Biophys Acta 93:445–461
Looman AC, Bodlaender J, Comstock LJ, Eaton D, Jhurani P, de Boer HA, van Knippenberg PH (1987) Influence of the codon following the AUG initiation codon on the expression of a modified lacZ gene in *Escherichia coli*. EMBO J 6:2489–2492
Lu AYH, Coon MJ (1968) Role of hemoprotein P450 in fatty acid ω-hydroxylation in a soluble enzyme system from liver microsomes. J Biol Chem 243:1331–1332
Lu AYH, Junk KW, Coon MJ (1969) Resolution of the cytochrome P-450-containing omega-hydroxylation system of liver microsomes into three components. J Biol Chem 244:3714–3721
Matsunaga I, Yamada A, Lee DS, Obayashi E, Fujiwara N, Kobayashi K, Ogura H, Shiro Y (2002) Enzymatic reaction of hydrogen peroxide-dependent peroxygenase cytochrome P450s: kinetic deuterium isotope effects and analyses by resonance Raman spectroscopy. Biochemistry 41:1886–1892
Miles JS, Munro AW, Rospendowski BN, Smith WE, McKnight J, Thomson AJ (1992) Domains of the catalytically self-sufficient cytochrome P450BM-3. Genetic construction, overexpression, purification and spectroscopic characterization. Biochem J 288:503–509
Miller TW, Shin I, Kagawa N, Evans DB, Waterman MR, Arteaga CL (2008) Aromatase is phosphorylated in situ at serine-118. J Steroid Biochem Mol Biol 112:95–101
Müller HG, Schunck WH, Kärgel E (1991) Cytochromes P450 in alkane assimilating yeasts. In: Ruckpaul K, Rein H (eds) Microbial and plant cytochromes P450: biochemical characteristics, genetic engineering and practical implications. Akadenic Verlag, Berlin, pp 87–126
Murakami H, Yabusaki Y, Sakaki T, Shibata M, Ohkawa H (1987) A genetically engineered P450 monooxygenase: construction of the functional fused enzyme between rat cytochrome P450c and NADPH-cytochrome P450 reductase. DNA 6:189–197
Nagano S, Cupp-Vickery JR, Poulos TL (2005) Crystal structures of the ferrous dioxygen complex of wild-type cytochrome P450eryF and its mutants, A245S and A245T: investigation of the proton transfer system in P450eryF. J Biol Chem 280:22101–22107
Nakahara K, Tanimoto T, Hatano K, Usuda K, Shoun H (1993) Cytochrome P-450 55A1 (P-450dNIR) acts as nitric oxide reductase employing NADH as the direct electron donor. J Biol Chem 268:8350–8355
Nakayama N, Takemae A, Shoun H (1996) Cytochrome P450foxy, a catalytically self-sufficient fatty acid hydroxylase of the fungus *Fusarium oxysporum*. J Biochem 119:435–440
Narhi LO, Fulco AJ (1986) Characterization of a catalytically self-sufficient 119,000-dalton cytochrome P450 monooxygenase induced by barbiturates in *Bacillus megaterium*. J Biol Chem 261:7160–7169
Narhi LO, Wen LP, Fulco AJ (1988) Characterization of the protein expressed in *Escherichia coli* by a recombinant plasmid containing the *Bacillus megaterium* cytochrome P-450BM-3 gene. Mol Cell Biochem 79:63–71
Nebert DW, Adesnik M, Coon MJ, Estabrook RW, Gonzalez FJ, Guengerich FP, Gunsaius IC, Johnson EF, Kemper B, Levin W, Phillips IR, Sato R, Waterman MR (1987) The P450 gene superfamily: recommended nomenclature. DNA 6:1–11
Nelson DR (2009) The cytochrome P450 homepage. Hum Genomics 4:59–65
Oeda K, Sakaki T, Ohkawa H (1985) Expression of rat liver cytochrome P-450MC cDNA in *Saccharomyces cerevisiae*. DNA 4:203–210
Oesch-Bartlomowicz B, Oesch F (2003) Cytochrome-P450 phosphorylation as a functional switch. Arch Biochem Biophys 409:228–234
Omura T (2010) Structural diversity of cytochrome P450 enzyme system. J Biochem (Tokyo) 147:297–306
Omura T (2011) Recollection of the early years of the research on cytochrome P450. Proc Jpn Acad Series B 87:617–640

Omura T, Sato R (1962) A new cytochrome in liver microsomes. J Biol Chem 237:1375–1376
Omura T, Sato R, Cooper DY, Rosenthal O, Estabrook RW (1965) Function of cytochrome P-450 of microsomes. Fed Proc 24:1181–1189
Omura T, Sanders E, Estabrook RW, Cooper DY, Rosenthal O (1966) Isolation from adrenal cortex of a non-heme iron protein and a flavoprotein functional as a reduced triphosphopyridine nucleotide-cytochrome P450 reductase. Arch Biochem Biophys 117:660–673
Ortiz de Montellano PR (ed) (2005) Cytochrome P450: structure, mechanism and biochemistry, 3rd edn. Kluwer Academic/Plenum Press, New York
Oster T, Boddupalli SS, Peterson JA (1991) Expression, purification, and properties of the flavoprotein domain of cytochrome P-450BM-3. Evidence for the importance of the amino-terminal region for FMN binding. J Biol Chem 266:22718–22725
Podust LM, Kim Y, Arase M, Neely BA, Beck BJ, Bach H, Sherman DH, Lamb DC, Kelly SL, Waterman MR (2003) The 1.92-Å structure of *Streptomyces coelicolor* A3(2) CYP154C1. A new monooxygenase that functionalizes macrolide ring systems. J Biol Chem 278:12214–12221
Podust LM, Bach H, Kim Y, Lamb DC, Arase M, Sherman DH, Kelly SL, Waterman MR (2004) Comparison of the 1.85 Å structure of CYP154A1 from *Streptomyces coelicolor* A3(2) with the closely related CYP154C1 and CYPs from antibiotic biosynthetic pathways. Protein Sci 13:255–268
Poulos TL, Finzel BC, Gunsalus IC, Wagner GC, Kraut J (1985) The 2.6 Å crystal structure of *Pseudomonas putida* cytochrome P450. J Biol Chem 260:16122–16130
Raag R, Martinis SA, Sligar SG, Poulos TL (1991) Crystal structure of cytochrome P450CAM active site mutant Thr252Ala. Biochemistry 20:9252–9253
Roberts GA, Celik A, Hunter DJ, Ost TW, White JH, Chapman SK, Turner NJ, Flitsch SL (2003) A self-sufficient cytochrome p450 with a primary structural organization that includes a flavin domain and a [2Fe-2S] redox center. J Biol Chem 278:48914–48920
Robin MA, Anandatheerthavarada HK, Biswas G, Sepuri NB, Gordon DM, Pain D, Avadhani NG (2002) Bimodal targeting of microsomal CYP2E1 to mitochondria through activation of an N-terminal chimeric signal by cAMP-mediated phosphorylation. J Biol Chem 277:40583–40593
Rupasinghe S, Schuler MA, Kagawa N, Yuan H, Lei L, Zhao B, Kelly SL, Waterman MR, Lamb DC (2006) The cytochrome P450 gene family CYP157 does not contain EXXR in the K-helix reducing the absolute conserved P450 residues to a single cysteine. FEBS Lett 580:6338–6342
Rylott EL, Jackson RG, Edwards J, Womack GL, Seth-Smith HM, Rathbone DA, Strand SE, Bruce NC (2006) An explosive-degrading cytochrome P450 activity and its targeted application for the phytoremediation of RDX. Nat Biotechnol 24:216–219
Rylott EL, Budarina MV, Barker A, Lorenz A, Strand SE, Bruce NC (2011) Engineering plants for the phytoremediation of RDX in the presence of the co-contaminating explosive TNT. New Phytol 192:405–413
Sabbadin F, Jackson R, Haider K, Tampi G, Turkenburg JP, Hart S, Bruce NC, Grogan G (2009) The 1.5-Å structure of XplA-heme, an unusual cytochrome P450 heme domain that catalyzes reductive biotransformation of royal demolition explosive. J Biol Chem 284:28467–28475
Sakaguchi M, Mihara K, Sato R (1984) Signal recognition particle is required for co-translational insertion of cytochrome P450 into microsomal membranes. Proc Natl Acad Sci USA 81:3361–3364
Sanglard D, Fiechter A (1989) Heterogeneity within the alkane-inducible cytochrome P450 gene family of the yeast *Candida tropicalis*. FEBS Lett 256:128–134
Schunck WH, Kärgel E, Gross B, Wiedmann B, Mauersberger S, Köpke K, Kiessling U, Strauss M, Gaestel M, Müller HG (1989) Molecular cloning and characterization of the primary structure of the alkane hydroxylating cytochrome P-450 from the yeast *Candida maltosa*. Biochem Biophys Res Commun 161:843–850

Seo JA, Proctor RH, Plattner RD (2001) Characterization of four clustered and coregulated genes associated with fumonisin biosynthesis in *Fusarium verticillioides*. Fungal Genet Biol 34:155–165

Shimada T, Misono KS, Guengerich FP (1986) Human liver microsomal cytochrome P-450 mephenytoin 4-hydroxylase, a prototype of genetic polymorphism in oxidative drug metabolism. Purification and characterization of two similar forms involved in the reaction. J Biol Chem 261:909–921

Shimozawa O, Sakaguchi M, Ogawa H, Harada N, Mihara K, Omura T (1993) Core glycosylation of cytochrome P-450(arom). Evidence for localization of N terminus of microsomal cytochrome P-450 in the lumen. J Biol Chem 268:21399–21402

Shyadehi AZ, Lamb DC, Kelly SL, Kelly DE, Schunck WH, Wright JN, Corina D, Akhtar M (1996) The mechanism of the acyl-carbon bond cleavage reaction catalyzed by recombinant sterol 14 alpha-demethylase of *Candida albicans* (other names are lanosterol 14 alpha-demethylase, P-45014DM, and CYP51). J Biol Chem 271:12445–12450

Sladek NE, Mannering GJ (1966) Evidence for a new P-450 hemoprotein in hepatic microsomes from methylcholanthrene treated rats. Biochem Biophys Res Commun 24:668–674

Song WC, Brash AR (1991) Purification of an allene oxide synthase and identification of the enzyme as a cytochrome P-450. Science 253:781–783

Stern JO, Peisach J (1974) A model compound study of the CO-adduct of cytochrome P-450. J Biol Chem 249:7495–7498

Stiles AR, McDonald JG, Bauman DR, Russell DW (2009) CYP7B1: one cytochrome P450, two human genetic diseases and multiple physiological functions. J Biol Chem 284:28485–28489

Suzuki K, Kimura T (1965) An iron protein as a component of steroid 11β-hydroxylase complex. Biochem Biophys Res Commun 19:340–345

Vanden Bossche H, Koymans L (1998) Cytochromes P450 in fungi. Mycoses 41:32–38

Vilgrain I, Defaye G, Chambaz EM (1984) Adrenocortical cytochrome P-450 responsible for cholesterol side chain cleavage (P-450scc) is phosphorylated by the calcium-activated, phospholipid-sensitive protein kinase (protein kinase C). Biochem Biophys Res Commun 125:554–561

Waterman MR, Mason HS (1970) The redox potential of liver cytochrome P-450. Biochem Biophys Res Commun 39:450–454

Williams PA, Cosme J, Sridhar V, Johnson EF, McRee DE (2000) Mammalian microsomal cytochrome P450 monooxygenase: structural adaptations for membrane binding and functional diversity. Mol Cell 5:121–131

Zhang LH, Rodriguez H, Ohno S, Miller WL (1995) Serine phosphorylation of human P450c17 increases 17,20-lyase activity: implications for adrenarche and the polycystic ovary syndrome. Proc Natl Acad Sci USA 92:10619–10623

Zhang L, Takaya N, Kitazume T, Kondo T, Shoun H (2001) Purification and cDNA cloning of nitric oxide reductase cytochrome P450nor (CYP55A4) from *Trichosporon cutaneum*. Eur J Biochem 268:3198–3204

Zhao B, Waterman MR (2011) Moonlighting cytochrome P450 monooxygenases. IUBMB Life 63:473–477

Zhao B, Guengerich FP, Bellamine A, Lamb DC, Izumikawa M, Funa N, Lei L, Podust LM, Sundamoorthy M, Reddy LM, Kelly SL, Moore BS, Stec D, Voehler M, Falck JR, Shimada T, Waterman MR (2005a) Bindings of two flaviolin substrate molecules, oxidative coupling, and crystal structure of *Streptomyces coelicolor* A3(2) cytochrome P450 158A2. J Biol Chem 280:11599–11607

Zhao B, Guengerich FP, Voehler M, Waterman MR (2005b) Role of active site water molecules and substrate hydroxyl groups in oxygen activation by cytochrome P450 158A2. A new mechanism of proton transfer. J Biol Chem 280:42188–42197

Zhao B, Lamb DC, Lei L, Kelly SL, Yuan H, Hachey DL, Waterman MR (2007) Different binding modes of two flaviolin substrate molecules in cytochrome P450 158A1 (CYP158A1) compared to CYP158A2. Biochemistry 46:8725–8733

Zhao B, Lin X, Lei L, Lamb DC, Kelly SL, Waterman MR, Cane DE (2008) Biosynthesis of the sesquiterpene antibiotic albaflavenone in *Streptomyces coelicolor* A3(2). J Biol Chem 283:8183–8189
Zhao B, Lei L, Vassylyev DG, Lin X, Cane DE, Kelly SL, Yuan H, Lamb DC, Waterman MR (2009) Crystal structure of albaflavenone monooxygenase containing a moonlighting terpene synthase active site. J Biol Chem 284:36711–36719
Zuber MX, Simpson ER, Waterman MR (1986) Expression of bovine 17 alpha-hydroxylase cytochrome P-450 cDNA in nonsteroidogenic (COS 1) cells. Science 234:1258–1261

Part II
Structure, Function, and Practical Applications of P450

Chapter 4
Cytochrome P450 Dynamics

Thomas L. Poulos

Abstract The unique P450 architecture is designed, in part, to enable the structure to adopt open and closed conformations. These changes are required to provide substrate access to the active site. Some of these motions also appear to be important in proton-coupled electron transfer associated with O_2 activation. Solvent protons must have access to the active site to properly protonate the iron-linked O_2 molecule, so a shift from the totally closed to at least partially open conformations is required. Such dynamics are probably shared by all P450s. Unique to drug-metabolizing P450s is the requirement that the active site adapt to substrates of very different sizes and shapes. Crystal structures, biophysical studies, and molecular dynamics have provided important insights into how drug-metabolizing P450s, especially CYP3A4, structurally adapt to a variety of inhibitors and substrates.

Keywords Conformational changes • Crystallography • Dynamics • Electron transfer • Inhibitor binding • Oxygen activation • Redox partners • Substrate binding

4.1 Introduction

The role of dynamics in enzyme action has a long and controversial history (Kamerlin and Warshel 2010). Part of the problem derives from confusing the difference between the roles that structural changes have in substrate binding/product release and the actual catalytic event of making and breaking bonds. Very often the rate-limiting step under steady-state conditions is not the actual chemical steps but rather substrate binding or release and associated structural adjustments of the protein. Such

T.L. Poulos (✉)
Departments of Molecular Biology and Biochemistry, Pharmaceutical Sciences, and Chemistry, University of California Irvine, Irvine, CA 92697-3900, USA
e-mail: poulos@uci.edu

H. Yamazaki (ed.), *Fifty Years of Cytochrome P450 Research*,
DOI 10.1007/978-4-431-54992-5_4,

changes may contribute very little to the strict definition of catalysis, which is lowering the energy of the transition state. It could well be that structural changes that result from substrate binding may be required to enable the active site to provide the appropriate electrostatic stabilization of the transition state, but once in the proper transition state configuration, it has been argued that subsequent dynamics is of little importance (Kamerlin and Warshel 2010).

With P450s, dynamics refers primarily to substrate binding and product release and to a lesser extent to changes associated with catalysis. P450 flexibility with respect to substrate binding became a problem immediately after the first P450 structure was solved (Poulos et al. 1985, 1987). The substrate, camphor, in P450cam was effectively sealed off from the outside world, which meant that there must be at least some small structural change that enables the protein to open up, thus allowing the substrate to enter and product to leave. Since these early days several more P450 crystal structures have shown that P450s indeed can undergo rather large motions that open the active site to bulk solvent. The breakthrough in solving mammalian P450 structures (Williams et al. 2000), and especially the drug-metabolizing CYP3A4 (Yano et al. 2004), presented another problem. In contrast to prokaryotic P450s that exhibit high substrate specificity, drug-metabolizing P450s such as CYP3A4 have relaxed selectivity and are able to bind and metabolize a large array of substrates of different size, shapes, and chemical properties. This capability raises interesting questions on how the active site adapts and suggests that protein dynamics may play a critical role. It also has become clear that conformational dynamics is important in not just substrate binding but also in the O_2 activation mechanism. In this chapter, we summarize recent advances in these areas that illustrate the critical and varied roles conformational dynamics plays in P450 catalysis.

4.2 Open and Closed P450 Structures

4.2.1 P450cam and P450BM3

The first hint on which regions of P450cam might be most flexible derived from the substrate-free P450cam crystal structure (Poulos et al. 1986). There was no change in structure, although the thermal or B factors, a measure of mobility, increased in the loop connecting the F and G helices on the surface of the protein (Fig. 4.1). This observation suggested that this region may provide the required flexibility to allow substrates to enter. P450BM3 provided the first crystallographic comparison between a substrate-free partially open and partially bound closed structure. Initially, the P450BM3 structure was solved in the substrate-free form (Ravichandran et al. 1993), followed by the substrate-bound form (Li and Poulos 1997). As shown in Fig. 4.2, the F and G helices are the main elements of structure that move to open up the access channel. More recently it has been possible to solve the structure of open

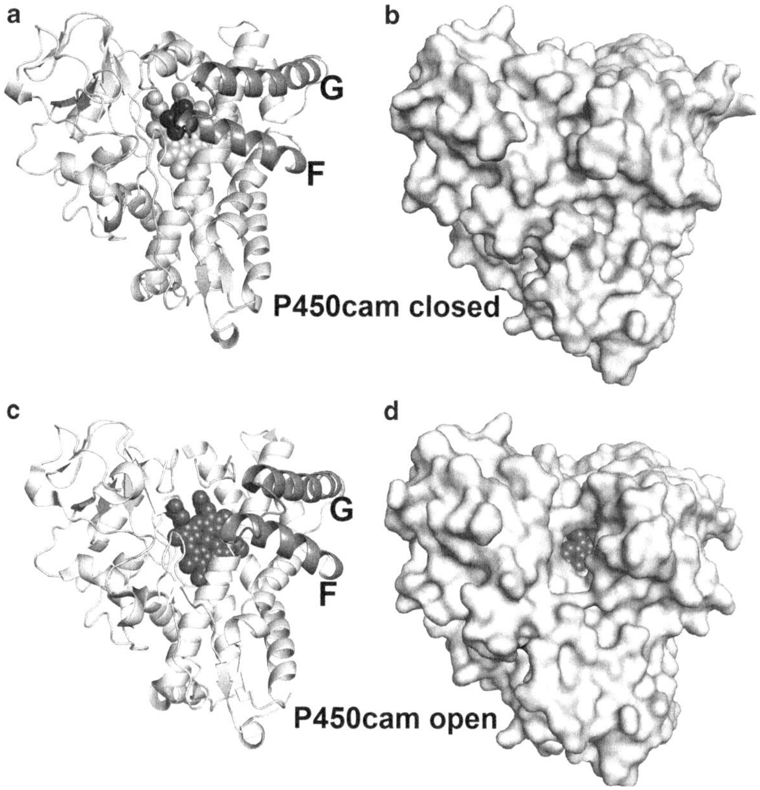

Fig. 4.1 Structure of P450cam in the open (Lee et al. 2010b) and closed states: cartoon representations (**a**, **c**) and surface representations (**b**, **d**). The F and G helices move the most, opening an access channel that leads directly to the substrate-binding site

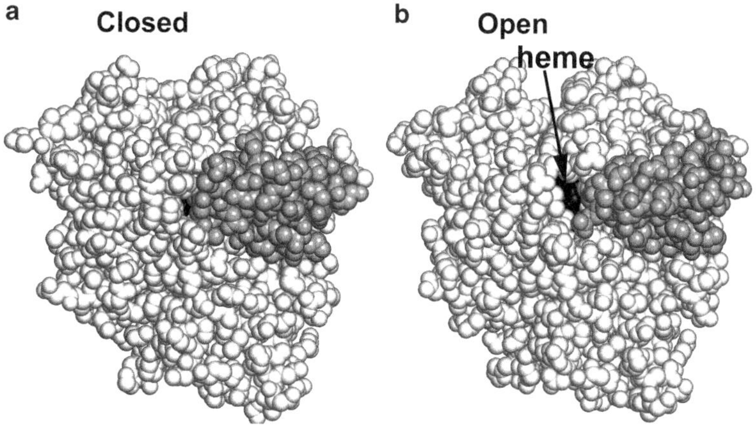

Fig. 4.2 Structure of P450BM3 in the closed and partially open conformation. The F/G helical segment is shaded *gray*

P450cam (Lee et al. 2010b) (Fig. 4.1). As with P450BM3, the open/close transition involves large movements of the F and G helices and the loop connecting these helices.

With P450cam it has been possible to capture the open/close transition using substrates tethered to a long linker that traps the substrate access channel in various stages along the open/close path. These structures together with principal component analysis show that P450cam has at least three distinct conformational states (Lee et al. 2010a). More recently, a double-labeling EPR approach has been used to study the effects of substrate binding on conformation. The basic idea is to engineer specific and well-defined spin-labeled sites and then use double electron–electron resonance (DEER) to measure distances (Stoll et al. 2012). Consistent with the crystal structures, these types of experiments show that P450cam shifts from the open state in the absence of substrate to the closed form when substrate binds. These types of open/close motions involving primarily the F and G helices now have been observed in other P450s, with the largest motions having been observed with CYP2B4 (Scott et al. 2003, 2004). Molecular dynamics simulations agree well with the crystal structures in that the F/G loop and helices provide the most flexible region for substrate binding and product egress (Lüdemann et al. 2000; Wade et al. 2004). These movements, however, are not limited to just the F and G helices. In P450cam, for example, the C and I helices also change (Lee et al. 2010b). This point becomes important further on when we consider structural changes involved in O_2 activation and electron transfer because the I helix is involved with O_2 activation whereas the C helix is involved with redox partner binding.

4.2.2 CYP119

Although the number of crystal structures of open and closed P450s is limited, the F/G helical movements are a common feature in all. CYP119, the first P450 from a thermophile to be described (Wright et al. 1996), also experiences a fairly large open/close motion, but in this case the F and G helices remain rigid and the main movement is confined to the long F/G loop (Fig. 4.3). The structure of CYP119 has been solved with ligands, imidazole and phenylimidazole, bound to the heme iron (Yano et al. 2000) and with no ligand bound (Park et al. 2002). In the absence of any ligand or substrate, the F/G loop extends away from the heme and Phe153 is positioned just over the heme buried in the active site (Fig. 4.3). To make room for ligands, Phe153 is displaced toward the surface. Phe153 is part of the C-terminal end of the F helix and, when a ligand binds, the C-terminal end of the F helix forms an additional turn of helix, which requires that Phe153 be expelled from the active site. When ligand binds, the Arg154-Glu 212 ion pair is disrupted. The resulting motion of Arg154 (Fig. 4.3) requires that the N-terminal end of the F helix unfolds in going from the ligand-free to the ligand-bound state. Therefore, the ligand/no ligand switch results in the loss of helical H-bonds in one state and the gain of helical H-bonds in the other. In addition, the unfavorable energetic loss of the Arg154-Glu 212 ion pair and removal of Phe153 from the active site in the ligand-free to ligand-bound

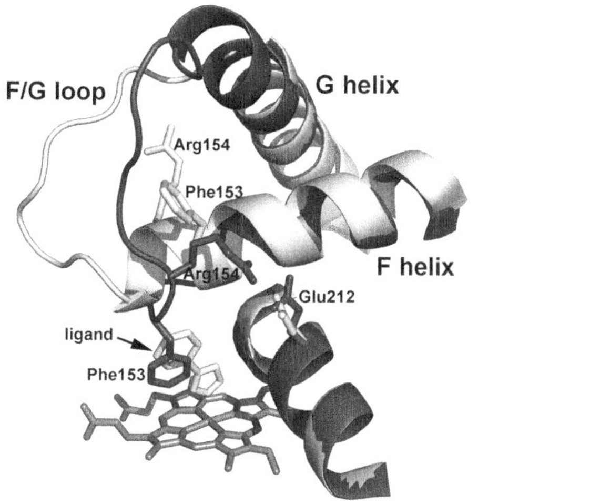

Fig. 4.3 Comparison between CYP119 with ligand bound (*white model*) and with no ligand bound (*dark model*). When the ligand binds, Phe153 moves out of the active site, which results in an extension of the F helix. The F/G loop moves into the active site, which requires a disruption of helical H-bonds in the G helix

transition is probably compensated for by favorable protein–ligand interactions and, of course, the ligand N–Fe bond. Thus, the new intramolecular protein–ligand contacts balance the energetic cost of disrupting nonbonded contacts required for ligand binding. Most noteworthy with CYP119 is that the observed changes are highly localized to the F/G loop and the ends of the F and G helices with very little change in the rest of the structure: this is in contrast to the small handful of examples for other P450s where the open/close motion requires more global changes.

4.3 Drug-Metabolizing P450s

Although we have a fairly clear picture on which regions of P450 are most dynamic with respect to substrate entry and product release, a far more challenging problem is in understanding how drug-metabolizing P450s can bind so many different types of substrates. Is the active site exceptionally large so it can accommodate a wide ranges of substrates, or does the active site "shape" itself around substrates? Intuitively one would think that those P450s that are very promiscuous should exhibit greater flexibility, and this led to the concept of a promiscuity index (Foti et al. 2011). Not too surprisingly, the most promiscuous P450, CYP3A4, has the highest promiscuity index whereas the more specific P450s have a lower promiscuity index. An important question raised by Foti et al. (2011) is the price to be paid for such enhanced promiscuity. For example, are those P450s such as CYP3A4 with a high promiscuity index more susceptible to P450 inhibitors? The answer appears to be no, which is perhaps a little surprising but in the context of evolution is clearly advantageous.

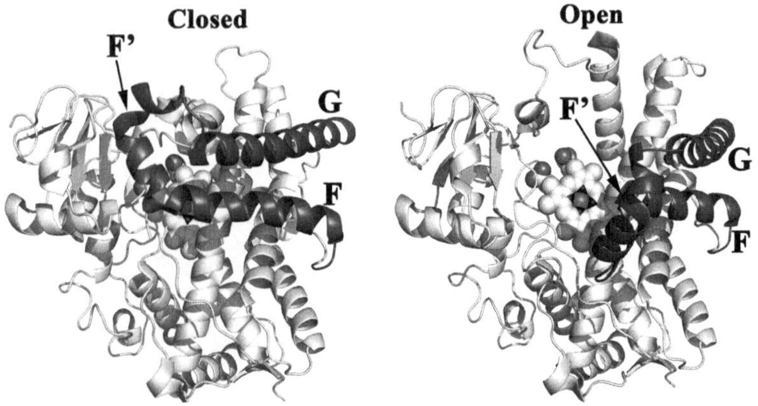

Fig. 4.4 CYP2B4 in the open and closed conformations (Scott et al. 2003, 2004). Note the very large movement of the F, F′, and G helices

4.3.1 CYP2B4

Although CYP2B4 was not calibrated on the promiscuity index, the crystal structure of the open and closed forms illustrates the very large motion that P450s can undergo (Scott et al. 2003). In the open form, CYP2B4 crystallizes as a dimer, and a His residue from one monomer penetrates into the active site of the second molecule where it coordinates the heme iron to give a low-spin complex; this requires a very large opening of the active site (Fig. 4.4). It has been argued that this is perhaps an artifact of crystallization. However, forces that hold crystals together are relatively weak, and crystallization cannot force a protein to adopt a conformation that it cannot also adopt in solution. Thus, it is very likely that the open conformation (Fig. 4.4) can occur in solution. Moreover, a similar dimer forms in solution (Scott et al. 2003). These results with CYP2B4 illustrate the extremely large range of flexibility of the F/G helical segments without disrupting much secondary structure.

4.3.2 CYP3A4

Large open/close motions can explain how substrates enter and products are released, but this does not explain how the active site adapts to substrates of different sizes and shapes. Here we focus on CYP3A4 because this is the most important drug-metabolizing P450, is responsible for the metabolism of more than half of currently used drugs, and is number one on the promiscuity index of the ten P450s examined (Foti et al. 2011). In contrast to some of the prokaryotic P450s in which it has been possible to solve many structures with ligands bound, CYP3A4 has proven less cooperative. Only ligands and substrates that bind tightly give

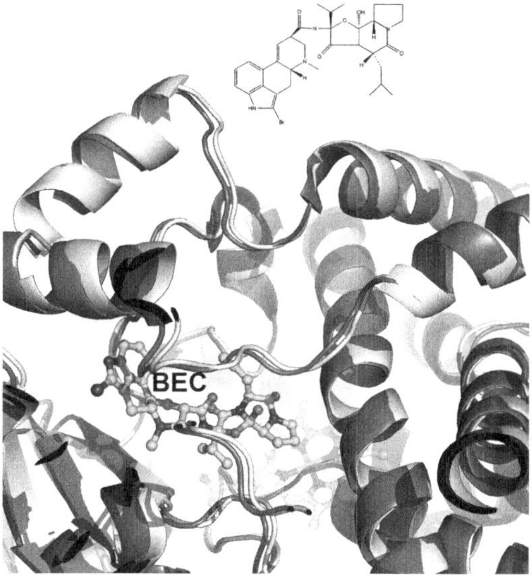

Fig. 4.5 Crystal structure of substrate-bound (*white*) and substrate-free (*gray*) CYP3A4. Bromoergocryptin (BEC) binding results in very few changes in the overall polypeptide conformation, although some adjustments of side chains occur

well-diffracting crystals, which is the reason why there is so far only one CYP3A4–substrate complex crystal structure where the substrate binds in a productive mode. Bromoergocryptine (BEC) is an excellent CYP3A4 substrate and binds tightly, which enabled the crystal structure to be solved (Sevrioukova and Poulos 2012). What was somewhat of a surprise is that there is very little change in the backbone between the substrate-free (Yano et al. 2004) and the BEC-bound structures (Fig. 4.5) and thus both are in the "closed" state; this, however, does not mean that CYP3A4 cannot experience the kinds of motions observed in other P450s but, rather, the closed form favors crystallization.

A clearer idea on which regions are most flexible derives from comparing a series of substrate and inhibitor complexes (Sevrioukova and Poulos 2013) even though some of the substrate complexes are not close enough to the heme iron to be considered binding in a productive mode. Two of the most informative structures are CYP3A4 bound to the inhibitors ketoconazole (Ekroos and Sjögren 2006) and ritonavir (Sevrioukova and Poulos 2010). In these cases there is a significant difference between the inhibitor-free and -bound structures. Most notable is the segment of polypeptide connecting the F and F′ helices that moves away from the heme to accommodate the inhibitor (Fig. 4.6). There also are associated changes with surrounding helices including the F, G, and I helices. In the CYP3A4–erytrhomycin complex (Ekroos and Sjögren 2006) the 214–219 segment that connects the F and F′ helices (Fig. 4.6) is disordered. These structures clearly indicate that the same F/G helical region that is flexible in other P450s also is quite flexible in CYP3A4. Particularly interesting is the 214–219 region, which is very flexible to the extent that it disappears in some electron density maps, raising the possibility that CYP3A4 may experience a more localized open/close motion similar to CYP119 rather than the more global motion observed in some other

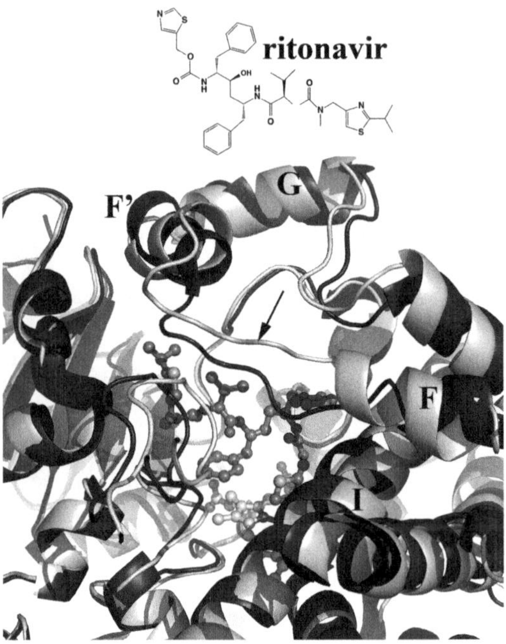

Fig. 4.6 Crystal of the CYP3A4–ritonavir complex (*dark model*) superimposed on ligand-free CYP3A4 (*white model*). The largest change is in the extended section of polypeptide between the F and F′ helices (*arrow*)

P450s. In P450cam, for example, the opening of the active site is associated with changes on the opposite side of the heme and especially the C helix (Fig. 4.6). In the CYP3A4 structures so far solved, the C helix does not change but only the substrate entry channel near the F/G helical region.

4.3.2.1 Ritonavir and Related Inhibitors

Ritonavir is a well-known human immunodeficiency virus (HIV) protease inhibitor but is used therapeutically in HIV drug cocktails as a "booster" because of its high inhibitory potency for CYP3A4 (Zeldin and Petruschke 2004). By inhibiting CYP3A4, ritonavir enables other drugs in the cocktail to escape CYP3A4 metabolism. Partly because ritonavir binds so tightly, it has been possible to obtain crystal structures of various ritonavir analogues, which has provided further insights into how CYP3A4 adapts structurally to various inhibitors. A list of ritonavir-like molecules in which crystal structures of CYP3A4-inhibitor complexes are known (Sevrioukova and Poulos 2013) is shown in Fig 4.7a. Figure 4.7b illustrates the large variation in the 206–216 stretch of polypeptide that connects the F and F′ helices, and Fig. 4.7c shows how two key Phe residues in this region adapt to different inhibitors. Clearly this region is quite flexible, and Phe213 and Phe215 as well as other nonpolar groups have substantial freedom to shape and adapt to inhibitors of different sizes and shapes. Molecular dynamics simulations (Park et al. 2005) also show that residues 211–218 are especially flexible and play a

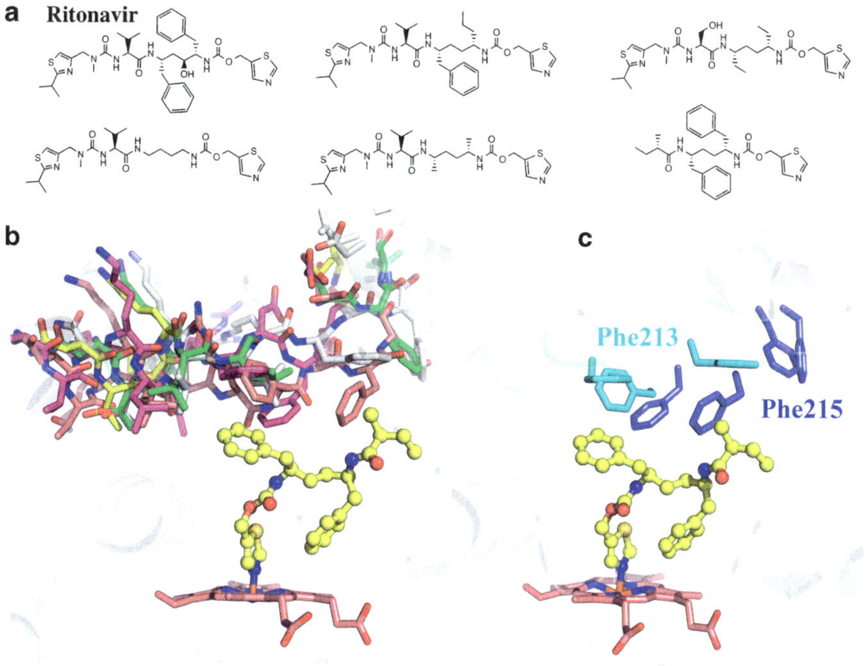

Fig. 4.7 (**a**) Structure of ritonavir and ritonavir analogues for which crystal structures in a complex with CYP3A4 have been determined. (**b**) Superposition of all the structures. (**c**) Various positioning of Phe213 and Phe215 illustrating how the Phe cluster can adapt to different inhibitors

major role in the ability of CYP3A4 to bind a wide range of substrates. Of the human P450 crystal structures that have been solved, CYP3A4 has the shortest F helix and the longest segment (residues 208–217) of extended polypeptide connecting the F and F′ helices. It is tempting to speculate that this structural feature imparts greater adaptability of CYP3A4 compared to other P450s.

4.3.2.2 Biophysical Methods and Kinetics as a Probe of Dynamics

Another area where CYP3A4 dynamics becomes important is in the ability of this P450 to bind more than one substrate at a time in addition to having peripheral binding sites. In the CYP3A4-progesterone structure (Williams et al. 2004) a molecule of progesterone binds near the "Phe cluster" shown in Fig. 4.8. It is generally thought that the F′-helical region and entry to the active site is oriented toward the membrane. If true, then nonpolar substrates concentrated in the membrane could first translocate to the Phe cluster region before entering the active site. Fluorescence energy transfer (FRET) experiments (Davydov et al. 2012) are consistent with a secondary binding site located in the flexible 211–218 segment between the F′ and G′ helices. In these experiments the distance between a

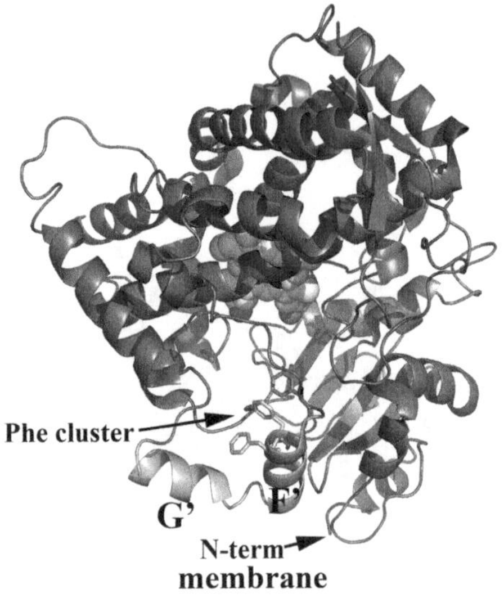

Fig. 4.8 CYP3A4 showing consensus view on how this P450 is oriented relative to the membrane. The F′ helix and Phe cluster are oriented toward the membrane, thus providing an initial docking site for nonpolar substrates that partition from the lipid bilayer to the P450 active site

fluorophore-labeled thiol and the fluorescent substrate, flurol-7GA, was determined and the binding site mapped out to the 217–220 region of CYP3A4 consistent with the CYP3A4-progesterone structure. This region, however, is likely not to be very specific given the established flexibility. Indeed this region was observed to be quite flexible when using CYP3A4-membrane molecular dynamics (MD) simulations (Denisov et al. 2012). A comparative 5–10 ns MD analysis of several P450s shows that CYP3A4 was found to be the most flexible (Hendrychová et al. 2010). Thus, the 211–218 region may represent a highly mobile and generic initial binding site to attract nonpolar substrates from the membrane toward the active site. A recent review (Otyepka et al. 2012) summarizes additional studies that demonstrate CYP3A4 is one of the more flexible mammalian P450s.

The biphasic kinetics often observed in the binding of both substrates and ligands/inhibitors (Isin and Guengerich 2006; Sevrioukova and Poulos 2010, 2012) is consistent with a peripheral binding site that does not influence the heme spectroscopic properties. The kinetics of BEC binding to CYP3A4 are biphasic. The fast phase increases with increasing BEC concentration but the slow phase decreases with increasing BEC concentration, which can best be explained by a two-binding-step model (Fig. 4.9). In the first step, BEC binds at the peripheral binding site, which results in no spectral change. In the second slower step, BEC moves closer to the heme, which does perturb the heme spectral properties, followed by a fast phase that positions BEC for productive catalysis. The structural models predict that Thr224 and Arg212 participate differently in the various kinetic phases, and altered kinetics of the Thr224 and Arg212 mutants support the view that Thr224 is involved in the peripheral binding site whereas Arg212 helps to fine tune the positioning of BEC in the active site (Sevrioukova and Poulos 2012).

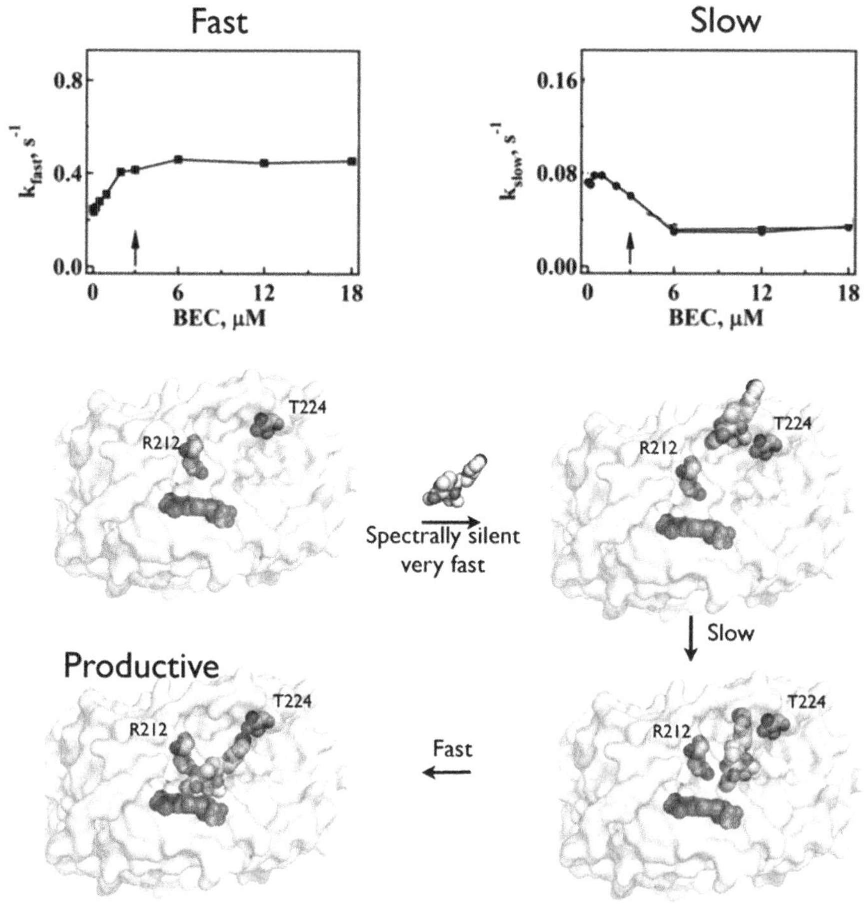

Fig. 4.9 Kinetics of BEC binding to CYP3A4. The low- to high-spin shift is used to follow the rate of BEC binding. The fast phase behaves normally and increases with increasing BEC concentration. However, the slow phase decreases with increasing BEC concentration, which can best be explained by a multi-step binding process. BEC first associates with a peripheral site, which does not result in any spectral change, followed by a second faster step wherein the BEC moves closer to the heme, resulting in spectral changes. In the third step the BEC adjusts to the final lowest free energy state, giving the complex observed in the crystal structure

4.4 Flexibility and O_2 Activation

Another area in which P450 flexibility is important is in the O_2 activation reaction. Unfortunately, there are only two crystal structures of P450-oxy complexes: P450eryF (Nagano et al. 2005) and P450cam (Nagano and Poulos 2005; Schlichting et al. 2000). This condition is no doubt because the limited stability of P450-oxy complexes precludes, in most cases, the determination of crystal structures.

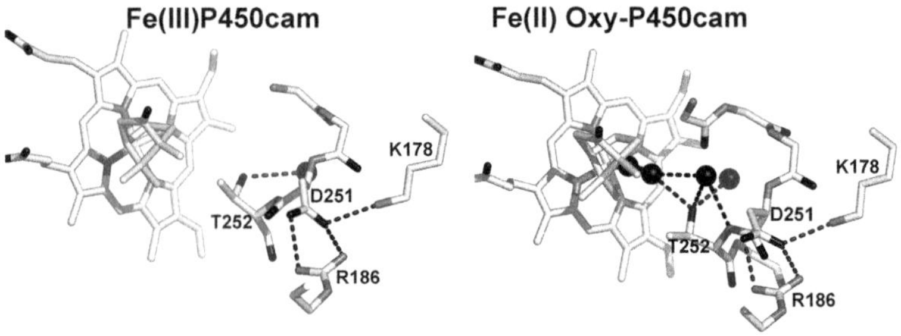

Fig. 4.10 Comparison of the Fe(III) and Fe(II)-oxy P450cam active sites. When O_2 binds, the I helix opens up, which enables water molecules (*large dark spheres*) to enter the active site. This H-bonded network of water molecules is thought to be important for O_2 activation

There is very little change between the ferric and oxy-P450eryF structures, so we focus here on P450cam. When O_2 binds, there is an opening of the I helix. This opening provides sufficient room for additional water molecules to move into the active site, one of which H-bonds to the distal oxygen atom of the iron-linked O_2 molecule (Fig. 4.10). This network of water-mediated H-bonds provides the proton relay network required to protonate the distal O atom, which promotes heterolysis of the O–O bond, thus generating the compound I Fe(IV)O center. The only other ligand that leads to the same changes is CN^- (Fedorov et al. 2003), but not CO or NO. One possible reason is that if the P450cam-oxy complex is best represented as the Fe(III)-O-O^- superoxide complex, then both the oxy and CN^- complexes have a negative charge on the distal ligand atom. The structural change in the I helix is required to properly stabilize this additional negative charge on the superoxide and CN^- via H-bonding with Thr252 and the new active site waters.

4.5 Redox Partner-Induced Structural Changes

Most simple redox complexes require very little structural change. In general, redox partners have complementary electrostatic surfaces, which enable a transient complex to form followed by electron transfer and rapid dissociation. Although P450s and other protein acceptors work best with their own natural partners, it is generally possible to achieve turnover with other "foreign" redox donors so long as the redox potential of the donor is low enough to reduce ferric P450. P450cam, however, behaves quite differently. P450cam has a strict requirement for its own redox partner, the Fe_2S_2 ferredoxin, putidaredoxin or Pdx. Any electron donor with a suitable redox potential can reduce ferric P450cam to ferrous but Pdx is strictly required for reduction of oxy-P450cam to generate product (Lipscomb et al. 1976). This observation has led to the idea that Pdx serves an effector role (Lipscomb et al. 1976; Sligar et al. 1974; Tyson et al. 1972) by inducing a structural change in P450cam required for electron transfer and O_2 activation. This initial observation

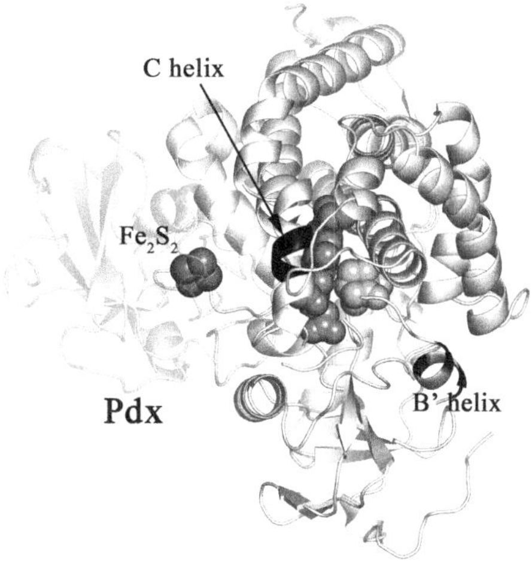

Fig. 4.11 Hypothetical model of the P450cam–Pdx complex (Pochapsky et al. 1996). The more darkly shaded regions in P450cam (B′ and C helices) exhibit some of the larger nuclear magnetic resonance (NMR) changes when Pdx binds (Pochapsky et al. 2003; Zhang et al. 2008)

was followed by a number of studies to track down the nature of the expected conformational change. Several studies have shown that Pdx binding results in changes that alter the spectral properties of various diatomic ligand complexes and nearby groups in the distal substrate-binding pocket (Nagano et al. 2003; Shimada et al. 2001; Shiro et al. 1989; Tosha et al. 2003; Unno et al. 2002). One of the more puzzling observations was that the binding of Pdx to P450cam shifts substrate-bound P450cam from high spin to low spin (Unno et al. 1997). The redox potential of low-spin P450cam is too low to be efficiently reduced by Pdx and, thus, is effectively inactive in electron transfer. In addition, Pdx binding to oxy-P450cam destabilizes the oxy complex by 150 fold (Glascock et al. 2005). These results indicate that Pdx induces some rather dramatic changes in the distal substrate-binding pocket. The first detailed picture on which regions of P450cam change when Pdx binding was derived from NMR data (Fig. 4.11), which showed that that regions well removed from the expected Pdx binding site are involved in Pdx-induced structural change (Pochapsky et al. 2003; Zhang et al. 2008).

The first crystallographic insight on the structural effect of Pdx binding was derived from the P450cam L358P mutant (Fig. 4.12), which mimics the effects of Pdx binding. That is, the L358P mutant exhibits spectral features similar to those induced by Pdx binding (Tosha et al. 2004). The crystal structure of the CO complex of the L358P provided a structural basis for these effects (Nagano et al. 2004). In wild-type P450cam, the binding of CO results in very few changes other than a slight repositioning of the substrate to make room for CO (Raag and Poulos 1989). In the L358P–CO complex the I helix adopts the conformation observed in the wild-type oxy complex. It thus appears that the L358P mutant promotes the deoxy to oxy change in the I helix required for O_2 activation. The reason appears to be a "push" from the larger Pro358 side chain on the proximal surface of the heme, which transmits movements to the distal pocket, resulting in the deoxy to oxy conformational switch of the I helix. Catalysis in the L358P

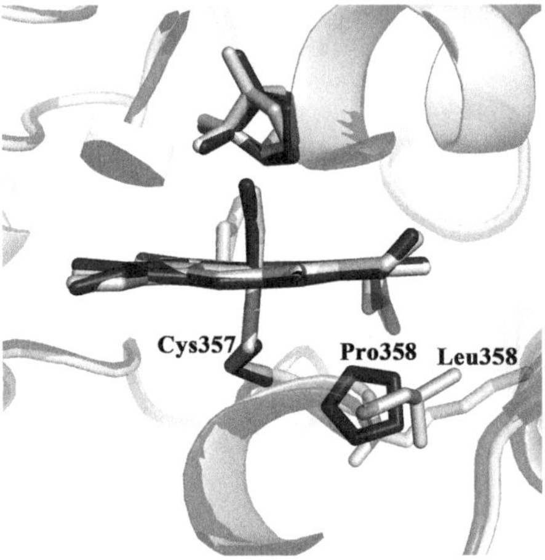

Fig. 4.12 The region around Leu358 in P450cam. The replacement of Leu358 with Pro "pushes" on the heme, resulting in a slight repositioning of the substrate in the active site. This movement enables the I helix to adopt the oxy-like conformation in the CO complex

mutant also can be supported by artificial reductants (Tosha et al. 2004), which indicates that the L358P mutant does indeed mimic the effects of Pdx binding.

The P450cam-Pdx crystal structure has recently been solved (Tripathi et al. 2013) and is basically the same as the NMR solution structure (Hiruma et al. 2013). Interactions at the interface are consistent with a wealth of mutagenesis data (Davies et al. 1990; Davies and Sligar 1992; Holden et al. 1997; Koga et al. 1993; Pochapsky et al. 1996; Shimada et al. 2001; Stayton and Sligar 1991; Unno et al. 1996). Pdx_{Asp38} interacts with $P450cam_{Arg112}$ (Fig. 4.13), which requires little movement in either protein in the vicinity of the ion pair. However, interactions involving Pdx_{Trp106} require substantial motion to optimize nonbonded and H-bonding interactions with P450cam (Fig. 4.13). The C helix in P450cam moves up toward Pdx by about 2–3 Å. This motion of the C helix is coupled to movements in the B′, I, F, and G helices, all of which are involved with substrate access or direct contacts with both substrates, camphor and O_2. This motion results in a large movement of the F and G helices and the F/G loop, which effectively opens the active site to bulk solvent. This open conformation is basically the same as that observed by Lee et al. (2010b). The main driving force in the closed-to-open switch appears to be Pdx_{Trp106}, which could not form tight interactions with P450cam without the structural switch. Mutagenesis and kinetic data (Fig. 4.13d) show that Pdx_{Trp106} is particularly important in the second electron-transfer step, reduction of oxy-P450cam (Kuznetsov et al. 2006).

The central question is how this opening of the active site is coupled to the effector role of Pdx. It seems most likely that the changes in the I helix when Pdx binds is the key to activating P450cam. The switch in the I helix in going from the closed to open state is in the same direction as the deoxy-to-oxy transition. Indeed, the oxy conformation of the I helix is about midway between the totally closed and open conformations (Fig. 4.13c). Thus, Pdx binding promotes formation of the H-bonded network required for O_2 activation. More than this, however, is an opening of the active site to bulk solvent that should be required to provide solvent

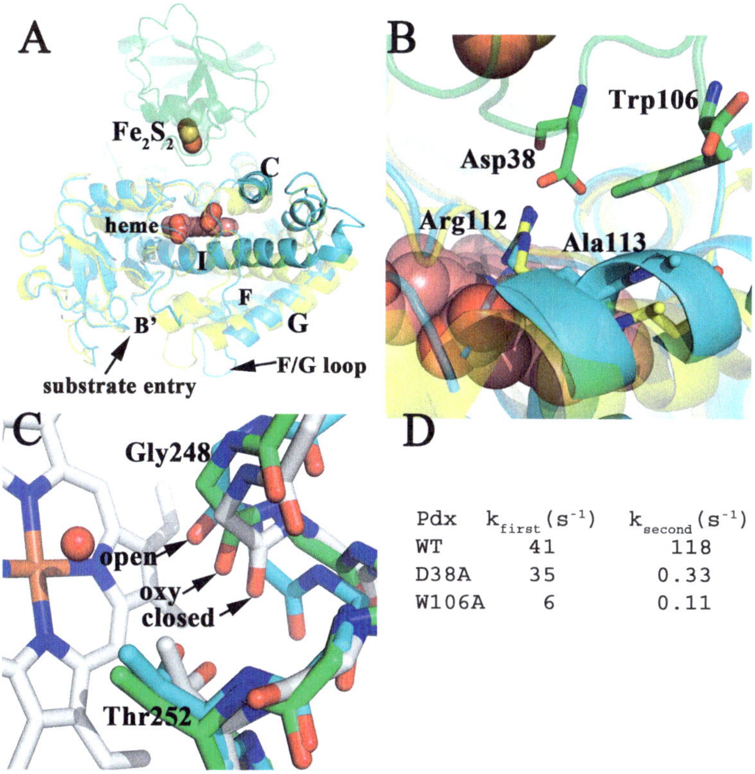

Pdx	k_{first} (s^{-1})	k_{second} (s^{-1})
WT	41	118
D38A	35	0.33
W106A	6	0.11

Fig. 4.13 (**a**) P450cam–Pdx structure (P450cam *yellow*, Pdx *green*) superimposed on the P450cam substrate-bound closed structure (*cyan*). When Pdx binds, P450cam adopts the open conformation, resulting in large movements of the F, G, and C helices plus the F/G loop. The C helix moves "up" to better interact with Pdx. (**b**) Close-up view of the P450cam–Pdx interface. Pdx_{Trp106} adopts a new rotamer conformation to interact with C-helical residues in P450cam. (**c**) Superposition of the closed, oxy, and open conformations of the I helix. The oxy conformation is about midway between the two extremes of the open and closed structures. It thus appears that Pdx promotes conversion to a more oxy-like conformation required for proton-coupled electron transfer. (**d**) Summary of kinetic results showing that Pdx_{Trp106} is critical for the second electron-transfer step and oxygen activation. It thus appears that the interactions between Pdx_{Trp106} and P450cam are critical for the conformational change required of O_2 activation

protons to dioxygen. In the closed state, there is no continuous H-bonded solvent network, so the active site must open up. Another key change that is important for catalysis involves Asp251. Asp251 is known to be critical for catalysis (Gerber and Sligar 1994) and participates in the proton relay network (Vidakovic et al. 1998). However, Asp251 is tied up with salt bridges to Lys178 and Arg186 and thus cannot participate in a proton shuttle network in the closed state. When Pdx binds, these salt bridges are broken, thus freeing Asp251 to serve its function in shuttling protons from bulk solvent to dioxygen. Because Asp251 is highly conserved (sometimes a Glu), it is likely that this residue is important in other P450s.

However, because many other P450s are not as specific for its redox partner as P450cam, redox partner-induced structural changes are probably not a general feature of P450 catalysis. Even so, similar to P450cam it will be necessary for all P450s to have access to bulk solvent to deliver solvent protons to dioxygen. The P450cam–Pdx structure implies that the open form of P450cam is the active form, but there are some problems with this scenario. The exposure of the active site to bulk solvent could result in uncoupling, and such loosening also could result in the loss of rigid substrate orientation and the loss of regio-selective hydroxylation. We have carried out a 6-ns MD simulation of the oxy-P450cam (open form)–Pdx complex, and indeed, bulk solvent moves into the active site and the camphor experiences substantial movement. It thus seems more likely that the truly active form is not the completely open structure but some intermediate state where Asp251 is free to serve its proton relay function, which may not require P450cam adopting the totally open conformation. The implication of such a model is that the active conformation is not the most stable conformation but is only an intermediate state along the way to the lowest energy open conformation observed in the crystal structure. This concept also requires the reasonable assumption that conformational changes are slow on the time scale of electron transfer and substrate hydroxylation.

4.6 Conclusions

There are a growing number of P450 structures that have been solved in the open and closed states. In all these structures, the most flexible region is in the F/G loop segment. In some cases there is concerted motion of the F and G helices that opens up the active site, and the hinge motion also involves changes on the opposite proximal side of the heme where redox partners bind, raising the possibility that the opening and closing of the access channel is coupled to redox partner binding. This likelihood certainly is the case with P450cam and can explain the specificity of P450cam for its own redox partner. However, this is not a general feature of all P450s. CYP119 and CYP3A4, for example, exhibit substantial dynamics in the long F/G loop region, but these changes do not involve other regions of the protein. The extreme selectivity of P450cam for its own redox partner is now better understood, but why P450cam is so selective while other P450s are not remains an interesting puzzle. One expects such differences to correlate with biological function. In contrast to many other P450s, P450cam is part of an operon that encodes genes required for using camphor as a source of energy. Other P450s, such as CYP3A4, are more "generic" and are not part of a critical biochemical pathway. Thus, P450cam may have evolved a higher level of conformational control that couples redox partner binding to proton-coupled electron transfer. If this scenario has a shred of truth, then one might anticipate that other P450s that are part of an operon critical to utilization of a specific substrate for energy might exhibit a P450cam-like level of selectivity.

References

Davies MD, Sligar SG (1992) Genetic variants in the putidaredoxin-cytochrome P-450cam electron-transfer complex: identification of the residue responsible for redox-state-dependent conformers. Biochemistry 31:11383–11389

Davies MD, Qin L, Beck JL, Suslick KS, Koga H, Horiuchi T, Sligar SG (1990) Putidaredoxin reduction of cytochrome P-450cam: dependence of electron-transfer on the identity of putidaredoxins C-terminal amino-acid. J Am Chem Soc 112:7396–7398

Davydov DR, Rumfeldt JA, Sineva EV, Fernando H, Davydova NY, Halpert JR (2012) Peripheral ligand-binding site in cytochrome P450 3A4 located with fluorescence resonance energy transfer (FRET). J Biol Chem 287:6797–6809

Denisov IG, Shih AY, Sligar SG (2012) Structural differences between soluble and membrane-bound cytochrome P450s. J Inorg Biochem 108:150–158

Ekroos M, Sjögren T (2006) Structural basis for ligand promiscuity in cytochrome P450 3A4. Proc Natl Acad Sci USA 103:13682–13687

Fedorov R, Ghosh DK, Schlichting I (2003) Crystal structures of cyanide complexes of P450cam and the oxygenase domain of inducible nitric oxide synthase-structural models of the short-lived oxygen complexes. Arch Biochem Biophys 409:25–31

Foti RS, Honaker M, Nath A, Pearson JT, Buttrick B, Isoherranen N, Atkins WM (2011) Catalytic versus inhibitory promiscuity in cytochrome P450s: implications for evolution of new function. Biochemistry 50:2387–2393

Gerber NC, Sligar SG (1994) A role for Asp-251 in cytochrome P-450cam oxygen activation . J Biol Chem 269:4260–4266

Glascock MC, Ballou DP, Dawson JH (2005) Direct observation of a novel perturbed oxyferrous catalytic intermediate during reduced putidaredoxin-initiated turnover of cytochrome P-450-CAM: probing the effector role of putidaredoxin in catalysis. J Biol Chem 280:42134–42141

Hendrychová T, Anzenbacherová E, Hudeček J, Skopalík J, Lange R, Hildebrandt P, Otyepka M, Anzenbacher P (2010) Flexibility of human cytochrome P450 enzymes: molecular dynamics and spectroscopy reveal important function-related variations. Biochim Biophys Acta 1814:58–68

Hiruma Y, Hass MA, Kikui Y, Liu WM, Olmez B, Skinner SP, Blok A, Kloosterman A, Koteishi H, Lohr F, Schwalbe H, Nojiri M, Ubbink M (2013) The structure of the cytochrome P450cam-putidaredoxin complex determined by paramagnetic NMR spectroscopy and crystallography . J Mol Biol 425:4353

Holden M, Mayhew M, Bunk D, Roitberg A, Vilker V (1997) Probing the interactions of putidaredoxin with redox partners in camphor P450 5-monooxygenase by mutagenesis of surface residues. J Biol Chem 272:21720–21725

Isin EM, Guengerich FP (2006) Kinetics and thermodynamics of ligand binding by cytochrome P450 3A4. J Biol Chem 281:9127–9136

Kamerlin SCL, Warshel A (2010) At the dawn of the 21st century: is dynamics the missing link for understanding enzyme catalysis? Proteins 78:1339–1375

Koga H, Sagara Y, Yaoi T, Tsujimura M, Nakamura K, Sekimizu K, Makino R, Shimada H, Ishimura Y, Yura K et al (1993) Essential role of the Arg112 residue of cytochrome P450cam for electron transfer from reduced putidaredoxin. FEBS Lett 331:109–113

Kuznetsov VY, Poulos TL, Sevrioukova IF (2006) Putidaredoxin-to-cytochrome P450cam electron transfer: differences between the two reductive steps required for catalysis. Biochemistry 45:11934–11944

Lee YT, Glazer EC, Wilson RF, Stout CD, Goodin DB (2010a) Three clusters of conformational states in p450cam reveal a multistep pathway for closing of the substrate access channel. Biochemistry 50:693–703

Lee YT, Wilson RF, Rupniewski I, Goodin DB (2010b) P450cam visits an open conformation in the absence of substrate. Biochemistry 49:3412–3419

Li H, Poulos TL (1997) The structure of the cytochrome p450BM-3 haem domain complexed with the fatty acid substrate, palmitoleic acid. Nat Struct Biol 4:140–146
Lipscomb JD, Sligar SG, Namtvedt MJ, Gunsalus IC (1976) Autooxidation and hydroxylation reactions of oxygenated cytochrome P-450cam. J Biol Chem 251:1116–1124
Lüdemann SK, Lounnas V, Wade RC (2000) How do substrates enter and products exit the buried active site of cytochrome P450cam? 2. Steered molecular dynamics and adiabatic mapping of substrate pathways. J Mol Biol 303:813–830
Nagano S, Poulos TL (2005) Crystallographic study on the dioxygen complex of wild-type and mutant cytochrome P450cam: implications for the dioxygen activation mechanism. J Biol Chem 280:31659–31663
Nagano S, Shimada H, Tarumi A, Hishiki T, Kimata-Ariga Y, Egawa T, Suematsu M, Park SY, Adachi S, Shiro Y, Ishimura Y (2003) Infrared spectroscopic and mutational studies on putidaredoxin-induced conformational changes in ferrous CO-P450cam. Biochemistry 42:14507–14514
Nagano S, Tosha T, Ishimori K, Morishima I, Poulos TL (2004) Crystal structure of the cytochrome p450cam mutant that exhibits the same spectral perturbations induced by putidaredoxin binding. J Biol Chem 279:42844–42849
Nagano S, Cupp-Vickery JR, Poulos TL (2005) Crystal structures of the ferrous dioxygen complex of wild-type cytochrome P450eryF and its mutants, A245S and A245T: investigation of the proton transfer system in P450eryF. J Biol Chem 280:22102–22107
Otyepka M, Berka K, Anzenbacher P (2012) Is there a relationship between the substrate preferences and structural flexibility of cytochromes P450? Curr Drug Metab 13:130–142
Park SY, Yamane K, Adachi S, Shiro Y, Weiss KE, Maves SA, Sligar SG (2002) Thermophilic cytochrome P450 (CYP119) from *Sulfolobus solfataricus*: high resolution structure and functional properties. J Inorg Biochem 91:491–501
Park H, Lee S, Suh J (2005) Structural and dynamical basis of broad substrate specificity, catalytic mechanism, and inhibition of cytochrome P450 3A4. J Am Chem Soc 127:13634–13642
Pochapsky TC, Lyons TA, Kazanis S, Arakaki T, Ratnaswamy G (1996) A structure-based model for cytochrome P450cam-putidaredoxin interactions. Biochimie 78:723–733
Pochapsky SS, Pochapsky TC, Wei JW (2003) A model for effector activity in a highly specific biological electron transfer complex: the cytochrome P450(cam)-putidaredoxin couple. Biochemistry 42:5649–5656
Poulos TL, Finzel BC, Gunsalus IC, Wagner GC, Kraut J (1985) The 2.6-Å crystal structure of *Pseudomonas putida* cytochrome P-450. J Biol Chem 260:16122–16130
Poulos TL, Finzel BC, Howard AJ (1986) Crystal structure of substrate-free *Pseudomonas putida* cytochrome P450. Biochemistry 25:5314–5322
Poulos TL, Finzel BC, Howard AJ (1987) High-resolution crystal structure of cytochrome P450cam. J Mol Biol 195:687–700
Raag R, Poulos TL (1989) Crystal structure of the carbon monoxide-substrate-cytochrome P-450CAM ternary complex. Biochemistry 28:7586–7592
Ravichandran KG, Boddupalli SS, Hasermann CA, Peterson JA, Deisenhofer J (1993) Crystal structure of hemoprotein domain of P450BM-3, a prototype for microsomal P450's. Science 261:731–736
Schlichting I, Berendzen J, Chu K, Stock AM, Maves SA, Benson DE, Sweet RM, Ringe D, Petsko GA, Sligar SG (2000) The catalytic pathway of cytochrome P450cam at atomic resolution. Science 287:1615–1622
Scott EE, He YA, Wester MR, White MA, Chin CC, Halpert JR, Johnson EF, Stout CD (2003) An open conformation of mammalian cytochrome P450 2B4 at 1.6-Å resolution. Proc Natl Acad Sci USA 100:13196–13201
Scott EE, White MA, He YA, Johnson EF, Stout CD, Halpert JR (2004) Structure of mammalian cytochrome P450 2B4 complexed with 4-(4-chlorophenyl)imidazole at 1.9-Å resolution: insight into the range of P450 conformations and the coordination of redox partner binding. J Biol Chem 279:27294–27301

Sevrioukova IF, Poulos TL (2010) Structure and mechanism of the complex between cytochrome P4503A4 and ritonavir. Proc Natl Acad Sci USA 107:18422–18427

Sevrioukova IF, Poulos TL (2012) Structural and mechanistic insights into the interaction of cytochrome P4503A4 with bromoergocryptine, a type I ligand. J Biol Chem 287:3510–3517

Sevrioukova IF, Poulos TL (2013) Dissecting cytochrome P450 3A4-ligand interactions using ritonavir analogues. Biochemistry 52:4474–4481

Shimada H, Nagano S, Hori H, Ishimura Y (2001) Putidaredoxin–cytochrome P450cam interaction. J Inorg Biochem 83:255–260

Shiro Y, Iizuka T, Makino R, Ishimura Y, Morishima I (1989) ^{15}N-NMR-study on cyanide (C-15N-) complex of cytochrome-P-450cam: effects of D-camphor and putidaredoxin on the iron ligand structure. J Am Chem Soc 111:7707–7711

Sligar SG, Debrunner PG, Lipscomb JD, Namtvedt MJ, Gunsalus IC (1974) A role of the putidaredoxin COOH-terminus in P-450cam (cytochrome m) hydroxylations. Proc Natl Acad Sci USA 71:3906–3910

Stayton PS, Sligar SG (1991) Structural microheterogeneity of a tryptophan residue required for efficient biological electron transfer between putidaredoxin and cytochrome P-450cam. Biochemistry 30:1845–1851

Stoll S, Lee Y-T, Zhang M, Wilson RF, Britt RD, Goodin DB (2012) Double electron-electron resonance shows cytochrome P450cam undergoes a conformational change in solution upon binding substrate. Proc Natl Acad Sci USA 109:12888–12893

Tosha T, Yoshioka S, Takahashi S, Ishimori K, Shimada H, Morishima I (2003) NMR study on the structural changes of cytochrome P450cam upon the complex formation with putidaredoxin. Functional significance of the putidaredoxin-induced structural changes. J Biol Chem 278:39809–39821

Tosha T, Yoshioka S, Ishimori K, Morishima I (2004) L358P mutation on cytochrome P450cam simulates structural changes upon putidaredoxin binding: the structural changes trigger electron transfer to oxy-P450cam from electron donors. J Biol Chem 279:42836–42843

Tripathi S, Li H, Poulos TL (2013) Structural basis for effector control and redox partner recognition in cytochrome P450. Science 340:1227–1230

Tyson CA, Lipscomb JD, Gunsalus IC (1972) The role of putidaredoxin and P450 cam in methylene hydroxylation. J Biol Chem 247:5777–5784

Unno M, Shimada H, Toba Y, Makino R, Ishimura Y (1996) Role of Arg112 of cytochrome p450cam in the electron transfer from reduced putidaredoxin. Analyses with site-directed mutants. J Biol Chem 271:17869–17874

Unno M, Christian JF, Benson DE, Gerber NC, Sligar SG, Champion PM (1997) Resonance Raman investigations of cytochrome P450cam complexed with putidaredoxin. J Am Chem Soc 119:6614–6620

Unno M, Christian JF, Sjodin T, Benson DE, Macdonald ID, Sligar SG, Champion PM (2002) Complex formation of cytochrome P450cam with putidaredoxin. Evidence for protein-specific interactions involving the proximal thiolate ligand. J Biol Chem 277:2547–2553

Vidakovic M, Sligar SG, Li H, Poulos TL (1998) Understanding the role of the essential Asp251 in cytochrome p450cam using site-directed mutagenesis, crystallography, and kinetic solvent isotope effect. Biochemistry 37:9211–9219

Wade RC, Winn PJ, Schlichting I, Sudarko (2004) A survey of active site access channels in cytochromes P450. J Inorg Biochem 98:1175–1182

Williams PA, Cosme J, Sridhar V, Johnson EF, McRee DE (2000) The crystallographic structure of a mammalian microsomal cytochrome P450 monooxygenase: structural adaptations for membrane binding and functional diversity. Mol Cell 5:121–132

Williams PA, Cosme J, Vinkovic DM, Ward A, Angove HC, Day PJ, Vonrhein C, Tickle IJ, Jhoti H (2004) Crystal structures of human cytochrome P450 3A4 bound to metyrapone and progesterone. Science 305:683–686

Wright RL, Harris K, Solow B, White RH, Kennelly PJ (1996) Cloning of a potential cytochrome P450 from the archaeon *Sulfolobus solfataricus*. FEBS Lett 384:235–239

Yano JK, Koo LS, Schuller DJ, Li H, Ortiz de Montellano PR, Poulos TL (2000) Crystal structure of a thermophilic cytochrome P450 from the archaeon *Sulfolobus solfataricus*. J Biol Chem 275:31086–31092

Yano JK, Wester MR, Schoch GA, Griffin KJ, Stout CD, Johnson EF (2004) The structure of human microsomal cytochrome P450 3A4 determined by X-ray crystallography to 2.05-Å resolution. J Biol Chem 279:38091–38094

Zeldin RK, Petruschke RA (2004) Pharmacological and therapeutic properties of ritonavir-boosted protease inhibitor therapy in HIV-infected patients. J Antimicrob Chemother 53:4–9

Zhang W, Pochapsky SS, Pochapsky TC, Jain NU (2008) Solution NMR structure of putidaredoxin-cytochrome P450cam complex via a combined residual dipolar coupling-spin labeling approach suggests a role for Trp106 of putidaredoxin in complex formation. J Mol Biol 384:349–363

Chapter 5
Structural and Functional Diversity of Cytochrome P450

Shingo Nagano

Abstract The elucidation of the crystal structure of P450cam has greatly advanced site-directed mutagenesis and other biochemical studies of P450. However, the proton supply system in the P450 active site was elusive for many years. In 1997, the crystal structure of the ferric substrate (6-deoxyerythronolide B)-bound P450eryF that is involved in erythromycin biosynthesis was solved. A water molecule that was presumed to be a proton donor to the heme-bound dioxygen molecule was found in the active site of P450eryF. Two water molecules were also found in the first crystal structure of P450cam-O_2. Two mutant P450cam-O_2 structures and other studies revealed that the I-helix Thr, which is a crucial structural feature for dioxygen activation, is not a proton donor, although it is a key residue for cleaving the O–O bond. No water molecules were found around the heme-bound O_2 in the P450eryF-O_2 structure. Therefore, the proton supply systems for P450eryF and P450cam are different and depend on the substrate and substrate cavity.

Although most P450s catalyze monooxygenation reactions, there are some exceptions. One example is P450 StaP, which is a key enzyme for forming the core structure of an anticancer agent, indolocarbazole, which catalyzes the intra-molecular C–C bond formation in chromopyrrolic acid (CPA). The crystal structure of CPA-bound P450 StaP and theoretical calculations of the active species of this enzyme suggest that C–C bond formation occurs via an indole cation radical intermediate that is equivalent to cytochrome *c* peroxidase compound I. In addition to the indole ring of CPA, water molecules and His250 are also important for the P450 StaP C–C coupling reaction.

Keywords C–C coupling • Dioxygen complex • P450 StaP • P450cam • P450eryF • Proton transfer

---

S. Nagano (✉)
Tottori University, Tottori, Japan
e-mail: snagano@bio.tottori-u.ac.jp

H. Yamazaki (ed.), *Fifty Years of Cytochrome P450 Research*,
DOI 10.1007/978-4-431-54992-5_5, © Springer Japan 2014

5.1 The First Crystal Structure of P450 and the Acid–Alcohol Pair in the I Helix

In 1985, 23 years after the discovery of P450, the first crystal structure of a P450 was determined by Poulos and coworkers at a resolution of 2.6 Å (Poulos et al. 1985). Their structure of P450cam from *Pseudomonas putida* revealed a triangular prism-shaped P450 fold. To date, more than 90 P450 structures have been solved and almost all of them, including eukaryotic and mitochondrial P450s, exhibit this P450 fold (Fig. 5.1). The substrate camphor fits into the pocket just above the heme cofactor. The substrate is hydrogen bonded to Tyr96, and several hydrophobic residues make contact with the substrate. Based on the substrate-binding mode of P450cam and the sequence alignment of P450cam and CYP2 family members, six regions were proposed as substrate recognition sites (SRS) for P450 (Gotoh 1992). Numerous subsequent crystallographic studies confirmed that many amino acid residues in SRS are involved in substrate binding and recognition. In the crystal structure of P450cam, one region of the substrate-binding pocket is part of the I helix, which runs through the middle of the enzyme. A region of the I helix that is close to the oxygen binding site, heme iron, and the substrate contains the hydrophilic residues, Asp251 and Thr252 (Fig. 5.2). The acid–alcohol pair in the I helix is highly conserved in P450s, and thus it has been proposed by two separate research groups that these residues may be involved in the proton supply to the iron-bound dioxygen. The role of Thr252 and Asp251 was investigated by site-directed mutagenesis (Imai et al. 1989; Martinis et al. 1989; Gerber and Sligar 1992; Gerber and Sligar 1994). The monooxygenase activity of the D251N mutant almost disappeared, and the T252A mutant consumed dioxygen and produced

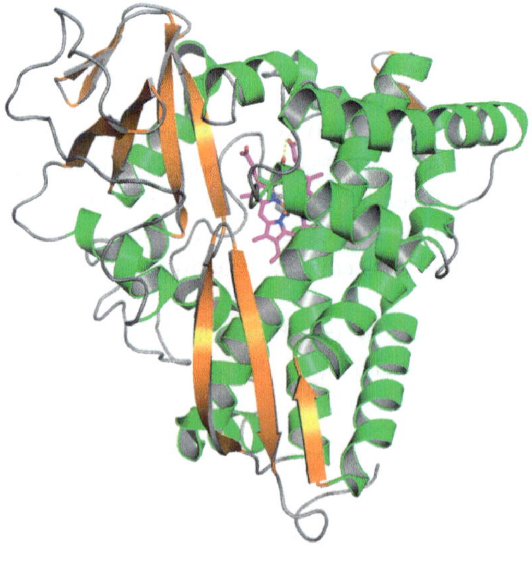

Fig. 5.1 Overall structure of cytochrome P450cam. Almost all P450s contain this triangular prism-shaped P450 fold

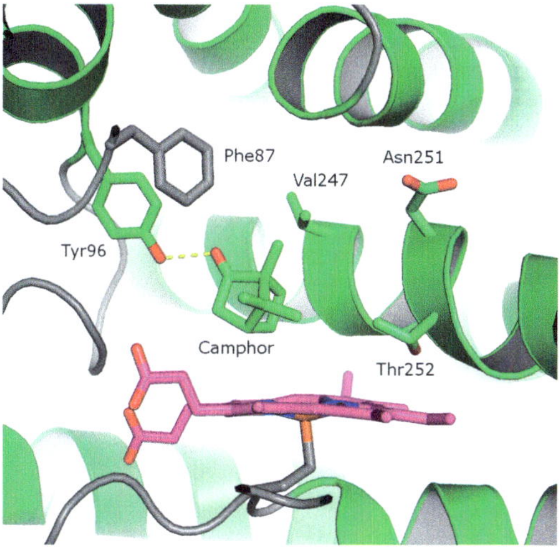

Fig. 5.2 Active site structure of camphor-bound P450cam. The substrate, camphor, is hydrogen bonded to Tyr96. Several hydrophobic residues interact with the substrate

hydrogen peroxide rather than hydroxycamphor via an uncoupling reaction. Based on these results, it was suggested that the highly conserved I helix acid–alcohol pair plays an important role in the proton supply and activation of molecular oxygen.

5.2 Dioxygen-Bound P450 Structures and Proton-Transfer Systems

Although mutagenesis studies indicated the importance of the acid–alcohol pair, its role in proton transfer and oxygen activation was not revealed until several crystal structures of dioxygen-bound P450 were solved. Schlichting et al. (Schlichting et al. 2000) reported that ferrous dioxygen-bound P450cam crystals can be prepared by high-pressure dioxygen gas treatment of dithionite-reduced P450cam crystals. The most important finding was that two water molecules that were not present in the ferric enzyme structure were present in the active site of dioxygen-bound P450cam (Fig. 5.3). One of these water molecules is hydrogen bonded to the Fe-bound dioxygen and to Thr252. Therefore, it is likely that these water molecules provide protons to the Fe-bound dioxygen. Direct evidence to prove the catalytic role of these active site water molecules came from the structure of the inactive mutant D251N P450cam-O_2, which was solved by Nagano and Poulos (Fig. 5.4) (Nagano and Poulos 2005). Although Asp251 does not interact directly with the active site water molecules, when negatively charged Asp251 is replaced with neutral Asn the catalytic water molecules are lost. The missing catalytic water molecules explain the very slow monooxygenation by the D251N mutant; these

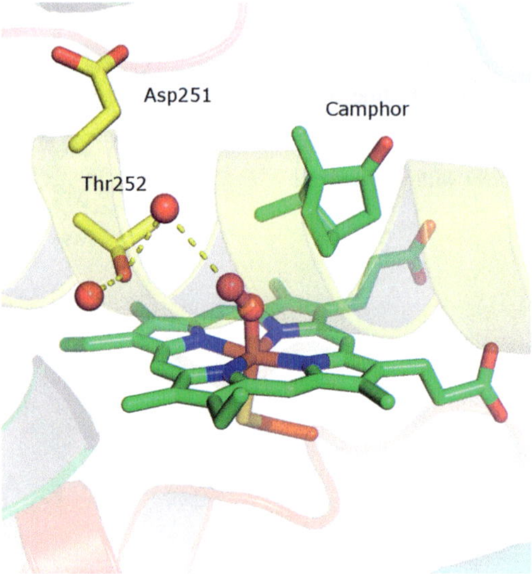

Fig. 5.3 Two water molecules and the hydrogen bond network in the active site of ferrous dioxygen-bound P450cam

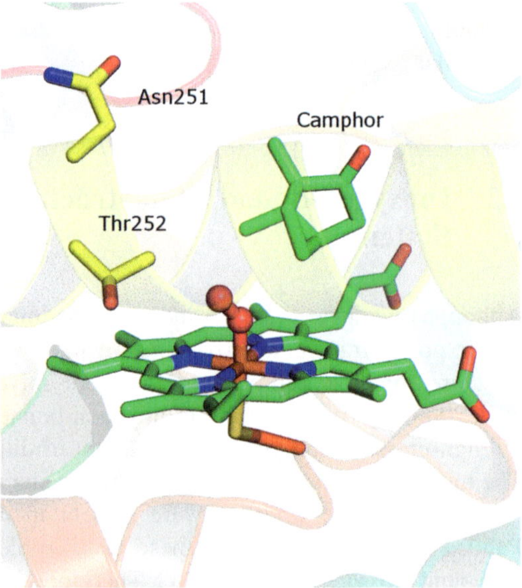

Fig. 5.4 Active site structure of ferrous dioxygen-bound D251N mutant P450cam. The impaired catalytic activity of this mutant is explained by the two missing water molecules found in the ferric enzyme

water molecules provide protons to the Fe-bound dioxygen. In contrast to the clear structure–function relationship in the dioxygen-bound structures of the wild-type and D251N mutant enzymes, it is not clear why hydrogen peroxide formation occurs instead of camphor hydroxylation in the T252A mutant. It had been proposed that water molecules in the active site of the T252A mutant may be relocated

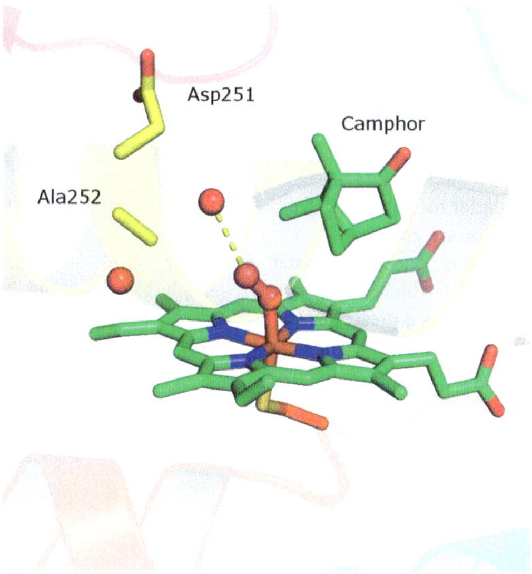

Fig. 5.5 Active site structure of ferrous dioxygen-bound T252A mutant P450cam. Although the hydroxyl group does not exist at the 252 position, two catalytic water molecules are found in a similar position to those in the wild-type enzyme

and that the relocation may result in inappropriate proton addition to the Fe-bound dioxygen to produce hydrogen peroxide. However, because the crystal structure of dioxygen-bound T252A revealed that catalytic water molecules reside in almost the same position as those in the wild-type dioxygen-bound structure (Fig. 5.5), the relocation of these water molecules cannot explain the uncoupling reaction of P450cam. The similar position of the two water molecules provided an important clue for explaining why the uncoupling reaction occurred in the T252A mutant. After the addition of one proton to the Fe-bound dioxygen, a hydroperoxy intermediate is formed. The hydroxyl oxygen atom of Thr252 accepts a hydrogen bond from the hydroperoxy intermediate, which assists O–O bond cleavage and the second proton addition to the terminal oxygen atom of the hydroperoxy species. This scenario is consistent with the results of the unnatural amino acid substitution study for P450cam. Shimada and coworkers prepared a mutant in which the hydroxyl oxygen atom of Thr252 was methylated (Kimata et al. 1995). The O-methylated Thr cannot donate but can accept a hydrogen bond from the hydroperoxy intermediate, and the mutant shows substantial hydroxylation activity. Therefore, the hydroxyl group of the acid–alcohol pair plays a crucial role as a hydrogen bond acceptor rather than as a hydrogen bond or proton donor.

Although the acid–alcohol pair is essential for P450 activity, it is missing in several unusual P450s. P450eryF, which is involved in the biosynthesis of erythromycin, is a unique P450 wherein the highly conserved I helix Thr is replaced with Ala. It was important to determine how molecular oxygen is activated in the active site of P450eryF in the absence of I helix Thr. The structure of substrate-bound P450eryF provided Cupp-Vickery and Poulos with an important insight (Cupp-Vickery and Poulos 1995). A water molecule is located close to the substrate hydroxyl group and Ala245, which is the position equivalent to the I helix Thr

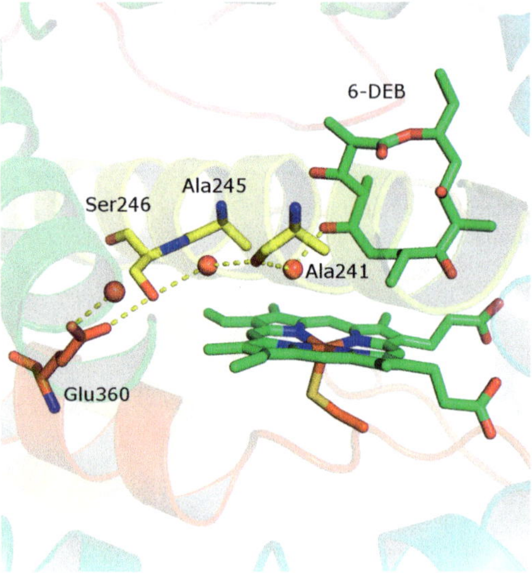

Fig. 5.6 Active site structure of ferric substrate (6-deoxyerythronolide B)-bound P450eryF. The I-helix Thr is substituted with Ala in this P450, and the substrate 5-OH is essential for catalysis. The 5-hydroxyl group is hydrogen bonded to a water molecule, which was proposed as a proton donor to the Fe-bound dioxygen

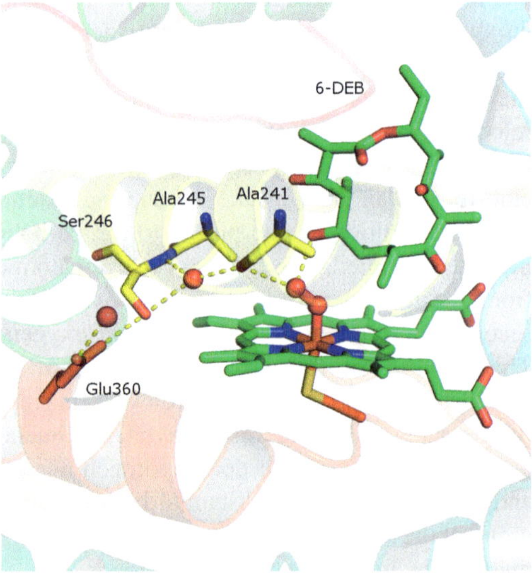

Fig. 5.7 Active site structure of ferrous dioxygen-bound P450eryF. The distal oxygen atom of the Fe-bound dioxygen is in a similar position to the water molecule that is proposed as a proton donor

(Fig. 5.6). It was proposed that this water molecule compensates for the I helix Thr and provides a proton to the Fe-bound dioxygen. However, the proton donor water molecule was not detected in P450cam when the P450eryF structure was first determined. Thus, it was important to ascertain whether the water molecule found in the active site of ferric substrate-bound P450eryF provides protons to the dioxygen ligand. To answer this question, we solved the P450eryF-O_2 structure (Fig. 5.7) (Nagano et al. 2005). Unexpectedly, the dioxygen bound to the heme expelled the

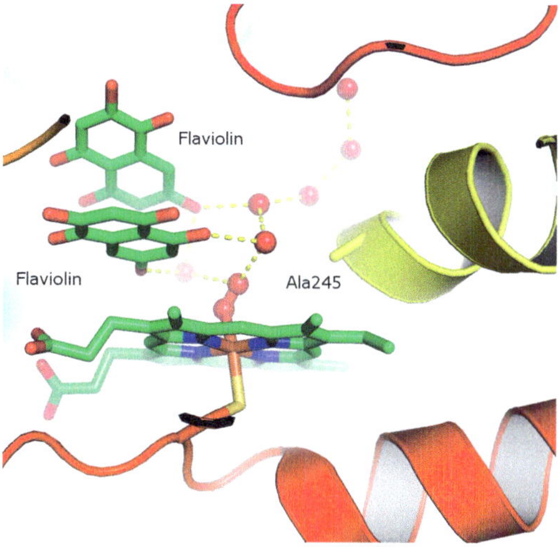

Fig. 5.8 Active site water molecules and extensive hydrogen bond network in the active site of ferrous dioxygen-bound CYP158A2

water molecule, and there were no other water molecules around the dioxygen ligand because the Fe-bound dioxygen ligand occupies almost the same position as the water molecule in the ferric enzyme and the space around the dioxygen ligand is too small (~77 Å^2) to accommodate a water molecule. Therefore, the proton supply system is not the same as for P450cam. Protons are delivered from the hydrogen bond network connecting Gly360, Ser246, water 63, and the Ala245 carbonyl. The substrate 5-OH group also plays a key role in dioxygen activation by P450eryF because the hydroxyl forms a direct hydrogen bond with the distal oxygen of Fe-bound dioxygen. Furthermore, conversion of the 5-OH group to a carbonyl eliminates enzyme activity. One problem with the proton supply is that there is no continuous hydrogen bond network connecting the bulk water, the ultimate proton source, and the active site in both P450eryF and P450cam. However, when the concerted motion of the side chains and water is taken into account, the proton delivery pathway in P450cam does connect to the molecular surface. Therefore, dynamics may well play a critical role in enabling bulk solvent protons to be delivered to the active site through the internal hydrogen bond network in P450eryF.

Important structural information is also provided by the proton transfer to Fe-bound dioxygen in CYP158A2, which catalyzes dimerization of flaviolin and also lacks the I helix Thr. The active site of this enzyme can accommodate two flaviolin molecules, which is consistent with the oligomerization of flaviolin by CYP158A2 (Zhao et al. 2005a). The substrate binding repositions the water molecules in the active site, and three water molecules form hydrogen bonds with the bound flaviolin molecules. Bio Zhao and Mike Waterman solved the structure of ferrous dioxygen-bound CYP158A2, and the three water molecules were found in similar positions in the ferric substrate-bound structure (Fig. 5.8) (Zhao et al. 2005b). Two of these water molecules also formed hydrogen bonds with the heme iron-bound dioxygen molecule. Therefore, they proposed that this hydrogen bond network is a new proton-transfer system in P450.

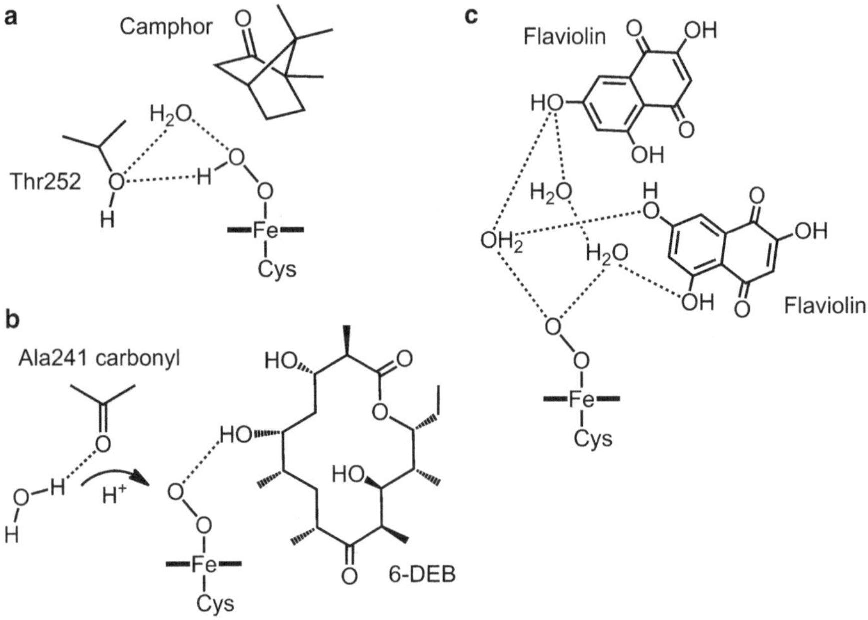

Fig. 5.9 Proton supply systems in P450cam (**a**), P450eryF (**b**), and CYP158A2 (**c**)

The structures of ferrous dioxygen-bound P450 (P450cam, P450eryF, and CYP158A2) reveal several features of proton transfer and dioxygen activation. Two protons are required to activate dioxygen in P450, and the arrangement of hydrogen bond network around the heme iron-bound dioxygen is critical because the T252A mutant P450cam also has two water molecules around the dioxygen ligand. However, this mutant cannot insert an oxygen atom into the substrate because of the missing hydroxyl group at the 252 position. P450 can catalyze monooxygenation and many other reactions for a wide variety of substrates with different shapes and volumes. Although the overall structure of P450 is common to almost all P450s, the shape and volume of the substrate-binding sites are unique to each P450 and depend on the natural substrate. The three different proton supply systems in P450cam, P450eryF, and CYP158A2 indicate that the proton-transfer mechanisms in the dioxygen-bound P450s depend on the substrate and type of reaction (Fig. 5.9).

5.3 Non-monooxygenase P450

P450s have a number of important physiological roles, such as drug and xenobiotic metabolism and the biosynthesis and metabolism of hormones and vitamins. Many P450s catalyze hydroxylation and monooxygenation, which are common reactions

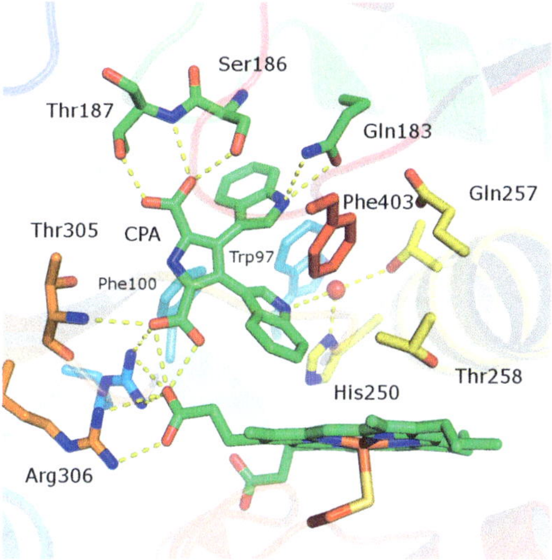

Fig. 5.10 Active site structure of P450 StaP. The substrate is bound through extensive hydrogen bonding and hydrophobic interactions

in drug metabolism (Ortiz de Montellano 1995). However, some P450 reactions produce non-monooxygenated products. These unusual P450 reactions are mainly found in secondary metabolite biosynthesis in plants and in drug metabolism.

Non-monooxygenase P450 is also found in the biosynthesis of indolocarbazole compounds, which have a unique 5-ring fused core structure and are lead compounds for anticancer drugs (Onaka et al. 2002). Tryptophan is oxidized by a flavin-containing enzyme to indole pyruvic acid, which is then dimerized to form chromopyrrolic acid (CPA). P450 StaP (StaP) forms the indolocarbazole core structure by creating a C–C bond between the C5 atoms of the indole rings in CPA. We solved the CPA-bound StaP structure and found that the enzyme forms eight hydrogen bonds with the substrate (Fig. 5.10) (Makino et al. 2007). The two indole rings are involved in C–H π and hydrophobic interactions. These extensive enzyme–substrate interactions hold CPA roughly perpendicular to the heme plane. Although the shortest distance between the heme iron and CPA is 4.7 Å, which is a reasonable distance for the hydroxylation reaction of P450, the C5 atom, which is oxidized by the C–C coupling reaction, is too far from the heme iron (8.6 Å). The substrate-binding mode suggests that the C5 atom may not be directly oxidized by the heme and the proximal indole ring may be directly involved in the C–C bond formation instead. Given that CPA has a number of π electrons, the removal of one electron from the proximal indole ring by compound I may occur during catalysis. The one-electron oxidation of indole rings is well known for cytochrome *c* peroxidase, where compound I is formed and then the indole ring of Trp191 is oxidized to form the Fe^{4+} porphyrin and tryptophan cation radical (CcP compound I). The tryptophan cation radical is stabilized by the surrounding electronegative side chains, such as Asp235, Met230, and Met231. In the crystal structure of substrate-bound StaP, CPA has two carboxyl groups and is hydrogen bonded to Ser186 and Gln183. The electronegative environment may promote the removal of one electron from CPA and indole radical formation.

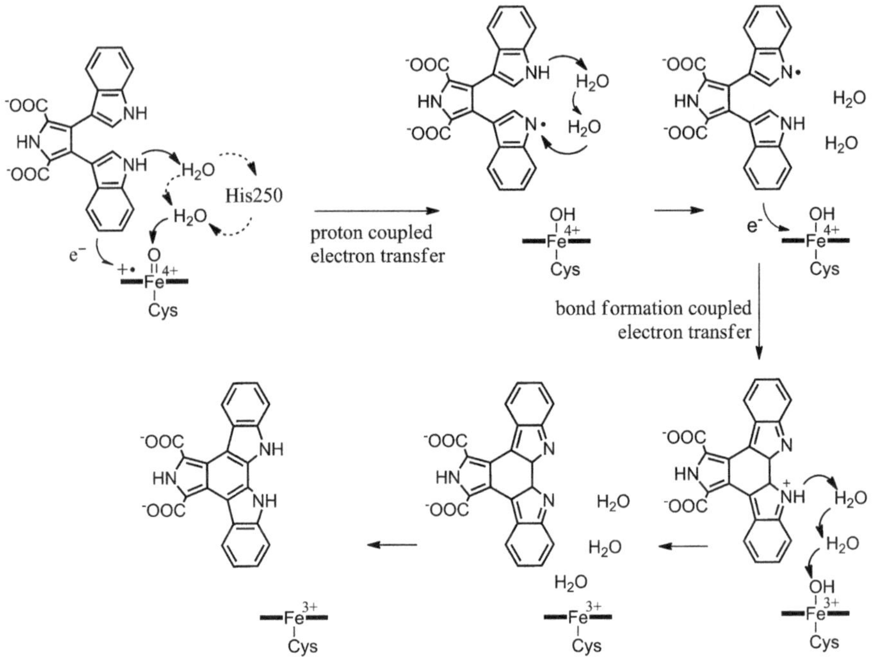

Fig. 5.11 Catalytic mechanism for C–C bond formation in CPA by StaP

The electron-transfer hypothesis for the substrate-bound StaP compound I species was examined by using density functional theory hybrid quantum mechanical/molecular mechanical calculations (Wang et al. 2008; Wang et al. 2009). The calculated orbital and spin density distribution shows that electron transfer from CPA to compound I is possible. An interesting consequence of this theory is that the hydrogen bond network, containing CPA, His250, and active site water molecules, assists the first electron transfer from the substrate to compound I. The electron transfer is favored when coupled with the proton transfer from the proximal indole nitrogen to compound I via the water molecules and His250 (Fig. 5.11). The ensuing mechanism involves C–C bond formation and tautomerization, both of which require proton transfer, which is also mediated by the water molecules and His250. The importance of these water molecules and His250 is exemplified by the crystal structure of Cl_2-CPA-bound StaP and its activity for Cl_2-CPA. The substitution of hydrogen with a bulky chlorine atom moves the imidazole ring of His250 slightly away from the substrate, and the imidazole ring rotates approximately 180°. The water molecule found in the CPA-bound form is missing, and the chlorine atom prevents water molecules from accessing the oxygen atom of the [FeOH] species. Furthermore, chlorine atoms increase the redox potential of the substrate, and the higher redox potential of Cl_2-CPA could also explain the very low turnover rate for Cl_2-CPA. However, an analogue of StaP, RebP, can catalyze C–C coupling reaction for Cl_2-CPA. Thus, the redox potential of the substrate does not explain the very low turnover rate of StaP for Cl_2-CPA. These theoretical and experimental results highlight the role of water molecules as biological catalysts that can enable P450 to catalyze peroxidase-type reactions.

5.4 Summary and Conclusions

Since the first crystal structure of P450cam was solved, there are more than 500 entries in the Protein Data Bank for P450s. These crystal structures have provided very important information about P450 catalysis, such as substrate binding, structural changes caused by substrate binding, oxygen activation, proton transfer, and electron transfer. The crystal structures of several ferrous dioxygen complex structures have revealed that there are different proton-transfer systems depending on the architecture of the substrate-binding site, implying that different P450s have many different proton supply systems. In addition to the diversity of the proton supply system, functional diversity is seen in many non-monooxygenase P450 reactions. Theoretical studies have shown that StaP involved in indole alkaloid biosynthesis has a catalytic intermediate that is equivalent to CcP compound I. Similar to many other non-monooxygenase C–C coupling reactions catalyzed by P450, StaP C–C bond formation proceeds via a radical mechanism. The charge transfer from the substrate and C–C coupling reaction are strongly affected by the hydrogen bond network, including the active site water molecules and His250.

References

Cupp-Vickery JR, Poulos TL (1995) Structure of cytochrome P450eryF involved in erythromycin biosynthesis. Nat Struct Biol 2:144–153

Gerber NC, Sligar SG (1992) Catalytic mechanism of cytochrome P-450: evidence for a distal charge relay. J Am Chem Soc 114:8742–8743

Gerber NC, Sligar SG (1994) A role for Asp-251 in cytochrome P-450cam oxygen activation. J Biol Chem 269:4260–4266

Gotoh O (1992) Substrate recognition sites in cytochrome P450 family 2 (CYP2) proteins inferred from comparative analyses of amino acid and coding nucleotide sequences. J Biol Chem 267:83–90

Imai M, Shimada H, Watanabe Y, Matsushima-Hibiya Y, Makino R, Koga H et al (1989) Uncoupling of the cytochrome P-450cam monooxygenase reaction by a single mutation, threonine-252 to alanine or valine: a possible role of the hydroxy amino-acid in oxygen activation. Proc Natl Acad Sci USA 86:7823–7827

Kimata Y, Shimada H, Hirose T, Ishimura Y (1995) Role of Thr-252 in cytochrome P450cam: a study with unnatural amino acid mutagenesis. Biochem Biophys Res Commun 208:96–102

Makino M, Sugimoto H, Shiro Y, Asamizu S, Onaka H, Nagano S (2007) Crystal structures and catalytic mechanism of cytochrome P450 StaP that produces the indolocarbazole skeleton. Proc Natl Acad Sci USA 104:11591–11596

Martinis SA, Atkins WM, Stayton PS, Sligar SG (1989) A conserved residue of cytochrome-P-450 is involved in heme-oxygen stability and activation. J Am Chem Soc 111:9252–9253

Nagano S, Poulos TL (2005) Crystallographic study on the dioxygen complex of wild-type and mutant cytochrome P450cam. Implications for the dioxygen activation mechanism. J Biol Chem 280:31659–31663

Nagano S, Cupp-Vickery JR, Poulos TL (2005) Crystal structures of the ferrous dioxygen complex of wild-type cytochrome P450eryF and its mutants, A245S and A245T: investigation of the proton transfer system in P450eryF. J Biol Chem 280:22102–22107

Onaka H, Taniguchi S, Igarashi Y, Furumai T (2002) Cloning of the staurosporine biosynthetic gene cluster from *Streptomyces* sp. TP-A0274 and its heterologous expression in *Streptomyces lividans*. J Antibiot (Tokyo) 55:1063–1071

Ortiz de Montellano PR (ed) (1995) Cytochrome P450: structure, mechanism, and biochemistry, 2nd edn. Plenum, New York

Poulos TL, Finzel BC, Gunsalus IC, Wagner GC, Kraut J (1985) The 2.6-Å crystal structure of *Pseudomonas putida* cytochrome P-450. J Biol Chem 260:16122–16130

Schlichting I, Berendzen J, Chu K, Stock AM, Maves SA, Benson DE et al (2000) The catalytic pathway of cytochrome p450cam at atomic resolution. Science 287:1615–1622

Wang Y, Hirao H, Chen H, Onaka H, Nagano S, Shaik S (2008) Electron transfer activation of chromopyrrolic acid by cytochrome P450 en route to the formation of an antitumor indolocarbazole derivative: theory supports experiment. J Am Chem Soc 130:7170–7171

Wang Y, Chen H, Makino M, Shiro Y, Nagano S, Asamizu S et al (2009) Theoretical and experimental studies of the conversion of chromopyrrolic acid to an antitumor derivative by cytochrome P450 StaP: the catalytic role of water molecules. J Am Chem Soc 131:6748–6762

Zhao B, Guengerich FP, Bellamine A, Lamb DC, Izumikawa M, Lei L et al (2005a) Binding of two flaviolin substrate molecules, oxidative coupling, and crystal structure of *Streptomyces coelicolor* A3(2) cytochrome P450 158A2. J Biol Chem 280:11599–11607

Zhao B, Guengerich FP, Voehler M, Waterman MR (2005b) Role of active site water molecules and substrate hydroxyl groups in oxygen activation by cytochrome P450 158A2: a new mechanism of proton transfer. J Biol Chem 280:42188–42197

Chapter 6
Oxygenation of Nonnative Substrates Using a Malfunction State of Cytochrome P450s

Osami Shoji and Yoshihito Watanabe

Abstract The substrate specificity of bacterial cytochrome P450s is very high. Therefore, their catalytic activities toward nonnative substrates are low, whereas their inherent catalytic activities are very high compared with P450s isolated from animals and plants. Using "decoy" molecules, whose structures are very similar to natural substrates, to trick their substrate recognition with decoy molecules, we can induce a malfunction state of cytochrome P450s. Decoy molecule binding under this malfunction state allows bacterial cytochrome P450s to catalyze the oxidation reaction of nonnative substrates. This system using decoy molecules does not require any substitution of amino acids to alter substrate specificity or any changes in the enantioselectivity of nonnative substrate oxidation.

Keywords Bacterial cytochrome P450s • Decoy molecules • Nonnative substrates • Substrate misrecognition • Substrate specificity

6.1 Introduction

Cytochrome P450s (P450s) are a family of hemoproteins that catalyze monooxygenation of inert substrates in conjunction with the biosynthesis of steroids, drug metabolism, and detoxification of xenobiotics (Ortiz de Montellano 2005; Sono et al. 1996). Because P450s efficiently catalyze the monooxygenation

O. Shoji (✉)
Department of Chemistry, Graduate School of Science, Nagoya University, Furo-cho, Chikusa-ku, Nagoya 464-8602, Japan
e-mail: shoji.osami@a.mbox.nagoya-u.ac.jp

Y. Watanabe
Research Center for Materials Science, Nagoya University, Furo-cho, Chikusa-ku, Nagoya 464-8602, Japan

H. Yamazaki (ed.), *Fifty Years of Cytochrome P450 Research*,
DOI 10.1007/978-4-431-54992-5_6, © Springer Japan 2014

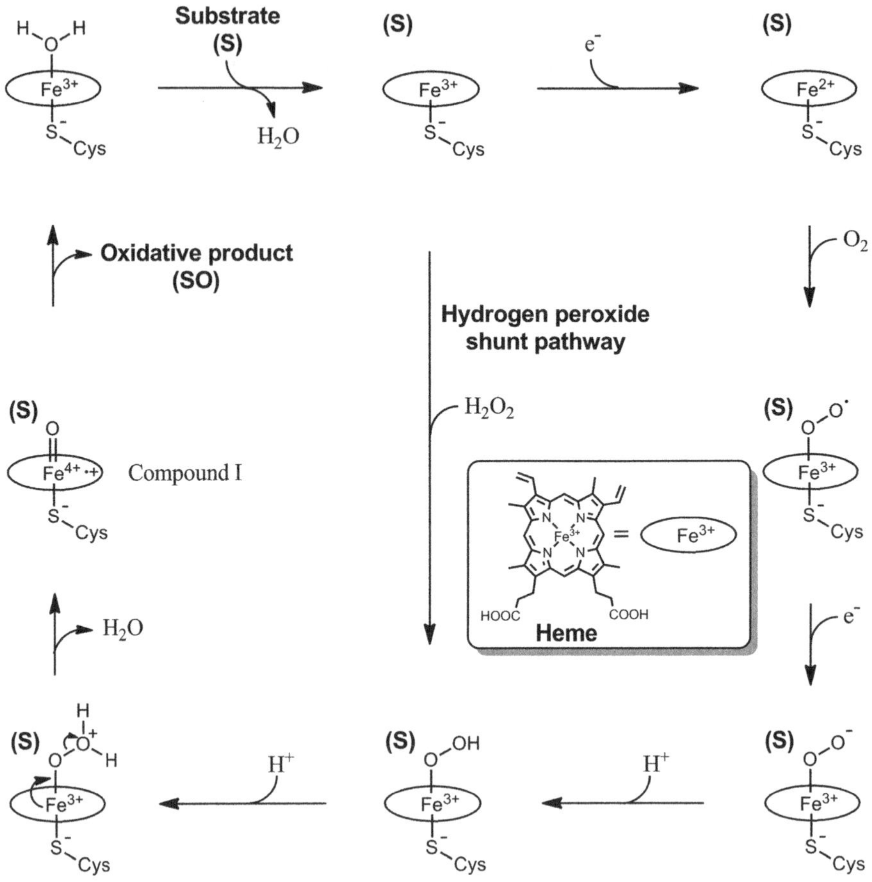

Fig. 6.1 The catalytic cycle of P450s including the hydrogen peroxide shunt pathway

of less reactive C–H bonds under mild conditions, P450s have attracted much attention as candidates for the construction of biocatalysts (Denisov et al. 2005; Fasan 2012; Whitehouse et al. 2012). Among the P450s reported thus far, bacterial P450s are regarded as a promising candidate because they have very high catalytic activity and because a practical amount of these enzymes can be obtained in soluble form using a typical *Escherichia coli* gene-expression system.

P450s activate molecular oxygen through a heme iron center with the thiolate of cysteine as the fifth ligand to generate the active oxidant species, oxoferryl (IV) porphyrin π cation radical, so-called Compound I (Dunford and Stillman 1976; Ost et al. 2003; Rittle and Green 2010). The catalytic cycle of P450 involves the following steps (Fig. 6.1): (1) substrate binding, which results in the removal of a water molecule ligated to the heme iron to cause a positive redox potential shift of the heme iron (Daff et al. 1997; Murataliev and Feyereisen 1996; Sligar 1976);

(2) reduction of Fe^{3+} (ferric) to Fe^{2+} (ferrous) by the first electron transfer from NADPH through the reductase domain; (3) binding of molecular oxygen to the ferrous heme; (4) production of compound I by reductive activation of the molecular oxygen through the second electron transfer from NADPH; and (5) oxidation of the bound substrate by compound I (Ortiz de Montellano 2005).

According to the reaction mechanism of P450s, appropriate binding of the substrate to the active site of P450s is crucial for initiating the catalytic cycle. Therefore, a molecule with a very different structure from that of native substrates cannot start the first step of the catalytic cycle, resulting in very low catalytic activity, especially for bacterial P450s having high substrate specificity. Because the substrate specificity in enzymatic reactions is governed by the local chemical environment of the enzyme active site, a variety of mutants of P450s, such as P450BM3 (Whitehouse et al. 2012), have been prepared by site-directed mutagenesis as well as by random mutagenesis to alter their substrate specificity. The resulting mutants have been found to catalyze the hydroxylation of gaseous alkanes such as ethane and propane (Chen et al. 2012; Fasan et al. 2007; Fasan et al. 2008; Meinhold et al. 2005). These mutants are assumed to provide space to accommodate gaseous alkanes in an appropriate manner because of the structural change in the active site. One of these mutants has been estimated to have a small active site that can accommodate gaseous alkanes (Fasan et al. 2008).

A similar strategy has been applied to P450cam to prepare a mutant to catalyze the hydroxylation of gaseous alkanes (Bell et al. 2003; Bell et al. 2002). Although the mutagenesis of P450s to construct a binding pocket suitable for nonnative substrates is regarded as a promising method, we proposed a simple and unique strategy to hydroxylate gaseous alkanes using wild-type P450s without replacing any amino acid residues. If the size and shape of the active site of P450s were reshaped by the addition of a small molecule that can bind to the active site, even wild-type P450BM3 would be able to provide a binding pocket for accommodating nonnative substrates and could oxidize them without the need for mutagenesis. This review focuses on the P450-catalyzed monooxygenation of nonnative substrates by inducing a malfunction state of P450s by adding "decoy" molecules whose chemical structures are similar to those of fatty acids. The second section of this review describes the hydroxylation of gaseous alkanes and benzene (and monosubstituted benzenes) by wild-type P450BM3 assisted by decoy molecules. The third section of this review describes the peroxygenation of nonnative substrates by hydrogen peroxide-dependent P450s with decoy molecules. In the final section, we conclude our discussion of the decoy molecule system for nonnative substrates oxidation by wild-type P450s.

6.2 Monooxygenation of Nonnative Substrates by P450BM3 Using Perfluorinated Carboxylic Acids as Decoy Molecules

P450BM3 (CYP102A1) isolated from *Bacillus megaterium* catalyzes the hydroxylation of long-alkyl-chain fatty acids at the ω-1, ω-2, and ω-3 positions (Fig. 6.2c) (Boddupalli et al. 1992; Narhi and Fulco 1986; Ravichandran et al. 1993). Because P450BM3 is a structurally self-sufficient P450 (i.e., the reductase domain is fused with the P450 domain on the same peptide chain), P450BM3 has an extremely high hydroxylation activity with a turnover number of more than 16,000 per minute (Girvan et al. 2004; Girvan et al. 2009; Noble et al. 1999); this is the highest oxidase activity among the P450s reported thus far. Because of its high catalytic activity, P450BM3 is expected to be a biocatalyst for the hydroxylation of inert C–H bonds. The crystal structure of P450BM3 with palmitoleic acid shows that palmitoleic acid is fixed by two major interactions (Fig. 6.2a, b): the ionic interaction of the substrate carboxylate group with Arg-47 and Tyr-51, and the hydrophobic interaction of the alkyl chain with amino acids at the substrate binding site (Li and Poulos 1997).

The catalytic cycle of P450BM3 begins with fatty acid binding to the heme cavity of P450BM3, which induces a structural change in P450BM3 and the removal of a water molecule coordinated to the heme iron atom (Fe^{3+}), resulting in a positive shift in the reduction potential of the heme iron atom, followed by an

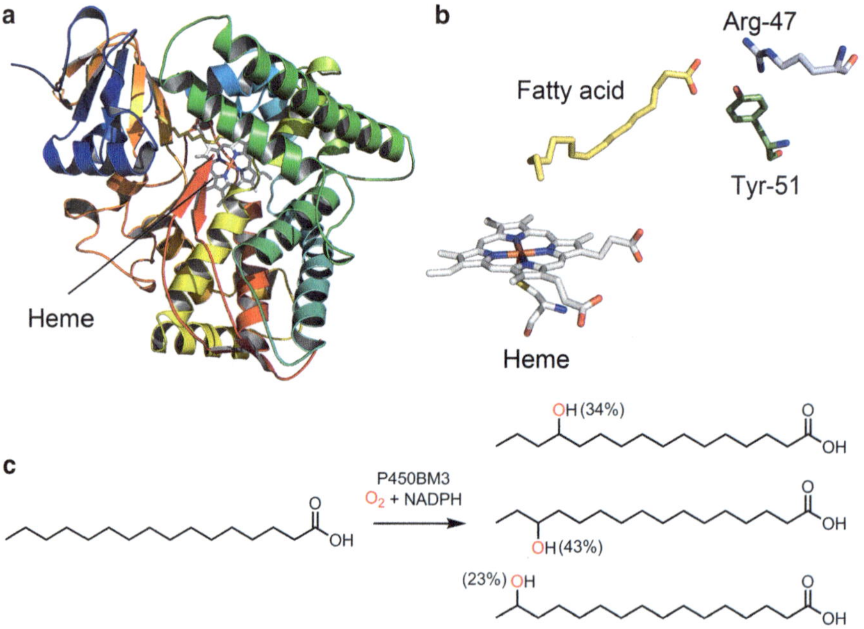

Fig. 6.2 The overall structure (**a**) and the active site structure (**b**) of P450BM3 (PDB code, 1FAG). (**c**) Fatty acid hydroxylation catalyzed by P450BM3

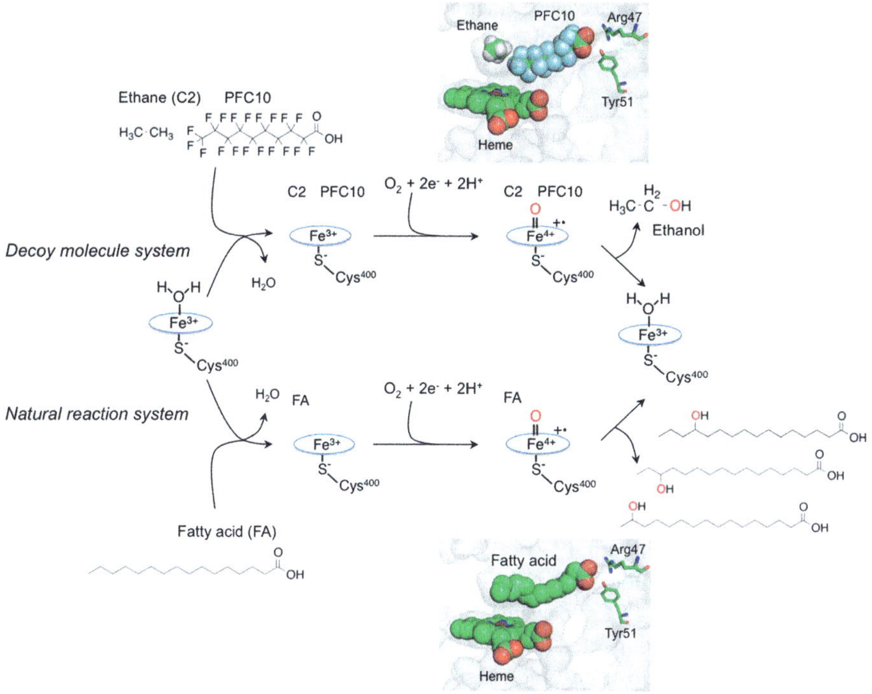

Fig. 6.3 Reaction mechanisms of the natural reaction system (*lower*) and the decoy molecule system (*upper*) of P450BM3. The subterminal carbons of fatty acids are hydroxylated in the natural reaction (*upper*). By simple addition of a decoy molecule such as PFC10, the hydroxylation reaction of a small alkane (e.g., ethane) is catalyzed by P450BM3 because of substrate misrecognition of P450BM3 (*upper*). The active site crystal structure of P450BM3 containing natural substrate (PDB code, 1FAG) and the plausible structure of P450BM3 containing PFC10 and ethane are shown in the respective reaction pathways

electron transfer from NADPH to reduce Fe^{3+} to Fe^{2+} (Figs. 6.1 and 6.3). After the generation of compound I, the bound fatty acid is oxidized. This reaction mechanism of P450BM3 shows that the fatty acid binding is crucial for initiating the catalytic cycle, and thus P450BM3 possesses very high substrate specificity. However, a simple addition of perfluorocarboxylic acid (PFCs) as inert dummy substrates (decoy molecules) can turn wild-type P450BM3 into a small alkane hydroxylase without the need to replace any amino acid residues (Kawakami et al. 2011; Zilly et al. 2011). It was assumed that the PFCs would bind to the active site of P450BM3 through an interaction of their carboxylic groups with Arg-47 and Tyr-51, similar to the natural substrate-binding manner, but PFCs are never oxidized by P450BM3 because of their inert C–F bonds (C–F bond dissociation energy, 116 kcal/mol). The decoy molecules initiate the activation of molecular oxygen in the same manner as do long-alkyl-chain fatty acids and induce

Table 6.1 Turnover rate and coupling efficiency of propane, butane, and cyclohexane hydroxylation by P450BM3 in the presence of perfluorinated carboxylic acids (PFCs)

	Propane hydroxylation		Butane hydroxylation		Cyclohexane hydroxylation	
Decoy molecule	Rates[a] (/min/P450)	Coupling efficiency (%)	Rates (/min/P450)	Coupling efficiency (%)	Rates (/min/P450)	Coupling efficiency (%)
None	Not detected	–	Not detected	–	Not detected	–
PFC8	Not detected	–	2 ± 3	1	63 ± 4	45
PFC9	24 ± 8	13	110 ± 30	55	110 ± 10	37
PFC10	67 ± 2	18	100 ± 30	17	72 ± 13	23
PFC11	40 ± 4	9	69 ± 20	11	54 ± 10	17
PFC12	25 ± 6	7	69 ± 15	10	44 ± 8	15
PFC13	11 ± 3	3	35 ± 10	4	26 ± 1	8
PFC14	Not detected	–	13 ± 3	2	16 ± 2	5

[a]Uncertainty is given as the standard deviation from at least three measurements

the generation of compound I (Fig. 6.3). Because the C–F bonds of PFCs are not oxidizable (Banks and Tatlow 1986), compound I can be used for nonnative substrate oxidation. PFCs bearing shorter alkyl chains of 8–14 carbon atoms (PFC8–PFC14) are expected to provide space for nonnative substrates because the structure of the substrate-binding site of P450BM3 is suitable for accommodating a fatty acid bearing 16 carbon atoms. This strategy for the hydroxylation of gaseous alkanes and a plausible active site structure of P450BM3 containing both perfluorodecanoic acid and gaseous alkane (ethane) are shown in Fig. 6.3, upper.

Upon the addition of PFCs bearing alkyl chains of 8 to 14 carbons, the ultraviolet and visible absorption spectra (UV–vis spectra) of P450BM3 were changed because of the binding of the decoy molecules. The dissociation constants (K_d) of the PFCs were estimated based on UV–vis spectral changes as 1,900 μM for PFC8, 980 μM for PFC9, 290 μM for PFC10, 91 μM for PFC11, 30 μM for PFC12, 1.8 μM for PFC13, and 13 μM for PFC14 (Kawakami et al. 2011). The consumption of NADPH upon the addition of decoy molecules was indicative of the elimination of the water molecule coordinated to the heme iron atom of P450BM3, which is the first step in the catalytic cycle for the formation of compound I (Figs. 6.1, 6.3).

The hydroxylation of gaseous alkane molecules (propane and butane) and cyclohexane in the presence of a series of PFCs was performed. The rate of product formation and the coupling efficiency [the ratio of product formation to NADPH consumption: (product)/(NADPH consumption) × 100] for a reaction time of 10 min were examined; the results are summarized in Table 6.1. Propane, butane, and cyclohexane were hydroxylated using 500 nM P450BM3 in the presence of 100 μM PFCs to yield 2-propanol, 2-butanol, and cyclohexanol, respectively. In the absence of PFCs, no product was detected in the reactions for any of the alkanes, showing that the addition of the decoy molecules to wild-type P450BM3 conferred the ability to catalyze the hydroxylation of gaseous alkanes. Interestingly, the rate of product formation was influenced significantly by the alkyl chain length of the PFCs.

Table 6.2 Turnover rate and coupling efficiency of benzene hydroxylation by P450BM3 in the presence of perfluorinated carboxylic acids (PFCs)

Decoy molecule	Rates[a] (/min/P450)	Coupling efficiency (%)
None	Not detected	–
PFC8	38 ± 6	17
PFC9	120 ± 9	24
PFC10	120 ± 5	19
PFC11	83 ± 6	15
PFC12	71 ± 5	12

[a]Uncertainty is given as the standard deviation from at least three measurements

PFC10 showed the fastest rate of product formation for propane hydroxylation (67/min/P450) and the highest coupling efficiency (18 %) among the PFCs examined. PFC9 and PFC10 were almost equally effective for butane hydroxylation and showed the fastest rate for formation of 2-butanol, 100–113/min/P450, whereas PFC9 showed the highest coupling efficiency of 57 %. In the hydroxylation of cyclohexane, PFC9 showed the fastest rate of cyclohexanol formation (110/min/P450) with a 35 % coupling efficiency. These results indicate that the larger alkanes tended to prefer the shorter alkyl chain PFCs for efficient reactions and that the combination of the alkyl chain length of the PFCs and the size of the alkane substrate governs the efficiency of the hydroxylation reaction catalyzed by P450BM3. The P450BM3–decoy molecule system also catalyzes the hydroxylation of octane, hexane, and branched alkanes (Zilly et al. 2011).

Hydroxylations of the small alkanes were also performed under a high-pressure condition by pressurizing the reaction vessel with 0.5 MPa (5 atm) using an autoclave (Kawakami et al. 2013). Pressurizing the reaction vessel improved the coupling efficiency for propane hydroxylation to 50 %. In addition, the primary carbon of propane was hydroxylated to give 1-propanol (100/h/P450). The primary carbons of ethane were also hydroxylated under the pressure condition of 0.5 MPa ethane using PFC10 as a decoy molecule (40/h/P450), whereas the hydroxylation of methane gas under the pressure condition of 0.5 MPa methane gave no methanol. These results indicate that even the primary carbon of ethane can be hydroxylated by this P450BM3–decoy molecule system when the concentration of ethane in the reaction mixture is increased under high pressure of ethane.

In addition to the hydroxylation of gaseous alkanes, the hydroxylation of benzene and monosubstituted benzenes has also been shown to be oxidized by the P450BM3–decoy molecule system (Shoji et al. 2013). Direct hydroxylation of benzene has attracted much attention as an alternative method for phenol production because phenol is currently produced by the cumene process (Schmidt 2005), which has a high energy consumption and leads to the formation of side products such as acetone. The hydroxylation of benzene by P450BM3 was examined by using a series of PFCs (PFC8–PFC12) (Table 6.2). PFC9 and PFC10 afforded the fastest turnover rate, 120/min/P450. The coupling efficiency of PFC9, 24 %, was the highest among the PFCs examined, indicating that the active site provided by

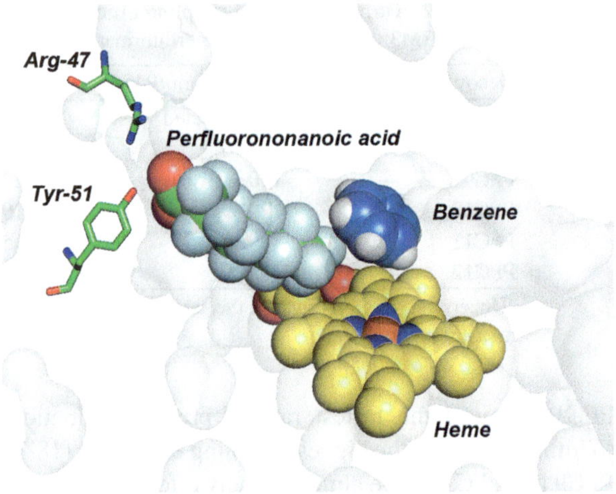

Fig. 6.4 The plausible active site structure of P450BM3 containing PFC9 and benzene

PFC9 is suitable for the accommodation of benzene. The catalytic turnover rate and the coupling efficiency were higher than those of the engineered P450BM3 mutants prepared by directed evolution (Farinas et al. 2004). Although the hydroxylation of benzene is generally accompanied by over-oxidation products (Zhai et al. 2005), selective formation of phenol without any over-oxidation products was observed.

As shown by high-performance liquid chromatography analysis, the selectivity toward phenol production was more than 99 %. Thus, phenol is expected to escape rapidly from the active site of P450BM3 because of the heme cavity comprising hydrophobic amino acid residues and the hydrophobic nature of PFC (Fig. 6.4). Benzene-d_6 hydroxylation had a faster turnover rate, 135/min/P450, indicating that the rate-determining step does not include C–H bond cleavage, but can be explained by assuming epoxide formation followed by rearrangement reactions, including a hydrid shift, known as the NIH shift (Fig. 6.5). The observed deuterium kinetic isotope effect is attributable to the change in carbon atoms from sp^2 to sp^3 for the formation of epoxide, which generally affords an inverse isotope effect. Toluene was also hydroxylated, and PFC9 gave the fastest turnover rate, 220/min/P450, for the selective *o*-position hydroxylation (Fig. 6.6). The coupling efficiency reached 56 %. The *o*-selective hydroxylation was also observed in the hydroxylation of anisole and chlorobenzene (Fig. 6.6). Furthermore, the hydroxylation of electron-poor benzene derivatives, nitrobenzene and acetophenone, gave the *o*-hydroxylated product despite their meta-preference in many reactions (Fig. 6.6). These results show that the *o*-position of monosubstituted benzenes was hydroxylated selectively, irrespective of the substituents.

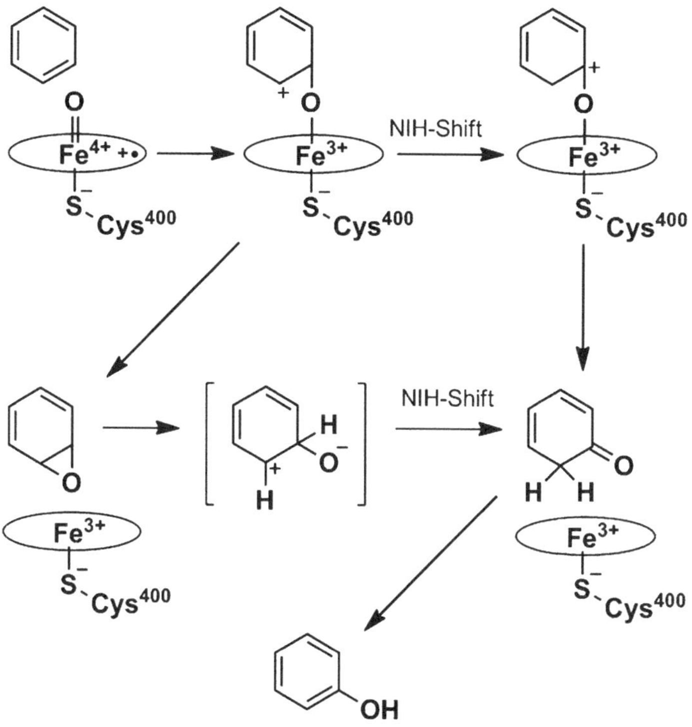

Fig. 6.5 A plausible reaction mechanism of benzene hydroxylation by P450BM3

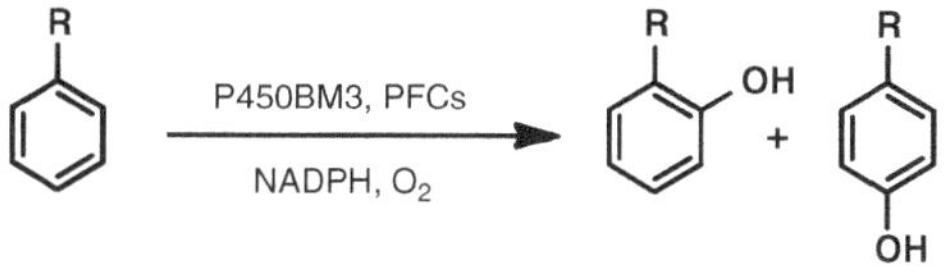

Substrate R =	Rates[a] [/min/P450]		
	o-	*m*-	*p*-
$-CH_3$	220 ± 7	-	5.3 ± 0.2
$-OCH_3$	260 ± 4	-	38 ± 0.3
-Cl	120 ± 4	-	14 ± 0.6
$-NO_2$	0.9 ± 0.05	-	-
$-COCH_3$	2.9 ± 0.1	-	-

[a] Uncertainty is given as the standard deviation from at least three measurements.

Fig. 6.6 Turnover rate of hydroxylation of monosubstituted benzenes by P450BM3 in the presence of perfluorononanoic acid (PFC9)

6.3 Hydrogen Peroxide-Dependent Monooxygenation of Nonnative Substrates Catalyzed by P450$_{BS\beta}$ and P450$_{SP\alpha}$

The use of hydrogen peroxide to generate the active species of P450s, the so-called "hydrogen peroxide-shunt pathway," is an attractive process for avoiding the consumption of expensive cofactors such as NADPH in monooxygenation reactions (Fig. 6.1). Because of its low cost, hydrogen peroxide is an oxidant applicable to industrial-scale processes. However, the catalytic activity of the shunt pathway is generally poor. In contrast to the fact that most P450s use the reductive molecular oxygen activation process for monooxygenation, P450$_{SP\alpha}$ (CYP152B1) from *Sphingomonas paucimobilis* (Imai et al. 2000; Matsunaga et al. 2000; Matsunaga et al. 1998a, b; Matsunaga et al. 1996; Matsunaga et al. 1997), P450$_{BS\beta}$ (CYP152A1) from *Bacillus subtilis* (Lee et al. 2002; Lee et al. 2003; Matsunaga and Shiro 2004; Matsunaga et al. 2002a, b; Matsunaga et al. 1999; Matsunaga et al. 2001), and P450$_{CLA}$ (CYP152A2) from *Clostridium acetobutylicum* (Girhard et al. 2007) use hydrogen peroxide as the oxidant and catalyze the hydroxylation of long-alkyl-chain fatty acids with high catalytic activities and high substrate specificity. These soluble enzymes work as a single component, and the amino acid identity of P450$_{SP\alpha}$ and P450$_{CLA}$ to P450$_{BS\beta}$ is 42 % and 57 %, respectively.

P450$_{BS\beta}$ efficiently catalyzes the exclusive hydroxylation of long-alkyl-chain fatty acids such as myristic acid to give a mixture of β-hydroxymyristic acid and α-hydroxymyristic acid. The X-ray crystal structure analysis of the palmitic acid-bound form of P450$_{BS\beta}$ revealed that it lacks general acid–base residues around the distal side of the heme, although general acid–base residues are highly conserved in peroxidases and peroxygenases. Instead of the general acid–base residue, the terminal carboxylate group of the bound fatty acid interacting with the guanidinium group of Arg-242 located near the heme group (Fig. 6.7a) is assumed to serve as the general acid–base catalyst. The location of the carboxylate is almost the same as the

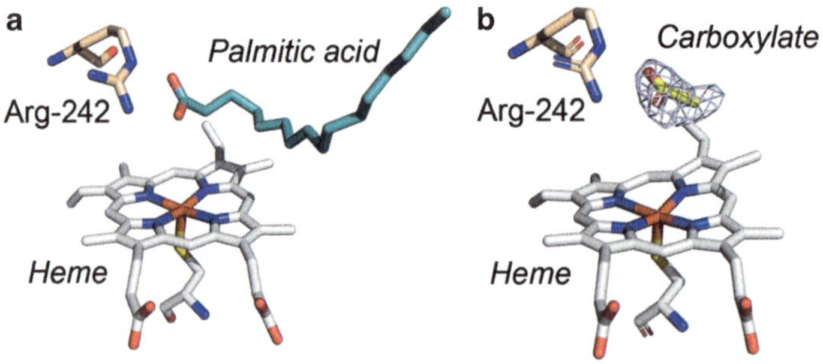

Fig. 6.7 The crystal structure of the palmitic acid-bound form (**a**) (PDB code, 1IZO) and the cocrystal with heptanoic acid (**b**) (PDB code, 2ZQX). An acetic acid molecule was placed in the electron density observed near the Arg-242 residue (*gray mesh*) in the cocrystal using a molecular graphics program

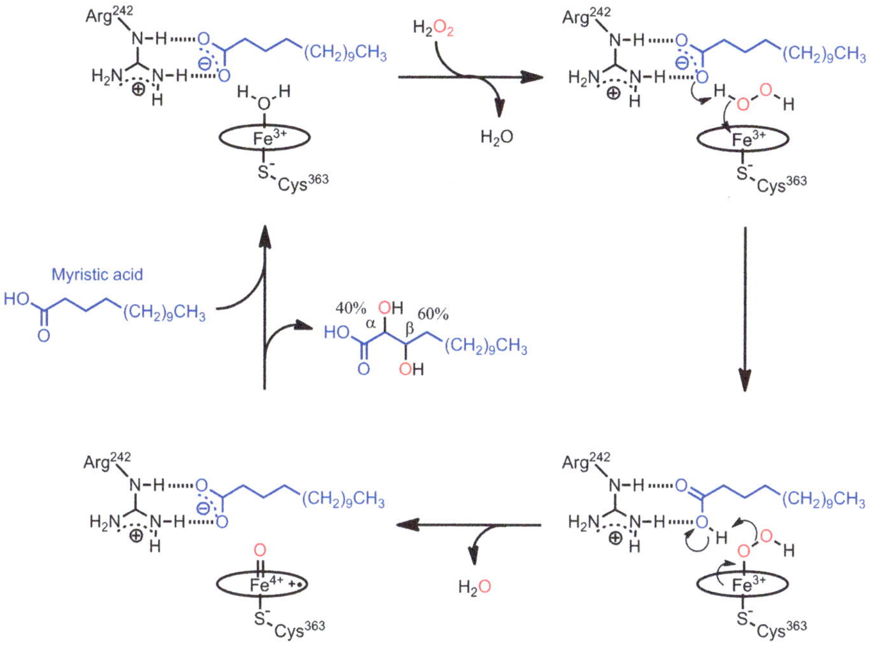

Fig. 6.8 A plausible catalytic hydroxylation mechanism of $P450_{BS\beta}$ and the roles of the substrate carboxylate–Arg-242 salt bridge

distal glutamate in chloroperoxidase (CPO), which is one of the most efficient hydrogen peroxide-dependent biocatalysts, and this suggests the following unique catalytic mechanism (Figs. 6.8, 6.9). (1) The catalytic reaction begins with the fixation of a substrate through interaction of the terminal carboxyl group of the fatty acid with Arg-242, located near the heme. (2) The general acid–base function of the fatty acid–Arg-242 salt bridge allows facile generation of the active species, compound I, to oxidize the substrate. Without the interaction of the carboxyl group, $P450_{BS\beta}$ does not start the reaction. Because this unique catalytic mechanism contributes to the high substrate specificity and regioselectivity of the hydroxylation, $P450_{BS\beta}$ never oxidizes substrates other than long-alkyl-chain fatty acids. For example, $P450_{BS\beta}$ never oxidizes tetradecane, 1-tetradecanol, or tetradecanal. However, $P450_{BS\beta}$ can oxidize a wide variety of nonnative substrates in the presence of a series of short-alkyl-chain carboxylic acids (C4–C10) as decoy molecules (Fig. 6.9) (Shoji et al. 2007).

Through this simple substrate-misrecognition trick using decoy molecules, peroxygenation of nonnative substrates oxidation such as the one-electron oxidation of guaiacol (Shoji et al. 2007), sulfoxidation of thioanisole (Fujishiro et al. 2010), epoxidation of styrene, C–H bond hydroxylation of ethylbenzene (Shoji et al. 2007), and aromatic C–H bond hydroxylation of 1-methoxynaphthalene (Shoji et al. 2010b) were catalyzed by $P450_{BS\beta}$ (Table 6.3, Fig. 6.10). Interestingly, the catalytic activities were highly dependent on the alkyl chain length of the

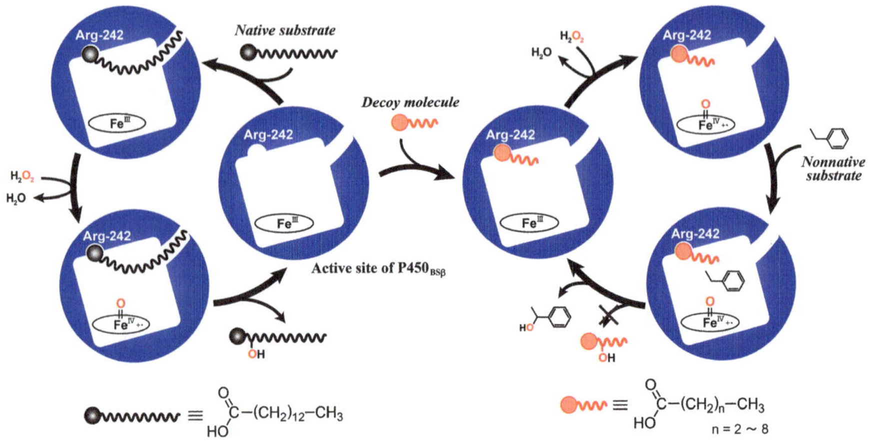

Fig. 6.9 Schematic representation of the hydroxylation of long-alkyl-chain fatty acid (*right*) and the substrate misrecognition system in the oxidation of a nonnative substrate (e.g., ethyl benzene) (*left*). The active site of $P450_{BS\beta}$ is depicted as *circles*

Table 6.3 Turnover rate of one-electron oxidation of guaiacol catalyzed by $P450_{BS\beta}$ in the presence of carboxylic acids

Decoy molecule (number of carbon atoms)	Turnover number/min
None	0
Acetic acid (C2)	0
Propionic acid (C3)	26
Butanoic acid (C4)	230
Pentanoic acid (C5)	1,900
Hexanoic acid (C6)	2,420
Heptanoic acid (C7)	3,750
Octanoic acid (C8)	2,490
Nonanoic acid (C9)	2,380
Decanoic acid (C10)	2,360
Myristic acid (C14)	14

decoy molecules (Table 6.3). The enantioselectivity in peroxygenation such as ethylbenzene hydroxylation is dependent on the structure of the decoy molecules used (Shoji et al. 2007).

The peroxygenation of nonnative substrates never proceeded without decoy molecules, and thus these reactions were catalyzed by $P450_{BS\beta}$ with the help of these short-alkyl-chain carboxylic acids. Apparently, $P450_{BS\beta}$ misrecognizes decoy molecules as its native substrates (long-alkyl-chain fatty acids), and the resulting salt bridge (Arg-242-carboxylate of fatty acid) serves as the general acid–base catalyst to allow the formation of compound I, followed by the oxidation of nonnative substrates. The role of decoy molecules was confirmed by the crystal structure analysis of a heptanoic acid (C7)-bound form of $P450_{BS\beta}$ (Fig. 6.7b)

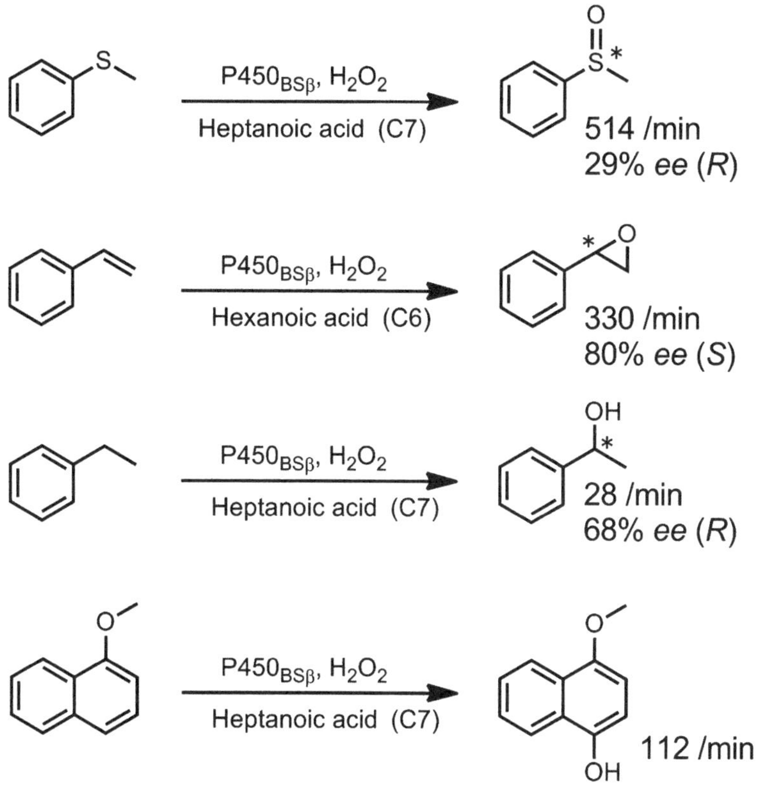

Fig. 6.10 The reactions catalyzed by $P450_{BS\beta}$ in the presence of carboxylic acids as a decoy molecule

(Shoji et al. 2010a). A clear electron density assignable to the carboxylate group of heptanoic acid was observed in the active site, indicating that the carboxylate group of the decoy molecules was involved in the formation of compound I.

The inversion of enantioselectivity by noncovalent modification of the active site of the enzyme was observed in $P450_{SP\alpha}$. Similar to $P450_{BS\beta}$, $P450_{SP\alpha}$ also catalyzed the hydroxylation of long-alkyl-chain fatty acids using hydrogen peroxide as an oxidant. The crystal structure of the palmitic acid-bound form of $P450_{SP\alpha}$ revealed that the key interactions between the carboxylate of palmitic acid and the guanidine group of arginine near the heme are conserved in $P450_{SP\alpha}$ (PDB code: 3AWM), indicating that the substrate-assisted reaction mechanism for the formation of compound I is the same as that of $P450_{BS\beta}$ (Fig. 6.8) (Fujishiro et al. 2011). By adding a series of short-alkyl-chain carboxylic acids as decoy molecules, epoxidation reactions of styrene have been catalyzed by $P450_{SP\alpha}$ (Fujishiro et al. 2012).

Interestingly, the enantioselectivity of styrene oxide was altered by the nature of the decoy molecules. The chirality at the α-carbons of the decoy molecules, in

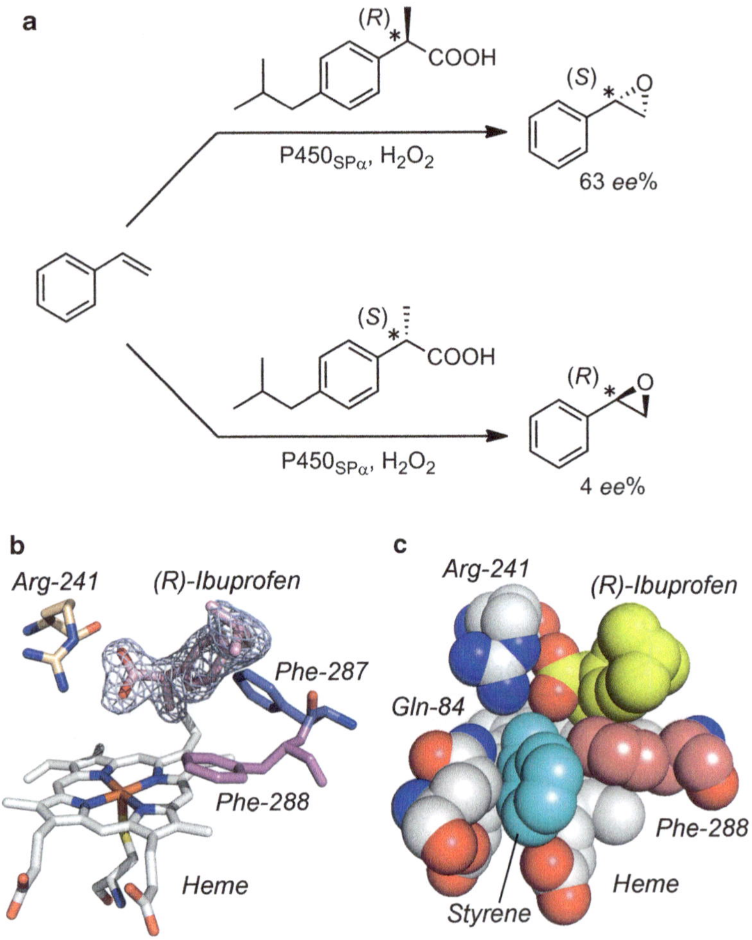

Fig. 6.11 (**a**) Stereoselective epoxidation of styrene by $P450_{SP\alpha}$ in the presence of (*R*)- or (*S*)-ibuprofen. (**b**) X-ray crystal structure of the (*R*)-ibuprofen-bound form of $P450_{SP\alpha}$ (PDB code, 3VM4). (**c**) Docking simulation of styrene toward $P450_{SP\alpha}$ with (*R*)-ibuprofen

particular, induces clear differences. For example, (*R*)-ibuprofen increased (*S*)-styrene oxide formation, whereas (*S*)-ibuprofen preferentially gave (*R*)-styrene oxide, showing that (*R*)-ibuprofen effectively increased (*S*)-selectivity and that the enantioselectivity could be controlled by simply selecting the (*R*)- or (*S*)-enantiomer of ibuprofen (Fig. 6.11a). The crystal structure of $P450_{SP\alpha}$ containing (*R*)-ibuprofen revealed that (*R*)-ibuprofen was accommodated similarly to the fatty acid and that the carboxylate group of (*R*)-ibuprofen interacted with Arg-241 (Fig. 6.11b). The docking simulation of styrene located in the active site of the

(*R*)-ibuprofen-bound form suggests the possible orientation of the vinyl group of styrene in the active site giving (*S*)-styrene oxide (Fig. 6.11c).

6.4 Conclusion

Although it has long been believed that bacterial P450s such as P450BM3, $P450_{BS\beta}$, and $P450_{SP\alpha}$ do not oxidize nonnative substrates efficiently, they can catalyze the oxidation of nonnative substrates after a simple addition of decoy molecules that induce substrate misrecognition of P450s. This malfunction state of P450s imposed by the decoy molecule binding allowed the oxidation of nonnative substrates without replacement of any amino acid residues. Interestingly, the catalytic activity and even enantioselectivity were clearly dependent on the structure of the decoy molecules. Accordingly, the catalytic activity of bacterial P450s for oxidation of nonnative substrates could be controlled by the structure of the decoy molecules. We believe that oxygenation of nonnative substrates using a malfunction state of P450s could be applied to most P450s and might replace repeated mutagenesis as a method to construct biocatalysts. The combination of the decoy molecules and mutants will lead to improvements in both the catalytic activity and the enantioselectivity of products.

Acknowledgments This work was supported by Grants-in-Aid for Scientific Research (S) to Y.W. (24225004) and a Grant-in-Aid for Young Scientists (A) to O.S. (21685018) from the Ministry of Education, Culture, Sports, Science, and Technology (Japan).

References

Banks RE, Tatlow JC (1986) A guide to modern organofluorine chemistry. J Fluor Chem 33:227–346

Bell SG, Stevenson JA, Boyd HD, Campbell S, Riddle AD, Orton EL, Wong LL (2002) Butane and propane oxidation by engineered cytochrome P450(cam). Chem Commun (5):490–491

Bell SG, Orton E, Boyd H, Stevenson JA, Riddle A, Campbell S, Wong LL (2003) Engineering cytochrome P450cam into an alkane hydroxylase. Dalton Trans 11:2133–2140

Boddupalli SS, Pramanik BC, Slaughter CA, Estabrook RW, Peterson JA (1992) Fatty-acid monooxygenation by P450bm-3: product identification and proposed mechanisms for the sequential hydroxylation reactions. Arch Biochem Biophys 292:20–28

Chen MMY, Snow CD, Vizcarra CL, Mayo SL, Arnold FH (2012) Comparison of random mutagenesis and semi-rational designed libraries for improved cytochrome P450 BM3-catalyzed hydroxylation of small alkanes. Protein Eng Des Sel 25:171–178

Daff SN, Chapman SK, Turner KL, Holt RA, Govindaraj S, Poulos TL, Munro AW (1997) Redox control of the catalytic cycle of flavocytochrome P-450 BM3. Biochemistry 36:13816–13823

Denisov IG, Makris TM, Sligar SG, Schlichting I (2005) Structure and chemistry of cytochrome P450. Chem Rev 105:2253–2277

Dunford HB, Stillman JS (1976) Function and mechanism of action of peroxidases. Coord Chem Rev 19:187–251

Farinas ET, Alcalde M, Arnold F (2004) Alkene epoxidation catalyzed by cytochrome P450 BM-3 139-3. Tetrahedron 60:525–528

Fasan R (2012) Tuning P450 enzymes as oxidation catalysts. ACS Catal 2:647–666

Fasan R, Chen MM, Crook NC, Arnold FH (2007) Engineered alkane-hydroxylating cytochrome P450(BM3) exhibiting native-like catalytic properties. Angew Chem Int Ed 46:8414–8418

Fasan R, Meharenna YT, Snow CD, Poulos TL, Arnold FH (2008) Evolutionary history of a specialized P450 propane monooxygenase. J Mol Biol 383:1069–1080

Fujishiro T, Shoji O, Watanabe Y (2010) Non-covalent modification of the active site of cytochrome P450 for inverting the stereoselectivity of monooxygenation. Tetrahedron Lett 52:395–397

Fujishiro T, Shoji O, Nagano S, Sugimoto H, Shiro Y, Watanabe Y (2011) Crystal structure of H_2O_2-dependent cytochrome P450SPα with its bound fatty acid substrate. J Biol Chem 286:29941–29950

Fujishiro T, Shoji O, Kawakami N, Watanabe T, Sugimoto H, Shiro Y, Watanabe Y (2012) Chiral-substrate-assisted stereoselective epoxidation catalyzed by H_2O_2-dependent cytochrome $P450_{SP\alpha}$. Chem Asian J 7:2286–2293

Girhard M, Schuster S, Dietrich M, Durre P, Urlacher VB (2007) Cytochrome P450 monooxygenase from *Clostridium acetobutylicum*: a new [alpha]-fatty acid hydroxylase. Biochem Biophys Res Commun 362:114–119

Girvan HM, Marshall KR, Lawson RJ, Leys D, Joyce MG, Clarkson J, Smith WE, Cheesman MR, Munro AW (2004) Flavocytochrome P450 BM3 mutant A264E undergoes substrate-dependent formation of a novel heme iron ligand set. J Biol Chem 279:23274–23286

Girvan HM, Toogood HS, Littleford RE, Seward HE, Smith WE, Ekanem IS, Leys D, Cheesman MR, Munro AW (2009) Novel haem co-ordination variants of flavocytochrome P450 BM3. Biochem J 417:65–76

Imai Y, Matsunaga I, Kusunose E, Ichihara K (2000) Unique heme environment at the putative distal region of hydrogen peroxide-dependent fatty acid [alpha]-hydroxylase from *Sphingomonas paucimobilis* (peroxygenase P450(SP[alpha])). J Biochem (Tokyo) 128:189–194

Kawakami N, Shoji O, Watanabe Y (2011) Use of perfluorocarboxylic acids to trick cytochrome P450BM3 into initiating the hydroxylation of gaseous alkanes. Angew Chem Int Ed 50:5315–5318

Kawakami N, Shoji O, Watanabe Y (2013) Direct hydroxylation of primary carbons in small alkanes by wild-type cytochrome P450BM3 containing perfluorocarboxylic acids as decoy molecules. Chem Sci 4:2344–2348

Lee D-S, Yamada A, Matsunaga I, Ichihara K, S-i A, Park S-Y, Shiro Y (2002) Crystallization and preliminary X-ray diffraction analysis of fatty-acid hydroxylase cytochrome P450BS[beta] from *Bacillus subtilis*. Acta Crystallogr D 58:687–689

Lee DS, Yamada A, Sugimoto H, Matsunaga I, Ogura H, Ichihara K, Adachi S, Park SY, Shiro Y (2003) Substrate recognition and molecular mechanism of fatty acid hydroxylation by cytochrome P450 from *Bacillus subtilis*. Crystallographic, spectroscopic, and mutational studies. J Biol Chem 278:9761–9767

Li HY, Poulos TL (1997) The structure of the cytochrome p450BM-3 haem domain complexed with the fatty acid substrate, palmitoleic acid. Nat Struct Biol 4:140–146

Matsunaga I, Shiro Y (2004) Peroxide-utilizing biocatalysts: structural and functional diversity of heme-containing enzymes. Curr Opin Chem Biol 8:127–132

Matsunaga I, Yamada M, Kusunose E, Nishiuchi Y, Yano I, Ichihara K (1996) Direct involvement of hydrogen peroxide in bacterial [alpha]-hydroxylation of fatty acid. FEBS Lett 386:252–254

Matsunaga I, Yokotani N, Gotoh O, Kusunose E, Yamada M, Ichihara K (1997) Molecular cloning and expression of fatty acid alpha-hydroxylase from *Sphingomonas paucimobilis*. J Biol Chem 272:23592–23596

Matsunaga I, Sumimoto T, Kusunose E, Ichihara K (1998a) Phytanic acid alpha-hydroxylation by bacterial cytochrome P450. Lipids 33:1213–1216

Matsunaga I, Yamada M, Kusunose E, Miki T, Ichihara K (1998b) Further characterization of hydrogen peroxide-dependent fatty acid {alpha}-hydroxylase from *Sphingomonas paucimobilis*. J Biochem (Tokyo) 124:105–110

Matsunaga I, Ueda A, Fujiwara N, Sumimoto T, Ichihara K (1999) Characterization of the ybdT gene product of *Bacillus subtilis*: novel fatty acid [beta]-hydroxylating cytochrome P450. Lipids 34:841–846

Matsunaga I, Sumimoto T, Ueda A, Kusunose E, Ichihara K (2000) Fatty acid-specific, regiospecific, and stereospecific hydroxylation by cytochrome P450 (CYP152B1) from *Sphingomonas paucimobilis*: substrate structure required for [alpha]-hydroxylation. Lipids 35:365–371

Matsunaga I, Ueda A, Sumimoto T, Ichihara K, Ayata M, Ogura H (2001) Site-directed mutagenesis of the putative distal helix of peroxygenase cytochrome P450. Arch Biochem Biophys 394:45–53

Matsunaga I, Sumimoto T, Ayata M, Ogura H (2002a) Functional modulation of a peroxygenase cytochrome P450: novel insight into the mechanisms of peroxygenase and peroxidase enzymes. FEBS Lett 528:90–94

Matsunaga I, Yamada A, Lee DS, Obayashi E, Fujiwara N, Kobayashi K, Ogura H, Shiro Y (2002b) Enzymatic reaction of hydrogen peroxide-dependent peroxygenase cytochrome P450s: kinetic deuterium isotope effects and analyses by resonance Raman spectroscopy. Biochemistry 41:1886–1892

Meinhold P, Peters MW, Chen MMY, Takahashi K, Arnold FH (2005) Direct conversion of ethane to ethanol by engineered cytochrome P450BM3. ChemBioChem 6:1765–1768

Murataliev MB, Feyereisen R (1996) Functional interactions in cytochrome P450BM3. Fatty acid substrate binding alters electron-transfer properties of the flavoprotein domain. Biochemistry 35:15029–15037

Narhi LO, Fulco AJ (1986) Characterization of a catalytically self-sufficient 119,000-Dalton cytochrome-P-450 monooxygenase induced by barbiturates in *Bacillus megaterium*. J Biol Chem 261:7160–7169

Noble MA, Miles CS, Chapman SK, Lysek DA, Mackay AC, Reid GA, Hanzlik RP, Munro AW (1999) Roles of key active-site residues in flavocytochrome P450 BM3. Biochem J 339:371–379

Ortiz de Montellano PR (2005) Cytochrome P450: structure, mechanism, and biochemistry. Plenum, New York

Ost TWB, Clark J, Mowat CG, Miles CS, Walkinshaw MD, Reid GA, Chapman SK, Daff S (2003) Oxygen activation and electron transfer in flavocytochrome P450BM3. J Am Chem Soc 125:15010–15020

Ravichandran KG, Boddupalli SS, Hasemann CA, Peterson JA, Deisenhofer J (1993) Crystal structure of hemoprotein domain of P450bm-3, a prototype for microsomal P450s. Science 261:731–736

Rittle J, Green MT (2010) Cytochrome P450 compound I: capture, characterization, and C–H bond activation kinetics. Science 330:933–937

Schmidt RJ (2005) Industrial catalytic processes—phenol production. Appl Catal A 280:89–103

Shoji O, Fujishiro T, Nakajima H, Kim M, Nagano S, Shiro Y, Watanabe Y (2007) Hydrogen peroxide dependent monooxygenations by tricking the substrate recognition of cytochrome $P450_{BS\beta}$. Angew Chem Int Ed 46:3656–3659

Shoji O, Fujishiro T, Nagano S, Tanaka S, Hirose T, Shiro Y, Watanabe Y (2010a) Understanding substrate misrecognition of hydrogen peroxide dependent cytochrome P450 from *Bacillus subtilis*. J Biol Inorg Chem 15:1331–1339

Shoji O, Wiese C, Fujishiro T, Shirataki C, Wünsch B, Watanabe Y (2010b) Aromatic C–H bond hydroxylation by P450 peroxygenases: a facile colorimetric assay for monooxygenation activities of enzymes based on Russig's blue formation. J Biol Inorg Chem 15:1109–1115

Shoji O, Kunimatsu T, Kawakami N, Watanabe Y (2013) Highly selective hydroxylation of benzene to phenol by wild-type cytochrome P450BM3 assisted by decoy molecules. Angew Chem Int Ed 52:6606–6610

Sligar SG (1976) Coupling of spin, substrate, and redox equilibria in cytochrome P450. Biochemistry 15:5399–5406

Sono M, Roach MP, Coulter ED, Dawson JH (1996) Heme-containing oxygenases. Chem Rev 96:2841–2888

Whitehouse CJC, Bell SG, Wong LL (2012) P450(Bm3) (Cyp102a1): connecting the dots. Chem Soc Rev 41:1218–1260

Zhai PM, Wang LQ, Liu CH, Zhang SC (2005) Deactivation of zeolite catalysts for benzene oxidation to phenol. Chem Eng J 111:1–4

Zilly FE, Acevedo JP, Augustyniak W, Deege A, Hausig UW, Reetz MT (2011) Tuning a P450 enzyme for methane oxidation. Angew Chem Int Ed 50:2720–2724

Chapter 7
Plant Cytochrome P450s in Triterpenoid Biosynthesis: Diversity and Application to Combinatorial Biosynthesis

Ery Odette Fukushima, Hikaru Seki, and Toshiya Muranaka

Abstract Plants produce a wide variety of specialized (secondary) metabolites, with which they interact in various environmental conditions for survival. Plant cytochrome P450s have a central function to enhance the diversity of the chemicals. Here we focus on the diversity of P450s in (tri)terpenoid biosynthesis and their application to combinatorial biosynthesis. A strategy combining a homology-based approach, gene coexpression analysis, and combinatorial biosynthesis with heterologous expression in yeast was successful in identifying enzymes involved in triterpenoid biosynthesis and also in generating natural and rare triterpenoids that do not accumulate in planta. Using this strategy is possible to construct a natural-unnatural triterpenoid library. The next steps are then to increase product yields as well as to diversify triterpenoids into novel synthetic entities with improved biological activities by combining enzymes from different sources.

Keywords Coexpression analysis • Glycosyltransferase • Glycyrrhizin • Oxidosqualene • Phytochemical • Saponin • Specialized metabolism • Terpenoid • Triterpenoid

7.1 Introduction

Plants produce a wide variety of phytochemicals. These chemicals, estimated to number approximately 1 million, are divided into two major classes: primary metabolites and specialized (secondary) metabolites. Primary metabolites such as sugars, fatty acids, amino acids, and nucleic acids are synthesized from water,

E.O. Fukushima • H. Seki • T. Muranaka (✉)
Osaka University, Suita, Japan
e-mail: muranaka@bio.eng.osaka-u.ac.jp

H. Yamazaki (ed.), *Fifty Years of Cytochrome P450 Research*,
DOI 10.1007/978-4-431-54992-5_7, © Springer Japan 2014

carbon dioxide, and minerals, which are essential to proper growth and development. Genes for biosynthesis of these metabolites are, in general, conserved across all the plant kingdom. On the other hand, specialized metabolites are required for plants to interact with both abiotic and biotic stresses and are also used to attract pollinators to the flowers. The interaction of specialized metabolites with the environment imparts an evolutionary pressure upon plants to create new specialized metabolites and, as such, the majority of these are specific to a limited number of plant species. In this section, we focus on the diversity of cytochrome P450 monooxidases hereafter P450s in (tri)terpenoid biosynthesis and their application to combinatorial biochemistry.

7.2 Triterpenoid Biosynthesis

Phytochemicals are classified into three basic groups: (1) terpenoids, (2) alkaloids, and (3) phenylpropanoids and allied phenolic compounds. Among these, terpenoids are the largest group, classified according to the number of 5-carbon units that they contain: hemi (C5), mono (C10), sesqui (C15), di (C20), tri (C30), tetra (C40), and poly (Cn) terpenoids.

Triterpenoids are a diverse group of phytochemicals produced by many plant species. Examples include medicinal plants that are exploited as drug sources, such as licorice and ginseng, as well as crop plants, such as legumes and oats (Haralampidis et al. 2002). Interest in triterpenoid saponins has increased recently because of data showing their diverse biological activities and beneficial properties, which include antifungal, antibacterial, antiviral, antitumor, molluscicidal, insecticidal, and antifeedant activities (Suzuki et al. 2002; Sparg et al. 2004; Huhman et al. 2005).

Triterpenoid saponins are derived from a common linear precursor, 2,3-oxidosqualene, which is synthesized through the mevalonate pathway. The biosynthesis of triterpenoid saponins involves the initial cyclization of 2,3-oxidosqualene to the triterpene skeleton triterpenols. The next process is oxidation, which seems to be catalyzed by P450s, although an unique P450 gene (*CYP93E1*) was isolated and characterized (Shibuya et al. 2006) before we started the project to identify P450 genes in triterpenoid biosynthesis (details are described in the following section; Fig. 7.1), which generate various triterpenoid sapogenins. Sapogenins are further glycosylated by UDP-dependent glycosyltransferases to produce triterpenoid saponins.

Among these chemicals, glycyrrhizin , a triterpenoid saponin derived from the underground parts of *Glycyrrhiza* (licorice) plants, is one of the most important crude drugs in the world. Glycyrrhizin and its aglycone glycyrrhetinic acid exhibit various pharmacological activities, including antiinflammatory (Matsui et al. 2004) and hepatoprotective activities (Shibata 2000). Their production largely depends on the collection of wild licorice plants, and this has caused a decrease in licorice

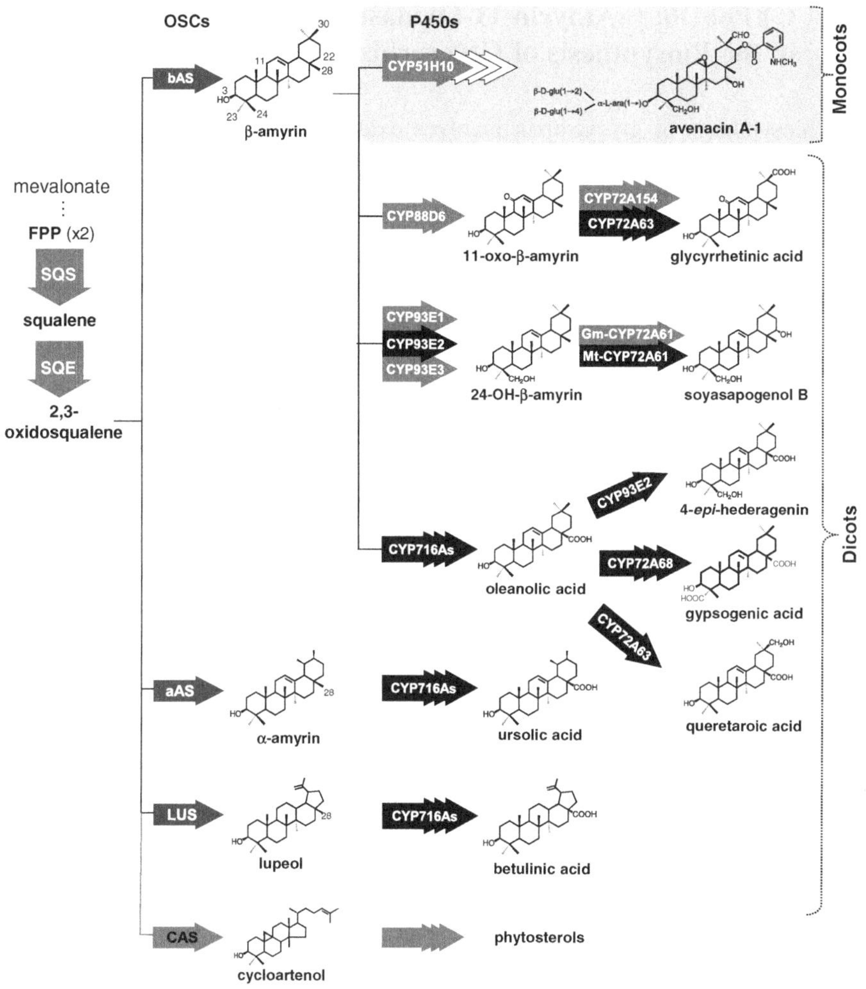

Fig. 7.1 Enzymes involved in triterpenoid biosynthesis in higher plants. Each reaction is indicated by one *arrow*. *Gray arrows* correspond to enzymes reported in previous studies. *Black arrows* correspond to enzymes identified through this work. *White arrows* correspond to yet-unidentified enzymes. **FPP** farnesyl pyrophosphate, **SQS** squalene synthase, **SQE** squalene epoxidase, **bAS** β-amyrin synthase, **aAS** α-amyrin synthase, **LUS** lupeol synthase, **CAS** cycloartenol synthase. (Original figure is from Fukushima et al. 2013b)

reserves and an increase in desertification where it is harvested. Glycyrrhizin is synthesized from β-amyrin, one of the most commonly occurring triterpenols in plants, followed by a series of oxidative reactions and glucuronydation. Although the gene encoding β-amyrin synthase was cloned from *Glycyrrhiza* species (Hayashi et al. 2001), information about genes for specific oxidases and glycosylases has not been obtained.

7.3 CYP88D6: β-Amyrin 11-Oxidase Committed in the Biosynthesis of Glycyrrhizin

The biosynthesis of glycyrrhizin involves oxidative reactions at positions C-11 (two-step oxidation) and C-30 (three-step oxidation) of β-amyrin. We identified a gene encoding β-amyrin 11-oxidase (CYP88D6) from *Glycyrrhiza uralensis*, by EST-based approach (Sudo et al. 2009), as the initial P450 gene of glycyrrhizin biosynthesis (Seki et al. 2008). CYP88D6 was shown in vitro and in vivo [in *β-amyrin synthase* (*bAS*)-expressing yeast; Fig. 7.2] to catalyze two sequential oxidation steps in the glycyrrhizin pathway: the conversion of β-amyrin to 11-oxo-β-amyrin via 11α-hydroxy-β-amyrin. Transcripts of *CYP88D6* were detected in the roots and stolons, whereas no transcripts were observed in the leaves or stems. These results are similar to those for *bAS* and are consistent with the fact that glycyrrhizin accumulates predominantly in the underground parts of *Glycyrrhiza* plants.

In terms of enzymes of known function, CYP88D6 is most similar to the *ent*-kaurenoic acid oxidases CYP88A6 (50 % identity) from pea (named PsKAO1; Davidson et al. 2003) and CYP88A4 (49 % identity) from *Arabidopsis* (named AtKAO2; Helliwell et al. 2001), both of which function in the biosynthesis of gibberellins (diterpenoid). Differing from the CYP88A subfamily of enzymes, which are distributed widely among plants, the CYP88D subfamily appears to be restricted to the Fabaceae and is not found in *Arabidopsis* or rice. The functional identification of CYP88D6 hints at the involvement of other CYP88D subfamily members, CYP88D2 (62 % sequence identity with CYP88D6) and CYP88D3 (79 % identity) of *Medicago truncatula*, and CYP88D4 (71 % identity) and CYP88D5 (72 % identity) of *Lotus japonicus*, in the biosynthesis of triterpenoid saponins with unique reaction specificity.

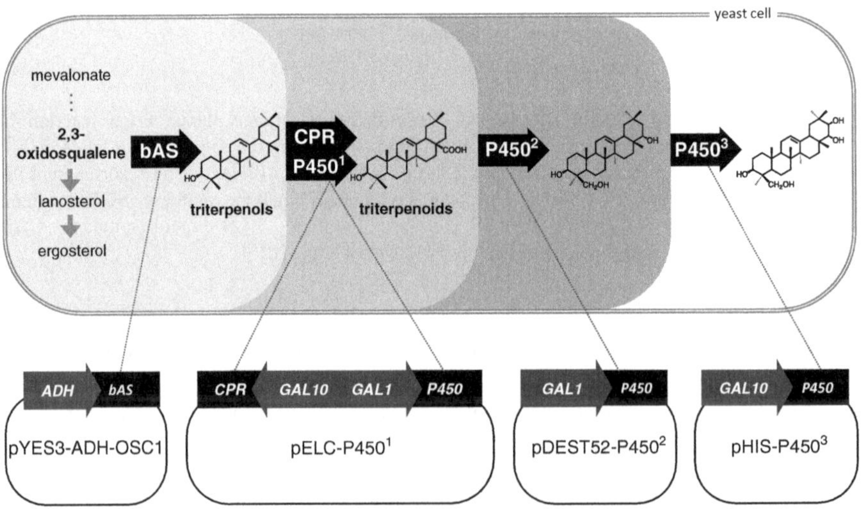

Fig. 7.2 Design of the yeast expression system. (Original figure is from Fukushima et al. 2013b)

7.4 CYP72A154 Is Responsible for C-30 Oxidation in the Biosynthesis of Glycyrrhizin

Among the P450 candidates isolated from *G. uralensis* that did not show β-amyrin-oxidizing activity in in vitro enzyme activity assays, we tested their activity against three additional substrates located between β-amyrin and glycyrrhetinic acid. This approach successfully identified a second relevant P450 (CYP72A154), which is responsible for C-30 oxidation in the glycyrrhizin pathway (Seki et al. 2011). We generated *bAS/CYP88D6/CYP72A154*-expressing yeast and confirmed the production of glycyrrhetinic acid, a glycyrrhizin aglycone. However, in this in vivo yeast production system, presumably a C-29 carboxylated form, an isomer of glycyrrhetinic acid was also detected. To test the further suggestion of the possible involvement of CYP72As in triterpenoind saponin biosynthesis, additional CYP72As identified from *M. truncatula*, a model Fabaceae plant, were tested for potential β-amyrin-oxidizing activity. Among all tested eight homologues, just one showed a β-amyrin-oxidizing activity; CYP72A63 was able to oxidize β-amyrin at C-30 and produced 30-OH-β amyrin and 11-deoxo-glycyrrhetinic acid. Interestingly, C-30-oxidized triterpenoids were not reported from *Medicago truncatula* but were from *Medicago arabica*, suggesting that they could be also present in *M. truncatula* but in a trace amount (Seki et al. 2011). Furthermore, bAS/CYP88D6/CYP72A63-expressing yeast did not produce the C-29 carboxylated form but produced glycyrrhetinic acid, suggesting that combinatorial biosynthesis using P450s from different plant species, namely CYP88D6 from *G. uralensis* and CYP72A63 from *M. truncatula* is suited to producing glycyrrhetinic acid in a yeast culture system (Seki et al. unpublished results).

7.5 β-Amyrin 24-Hydroxylase Committed in the Biosynthesis of Soyasaponins

CYP93E1, a P450 that participates in soyasaponin biosynthesis as a β-amyrin and sophoradiol 24-hydoxylase, was identified from elicitor-inducible P450s of soybean (Shibuya et al. 2006) before we started the project to identify P450 genes in triterpenoid biosynthesis. We isolated a cDNA encoding CYP93E3, a homologue of CYP93E1 from *G. uralensis*, by a homology-based polymerase chain reaction (PCR) cloning method. The amino acid sequence of CYP93E3 showed 82 % identity to that of CYP93E1. Moreover, it was shown that CYP93E3 also possesses β-amyrin 24-hydroxylase activity (Seki et al. 2008). P450s with more than 85 % amino acid sequence identity with CYP93E3 are identified also in *M. truncatula* (CYP93E2) and *Lotus japonicus* (Fukushima et al. 2011). These observations suggest that, in legumes, the P450 enzymes involved in saponin biosynthesis are recruited in at least three very distant CYP families, CYP93 (CYP71 clan) of the A-type P450 and CYP88 (CYP85 clan) and CYP72 (CYP72 clan) of the non-A type P450s. Plant P450s are generally assigned to two major clades: A-type and non-A-type (Durst and Nelson 1995). This example clearly shows that P450s involved in the same pathway or that catalyze similar reactions are not always phylogenetically related.

7.6 CYP716As: Multifunctional Oxidases in Triterpenoid Biosynthesis

As shown in the previous section, P450s so far identified are classified into different subgroups, namely, CYP93E, CYP88D, and CYP72. For this reason, based on homology search strategy, it is difficult to predict whether other types of P450 subgroups for triterpenoid biosynthesis exist. "Gene coexpression analysis" is based on the following principle: genes whose expression patterns are strongly correlated with each other and are expected to be involved in the same biological processes. For example, enzymes involved in the glycyrrhetinic acid biosynthetic pathway bAS, CYP88D6, and CYP72A154 showed similar gene expression patterns in planta. Using the gene coexpression analysis tool of *Medicago truncatula*, we found that the expression of a member of a new P450 subfamily, *CYP716A12*, was highly correlated with that of *bAS*. Through a yeast expression system, CYP716A12 was characterized as a β-amyrin 28-hydroxylase, able to modify β-amyrin to oleanolic acid through erythrodiol. Similarly, CYP716A12 was evaluated for its potential α-amyrin- and lupeol-oxidizing activities, being able to modify them to ursolic acid (through uvaol) and betulinic acid (through betulin), respectively. Therefore, CYP716A12 was characterized as a multifunctional enzyme having β-amyrin 28-hydroxylase, α-amyrin 28-hydroxylase, and lupeol 28-hydroxylase activities (Fukushima et al. 2011). Moreover, CYP716A12 homologues were also found in beet, coffee, grape, and olive. Testing the enzymatic activity of CYP716A12 homologues, these presented β-amyrin 28-hydroxylase, α-amyrin 28-hydroxylase, and lupeol 28-hydroxylase activities, respectively (Fukushima et al. unpublished results).

7.7 Combinatorial Biosynthesis

Combinatorial biosynthesis is a method that establishes novel enzyme–substrate combinations in vivo. Its ultimate goals are to construct de novo pathways by assembling enzymes and to produce novel chemical compounds that possess improved biological activities. As described in the sect. 7.4, among P450s in CYP72A subfamilies in *M. truncatula*, CYP72A63 alone oxidized the C-30 position of β-amyrin. On the other hand, based on coexpression analysis with bAS, we could identify the function of CYP716A12, which showed three-step oxidation activity in the C-28 position of β-amylin. We therefore hypothesized that these nonfunctional CYP72As were involved in the modification of oxidized forms of β-amyrin. We used the *M. truncatula* gene coexpression tool to determine the correlation between their gene expression and those of *CYP716A12* and *CYP93E2* that were encoding β-amyrin oxidases. As a result, *CYP72A68v2* was coexpressed with *CYP716A12*, and *CYP72A61v2* was with CYP93E2, respectively. Next, we generated *bAS/CYP716A12/CYP72A68v2* and *bAS/CYP93E2/CYP72A61v2*-expressing transgenic yeast and analyzed the product. Gypsogenic acid and soyasapogenol B were produced, respectively (Fukushima et al. 2013a). Furthermore, P450s that seemed not to coexpress in *M. truncatula* were combinatorially expressed in the transgenic yeast (*bAS/CYP72A63/CYP93E2* and *bAS/CYP72A63/CYP716A12*). The yeast strain produced rare triterpenoids that are not found in this legume, indicating the ability of combinatorial synthesis in diversifying triterpenoids with potential new biological activities, which leads to the discovery of novel drugs (Fukushima et al. 2013a).

7.8 Prospects

Medicago truncatula sapogenins have different functional groups at C-2, C-3, C-16, C-21, C-22, C-23, C-24, and C-28 of β-amyrin. We identified enzymes involved in C-22, C-23, C-24, and C-28 modifications. Enzymes involved in C-2, C-16, and C-21 modifications remain unidentified. *CYP89A2* expression correlated highly with *bAS* expression, making it potentially involved in these modifications. Other potential candidates are the CYP72A subfamily members, because of all the P450s identified, these enzymes had the widest range of catalytic activity. Improvements to our current expression system will help in identifying even more of these enzymes.

The P450 enzymes identified in triterpenoid biosynthesis phylogenetically belong to distinct and distant clans. CYP93E and CYP705A belong to clan CYP71 (A-type); CYP72A belongs to clan CYP72; CYP88D, CYP716A, and CYP708 belong to clan CYP85; and CYP51H (Kunii et al. 2012) belongs to clan CYP51 (non-A type; Bak et al. 2011). This low degree of relatedness supports the hypothesis that enzyme recruitment for triterpenoid biosynthesis occurred multiple times during evolution and that it is not restricted to a specific origin (Augustin et al. 2011). Convergent evolution in this case could explain the presence of the same specialized metabolites in polyphyletic plant lineages but also explains how selection to achieve similar physiological functions in different lineages can lead, paradoxically, to greater chemical diversity in the plant kingdom.

The strategy combining a homology-based approach, gene coexpression analysis, and combinatorial biosynthesis with heterologous expression in yeast was successful in identifying enzymes involved in triterpenoid biosynthesis and also in generating natural and rare triterpenoids that do not accumulate in planta.

This strategy opens the possibility to construct a natural-unnatural triterpenoid library. To achieve the ultimate goals of combinatorial biosynthesis, the next steps will be to increase product yields as well as to diversify triterpenoids into novel synthetic entities with improved biological activities by combining enzymes from different sources.

References

Augustin JM, Kuzina V, Andersen SB, Bak S (2011) Molecular activities, biosynthesis and evolution of triterpenoid saponins. Phytochemistry 72:435–457

Bak S, Beisson F, Bishop G, Hamberger B, Hofer R, Paquette S et al (2011) Cytochromes p450. Arabidopsis Book 9:e0144

Davidson SE, Elliott RC, Helliwell CA, Poole AT, Reid JB (2003) The pea gene NA encodes *ent*-kaurenoic acid oxidase. Plant Physiol 131:335–344

Durst F, Nelson DR (1995) Diversity and evolution of plant P450 and P450 reductases. Drug Metab Drug Interact 12:189–206

Fukushima EO, Seki H, Ohyama K, Ono E, Umemoto N, Mizutani M, Saito K, Muranaka T (2011) CYP716A subfamily members are multifunctional oxidases in triterpenoid biosynthesis. Plant Cell Physiol 52:2050–2061

Fukushima EO, Seki H, Sawai S, Suzuki M, Ohyama K, Saito K, Muranaka T (2013a) Combinatorial biosynthesis of legume natural and rare triterpenoids in engineered yeast. Plant Cell Physiol 54:740–749

Fukushima EO, Seki H, Muranaka T (2013b) Metabolic diversity in plant terpenoids and their combinatorial biosynthesis. Seibutsu-kogaku 91:337–341 (in Japanese)

Haralampidis K, Trojanowska M, Osbourn A (2002) Biosynthesis of triterpenoid saponins in plants. Adv Biochem Eng Biotechnol 75:32–49

Hayashi H, Huang P, Kirakosyan A, Inoue K, Hiraoka N, Ikeshiro Y, Kushiro T, Shibuya M, Ebizuka Y (2001) Cloning and characterization of a cDNA encoding β-amyrin synthase involved in glycyrrhizin and soyasaponin biosyntheses in licorice. Biol Pharm Bull 24:912–916

Helliwell CA, Chandler PM, Poole A, Dennis ES, Peacock WJ (2001) The CYP88A cytochrome P450, *ent*-kaurenoic acid oxidase, catalyzes three steps of the gibberellin biosynthesis pathway. Proc Natl Acad Sci USA 98:2065–2070

Huhman DV, Berhow MA, Sumner LW (2005) Quantification of saponins in aerial and subterranean tissues of *Medicago truncatula*. J Agric Food Chem 53:1914–1920

Kunii M, Kitahama Y, Fukushima EO, Seki H, Muranaka T, Yoshida Y, Aoyama Y (2012) β-Amyrin oxidation by oat CYP51H10 expressed heterologously in yeast cells: the first example of CYP51-dependent metabolism other than the 14-demethylation of sterol precursors. Biol Pharm Bull 35:801–804

Matsui S, Matsumoto H, Sonoda Y, Ando K, Aizu-Yokota E, Sato T, Kasahara T (2004) Glycyrrhizin and related compounds down-regulate production of inflammatory chemokines IL-8 and eotaxin 1 in a human lung fibroblast cell line. Int Immunopharmacol 4:1633–1644

Seki H, Ohyama K, Sawai S, Mizutani M, Ohnishi T, Sudo H, Akashi T, Aoki T, Saito K, Muranaka T (2008) Licorice beta-amyrin 11-oxidase, a cytochrome P450 with a key role in the biosynthesis of the triterpene sweetener glycyrrhizin. Proc Natl Acad Sci USA 105:14204–14209

Seki H, Sawai S, Ohyama K, Mizutani M, Ohnishi T, Sudo H, Fukushima EO, Akashi T, Aoki T, Saito K, Muranaka T (2011) Triterpene functional genomics in licorice for identification of CYP72A154 involved in the biosynthesis of glycyrrhizin. Plant Cell 23:4112–4123

Shibata S (2000) A drug over the millennia: pharmacognosy, chemistry, and pharmacology of licorice. Yakugaku Zasshi 120:849–862

Shibuya M, Hoshino M, Katsube Y, Hayashi H, Kushiro T, Ebizuka Y (2006) Identification of β-amyrin and sophoradiol 24-hydroxylase by expresssed sequence tag mining and functional expression assay. FEBS J 273:948–959

Sparg SG, Light ME, van Staden J (2004) Biological activities and distribution of plant saponins. J Ethnopharmacol 94:219–243

Sudo H, Seki H, Sakurai N, Suzuki H, Shibata D, Toyoda A, Totoki Y, Sakaki Y, Iida O, Shibata T, Kojoma M, Muranaka T, Saito K (2009) Expressed sequence tags from rhizomes of *Glycyrrhiza uralensis*. Plant Biotechnol 26:105–108

Suzuki H, Achnine L, Xu R, Matsuda SP, Dixon RA (2002) A genomics approach to the early stages of triterpene saponin biosynthesis in *Medicago truncatula*. Plant J 32:1033–1048

Chapter 8
Mammalian and Bacterial Cytochromes P450 Involved in Steroid Hydroxylation: Regulation of Catalysis and Selectivity, and Potential Applications

Rita Bernhardt

Abstract In this review, recent results concerning the function and potential applications of mammalian steroid hydroxylases, which catalyze the important tailoring of steroid molecules, are discussed. A better understanding of the mechanism and modulation of these enzymes opens up new perspectives and innovative possibilities for the treatment of diseases caused by misfunction of the steroidogenic enzymes such as overproduction of aldosterone leading to hypertension and congestive heart failure. In this chapter, special attention is given to the role of protein–protein interactions on the activity of mitochondrial steroid hydroxylase systems. In addition, the role of steroids themselves as important drugs is considered. The progress in recombinant protein expression and in genome sequencing (leading to the identification of novel cytochrome P450 systems) as well as the application of engineering of mammalian and bacterial steroid hydroxylases opens up a tremendous reservoir of possibilities for the application of the corresponding strains and enzymes for the sustainable production of steroidal drugs and products. It can be expected that the extensive use of methods of both enzyme and strain engineering will further promote and increase the application of steroid hydroxylases in biotechnological processes.

Keywords Adrenodoxin • Biotechnological application • Cytochrome P450 • Electron transfer • Steroid hormone • Steroid hydroxylase

---

R. Bernhardt (✉)
Saarland University, Saarbrücken, Germany
e-mail: ritabern@mx.uni-saarland.de

H. Yamazaki (ed.), *Fifty Years of Cytochrome P450 Research*,
DOI 10.1007/978-4-431-54992-5_8, © Springer Japan 2014

8.1 Introduction

The human cytochrome P450 (CYP) superfamily comprises 57 members that by sequence homology can be grouped into 18 families and 43 subfamilies (http://drnelson.uthsc.edu/CytochromeP450.html). However, they may also be categorized according to their subcellular localization, their electron-transport chains, or their biological functions (Guengerich 2005; Hannemann et al. 2007). Regarding the last aspect, a distinction has often been made between those CYPs that have physiological substrates and those which do not, but act as a protective system against foreign compounds. However, it is becoming increasingly clear that this differentiation is not a strict one (Omura 2013). A prominent group of CYPs that are known to fulfill important physiological functions are the steroid hydroxylases. In humans, there are six of these enzymes: CYP11A1 (cytochrome P450scc, side-chain cleavage enzyme), CYP17A1 (CYP17, cytochrome P45017α, steroid 17α-hydroxylase and 17, 20-lyase), CYP21A2 (CYP21, cytochrome P450c21, steroid 21-hydroxylase), CYP11B1 (cytochrome P45011β, steroid 11β-hydroxylase), CYP11B2 (cytochrome P450aldo, aldosterone synthase), and CYP19A1 (cytochrome P450aro, aromatase) (Bernhardt and Waterman 2007; Bureik et al. 2002a; Lisurek and Bernhardt 2004). It has to be mentioned here that it was Tsuneo Omura who not only discovered the cytochromes P450 50 years ago (Omura and Sato 1962) but also was the first to isolate CYP11B1 (Omura et al. 1966) and to clone bovine CYP11A1 as well as CYP11B1 (Morohashi et al. 1984, 1987).

The physiological substrates or products of steroid hydroxylases, the steroid hormones, constitute a class of very important molecules, which are derived from cholesterol and involved in the regulation of essential processes in mammals. They can be divided into glucocorticoids—regulating the sugar metabolism and stress response of the body, mineralocorticoids—which are responsible for the salt and water household, thus affecting the blood pressure, and sexual hormones—determining the sexual characteristics. Dysregulation of steroid hormone biosynthesis can lead to severe implications in metabolic regulation and thus pathological changes. Deficiencies in steroid hormone biosynthesis were described to cause congenital adrenal hyperplasia as well as salt waste, reduced stress response (CYP21A2 and CYP11B1 deficiency), or hypotension (CYP11B2 deficiency) (Bureik et al. 2002a; Lisurek and Bernhardt 2004). In contrast, overproduction of aldosterone can cause severe hypertension and congestive heart failure (for reviews, see Bernhardt and Waterman 2007; Funder 2007). Therefore, investigation of the steroid hormone biosynthesis and its regulation on different levels is needed to fully understand these complex processes and to be able to apply the obtained knowledge for the development of rational treatment strategies (Hakki and Bernhardt 2006; Schuster and Bernhardt 2007).

The biosynthesis of steroid hormones is realized in two subcellular compartments: (1) the endoplasmic reticulum (ER) and (2) the mitochondria. Being external monooxygenases, CYPs need electrons for oxygen activation and substrate

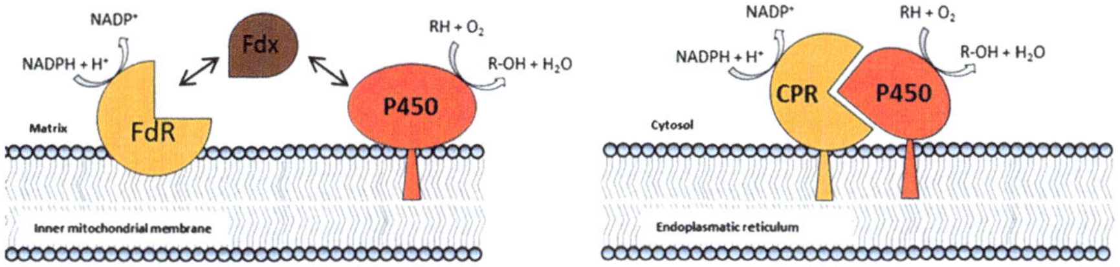

Fig. 8.1 Schematic organization of mitochondrial and microsomal cytochrome P450 systems. *P450* cytochrome P450, *FdR* ferredoxin reductase, *Fdx* ferredoxin, *CPR* NADPH-dependent cytochrome P450 reductase. *Left side*: class 1 CYP system; *right side*: class 2 system

hydroxylation. While the cytochrome P450 systems present in the ER belong to class 2 CYPs, those of the mitochondria are class 1 systems (Hannemann et al. 2007). Class 1 systems receive the necessary electrons from NADPH via the action of a NADPH-dependent ferredoxin reductase called adrenodoxin reductase (AdR) and a ferredoxin, called adrenodoxin (Adx). The mitochondrial CYPs are bound to the inner mitochondrial membrane, whereas AdR is associated to it, and Adx is soluble in the matrix. In contrast, class 2 CYPs are bound at the cytoplasmic side of the ER and acquire their electrons from NADPH via a NADPH-dependent cytochrome P450 reductase (CPR) also bound there (Fig. 8.1).

The role of redox partners in modulating P450-dependent substrate conversions is often underestimated, one of the reasons being that some of the steroid-converting CYPs have only very recently been expressed with sufficient yields in bacteria to provide necessary amounts for in vitro studies (Hobler et al. 2012; Kagawa 2011; Zollner et al. 2008).

8.2 Regulation of Steroid Hormone Production in Mitochondrial P450 Systems by Protein–Protein Interactions

While the different levels of regulation of the steroid hormone biosynthesis have been described in detail in a recent review (Bernhardt and Waterman 2007), I focus in this review on the regulation of mitochondrial steroid hormone biosynthesis on the level of protein–protein interactions. Adx plays a key role as electron-transfer protein in mitochondrial P450 systems catalyzing the initial step of steroid hormone biosynthesis, the side-chain cleavage of cholesterol to pregnenolone, as well as the formation of cortisol and aldosterone. The interactions between Adx and AdR as well as Adx and CYP11A1 have been investigated in great detail during the past years. In contrast, the interactions of Adx with CYP11B1 and especially with CYP11B2 have only recently been started to be investigated, because heterologous expression of CYP11B proteins in sufficient amounts turned out to be very tedious and was obtained only recently (Hobler et al. 2012; Zollner et al. 2008).

Four questions are in the focus of our studies: (1) which are the residues of Adx besides the negatively charged ones involved in protein–protein interactions; (2) how does the presence of additional proteins affect CYP11B1 and CYP11B2 function; (3) is it possible to improve CYP11/Adx interactions, electron transfer and corresponding activities; (4) will changes of the Adx charge caused by phosphorylation affect substrate conversion.

Using site-directed mutagenesis it has been demonstrated that in addition to the well-described influence of negatively charged residues of Adx on redox partner interactions (Ewen et al. 2010; Ewen et al. 2012; Grinberg et al. 2000; Vickery 1997), also residues 109–128 of Adx (Uhlmann et al. 1994), tyrosine 82 (Beckert et al. 1994), as well as the loop region covering the iron–sulfur cluster (Hannemann et al. 2001; Zollner et al. 2002), are significantly affecting the recognition, interaction, and electron transfer between AdR and Adx and/or Adx and CYP11A1 or CYP11B1. Surprisingly, the dipole moment of Adx does not affect these interactions, although Adx as well as AdR and CYP11A1 show significant dipole moments and complementary distributions of charges (Hannemann et al. 2009).

Another important question is whether proteins that might be associated with a certain mitochondrial CYP could modulate their activity. The most obvious partner is another CYP, for example, CYP11A1 in case of CYP11B1 and CYP11B2. Concerning the effect of an additional CYP on the activity of a present one, it can be expected that at least during redox equivalent limitation a competition for electrons can occur. Such competitions could take place in the *zona glomerulosa*, where CYP11A1 and CYP11B2 are coexpressed, as well as in the *zona fasciculata/reticularis*, where CYP11A1 and CYP11B1 are colocalized. In fact, it was demonstrated that the activity of bovine CYP11B1 as well as human CYP11B1 and CYP11B2 decreased when CYP11A1 was coexpressed in COS-1 cells (Cao and Bernhardt 1999a). When the amount of Adx was increased, however, this decrease was overcome. Interestingly, in the case of bovine CYP11B1, a change in the product profile was observed (Cao and Bernhardt 1999a; Ikushiro et al. 1992), although this was not the case with the human counterparts, for example, when human CYP11A1 was coexpressed with human CYP11B1 or CYP11B2, indicating species-specific differences in modulating the product profile (Cao and Bernhardt 1999a).

A very interesting and challenging question is whether it would be possible to improve Adx–redox partner interactions and the following electron transfer. If this question can be solved, it does not only provide a deeper insight into the reaction mechanism of mitochondrial CYPs but also allows an application of the obtained knowledge for improving biotechnological processes with these steroid hydroxylases. It was shown very recently that the choice of the electron transfer partner in a recombinant host significantly influences the activity of human microsomal CYPs (Neunzig et al. 2013). Moreover, as indicated by the aforementioned fact that a higher expression of Adx would help to overcome competition for reducing equivalents and thus increase CYP11B1 and CYP11B2 activities, it has been investigated whether mutants of Adx can be created, accepting and supplying

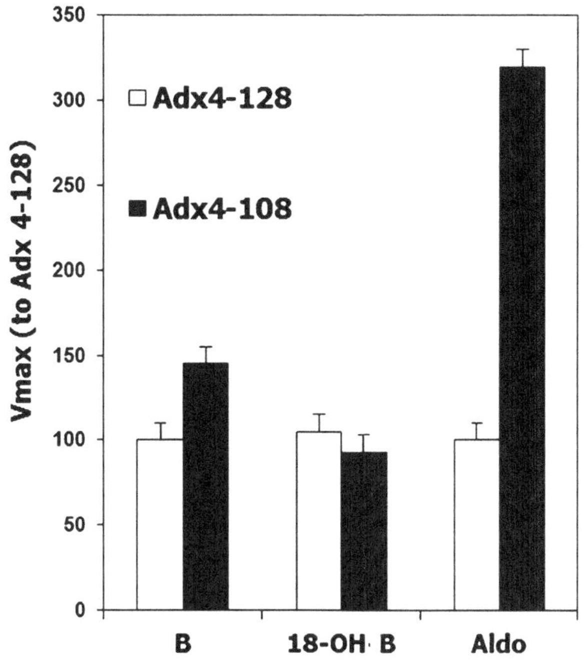

Fig. 8.2 Effect of Adx truncation onto bovine CYP11B1-catalyzed reaction (Cao and Bernhardt 1999b). *B* corticosterone, *18-OH B* 18-OH corticosterone, *Aldo* aldosterone, Adx4-128 serves as wild type

electrons more efficiently. A first and surprising result concerning this question was the observation that truncated Adx, Adx_{4-108}, lead to an increase of bovine CYP11B1-dependent cortisol formation while CYP11A1-dependent pregnenolone production remained unchanged under in vitro conditions (Uhlmann et al. 1994). Interestingly, when studying the influence of the truncated Adx in COS-1 cells on bovine CYP11B1, also a significant increase in 11β-hydroxylation as well as 18-oxidation was found (Fig. 8.2). Moreover, when using human CYP11B1 and CYP11B2 expressed in COS-1 cells, again an increase in the product formation from 11-deoxycorticosterone was observed, leading to increased formation of 18-OH corticosterone and aldosterone, which are both correlated with hypertension (Cao and Bernhardt 1999b). The reason for the increased product formation with CYP11B enzymes is still not clear but could be that the redox potential of the truncated Adx is much lower than that of the wild type (−344 vs. −270 mV) (Grinberg et al. 2000), thus facilitating electron transfer and overall activity. Expression and analysis of bovine CYP11B1 or respective studies using human CYP11B1 and CYP11B2 will be necessary to prove this hypothesis. Whether this observation is of physiological or pathophysiological impact also remains an open question. It seems quite reasonable to assume that under certain conditions certain proteinases might be activated that cause a truncation of Adx, which would then show the aforementioned changes in product formation. In addition to Adx_{4-108}, other mutants of Adx have also been created that displayed improved redox partner interaction, electron transfer, and substrate conversion. In one of these attempts, an "evolutionary" approach has been applied. Based on the observation that the bacterial putidaredoxin (Pdx) is able to transfer electrons significantly faster to its partner CYP101 than Adx to CYP11A1 or CYP11B1, the structures of both proteins have been compared thoroughly. Pdx is about 40 % similar in its primary structure to Adx. It was found that Pdx is shorter than Adx and contains a C-terminal tryptophan (Schiffler and Bernhardt 2003; Schiffler et al. 2001), which was shown to be critical for CYP101A1 recognition (Sligar et al. 1974). Consequently, corresponding changes (truncation of Adx to 112 amino acids and an exchange S112W) were introduced into Adx. This mutant in fact displayed a 3-fold-increased efficiency (k_{cat}/V_{max}) of CYP11B1-dependent corticosterone formation and an even 75-fold increase of CYP11A1-dependent cholesterol conversion. Introduction of a second replacement (Y82F) that had previously been shown to be important for CYP recognition (Beckert et al. 1994) further increased both efficiencies (Schiffler et al. 2001). All in all, this is the highest improvement of a steroid hydroxylase system reported to date and demonstrates the power of rational design on the basis of a detailed knowledge of the proteins of concern. Interestingly, when studying the efficiency of this mutant and a mutant of Adx obtained by directed evolution (Bichet et al. 2007) in the recombinant fission yeast *Schizosaccharomyces pombe*, it turned out that both mutants did not increase cortisol formation compared with the wild-type Adx, which was leading to a 3.4-fold increase in CYP11B1 activity compared with the strain using only autologous redox partners (Hakki et al. 2008). The reason for this observation is not yet clear but could be that other factors such as substrate uptake into yeast became rate limiting. In addition, it was shown that the effect of mutant AdxS112W upon CYP11A1 activity dramatically depends on the ionic strength of the solution (Schiffler et al. 2011). A prominent ionic-strength dependence was also observed for the electron transfer from Adx to CYP11A1 and, to a lesser extent, from AdR to Adx (Schiffler et al. 2004), underlining the role of salt bridges between the redox partners (Grinberg et al. 2000; Vickery 1997). Therefore, it was tempting to investigate whether physiological parameters causing changes of the charges such as accumulation of cations such as polyamines or phosphorylation of one of the participating proteins would also affect the rate of steroid hormone production. It could be shown that, in fact, naturally occurring polyamines (putrescine, spermidine, spermine), which can accumulate under certain circumstances to concentrations up to 6 mM (Schuster and Bernhardt 2011), can affect redox partner interactions, electron transfer, and product formation. Although AdR–Adx interaction was tightened upon polyamine addition, that between Adx and CYP11A1 was weakened, leading to a decrease in cholesterol conversion in the presence of spermine and spermidine but not putrescine (Berwanger et al. 2010). Because polyamine levels can vary significantly and change rapidly during the cell cycle, this opens up a novel possibility for the regulation of steroid hormone biosynthesis and possibly also other CYP reactions and metabolic pathways.

When studying the effect of Adx phosphorylation on steroid hormone biosynthesis, it was found that CK2-dependent phosphorylation of T71, or the replacement T71E mimicking phosphorylation, increases CYP11A1-dependent pregnenolone formation whereas no effect on CYP11B1-dependent 11-deoxycorticosterone conversion was observed (Bureik et al. 2005). Detailed biochemical and biophysical studies revealed that also in this case the changes in the charge affect the binding of CYP11A1 although no significant effect on AdR and CYP11B1 binding was found (Bureik et al. 2005; Zollner et al. 2007). These results demonstrate that overlapping but not identical binding sites occur between Adx/AdR and Adx/CYP11A1 or Adx/CYP11B1 and that charges play a prominent role, mainly in the recognition and interaction between Adx and CYP11A1.

8.3 Impact of Mammalian Steroid Hydroxylases for Biotechnological Applications

Steroids play an important role in biotechnology (Bernhardt 2006; Bureik and Bernhardt 2007; Donova and Egorova 2012; Megges et al. 1990). About 300 steroidal drugs have been approved so far and belong to the second most marketed category of drugs next to antibiotics (Tong and Dong 2009). For many years microbial biotransformations have been used for the generation of novel steroidal drugs (Donova and Egorova 2012). An approach using coexpression of bovine CYP17A1 and CYP21B1 in *Saccharomyces cerevisiae* leading to the conversion of progesterone to 11-deoxycortisol (Sakaki et al. 1991) demonstrated the principal applicability of steroid hydroxylases for biotechnological processes using recombinant microorganisms. In the 1990s Roussel–Uclaf started a visionary project to replace the semi-synthetic process for hydrocortisone production (using several chemical synthetic steps as well as a microorganism) by a biotechnological method using recombinant microorganisms. The project was aimed to use the steroid biosynthesis pathway present in mammalian adrenals for the production of hydrocortisone (cortisol), which is an important precursor of steroidal drugs such as prednisolone. It has to be stated that at that time this was an extraordinary challenge, because nearly all proteins involved are membrane bound and difficult to express in microorganisms. Furthermore, efficient protein–protein interactions (especially between the CYPs and their corresponding redox partners) and transfer of intermediates between different compartments are necessary to create an effective system for biotransformation. Finally, sufficient supply of redox equivalents is necessary as is the import of substrate into the cell as well as into mitochondria. The first challenge solved was the development of a progesterone-producing strain of *S. cerevisiae* that uses ergosterol, the inherent sterol of yeast cells, instead of cholesterol, which does not cross the yeast membrane to a remarkable extent. The recombinant strain carrying a deletion of the CYP61 gene and an introduction of a Δ7 reductase from *Arabidopsis thaliana* was able to produce progesterone from a simple carbon source such as glucose (Duport et al. 1998). As a next step, progesterone had to be modified to allow the production of hydrocortisone by expressing the genes for CYP17A1, CYP21A2, 3β-HSD, and CYP11B1 as well as the redox partners, CPR, Adx, and AdR (see Fig. 8.1). The resultant strain was in fact capable of producing hydrocortisone from glucose (Szczebara et al. 2003). It can be expected that the production of hydrocortisone using the described recombinant strain will open new horizons for the further enzymatic tailoring of the product using additional cytochromes P450 or other enzymes. Furthermore, individual blocks of this system can be used to produce different steroidal compounds from glucose by introducing other tailoring systems, such as enzymes catalyzing 11- or 16-hydroxylation of progesterone. This outstanding achievement was the starting point for the design of recombinant strains using cytochromes P450 for the sustainable biotechnological production of valuable compounds. A recent example based on these experiences and approaches is the successful and efficient semi-synthetic production of artemisinin using an engineered CYP102A1 (Dietrich et al. 2009; Paddon et al. 2013).

Another approach to produce hydrocortisone in a sustainable process is the replacement of the chemically most complicated and tedious step, the 11β-hydroxylation of 11-deoxycortisol to cortisol, by an enzymatic step. In classical biosynthesis, this step can be performed by a microbial reaction using *Curvularia lunata* (Megges et al. 1990). However, after many years of use of this microorganism, the possibility to further improve the strain appears be very limited. Therefore, expression of recombinant human CYP11B1 and CYP11B2 in the fission yeast *Schizosaccharomyces pombe* was used as an alternative route (Bureik et al. 2002b; Dragan et al. 2005; Hakki et al. 2008). Initial results using this expression host are promising, but further improvements are needed for a successful biotechnological application.

8.4 Bacterial Steroid Hydroxylases

From an industrial perspective, hydroxylations of the steroid scaffold in positions 7α, 9α, 11α, 11β, 16α, and 17α are most important for the production of novel hydroxysteroids. As described in the very informative review of (Donova and Egorova 2012), many different microorganisms have been found to catalyze these and other reactions on steroid molecules. However, in most cases the steroid-modifying enzymes have not been characterized so far. A well-known exception is CYP106A2, which was described already in the 1970s to be a potent steroid hydroxylase (Berg et al. 1975, 1976, 1979). In addition, very recently CYP509C12 from *Rhizopus oryzea* was used to efficiently produce 11α-OH progesterone from progesterone using the fission yeast *S. pombe* as recombinant host. Coexpression of the natural redox partner of CYP509C12, Ro CPR, led to an increase of product formation, thus again demonstrating the importance of redox partner recognition and protein–protein interaction for the cytochrome P450 reaction (Petric et al. 2010). Moreover, CYP154C5 from *Nocardia farcinica* IFM 10152 was shown to convert testosterone and similar steroids to 16α-OH steroids (Bracco et al. 2013). When studying the substrate specificity of CYP106A2, it was originally shown that it is able to catalyze a 15β-hydroxylation of 3-oxo-Δ^4-steroids such as progesterone, 11-deoxycorticosterone, testosterone, 11-deoxycortisol, etc. (Berg and Rafter 1981; Lisurek et al. 2008; Simgen et al. 2000; Virus et al. 2006). Later on it was found that, in addition to this, di- and triterpenes also can be highly selectively hydroxylated by CYP106A2 (Bleif et al. 2011, 2012; Schmitz et al. 2012). Taking into account the much higher expression rates of bacterial P450s in a recombinant host as well as the advantage of using soluble cytochromes P450 compared with membrane-bound ones, it can be expected that, on the one hand, bacterial cytochromes P450 are a challenging tool for studying basic questions of steroid hydroxylation and, on the other hand, they are of high biotechnological impact.

8.5 Engineering of Steroid Hydroxylases

When taking a closer look at the progesterone hydroxylation by CYP106A2, it was shown that in addition to the 15β-hydroxylation, hydroxylations at positions 11α, 9α, and 6β took place, although to a much lesser extent (Lisurek et al. 2004) (Fig. 8.3). Because hydroxylations in positions 11, 9, and 6 are of much higher biotechnological impact compared with position 15, and to get a deeper insight into

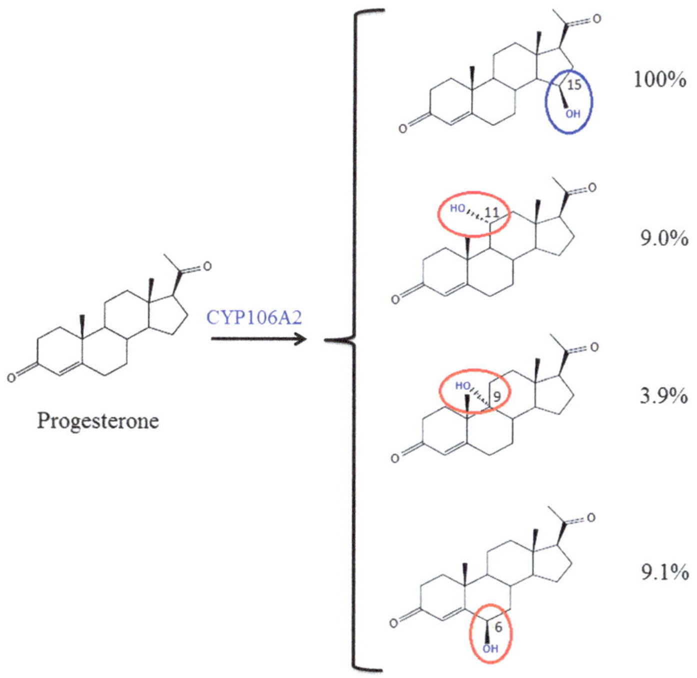

Fig. 8.3 Reactions catalyzed by CYP106A2 on progesterone

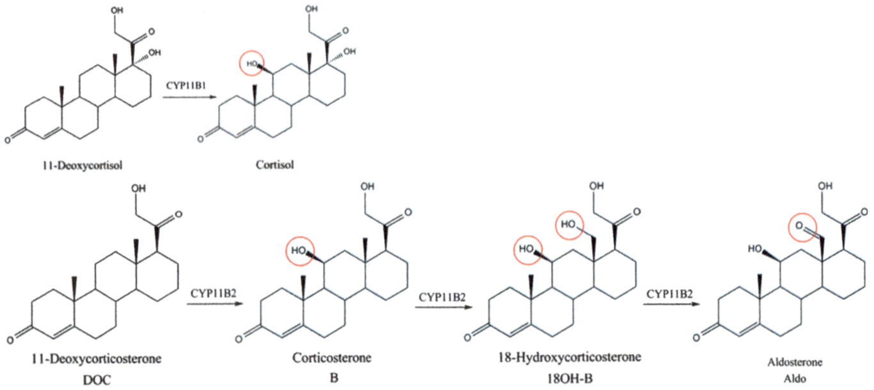

Fig. 8.4 Reactions catalyzed by human CYP11B1 and CYP11B2

Table 8.1 Change of the regioselectivity of hydroxylation of human CYP11B1 and CYP11B2 by site-directed mutagenesis of active site residues

	CYP11B2		CYP11B1	
	WT (%)	L301P/E302D/A320V (%)	WT (%)	V320A (%)
Cortisol	5–10	85	100	100
Aldosterone	100	10	0	20

CYP11B2 can be converted into a cortisol synthase by three-point mutations nearly completely losing its ability to produce aldosterone. CYP11B1 can be converted into an aldosterone synthase by only a one-point mutation

Source: Data taken from Bottner and Bernhardt 1996; Bottner et al. 1998; Bottner et al. 1996

the structural basis of steroid hydroxylation, we set out to rationally change the selectivity of steroid hydroxylation using CYP106A2. Our approach was based on the successful work performed on human CYP11B1 and CYP11B2 (Fig. 8.4), where we were able to change the mineralocorticoid-synthesizing CYP11B2 into a glucocorticoid-producing enzyme and, vice versa, the glucocorticoid-producing CYP11B1 into an aldosterone synthase. Based on a computer model (both proteins share 93 % amino acid identity) showing a "hot spot" at and around the I helix (Bottner and Bernhardt 1996), single, double, and triple mutants were produced, and it was demonstrated that by changing only three amino acids, CYP11B2 could be changed into a cortisol-producing enzyme while at the same time decreasing its ability to produce aldosterone to 10 % compared with the wild-type protein (Bottner et al. 1996; Bottner and Bernhardt 1996). After that we aimed to introduce a new function into CYP11B1: 18-hydroxylation and 18-oxidation ability. As shown in Table 8.1, a single amino acid replacement was able to convert CYP11B1 into an aldosterone synthase (Bottner and Bernhardt 1996; Bottner et al. 1998). In this way we were able to demonstrate that amino acids in the I helix and in the region

Table 8.2 Engineering of CYP106A2 to change the hydroxylation selectivity from C15 to C11 (Nguyen et al. 2012)

	K_m (μM)	V_{max} (nmol/min nmol P450)	k_{cat}/K_m (1/Ms) × 10^3
CYP106A2 WT	203.4 ± 66.9	16.8 ± 2.5	1.4 ± 0.6
A395I	31.6 ± 12.2	79.6 ± 7.8	42.0 ± 20.9
A395W/G397K	54.7 ± 19.8	71.3 ± 8.0	21.7 ± 10.1
T89N/A395I	83.7 ± 26.7	171.7 ± 20.0	34.2 ± 13.6
A106T/A395I	91.0 ± 23.6	300.9 ± 28.2	55.1 ± 16.8
A106T/A395I/R409L	45.7 ± 14.9	415.2 ± 39.2	151.4 ± 60.5

flanking this helix play a pivotal role in determining the regio-selectivity of steroid hydroxylation. This knowledge can be used for the design of novel drugs for the treatment of severe diseases such as hypertension, congestive heart failure, and metabolic syndrome (Bureik et al. 2002a; Ehmer et al. 2002; Hakki and Bernhardt 2006; Roumen et al. 2007). Based on this approach, we were able to change the hydroxylation selectivity of the bacterial steroid hydroxylase, CYP106A2, from the 15- to the 11-position. For this, we first made an alignment of CYP106A2 (performing 15-hydroxylation) and CYP11B1 (supporting 11-hydroxylation) and identified amino acid residues that could be responsible for the selectivity of hydroxylation in CYP106A2 (Lisurek et al. 2008). A prominent difference was found between residues 381–410 of CYP106A2 and the corresponding residues of CYP11B1 located in the substrate recognition site SRS6 (Lisurek et al. 2004) defined by Gotoh (Gotoh and Fujii-Kuriyama 1989). Using this result, five site-directed mutants have been produced changing the original amino acids from CYP106A2 to those from CYP11B1: S394I, A395L, T396R, G397P, and Q398S. As expected, some of the mutants (A395L, G397P) led to a significant increase (fourfold) of 11α-hydroxylation compared with the wild type, thus supporting the usefulness of our approach (Lisurek et al. 2008). To further improve 11α-hydroxylation, we used saturation mutagenesis at positions 395 and 397 to change the selectivity of hydroxylation to 11α (~13,100 transformants were screened). The best mutants, showing an 8.8- and 11.5-fold-higher 11α-hydroxylase activity compared to the wild type, were A395I and A395W/G397K (Table 8.2). Both showed a greater shift in regio-selectivity compared with the parental mutants (Nguyen et al. 2012). To improve the activity of these novel mutants, amino acid replacements were conducted that were shown in previous work to improve the activity of steroid hydroxylases. Forty mutants were analyzed in detail, and it was found that mutants A106T/A395I, A106T/A395I/R409L, and T89N/A395I displayed increased 11α-hydroxylase selectivity and activity compared with the wild type (14.3-, 12.6-, and 11.8-fold increase in selectivity and 39.3-, 108.1-, and 24.4-fold in k_{cat}/K_m) (Table 8.2). The best mutants were applied in a whole-cell biotransformation, and it was shown that the percentage of 15β-hydroxyprogesterone decreased from 50.4 % (wild-type) to 4.8 % (mutant T89N/A395I), whereas the percentage of 11α-hydroxyprogesterone increased from 27.7 % to 80.9 %, demonstrating an impressive alteration in regio-selectivity. This result indicates a nearly complete regio-selectivity change of the progesterone hydroxylation from the D-ring to the C-ring (Nguyen et al. 2012).

There are also other examples of engineering steroid hydroxylation with CYPs. CYP102A1 was changed using directed evolution with iterative saturation mutagenesis to selectively hydroxylate testosterone as well as progesterone (Kille et al. 2011; Rea et al. 2012; Venkataraman et al. 2012). In addition, human (or other mammalian) CYPs that do not belong to the CYP11, CYP17, CYP19, or CYP21 families may still be considered for specific steroid hydroxylation reactions, and, indeed, some isoenzymes of the other 14 human CYP families (most notably CYP3A4, CYP3A5, CYP3A7, CYP2D6) are well known to hydroxylate at least some steroids. Consequently, the improvement of CYP3A and CYP2D6 enzyme activity toward steroidal substrates has also been a subject of investigation (Geier et al. 2013; Neunzig et al. 2012). However, many studies from my laboratory and from many other colleagues indicate, that, generally speaking, successful engineering of human CYPs with respect to activity is much more demanding than that of bacterial CYPs. Interestingly, it was shown that coexpression of a CPR not only influences the reaction rate but, in some cases, also exerts an influence on the product pattern of human CYPs (Neunzig et al. 2013). Moreover, we observed in our studies that the product pattern of CYP106A2 mutants can vary in in vitro and in vivo studies (Nguyen et al. 2012; Zehentgruber et al. 2010). The reason for this observation is not yet clear.

Taken together, the discussed results clearly demonstrate that it is not only possible to interconvert the regio-selectivities of closely related human steroid hydroxylases such as CYP11B1 and CYP11B2, but also that of a more distantly related one (such as the bacterial CYP106A2) when taking into account the knowledge about important regions of the active site determining the substrate specificity and regio-selectivity and by applying a combination of rational design and directed evolution of the protein. Further studies will show whether this approach can be successfully applied to further rationally change the hydroxylation selectivity of CYP106A2. If so, this would be of great biotechnological importance and could be used to design bacterial steroid hydroxylases for the production of various hydroxyl steroids.

8.6 Epilogue

This review article is based on a lecture given on the occasion of the celebration of 50 years of the discovery of P450s by Tsuneo Omura and Ryo Sato. I was truly honored to be invited to this important event and to be able to meet again Tsuneo Omura and all the very productive Japanese and international colleagues and friends. The contribution of Tsuneo Omura to the field cannot be overestimated. And in addition to his extraordinary scientific input, Omura “sensei” always impresses by his modesty and his encyclopedic knowledge in very different fields such as Japanese and Asian (but also European) culture, botany, and literature. I feel, therefore, very fortunate to know him for many years.

Acknowledgments I thank Antje Eiden-Plach for technical support upon finalizing the manuscript, and Martin Litzenburger and Jens Neunzig for help with the figures, as well as Dr. Matthias Bureik, Dr. Frank Hannemann, Simon Janocha, and Dr. Daniela Schmitz for critical reading of the manuscript. The support of the EC INTERREG program to R.B. is acknowledged.

References

Beckert V, Dettmer R, Bernhardt R (1994) Mutations of tyrosine 82 in bovine adrenodoxin that affect binding to cytochromes P45011A1 and P45011B1 but not electron transfer. J Biol Chem 269:2568–2573

Berg A, Rafter JJ (1981) Studies on the substrate specificity and inducibility of cytochrome P-450meg. Biochem J 196:781–786

Berg A, Carlstrom K, Gustafsson JA, Ingelman-Sundberg M (1975) Demonstration of a cytochrome P-450-dependent steroid 15beta-hydroxylase in *Bacillus megaterium*. Biochem Biophys Res Commun 66:1414–1423

Berg A, Gustafsson JA, Ingelman-Sundberg M (1976) Characterization of a cytochrome P-450-dependent steroid hydroxylase system present in *Bacillus megaterium*. J Biol Chem 251:2831–2838

Berg A, Ingelman-Sundberg M, Gustafsson JA (1979) Isolation and characterization of cytochrome P-450meg. Acta Biol Med Ger 38:333–344

Bernhardt R (2006) Cytochromes P450 as versatile biocatalysts. J Biotechnol 124:128–145

Bernhardt R, Waterman MR (2007) Cytochrome P450 and steroid hormone biosynthesis. In: Sigel A, Sigel H, Sigel RKO (eds) The ubiquitous roles of cytochrome P450 proteins. Wiley, Chichester, pp 361–369

Berwanger A, Eyrisch S, Schuster I, Helms V, Bernhardt R (2010) Polyamines: naturally occurring small molecule modulators of electrostatic protein–protein interactions. J Inorg Biochem 104:118–125

Bichet A, Hannemann F, Rekowski M, Bernhardt R (2007) A new application of the yeast two-hybrid system in protein engineering. Protein Eng Des Sel 20:117–123

Bleif S, Hannemann F, Lisurek M, von Kries JP, Zapp J, Dietzen M, Antes I, Bernhardt R (2011) Identification of CYP106A2 as a regioselective allylic bacterial diterpene hydroxylase. Chembiochem 12:576–582

Bleif S, Hannemann F, Zapp J, Hartmann D, Jauch J, Bernhardt R (2012) A new *Bacillus megaterium* whole-cell catalyst for the hydroxylation of the pentacyclic triterpene 11-keto-beta-boswellic acid (KBA) based on a recombinant cytochrome P450 system. Appl Microbiol Biotechnol 93:1135–1146

Bottner B, Bernhardt R (1996) Changed ratios of glucocorticoids/mineralocorticoids caused by point mutations in the putative I-helix regions of CYP11B1 and CYP11B2. Endocr Res 22:455–461

Bottner B, Schrauber H, Bernhardt R (1996) Engineering a mineralocorticoid- to a glucocorticoid-synthesizing cytochrome P450. J Biol Chem 271:8028–8033

Bottner B, Denner K, Bernhardt R (1998) Conferring aldosterone synthesis to human CYP11B1 by replacing key amino acid residues with CYP11B2-specific ones. Eur J Biochem 252:458–466

Bracco P, Janssen DB, Schallmey A (2013) Selective steroid oxyfunctionalisation by CYP154C5, a bacterial cytochrome P450. Microb Cell Fact 12(95):1–11

Bureik M, Bernhardt R (2007) Steroid hydroxylation: microbial steroid biotransformations using cytochrome P450 enzymes. In: Schmid RD, Urlacher VB (eds) Modern biooxidation: enzymes, reactions and applications. Wiley, Weinheim

Bureik M, Lisurek M, Bernhardt R (2002a) The human steroid hydroxylases CYP1B1 and CYP11B2. Biol Chem 383:1537–1551

Bureik M, Schiffler B, Hiraoka Y, Vogel F, Bernhardt R (2002b) Functional expression of human mitochondrial CYP11B2 in fission yeast and identification of a new internal electron transfer protein, etp1. Biochemistry 41:2311–2321

Bureik M, Zollner A, Schuster N, Montenarh M, Bernhardt R (2005) Phosphorylation of bovine adrenodoxin by protein kinase CK2 affects the interaction with its redox partner cytochrome P450scc (CYP11A1). Biochemistry 44:3821–3830

Cao PR, Bernhardt R (1999a) Interaction of CYP11B1 (cytochrome P-45011 beta) with CYP11A1 (cytochrome P-450scc) in COS-1 cells. Eur J Biochem 262:720–726

Cao PR, Bernhardt R (1999b) Modulation of aldosterone biosynthesis by adrenodoxin mutants with different electron transport efficiencies. Eur J Biochem 265:152–159

Dietrich JA, Yoshikuni Y, Fisher KJ, Woolard FX, Ockey D, McPhee DJ, Renninger NS, Chang MC, Baker D, Keasling JD (2009) A novel semi-biosynthetic route for artemisinin production using engineered substrate-promiscuous P450(BM3). ACS Chem Biol 4:261–267

Donova MV, Egorova OV (2012) Microbial steroid transformations: current state and prospects. Appl Microbiol Biotechnol 94:1423–1447

Dragan CA, Zearo S, Hannemann F, Bernhardt R, Bureik M (2005) Efficient conversion of 11-deoxycortisol to cortisol (hydrocortisone) by recombinant fission yeast *Schizosaccharomyces pombe*. FEMS Yeast Res 5:621–625

Duport C, Spagnoli R, Degryse E, Pompon D (1998) Self-sufficient biosynthesis of pregnenolone and progesterone in engineered yeast. Nat Biotechnol 16:186–189

Ehmer PB, Bureik M, Bernhardt R, Muller U, Hartmann RW (2002) Development of a test system for inhibitors of human aldosterone synthase (CYP11B2): screening in fission yeast and evaluation of selectivity in V79 cells. J Steroid Biochem Mol Biol 81:173–179

Ewen KM, Kleser M, Bernhardt R (2010) Adrenodoxin: the archetype of vertebrate-type [2Fe-2S] cluster ferredoxins. Biochim Biophys Acta 1814:111–125

Ewen KM, Ringle M, Bernhardt R (2012) Adrenodoxin–a versatile ferredoxin. IUBMB Life 64:506–512

Funder JW (2007) The role of aldosterone and mineralocorticoid receptors in cardiovascular disease. Am J Cardiovasc Drugs 7:151–157

Geier M, Braun A, Fladischer P, Stepniak P, Rudroff F, Hametner C, Mihovilovic MD, Glieder A (2013) Double site saturation mutagenesis of the human cytochrome P450 2D6 results in regioselective steroid hydroxylation. FEBS J 280:3094–3108

Gotoh O, Fujii-Kuriyama Y (1989) Evolution, structure and gene regulation of cytochrome P-450. In: Ruckpaul K, Rein H (eds) Basis and mechanisms of regulation of cytochrome P-450. Akademie-Verlag, Berlin, pp 195–243

Grinberg AV, Hannemann F, Schiffler B, Muller J, Heinemann U, Bernhardt R (2000) Adrenodoxin: structure, stability, and electron transfer properties. Proteins 40:590–612

Guengerich FP (2005) Human cytochrome P450 enzymes. In: Ortiz de Montellano PR (ed) Cytochrome P450: structure, mechanism and biochemistry. Kluwer Academic/Plenum, New York, pp 377–530

Hakki T, Bernhardt R (2006) CYP17- and CYP11B-dependent steroid hydroxylases as drug development targets. Pharmacol Ther 111:27–52

Hakki T, Zearo S, Dragan CA, Bureik M, Bernhardt R (2008) Coexpression of redox partners increases the hydrocortisone (cortisol) production efficiency in CYP11B1 expressing fission yeast *Schizosaccharomyces pombe*. J Biotechnol 133:351–359

Hannemann F, Rottmann M, Schiffler B, Zapp J, Bernhardt R (2001) The loop region covering the iron-sulfur cluster in bovine adrenodoxin comprises a new interaction site for redox partners. J Biol Chem 276:1369–1375

Hannemann F, Bichet A, Ewen KM, Bernhardt R (2007) Cytochrome P450 systems–biological variations of electron transport chains. Biochim Biophys Acta 1770:330–344

Hannemann F, Guyot A, Zollner A, Muller JJ, Heinemann U, Bernhardt R (2009) The dipole moment of the electron carrier adrenodoxin is not critical for redox partner interaction and electron transfer. J Inorg Biochem 103:997–1004

Hobler A, Kagawa N, Hutter MC, Hartmann MF, Wudy SA, Hannemann F, Bernhardt R (2012) Human aldosterone synthase: recombinant expression in *E. coli* and purification enables a detailed biochemical analysis of the protein on the molecular level. J Steroid Biochem Mol Biol 132:57–65

Ikushiro S, Kominami S, Takemori S (1992) Adrenal P-450scc modulates activity of P-45011 beta in liposomal and mitochondrial membranes. Implication of P-450scc in zone specificity of aldosterone biosynthesis in bovine adrenal. J Biol Chem 267:1464–1469

Kagawa N (2011) Efficient expression of human aromatase (CYP19) in *E. coli*. Methods Mol Biol 705(109-22):109–122

Kille S, Zilly FE, Acevedo JP, Reetz MT (2011) Regio- and stereoselectivity of P450-catalysed hydroxylation of steroids controlled by laboratory evolution. Nat Chem 3:738–743

Lisurek M, Bernhardt R (2004) Modulation of aldosterone and cortisol synthesis on the molecular level. Mol Cell Endocrinol 215:149–159

Lisurek M, Kang MJ, Hartmann RW, Bernhardt R (2004) Identification of monohydroxy progesterones produced by CYP106A2 using comparative HPLC and electrospray ionisation collision-induced dissociation mass spectrometry. Biochem Biophys Res Commun 319:677–682

Lisurek M, Simgen B, Antes I, Bernhardt R (2008) Theoretical and experimental evaluation of a CYP106A2 low homology model and production of mutants with changed activity and selectivity of hydroxylation. Chembiochem 9:1439–1449

Megges R, Müller-Frohne M, Pfeil D, Ruckpaul K (1990) Microbial steroid hydroxylation enzymes in glucocorticoid production. In: Ruckpaul K, Rein H (eds) Molecular mechanisms of adrenal steroidogenesis and aspects of regulation and application. Akademie-Verlag, Berlin, pp 204–243

Morohashi K, Fujii-Kuriyama Y, Okada Y, Sogawa K, Hirose T, Inayama S, Omura T (1984) Molecular cloning and nucleotide sequence of cDNA for mRNA of mitochondrial cytochrome P-450(SCC) of bovine adrenal cortex. Proc Natl Acad Sci USA 81:4647–4651

Morohashi K, Yoshioka H, Gotoh O, Okada Y, Yamamoto K, Miyata T, Sogawa K, Fujii-Kuriyama Y, Omura T (1987) Molecular cloning and nucleotide sequence of DNA of mitochondrial cytochrome P-450(11 beta) of bovine adrenal cortex. J Biochem 102:559–568

Neunzig I, Widjaja M, Dragan CA, Peters FT, Maurer HH, Bureik M (2012) Engineering of human CYP3A enzymes by combination of activating polymorphic variants. Appl Biochem Biotechnol 168:785–796

Neunzig I, Widjaja M, Peters FT, Maurer HH, Hehn A, Bourgaud F, Bureik M (2013) Coexpression of CPR from various origins enhances biotransformation activity of human CYPs in *S. pombe*. Appl Biochem Biotechnol 170:1751–1766

Nguyen KT, Virus C, Gunnewich N, Hannemann F, Bernhardt R (2012) Changing the regioselectivity of a P450 from C15 to C11 hydroxylation of progesterone. Chembiochem 13:1161–1166

Omura T (2013) Contribution of cytochrome P450 to the diversification of eukaryotic organisms. Biotechnol Appl Biochem 60:4–8

Omura T, Sato R (1962) A new cytochrome in liver microsomes. J Biol Chem 237:1375–1376

Omura T, Sanders E, Estabrook RW, Cooper DY, Rosenthal O (1966) Isolation from adrenal cortex of a nonheme iron protein and a flavoprotein functional as a reduced triphosphopyridine nucleotide-cytochrome P-450 reductase. Arch Biochem Biophys 117:660–672

Paddon CJ, Westfall PJ, Pitera DJ, Benjamin K, Fisher K, McPhee D, Leavell MD, Tai A, Main A, Eng D, Polichuk DR, Teoh KH, Reed DW, Treynor T, Lenihan J, Fleck M, Bajad S, Dang G, Dengrove D, Diola D, Dorin G, Ellens KW, Fickes S, Galazzo J, Gaucher SP, Geistlinger T, Henry R, Hepp M, Horning T, Iqbal T, Jiang H, Kizer L, Lieu B, Melis D, Moss N, Regentin R, Secrest S, Tsuruta H, Vazquez R, Westblade LF, Xu L, Yu M, Zhang Y, Zhao L, Lievense J, Covello PS, Keasling JD, Reiling KK, Renninger NS, Newman JD (2013) High-level semi-synthetic production of the potent antimalarial artemisinin. Nature (Lond) 496:528–532

Petric S, Hakki T, Bernhardt R, Zigon D, Cresnar B (2010) Discovery of a steroid 11alpha-hydroxylase from *Rhizopus oryzae* and its biotechnological application. J Biotechnol 150:428–437

Rea V, Kolkman AJ, Vottero E, Stronks EJ, Ampt KA, Honing M, Vermeulen NP, Wijmenga SS, Commandeur JN (2012) Active site substitution A82W improves the regioselectivity of steroid hydroxylation by cytochrome P450 BM3 mutants as rationalized by spin relaxation nuclear magnetic resonance studies. Biochemistry 51:750–760

Roumen L, Sanders MP, Pieterse K, Hilbers PA, Plate R, Custers E, de Gooyer M, Smits JF, Beugels I, Emmen J, Ottenheijm HC, Leysen D, Hermans JJ (2007) Construction of 3D models of the CYP11B family as a tool to predict ligand binding characteristics. J Comput Aided Mol Des 21:455–471

Sakaki T, Akiyoshi-Shibata M, Yabusaki Y, Manabe K, Murakami H, Ohkawa H (1991) Progesterone metabolism in recombinant yeast simultaneously expressing bovine cytochromes P450c17 (CYP17A1) and P450c21 (CYP21B1) and yeast NADPH-P450 oxidoreductase. Pharmacogenetics 1:86–93

Schiffler B, Bernhardt R (2003) Bacterial (CYP101) and mitochondrial P450 systems: how comparable are they? Biochem Biophys Res Commun 312:223–228

Schiffler B, Kiefer M, Wilken A, Hannemann F, Adolph HW, Bernhardt R (2001) The interaction of bovine adrenodoxin with CYP11A1 (cytochrome P450scc) and CYP11B1 (cytochrome P45011beta). Acceleration of reduction and substrate conversion by site-directed mutagenesis of adrenodoxin. J Biol Chem 276:36225–36232

Schiffler B, Zollner A, Bernhardt R (2004) Stripping down the mitochondrial cholesterol hydroxylase system, a kinetics study. J Biol Chem 279:34269–34276

Schiffler B, Zollner A, Bernhardt R (2011) Kinetic and optical biosensor study of adrenodoxin mutant AdxS112W displaying an enhanced interaction towards the cholesterol side chain cleavage enzyme (CYP11A1). Eur Biophys J 40:1275–1282

Schmitz D, Zapp J, Bernhardt R (2012) Hydroxylation of the triterpenoid dipterocarpol with CYP106A2 from *Bacillus megaterium*. FEBS J 279:1663–1674

Schuster I, Bernhardt R (2007) Inhibition of cytochromes p450: existing and new promising therapeutic targets. Drug Metab Rev 39:481–499

Schuster I, Bernhardt R (2011) Interactions of natural polymanines with mammalian proteins. Biomol Concepts 2:79–94

Simgen B, Contzen J, Schwarzer R, Bernhardt R, Jung C (2000) Substrate binding to 15beta-hydroxylase (CYP106A2) probed by FT infrared spectroscopic studies of the iron ligand CO stretch vibration. Biochem Biophys Res Commun 269:737–742

Sligar SG, Debrunner PG, Lipscomb JD, Namtvedt MJ, Gunsalus IC (1974) A role of the putidaredoxin COOH-terminus in P-450cam (cytochrome *m*) hydroxylations. Proc Natl Acad Sci USA 71:3906–3910

Szczebara FM, Chandelier C, Villeret C, Masurel A, Bourot S, Duport C, Blanchard S, Groisillier A, Testet E, Costaglioli P, Cauet G, Degryse E, Balbuena D, Winter J, Achstetter T, Spagnoli R, Pompon D, Dumas B (2003) Total biosynthesis of hydrocortisone from a simple carbon source in yeast. Nat Biotechnol 21:143–149

Tong WY, Dong X (2009) Microbial biotransformation: recent developments on steroid drugs. Recent Pat Biotechnol 3:141–153

Uhlmann H, Kraft R, Bernhardt R (1994) C-terminal region of adrenodoxin affects its structural integrity and determines differences in its electron transfer function to cytochrome P-450. J Biol Chem 269:22557–22564

Venkataraman H, Beer SB, Bergen LA, Essen N, Geerke DP, Vermeulen NP, Commandeur JN (2012) A single active site mutation inverts stereoselectivity of 16-hydroxylation of testosterone catalyzed by engineered cytochrome P450 BM3. Chembiochem 13:520–523

Vickery LE (1997) Molecular recognition and electron transfer in mitochondrial steroid hydroxylase systems. Steroids 62:124–127

Virus C, Lisurek M, Simgen B, Hannemann F, Bernhardt R (2006) Function and engineering of the 15beta-hydroxylase CYP106A2. Biochem Soc Trans 34:1215–1218

Zehentgruber D, Hannemann F, Bleif S, Bernhardt R, Lutz S (2010) Towards preparative scale steroid hydroxylation with cytochrome P450 monooxygenase CYP106A2. Chembiochem 11:713–721

Zollner A, Hannemann F, Lisurek M, Bernhardt R (2002) Deletions in the loop surrounding the iron–sulfur cluster of adrenodoxin severely affect the interactions with its native redox partners adrenodoxin reductase and cytochrome P450(scc) (CYP11A1). J Inorg Biochem 91:644–654

Zollner A, Pasquinelli MA, Bernhardt R, Beratan DN (2007) Protein phosphorylation and intermolecular electron transfer: a joint experimental and computational study of a hormone biosynthesis pathway. J Am Chem Soc 129:4206–4216

Zollner A, Kagawa N, Waterman MR, Nonaka Y, Takio K, Shiro Y, Hannemann F, Bernhardt R (2008) Purification and functional characterization of human 11beta hydroxylase expressed in *Escherichia coli*. FEBS J 275:799–810

Chapter 9
Neurosteroids: Regional Steroidogenesis

Takeshi Yamazaki and Yasuhiro Ishihara

Abstract An enzymatic activity of cytochrome P450 was first identified as a component of adrenal steroidogenesis by Estabrook and his coworkers in 1963. The adrenal glands and gonads synthesize large amounts of steroid hormones and secrete them for systemic circulation. Two decades later, regional steroidogenesis in the brain was well established by Baulieu and his coworkers. Brain-synthesized steroids were termed neurosteroids. Recently, regional steroidogenesis has been reported in various organs, including the heart, lung, pancreas, and intestine. The locally synthesized steroids are present at low levels and may function as paracrine modulators.

Neurosteroids exert various important modulatory effects on brain functions and diseases. This chapter presents evidence for the regulation of neurosteroidogenesis and the protective functions of brain-synthesized steroids in the rat hippocampus.

(1) Rat hippocampal cultured slices synthesized testosterone and 17β-estradiol as neurosteroids. Retinoic acid stimulated the synthesis of sex steroids via transcriptional activation of cytochromes P450(17α) and P450arom. (2) The housing conditions of immature rats strongly affected the hippocampal steroid synthesis by alteration of mRNA levels for steroidogenic enzymes. (3) De novo synthesized estradiol self-protected rat hippocampal cultured slices from methylmercury toxicity by an estrogen receptor-mediated mechanism. (4) The neuroprotective effect of progesterone against tributyltin in the rat hippocampal slices was not mediated by a progesterone receptor but by allopregnanolone, which was synthesized from progesterone.

Keywords 17β-Estradiol • Allopregnanolone • Environmental enrichment • Hippocampus • Methylmercury • Neurosteroids • Progesterone • Social isolation • Testosterone • Tributyltin

---

T. Yamazaki (✉) • Y. Ishihara
Graduate School of Integrated Arts and Sciences, Hiroshima University, Hiroshima, Japan
e-mail: takey@hiroshima-u.ac.jp

H. Yamazaki (ed.), *Fifty Years of Cytochrome P450 Research*,
DOI 10.1007/978-4-431-54992-5_9,

9.1 Regional Steroidogenesis

Classically, steroid hormones have been thought to be secreted exclusively by the endocrine organs, such as gonads, adrenal glands, and placenta. These organs possess higher levels of steroidogenic cytochrome P450, electron-transfer proteins for the cytochromes, and hydroxysteroid dehydrogenases to synthesize large amounts of active steroid hormones from cholesterol and secrete them for systemic circulation (Norman et al. 1997). The first identified physiological function of cytochrome P450 was the C21-hydroxylation of steroids for corticoid synthesis in adrenal glands (Estabrook et al. 1963).

De novo synthesis of steroid hormones from cholesterol in extra-endocrine organs was reported 1969 in the rat submandibular glands, which convert acetate into cholesterol, pregnenolone, and dehydroepiandrosterone (DHEA) (Rosner et al. 1969). In the 1980s, regional steroidogenesis in the brain was well established by Baulieu and coworkers (Baulieu 1998). Recent evidence has shown that various steroid hormones can also be locally synthesized in many other tissues. These organs contain relatively small amounts of steroidogenic or steroid-converting enzymes. Because the locally synthesized steroids, other than those from gonads, adrenal glands, and the placenta, are present at low levels, most of the local steroids exert autocrine and paracrine functions in their tissues of origin. Some examples of regional steroidogenesis are shown here.

9.1.1 Regional Corticoid Synthesis

Corticoids, glucocorticoids and mineral corticoids, are synthesized in the adrenals and secreted for systemic circulation. Corticoid synthesis in adrenals is regulated by the hypothalamic–pituitary–adrenal axis or the renin-angiotensin system (Norman et al. 1997). Hypothalamic release of corticotropin-releasing hormone triggers pituitary release of adrenocorticotropic hormone, which stimulates glucocorticoid production by the zona fasciculata of the adrenals and an increase in circulating glucocorticoids. The concentration of a circulating mineral corticoid, aldosterone, was responsive to the renin-angiotensin system. The kidneys release renin, which converts angiotensinogen to angiotensin I. Angiotensin I is then cleaved by angiotensin-converting enzyme to produce active angiotensin II, which stimulates mineralocorticoid production in the zona glomerulosa of the adrenals.

Local synthesis of corticosteroids in various organs other than adrenals was recently reviewed (Taves et al. 2011). Corticosteroidogenic tissues are primary lymphoid organs, intestine, skin, brain, heart, lung, retina, and kidney. These tissues possess steroidogenic enzymes and high local corticosteroid levels, even after adrenalectomy. Local synthesis of steroids is regulated by locally expressed mediators of the hypothalamic–pituitary–adrenal axis or of the renin-angiotensin system. For example, the local control pathways in skin might form miniature analogues of

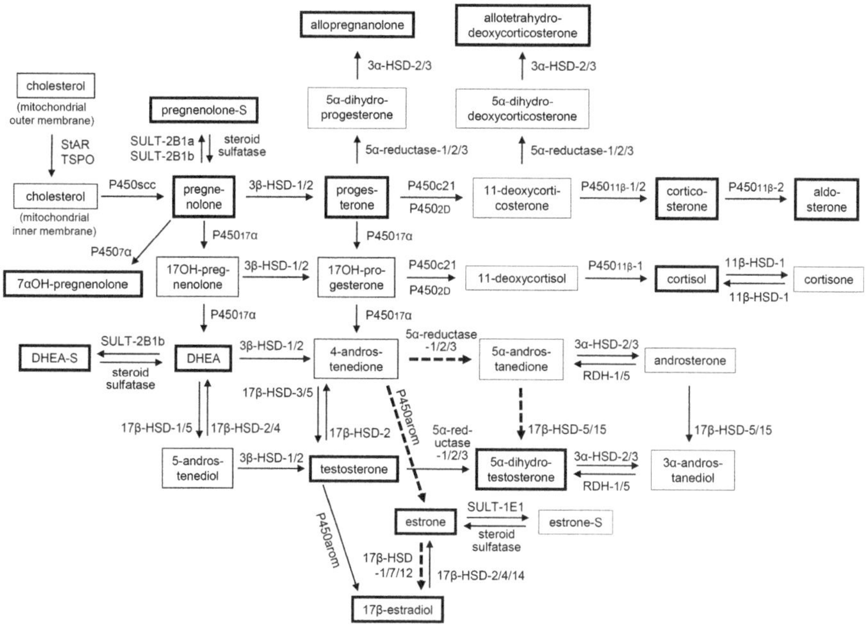

Fig. 9.1 Pathway of neurosteroidogenesis. Typical neuroactive steroids are shown in *rectangles* with *thick lines*. *Dashed arrows* indicate back-door pathway in humans. *estrone-S* estrone sulfate; P4502D, cytochrome P4502D4, CYP2D4, in rat and cytochrome P4502D6, CYP2D6, in human; *RDH* retinol dehydrogenase

the pathways that stimulate adrenal corticosteroid production (Slominski et al. 2005). Regional corticoid synthesis regulates immune cell activation, blood volume, and blood pressure. The physiological importance of regional corticoid synthesis has been shown by inhibition of local steroidogenic activity, which produced significant effects even in adrenal-intact subjects. The locally synthesized corticosteroids exert steroid action in specific regions, whereas systemic corticoids from the adrenal glands coordinate multiple organ systems.

In addition to de novo synthesis of corticoids from cholesterol, some tissues, such as bones, joints, liver, muscle, and fat, can convert circulating inactive glucocorticoid metabolites (e.g., cortisone) into active glucocorticoids by 11β-hydroxysteroid dehydrogenase (11β-HSD) type I activity. Circulating precursors for corticoids can also be converted to glucocorticoids or mineralocorticoids by the activity of regional steroidogenic cytochromes or 3β-hydroxysteroid dehydrogenase/Δ^5-Δ^4 isomerase (3β-HSD) (Fig. 9.1). Tissue levels of corticoids in target organs need not (and very often do not) reflect the concentration of the steroids in systemic circulation (Slominski et al. 2005).

9.1.2 Regional Synthesis of Sex Steroids

Sex steroids, androgens and estrogens, in peripheral organs can be derived from gonads via systemic circulation and also from de novo synthesis at the peripheral organs. In humans, a large proportion of androgens and estrogens are synthesized locally in peripheral target tissues from the inactive adrenal precursor DHEA and its sulfate, DHEA-S (Labrie et al. 2000). High concentrations of circulating DHEA and DHEA-S provide sufficient substrates for conversion into potent androgens and estrogens in peripheral tissues. In fact, plasma DHEA-S levels in adult men and women are 100 to 500 times higher than those of testosterone and 1,000 to 10,000 times higher than those of 17β-estradiol (estradiol). In postmenopausal women, almost 100 % of sex steroids are synthesized in peripheral tissues from precursors of adrenal origin or cholesterol, except for a small contribution from ovarian or adrenal testosterone and 4-androstenedione (Labrie et al. 2000). The term "intracrinology" was coined by Labrie in 1988 to describe the synthesis of active steroids in peripheral target tissues, where steroid action is exerted in the same cells in which synthesis takes place (Labrie et al. 1988). It has been claimed that estradiol and 5α-dihydrotestosterone in some intracrine peripheral tissues might be synthesized by a "back-door pathway" from 4-androstenedione or other steroid precursors, rather than the classical "front-door pathway" from testosterone (Luu-The 2013) (Fig. 9.1).

In rodents, sex steroid synthesis from DHEA should be less important, given the lack of adrenal androgens and the low level (below 0.01 ng/ml in rat) of plasma DHEA in the intact animal (Hojo et al. 2009). However, steroidogenic and steroid-converting enzymes have been detected in various organs in rodents (Dalla Valle et al. 1992). Rat pancreas β-cells, retina, cochlea, and brain express mRNA or proteins for cytochrome P450scc (CYP11A1), P450(17α) (CYP17) and/or P450arom (CYP19), and 3β-HSD (Kawato et al. 2003; Lecain et al. 2003; Cascio et al. 2007; Ogishima et al. 2008) (Fig. 9.1). These organs may synthesize sex steroids from cholesterol as paracrine modulators. Regional synthesis of sex steroids from cholesterol in the brain is described in the next section.

9.2 Neurosteroid Synthesis

It is well known that the brain is a target organ for steroid hormones. On the other hands, the brain regulates biosynthesis of gonadal sex steroids and adrenal glucocorticoids via the hypothalamic–pituitary–gonadal axis and the hypothalamic–pituitary–adrenal axis (Norman et al. 1997).

The brain is also a steroidogenic organ. De novo synthesis of steroids in the brain was demonstrated in the 1980s by Baulieu and colleagues (Baulieu 1998). They discovered that the concentrations of several steroids, including pregnenolone, DHEA, and their sulfate esters, pregnenolone-S and DHEA-S, are much higher in

Table 9.1 mRNA levels for steroidogenic acute regulatory (StAR) protein and steroidogenic enzymes in rat hippocampus

mRNA	10 days old (molecules per mg protein)	8 weeks old (molecules per mg protein)
StAR	$220 \times 10^3 \pm 60 \times 10^3$	$290 \times 10^3 \pm 60 \times 10^3$
P450scc	110 ± 0.15	>50
P450(17α)	$57 \times 10^3 \pm 1.8 \times 10^3$	$71 \times 10^3 \pm 57 \times 10^3$
3β-HSD-1/2	$9.7 \times 10^3 \pm 3.9 \times 10^3$	$10 \times 10^3 \pm 9.7 \times 10^3$
17β-HSD-1	$49 \times 10^3 \pm 11 \times 10^3$	$48 \times 10^3 \pm 7 \times 10^3$
17β-HSD-3	$28 \times 10^6 \pm 3.7 \times 10^6$	$50 \times 10^6 \pm 7 \times 10^6$
17βHSD-4	$2.9 \times 10^6 \pm 0.4 \times 10^6$	$9.5 \times 10^6 \pm 0.6 \times 10^6$
P450arom	$42 \times 10^3 \pm 3.1 \times 10^3$	$69 \times 10^3 \pm 44 \times 10^3$

Hippocampi were isolated from 10-day-old and 8-week-old male rats. mRNA levels were determined by real-time RT-PCR as described in (Munetsuna et al. 2009b). mRNA levels are given as molecules per mg protein in the hippocampus. Values are means ± SE of the results from three separate experiments

the brain than in the systemic circulation. The levels of these steroids in the brain remain high long after adrenalectomy and gonadectomy. In addition, the circadian variations of steroid concentrations in brain tissue are not synchronized with those of circulating steroids. These observations suggested that steroids either might be synthesized de novo in the central and peripheral nervous systems or might accumulate in those structures. Such steroids were named "neurosteroids" to refer to their unusual origin and to differentiate them from steroids derived from classical steroidogenic organs such as the gonads and adrenals (Baulieu 1997, 1998). More recently, de novo synthesis of the other bioactive steroids in neuronal systems has been demonstrated, and the term neurosteroids has been expanded to include sex steroids, corticoids, allopregnanolone, 7α-hydroxysteroids, etc. (Tsutsui et al. 1999, 2003; Compagnone and Mellon 2000; Mellon and Vaudry 2001; Do Rego et al. 2009) (Fig. 9.1).

9.2.1 Steroidogenic Enzymes in the Brain

Most of the enzymes present in the adrenals and gonads have been found in the brain by measuring some combination of their enzymatic activity, their mRNA transcript level, and their protein expression (Compagnone and Mellon 2000; Do Rego et al. 2009). Steroidogenic acute regulatory protein (StAR), translocator protein (TSPO, peripheral benzodiazepine receptor), cytochromes P450scc, P450 (17α), P450arom, and multiple subtypes of 3β-HSD and 17β-HSD have been observed in human and rodent brains (Compagnone and Mellon 2000; Do Rego et al. 2009; Munetsuna et al. 2009b). These proteins participate in de novo synthesis of sex steroids from cholesterol (Fig. 9.1). Copy numbers of transcripts for these proteins in hippocampus of neonatal and adult rats are shown in Table 9.1.

These numbers were lower than for samples from endocrine tissues. The relative number of transcripts was of the order of 1/100 to 1/200 of that in the bovine adrenal gland for StAR, less than 1/200,000 for P450scc, and of the order of 1/10,000 to 1/20,000 for P450(17α) and 3β-HSD (Yamazaki et al. 2005).

Activities of cytochromes P450c21 (CYP21), P450(11β)-1 (CYP11B1), and P450(11β)-2 (P450aldo, CYP11B2) are required for corticoid synthesis (Fig. 9.1). Small amounts of these enzymes were also detected in the brain (Higo et al. 2011). Cytochrome P450(2D) isoforms, drug-metabolizing enzymes, are expressed abundantly in the brain (Hiroi et al. 1998; McFayden et al. 1998). Funae and coworkers revealed that cytochromes P450(2D4) in rat and P450(2D6) in human mediate the 21-hydroxylation of steroids (Kishimoto et al. 2004). These cytochrome P450 (2D) isoforms might contribute to corticoid synthesis in the brain along with cytochrome P450c21 (Kishimoto et al. 2004; Higo et al. 2011). The relative number of transcripts expressed in the hippocampus of adult male rats was of the order of 1/20,000 of that in the adrenal gland for P450c21, almost the same level as that in the liver for P450(2D4), and of the order of 1/5,000 to 1/10,000 of that in the adrenal gland for P450(11β)-1 and P450(11β)-2 (Higo et al. 2011).

An isoform of 7α-hydroxylase, cytochrome P450(7α) (CYP7B) is predominantly found in the brain and is specifically expressed in the hippocampus (Do Rego et al. 2009). Recently, physiological functions of the 7α-hydroxysteroids in amphibian brains have been demonstrated (Tsutsui et al. 2013).

Activities of cytochrome P450 were supported by electron-transfer proteins, which transport electrons from NADPH to the P450. In mitochondria, ferredoxin-1, so-called adrenodoxin or hepatoredoxin, and the ferredoxin reductase are required for activities of cytochromes P450scc, P450(11β)-1, and P450(11β)-2. NADPH-hemoprotein reductase supplies electrons to the microsomal cytochromes. These electron-transport proteins have been detected in the brain (Compagnone and Mellon 2000). Cytochrome b_5 modulates the activity of microsomal cytochrome P450. Cytochrome b_5 and cytochrome b_5 reductase have also been detected in the brain (Du et al. 1997).

The 5α-reductase converts progesterone, testosterone, and 11-deoxycorticosterone to their respective 5α-dihydrosteroids (Fig. 9.1). Some 5α-dihydrosteroids are further metabolized by 3α-hydroxysteroid dehydrogenase (3α-HSD) (Fig. 9.1). Isoforms of the 5α-reductase and 3α-HSD have been detected in the brain (Compagnone and Mellon 2000; Do Rego et al. 2009).

Sulfate esters of active steroids have been considered neurosteroids (Baulieu 1998). Although the presence of pregnenolone-S in rodent brain was controversial, the human brain contains significant amounts of pregnenolone-S and DHEA-S (Schumacher et al. 2008). Several subtypes of steroid sulfotransferase (SULT) and steroid sulfatase have been detected in the brain (Compagnone and Mellon 2000; Schumacher et al. 2008; Do Rego et al. 2009).

9.2.2 *Regulation of Neurosteroid Synthesis*

Although there is a growing body of evidence of the biosynthesis and physiological importance of neurosteroids, little is known regarding the neural mechanisms regulating neurosteroid biosynthesis in the mammalian brain. The regulation of activity of several neurosteroidogenic enzymes by neurotransmitters and neuropeptides in vertebrates has been reviewed by Do Rego et al. (2009). In this section, we describe examples of the regulation of estradiol synthesis in rat hippocampus.

9.2.2.1 Retinoic Acid Stimulates Sex Steroid Synthesis in Rat Hippocampal Slice Cultures: In Vitro Experiments

The adult rat hippocampus possesses the active steroidogenic enzymes and proteins required for de novo synthesis of testosterone and estradiol (Munetsuna et al. 2009b). mRNA levels of these enzymes in the rat hippocampus are listed in Table 9.1.

Retinoic acids, which are metabolites of vitamin A, play important roles in the central nervous system during development and in adulthood. The retinoic acids and sex steroids in the hippocampus have similar functions. Both signaling pathways are involved in hippocampal plasticity and spatial memory, as well as in neuroprotection (Chiang et al. 1998; Shinozaki et al. 2007). However, steroid hormone synthesis is stimulated by retinoic acid in peripheral steroidogenic organs, such as mouse Leydig cells, the K9 mouse Leydig cell line, and in human ovarian thecal cells (Lefevre et al. 1994; Lee et al. 1999; Wickenheisser et al. 2005). Based on these findings, we hypothesized that retinoic acid modified sex steroid synthesis in the hippocampus. This hypothesis was tested with organotypic slice culture of rat hippocampus.

Rat hippocampal slices were prepared from 10- to 12-day-old male Wistar rats and cultured as described previously (Munetsuna et al. 2009b). After the 24-h pre-culture, the slices were incubated for 48 h in serum-free medium with or without 9-*cis*-retinoic acid. Testosterone and estradiol levels in the slices were increased during the 48-h incubation with serum-free medium, indicating de novo synthesis of the sex steroid (black and gray bars in Fig. 9.2). Incubation of the slices with 0.1 and 1 μM 9-*cis*-retinoic acid induced a further increase in the amount of de novo synthesized sex steroids, about twofold (Fig. 9.2).

After the 48-h incubation, mRNA levels for steroidogenic enzymes in the slices were determined, and the levels in retinoic acid-treated slices normalized to those in nontreated slices (Fig. 9.3). Relative mRNA levels for of P450(17α) and P450arom were significantly increased by the treatment with 9-*cis*-retinoic acid. These increases were concomitant with an increase in the level of mRNA for cellular retinoid-binding protein type 2 (CRBP-2), transcription of which is activated by retinoid X receptors (RXR) signaling (Nakshatri and Chambon 1994). Stimulation of gene expression for the cytochromes P450 might be mediated by the RXR signaling pathway.

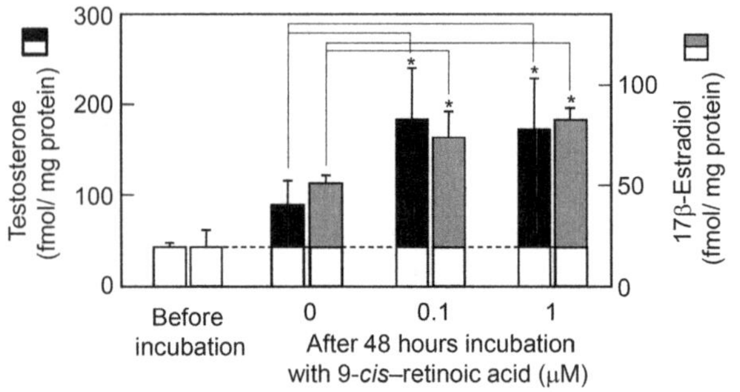

Fig. 9.2 Testosterone and estradiol levels in cultured hippocampal slices. Steroids were extracted from slices after 24-h pre-culture (*Before incubation*), and from slices and medium after 48-h incubation with 0, 0.1 or 1 μM 9-*cis*-retinoic acid. Levels of extracted testosterone and 17β-estradiol were quantified by specific EIA and RIA, as described previously (Munetsuna et al. 2009b). Values are means ± SE of the results from four separate experiments. $^*P < 0.05$ versus nontreated slices and medium

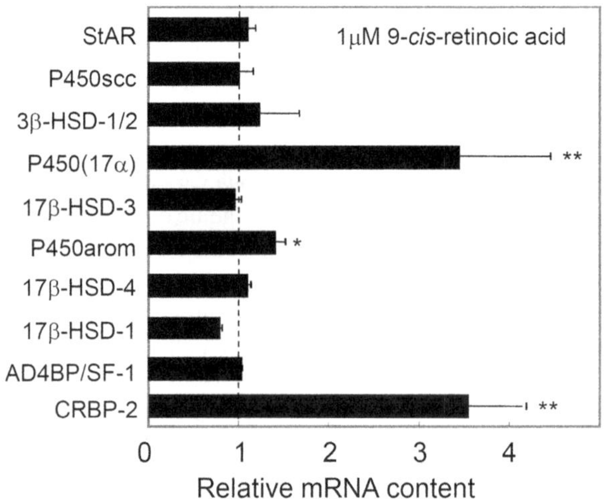

Fig. 9.3 Relative mRNA levels for steroidogenic enzymes and selected proteins in cultured hippocampal slices. Slices were incubated with or without 1 μM 9-*cis*-retinoic acid for 48 h. The levels of mRNAs for steroidogenic enzymes and StAR proteins, Ad4 binding protein/steroidogenic factor-1 (AD4BP/SF-1), and cellular retinol binding protein type-2 (CRBP-2) were determined by real-time RT-PCR as described previously (Munetsuna et al. 2009b). The mRNA levels in retinoic acid-treated slices are normalized to those in nontreated slices (*dashed line*). Values are means ± SE of the results from three or four separate experiments. $^*P < 0.05$ versus nontreated slices

After the 48-h incubation with 1 μM 9-*cis*-retinoic acid, the P450(17α) protein level in the slices increased approximately 1.7 fold (Munetsuna et al. 2009b). The enzymatic activity of P450(17α) produces DHEA and 4-androstenedione, both of which are precursors for testosterone and estradiol (Fig. 9.1). Testosterone and estradiol synthesis in the hippocampus was stimulated about 2 fold by 1 μM 9-*cis*-retinoic acid (Fig. 9.2), consistent with the increase in the level of P450 (17α) protein. The mRNA level for P450arom was also increased by the same treatment. This increase might contribute to the stimulation of estradiol synthesis.

It can be concluded that the treatment of slices with 9-*cis*-retinoic acid induced an increase in sex steroid synthesis by activation of the gene expression of P450 (17α) and P450arom. This stimulation of gene expression might be mediated by the RXR signaling pathway. Transcriptional regulation by retinoids is a candidate regulatory mechanism for neurosteroid synthesis.

9.2.2.2 Effect of Housing Conditions on Hippocampal Neurosteroidogenesis: In Vivo Experiments

Rats deprived of social contact with other rats at a young age experience a form of prolonged stress that leads to long-lasting alterations in their behavioral profiles (Fone and Porkess 2008). Isolated rats are aggressive and neophobic. The hippocampus was reported to be sensitive to the social isolation stress (Blanchard et al. 2001). Isolation stress also has effects on brain steroids. Allopregnanolone level and mRNA expression of 5α-reductase type 1 were decreased in specific neurons of the medial prefrontal cortex, basolateral amygdala, and hippocampus (Agis-Balboa et al. 2007; Pibiri et al. 2008).

Environmental enrichment is a reliable and well-characterized paradigm of experience-dependent plasticity in rodents (van Praag et al. 2000; Mora et al. 2007). Enriched environment can be provided to rodents by rearing in large groups and in wide and stimulating environments including a variety of toys and objects (Sale et al. 2009). Enriched rats have shown morphological changes in the brain, including neurogenesis and increased synaptic contacts, as well as improvements in cognitive performance (Sale et al. 2009). These changes might be related to the level of neuroactive steroids in the brain. The effect of environmental enrichment on neurosteroidogenesis had not been well studied.

It has been reported the effect of housing conditions on the transcript levels of the neurosteroidogenic enzymes in rat hippocampus (Munetsuna et al. 2009a, 2011). The results are briefly described next.

On postnatal day 28, male Wistar rats were randomly divided into three groups: (1) social isolation, (2) pair-housed, and (3) environmental enrichment. Socially isolated rats were housed individually in clear plastic cages (26.5 × 42 × 18 cm), and pair-housed rats were housed two per cage in the same type of cage. Environmental enrichment rats were housed in a group of nine in a large cage (100 × 50 × 70 cm) containing a running wheel, a branched tunnel, a wooden shelter, and several plastic toys.

After 8 weeks in these housing conditions, mRNA levels of neurosteroidogenic enzymes and proteins in the rat hippocampus were determined. The mRNA levels from isolated or enriched rats are normalized to those from pair-housed rats (Fig. 9.4). In the socially isolated rats, the mRNA level for cytochrome P450arom

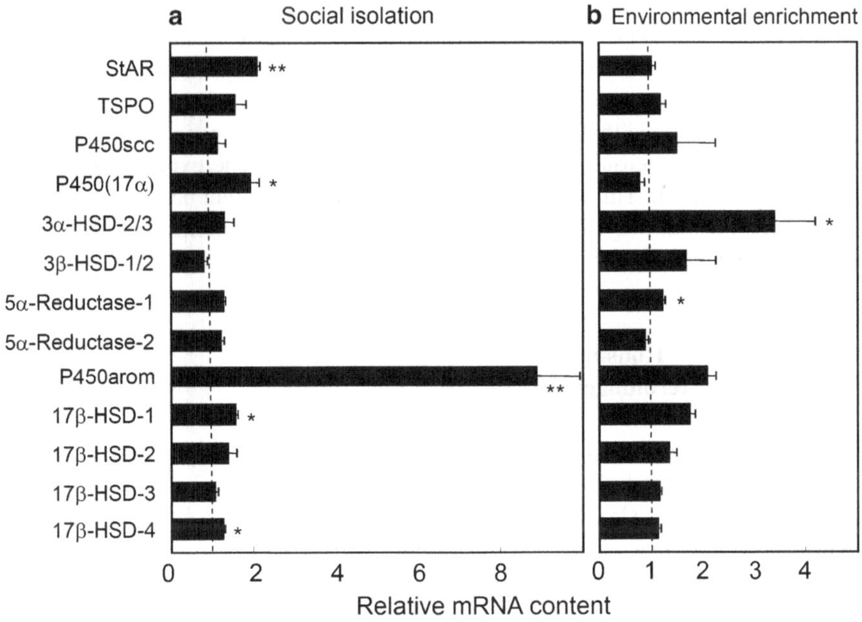

Fig. 9.4 Relative mRNA contents for steroidogenic enzymes and proteins in rat hippocampus. Hippocampi of socially isolated, pair-housed, and environmentally enriched rats were isolated after 8 weeks of housing. Levels of mRNAs for StAR, TSPO, and steroidogenic enzymes in the hippocampi were determined by real-time PCR as described (Munetsuna et al. 2009a, 2011). The amounts of mRNA in socially isolated rats (**a**) and environmentally enriched rats (**b**) are normalized to those in pair-housed rats (*dashed line*). Values are means ± SE of the results from nine to ten rats. $^*P < 0.05$, $^{**}P < 0.01$ versus pair-housed rats

was increased more than eightfold. mRNA levels for StAR, cytochrome P450(17α), 17β-HSD-1, and 17β-HSD-4 were also increased significantly (Fig. 9.4a). The estradiol levels in the hippocampi of isolated and pair-housed rats were determined by enzyme immunoassay (EIA) (Munetsuna et al. 2009a), and they were 1.5 fold higher in the socially isolated rats (Fig. 9.5). The estradiol levels in individual rats were significantly correlated with the mRNA levels for StAR protein and cytochrome P450arom in isolated and pair-housed rats (Munetsuna et al. 2009a).

Effects of environmental enrichment on mRNA levels of neurosteroidogenic enzymes in the rat hippocampus are shown in Fig. 9.4b. A significant increase in mRNA for 3α-HSD and 5α-reductase type 1 was observed. These enzymes mediate allopregnanolone synthesis from progesterone (Fig. 9.1). The levels of mRNAs for P450arom and 17β-HSD-1 were elevated but not significantly. These data indicated that environmental enrichment induces the transcriptional activation of enzymes involved in hippocampal allopregnanolone synthesis (Munetsuna et al. 2011).

It can be concluded that housing conditions during rearing strongly affected neurosteroid synthesis in the rat hippocampus.

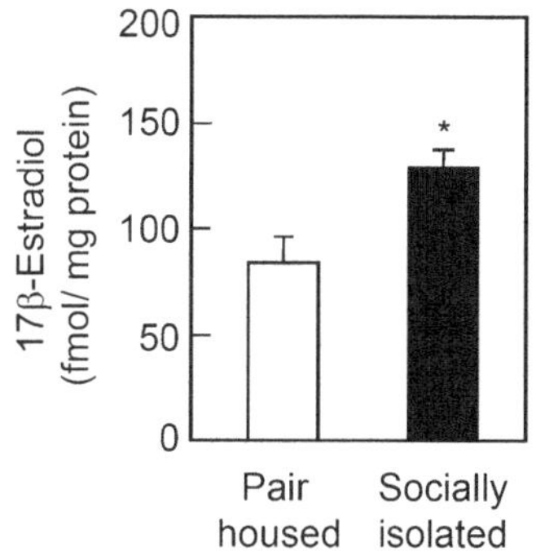

Fig. 9.5 Hippocampal estradiol levels in isolated and pair-housed rats. Hippocampi of socially isolated and pair-housed rats were isolated after 8 weeks of housing. Estradiol levels in the hippocampi were determined by specific EIA (Munetsuna et al. 2009a). Values are means ± SE of three separate experiments. *$P < 0.05$ versus pair-housed rats

9.2.2.3 Effect of Gonadectomy on Brain Sex Steroid Levels

Brain sex steroids are derived from gonads via systemic circulation and from de novo synthesis from precursors or cholesterol. To evaluate the contributions of these sources, sex steroid levels in the brain and serum of adult male rats were determined after castration or sham operation.

Testosterone levels in cerebrum and cerebellum were lower than in plasma (Fig. 9.6a). Ten days after the castration, blood testosterone level was decreased in one-fifth of sham-operated rats. Brain testosterone levels were decreased to about one-half or one-third of those in sham-operated rats, indicating de novo synthesis of small amounts of testosterone in the brain. Estradiol levels in cerebrum, cerebellum, and hippocampus were higher than in plasma (Fig. 9.6b). Brain estradiol is thus likely to be synthesized in the brain. Surprisingly, the brain estradiol levels were not affected by the decrease in brain testosterone content after the castration. A similar result has been observed in the hippocampus of 12-week-old male rats (Hojo et al. 2009). According to the data from Hojo et al., hippocampal estradiol levels did not correlate with testosterone levels, although 5α-dihydrotestosterone levels in the hippocampus were well correlated with testosterone. Testosterone is a direct precursor for both estradiol and 5α-dihydrotestosterone (Fig. 9.1). These data indicate that brain estradiol levels might be regulated by a mechanism that is independent of brain testosterone levels.

One hypothetical feedback mechanism for brain estradiol was proposed by Atwood and coworkers (Meethal et al. 2009). In female mice, hypothalamic–pituitary gonadal axis components in the extrahypothalamic brain might comprise a feedback loop to regulate brain sex-steroid synthesis. Processing of StAR protein in the brain was increased by the gonadectomy-induced conditions of high serum gonadotropins and low serum sex steroids. Processing of StAR protein, from the 37-kDa form to the 30-kDa form, indicated cholesterol transport from the mitochondrial outer membrane to the inner membrane, where cholesterol is converted to

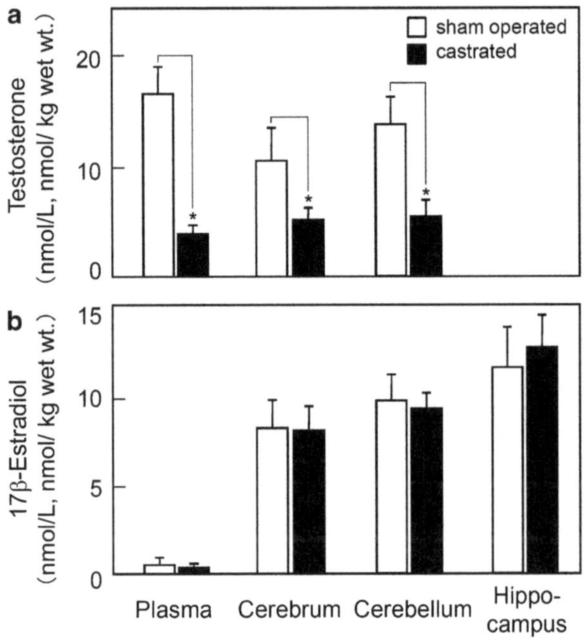

Fig. 9.6 Sex steroid levels in the plasma and brains of castrated and sham-operated rats. Male Wistar rats (3–4 months old) were castrated or sham-operated. Ten days after the operations, plasma and brain samples were collected, and steroids were extracted. Testosterone (**a**) and estradiol (**b**) levels were quantified by specific EIA and RIA, respectively (Munetsuna et al. 2009b). Values are means ± SE of five to six rats. *$P < 0.05$ versus sham-operated rats

a precursor of neurosteroids. This processing was decreased by administration of estradiol or progesterone, or the GnRH agonist leuprolide acetate. These data may indicate the existence of endocrine and autocrine/paracrine feedback loops that regulate neurosteroid synthesis.

9.3 Physiological Functions of Brain Steroids

Neurosteroids play a crucial role in the development and functioning of the central nervous system, as reviewed by several authors (Baulieu 1997; Compagnone and Mellon 2000; Kawato et al. 2003; Mellon 2007; Dean and McCarthy 2008; Tsutsui 2012). For example, brain-synthesized sex steroids are essential for neurodevelopment and for the establishment of the neural network at neonatal stages because sex steroid concentrations in circulation are not high enough at these developmental stages (Tsutsui 2006; Konkle and McCarthy 2011). In adults, the physiological functions of neurosteroids include various behavioral processes, such as cognition, arousal, stress, depression, anxiety, and sleep, as well as in sexual- and feeding-related behaviors and locomotion (Do Rego et al. 2009). The list of physiological functions of the neurosteroids is still expanding.

The biological functions of the neurosteroids and steroids from circulation are exerted either through a conventional genomic process via estrogen receptors, androgen receptors, progesterone receptors, mineral corticoid receptors, and glucocorticoid receptors, or through interaction with membrane receptors as allosteric modulators of the $GABA_A$/central-type benzodiazepine receptor complex, NMDA receptors, kainate receptors, AMPA receptors, sigma receptors, glycine receptors, serotonin receptors, nicotinic receptors, and muscarinic receptors. Neurosteroids may directly activate G protein-coupled transmembrane receptors or indirectly modulate the binding of neuropeptides to their receptors. Neurosteroids have also been shown to bind to microtubule-associated protein 2 and to stimulate tubulin polymerization in cultured neurons (Do Rego et al. 2009).

9.3.1 Neuroprotection by Steroids

Many lines of evidence have revealed neurotrophic and neuroprotective properties of steroid hormones, including pregnenolone, pregnenolone-S, progesterone, allopregnanolone, DHEA, DHEA-S, deoxycorticosterone, allotetrahydrodeoxycorticosterone, testosterone, and estradiol (Wojtal et al. 2006; Liu et al. 2010; Faroni and Magnaghi 2011; Melcangi et al. 2011; Traish et al. 2011; Scott et al. 2012; Singh and Su 2013).

The neuroprotective steroids in the brain are derived both from gonads via systemic circulation and from de novo synthesis from precursors or cholesterol. Two examples of receptor-mediated neuroprotective functions of brain steroids are described next.

9.3.2 Neuroprotection by Neurosteroids

Methylmercury is a neurotoxin that induces neuronal degeneration in the central nervous system. Anthropogenic exposure to methylmercury in Japan, Minamata disease, and the methylmercury poisoning in Iraq have established the toxicity of methylmercury in the nervous system (Bakir et al. 1973; Takeuchi 1982; Eto 2000).

The toxic effect of methylmercury has been shown to be attenuated by exogenous estradiol in primary cultured rat cerebellar granule cells. In these cultured cells, estradiol protected against methylmercury toxicity by acting as an antioxidant without stimulating estrogen receptors (Dare et al. 2000). In male mice, the administration of estradiol partially prevented the methylmercury-induced motor activity deficits and modification of cerebellar glutathione metabolism (Malagutti et al. 2009).

The protective effect of de novo synthesized estradiol on methylmercury-induced neurotoxicity in rat hippocampus is described next (Yamazaki et al. 2013).

The hippocampal organotypic slices of male rats were prepared as described earlier. Damaged cells were stained with propidium iodide (PI), which enters cells with damaged membranes and becomes brightly red fluorescent after binding

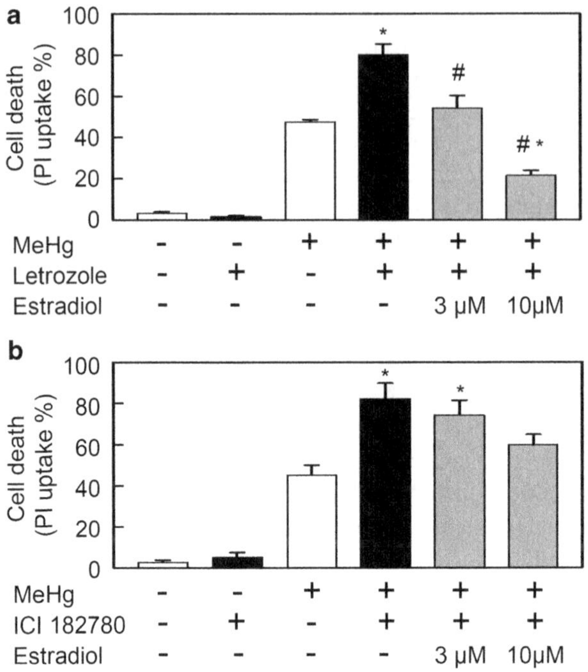

Fig. 9.7 Involvement of de novo synthesized estradiol and estrogen receptors on neuroprotection for methylmercury-induced cell death. Rat hippocampal slices were treated for 48 h with letrozole, an estradiol synthesis blocker, or vehicle, with 24-h treatment with 1 μM methylmercury (MeHg) or vehicle (**a**). Slices were treated for 30 h with 100 μM ICI 182,780, an estrogen receptor antagonist, or vehicle, with 24-h treatment with 1 μM methylmercury or vehicle (**b**). Methylmercury was added 24 or 6 h after the administration of letrozole or ICI 182,780, respectively. Selected slices were given the indicated concentrations of estradiol 2 h before the addition of methylmercury. Cell death in the neuronal cell-dense region of the slices was evaluated by PI uptake. Values are means ± SE of five or four different cultures. *$P < 0.05$ versus methylmercury-treated slices (*white bars*); #$P < 0.05$ versus methylmercury and letrozole-treated slices (*black bar* in **a**)

nucleic acids (Macklis and Madison 1990; Vitale et al. 1993). Approximately 50 % of the cells in the neuronal cell-dense regions of hippocampal slices were damaged and stained by PI after 24-h treatment with 1 μM methylmercury (Fig. 9.7).

Letrozole is a specific inhibitor for P450arom. De novo synthesis of estradiol in the hippocampal slices was completely blocked by pre-incubation with letrozole (Rune et al. 2006; Yamazaki et al. 2013). The letrozole itself showed no neurotoxic effect (Fig. 9.7a). After the slices were incubated for 24 h with letrozole followed by 24 h with 1 μM methylmercury and letrozole, the cell death rate was approximately 80 %, which was 30 % higher than the slices treated with methylmercury only (Fig. 9.7a). The increase in the cell death was reversed by the co-addition of 3 μM estradiol 2 h before the administration of methylmercury (Fig. 9.7a). These data indicate that the hippocampus-synthesized estradiol protected the slices from the toxic effects of methylmercury. Pre-treatment of the slices with 10 μM estradiol for 2 h attenuated methylmercury-induced cell death (Fig. 9.7a). The exogenously added estradiol also protected hippocampal cells from the toxic effect of methylmercury.

ICI 182,780 is an antagonist for both estrogen receptor (ER)-α and ER-β (Dahlman-Wright et al. 2006). Treatment of the slices with 100 μM ICI 182,780 alone had no neurotoxic effect (Fig. 9.7b). In the presence of ICI 182,780, cell death after 1 μM methylmercury treatment was increased to 82 %, the same level as that of the letrozole-treated slices (Fig. 9.7a, b). These data indicate that the protective function of de novo synthesized estradiol was mediated by ERs in the hippocampal slices. The protective effect of exogenous estradiol was also mediated by ERs because ICI 182,780 blocked the protective effect of co-added estradiol at both 3 and 10 μM (Fig. 9.7b). The protective effect of estradiol was mimicked by the administration of specific agonists for ERα or ERβ, 1,3,5-tris(4-hydroxyphenyl)-4-propyl-1*H*-pyrazole (PPT) or 2,3-bis(4-hydroxyphenyl) propionitrile (DPN), respectively (Yamazaki et al. 2013).

It can be concluded that hippocampal de novo synthesized estradiol protected hippocampal cells from methylmercury-induced neurotoxicity via ER-α- and ER-β-mediated pathways.

9.3.3 Neuroprotection by Brain-Converted Steroids

Some steroid hormones can be converted to other active steroids in the brain. The physiological functions of these steroids may be exerted by metabolic product(s) rather than by the steroids themselves. Progesterone is known to bind to the progesterone receptor to protect neuronal cells via both genomic and nongenomic pathways, and it is also known as a precursor for neuroprotective metabolites (Brinton et al. 2008). It has been analyzed the neuroprotective mechanism of progesterone against the toxicity of tributyltin in hippocampal slice culture (Ishihara et al. 2013). The results are briefly described next.

Tributyltin is an environmental pollutant that has been widely used in paint formulations to prevent marine fouling on ships, boats, and fish-farming nets. The central nervous system is a primary target of tributyltin toxicity (Ishihara et al. 2012, 2013).

Hippocampal organotypic slices of male rats were prepared as described previously and treated with tributyltin. After a 24-h treatment with 3 μM tributyltin, approximately 50 % of the hippocampal cells were damaged (Fig. 9.8). The neurotoxic effect of tributyltin was significantly protected by a 2-h pre-incubation with 1 μM progesterone (Fig. 9.8). In these hippocampal slices, the protective effect of progesterone was not mediated by the progesterone receptor because co-addition of mifepristone, a progesterone receptor antagonist, had no effect on the neuroprotection of progesterone (Fig. 9.8).

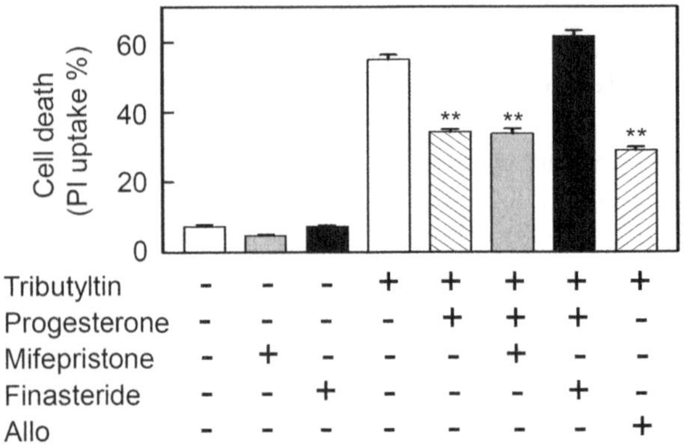

Fig. 9.8 Protective effects of progesterone and its metabolite on tributyltin-induced cell death. Rat hippocampal slices were pretreated with 1 μM progesterone, 1 μM allopregnanolone (*Allo*), or vehicle 2 h before a 24-h treatment with or without 3 μM tributyltin. Some slices were pretreated with 10 μM mifepristone, a progesterone receptor antagonist, or 100 μM finasteride, a 5α-reductase inhibitor, 20 min before the addition of progesterone. Cell death was evaluated by PI uptake. Values are mean ± SE of five separate experiments. $^{**}P < 0.01$ versus 3 μM tributyltin-treated slices (*white bar*)

Allopregnanolone, a metabolite of progesterone, was also reported to provide neuroprotective activity (Shibuya et al. 2003). Finasteride is a 5α-reductase inhibitor that completely blocks the conversion of progesterone to allopregnanolone in hippocampal slices (Ishihara et al. 2013). Co-addition of finasteride blocked the protective effects of progesterone (Fig. 9.8). Pre-treatment with allopregnanolone protected the hippocampal slices from the neurotoxicity of tributyltin (Fig. 9.8). These results indicate that allopregnanolone converted from progesterone protected hippocampal neurons from tributyltin, but progesterone itself via the progesterone receptor did not.

It has been shown that allopregnanolone acts on $GABA_A$ receptors as an allosteric modulator (Belelli and Lambert 2005). The neuroprotective effects of progesterone and allopregnanolone were mediated by $GABA_A$ receptors. Pretreatment with bicuculline, a potent $GABA_A$ receptor antagonist, significantly abrogated the neuroprotective actions of progesterone or allopregnanolone for the toxic effect of tributyltin (Fig. 9.9). Treatment with bicuculline alone showed no toxicity in the hippocampal slices. Furthermore, pretreatment with muscimol, a potent $GABA_A$ receptor agonist, partially but significantly suppressed neuronal cell death induced by tributyltin (Fig. 9.9). These data suggest that the $GABA_A$ receptor is involved in the protective effects of progesterone and allopregnanolone on neuronal injury induced by tributyltin.

Taken together, allopregnanolone converted from progesterone in hippocampal slices may protect neurons from tributyltin-induced neurotoxicity via a $GABA_A$ receptor-dependent mechanism.

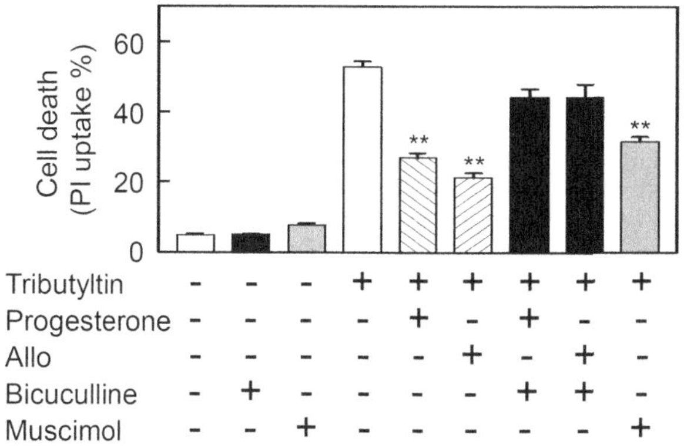

Fig. 9.9 Involvement of $GABA_A$ receptor in the protective effects of progesterone and allopregnanolone on tributyltin-induced cell death. Rat hippocampal slices were pretreated with 100 μM bicuculline, a $GABA_A$ receptor antagonist, for 20 min. Progesterone (1 μM) or allopregnanolone (1 μM) was added, and slices were cultured for 2 h. Slices were exposed to 3 μM tributyltin for 24 h, and cell death was then evaluated by PI uptake. Some slices were pretreated with 10 μM muscimol, a $GABA_A$ receptor agonist, for 20 min and treated with or without 3 μM tributyltin for 24 h. Values are mean ± SE of five separate experiments. $^{**}P < 0.01$ versus 3 μM tributyltin-treated slices (*white bar*)

9.4 Conclusion

Various organs possess the steroidogenic cytochrome P450 and related steroidogenic enzymes, which synthesize active steroid hormones from cholesterol or precursor steroids. These organs, except gonads, adrenal glands, and the placenta, synthesize small amounts of steroids, which exert autocrine and paracrine functions in these tissues, although systemic steroids from the gonads and adrenals coordinate multiple organ systems. There is a growing body of evidence about the biosynthesis and physiological importance of locally synthesized steroids.

Synthesis of steroids in the brain and peripheral nervous system is the most well characterized local steroidogenesis (neurosteroidogenesis). Approximately three decades of studies of neurosteroids have revealed a wide range of important physiological functions and functional mechanisms in the development and maintenance of the nervous system; however, novel and unexpected functions for neurosteroids are still being discovered. This chapter includes descriptions of two examples of neuroprotective functions, by a neurosteroid and a brain-converted steroid.

Despite the importance of neurosteroids in the brain, the mechanisms regulating the production of neurosteroids are not well understood. Some of the regulation of neurosteroidogenesis was described in this chapter, but further experiments are required for full understanding of thc regulatory mechanisms.

References

Agis-Balboa RC, Pinna G, Pibiri F, Kadriu B, Costa E, Guidotti A (2007) Down-regulation of neurosteroid biosynthesis in corticolimbic circuits mediates social isolation-induced behavior in mice. Proc Natl Acad Sci USA 104:18736–18741

Bakir F, Damluji SF, Amin-Zaki L, Murtadha M, Khalidi A, al-Rawi NY et al (1973) Methylmercury poisoning in Iraq. Science 181:230–241

Baulieu EE (1997) Neurosteroids: of the nervous system, by the nervous system, for the nervous system. Recent Prog Horm Res 52:1–32

Baulieu EE (1998) Neurosteroids: a novel function of the brain. Psychoneuroendocrinology 23:963–987

Belelli D, Lambert JJ (2005) Neurosteroids: endogenous regulators of the GABA(A) receptor. Nat Rev Neurosci 6:565–575

Blanchard RJ, McKittrick CR, Blanchard DC (2001) Animal models of social stress: effects on behavior and brain neurochemical systems. Physiol Behav 73:261–271

Brinton RD, Thompson RF, Foy MR, Baudry M, Wang J, Finch CE et al (2008) Progesterone receptors: form and function in brain. Front Neuroendocrinol 29:313–339

Cascio C, Russo D, Drago G, Galizzi G, Passantino R, Guarneri R et al (2007) 17beta-estradiol synthesis in the adult male rat retina. Exp Eye Res 85:166–172

Chiang MY, Misner D, Kempermann G, Schikorski T, Giguere V, Sucov HM et al (1998) An essential role for retinoid receptors RARbeta and RXRgamma in long-term potentiation and depression. Neuron 21:1353–1361

Compagnone NA, Mellon SH (2000) Neurosteroids: biosynthesis and function of these novel neuromodulators. Front Neuroendocrinol 21:1–56

Dahlman-Wright K, Cavailles V, Fuqua SA, Jordan VC, Katzenellenbogen JA, Korach KS et al (2006) International union of pharmacology. LXIV. Estrogen receptors. Pharmacol Rev 58:773–781

Dalla Valle L, Belvedere P, Simontacchi C, Colombo L (1992) Extraglandular hormonal steroidogenesis in aged rats. J Steroid Biochem Mol Biol 43:1095–1098

Dare E, Gotz ME, Zhivotovsky B, Manzo L, Ceccatelli S (2000) Antioxidants J811 and 17beta-estradiol protect cerebellar granule cells from methylmercury-induced apoptotic cell death. J Neurosci Res 62:557–565

Dean SL, McCarthy MM (2008) Steroids, sex and the cerebellar cortex: implications for human disease. Cerebellum 7:38–47

Do Rego JL, Seong JY, Burel D, Leprince J, Luu-The V, Tsutsui K et al (2009) Neurosteroid biosynthesis: enzymatic pathways and neuroendocrine regulation by neurotransmitters and neuropeptides. Front Neuroendocrinol 30:259–301

Du M, Shirabe K, Takeshita M (1997) Identification of alternative first exons of NADH-cytochrome b5 reductase gene expressed ubiquitously in human cells. Biochem Biophys Res Commun 235:779–783

Estabrook RW, Cooper DY, Rosenthal O (1963) The light reversible carbon monoxide inhibition of the steroid C21-hydroxylase system of the adrenal cortex. Biochem Z 338:741–755

Eto K (2000) Minamata disease. Neuropathology 20(Suppl):S14–S19

Faroni A, Magnaghi V (2011) The neurosteroid allopregnanolone modulates specific functions in central and peripheral glial cells. Front Endocr (Lausanne) 2:103

Fone KC, Porkess MV (2008) Behavioural and neurochemical effects of post-weaning social isolation in rodents-relevance to developmental neuropsychiatric disorders. Neurosci Biobehav Rev 32:1087–1102

Higo S, Hojo Y, Ishii H, Komatsuzaki Y, Ooishi Y, Murakami G et al (2011) Endogenous synthesis of corticosteroids in the hippocampus. PLoS One 6:e21631

Hiroi T, Imaoka S, Chow T, Funae Y (1998) Tissue distributions of CYP2D1, 2D2, 2D3 and 2D4 mRNA in rats detected by RT-PCR. Biochim Biophys Acta 1380:305–312

Hojo Y, Higo S, Ishii H, Ooishi Y, Mukai H, Murakami G et al (2009) Comparison between hippocampus-synthesized and circulation-derived sex steroids in the hippocampus. Endocrinology 150:5106–5112
Ishihara Y, Kawami T, Ishida A, Yamazaki T (2012) Tributyltin induces oxidative stress and neuronal injury by inhibiting glutathione S-transferase in rat organotypic hippocampal slice cultures. Neurochem Int 60:782–790
Ishihara Y, Kawami T, Ishida A, Yamazaki T (2013) Allopregnanolone-mediated protective effects of progesterone on tributyltin-induced neuronal injury in rat hippocampal slices. J Steroid Biochem Mol Biol 135:1–6
Kawato S, Yamada M, Kimoto T (2003) Brain neurosteroids are 4th generation neuromessengers in the brain: cell biophysical analysis of steroid signal transduction. Adv Biophys 37:1–48
Kishimoto W, Hiroi T, Shiraishi M, Osada M, Imaoka S, Kominami S et al (2004) Cytochrome P450 2D catalyze steroid 21-hydroxylation in the brain. Endocrinology 145:699–705
Konkle AT, McCarthy MM (2011) Developmental time course of estradiol, testosterone, and dihydrotestosterone levels in discrete regions of male and female rat brain. Endocrinology 152:223–235
Labrie C, Belanger A, Labrie F (1988) Androgenic activity of dehydroepiandrosterone and androstenedione in the rat ventral prostate. Endocrinology 123:1412–1417
Labrie F, Luu-The V, Lin SX, Simard J, Labrie C, El-Alfy M et al (2000) Intracrinology: role of the family of 17 beta-hydroxysteroid dehydrogenases in human physiology and disease. J Mol Endocrinol 25:1–16
Lecain E, Yang TH, Tran Ba Huy P (2003) Steroidogenic enzyme expression in the rat cochlea. Acta Otolaryngol 123:187–191
Lee HK, Yoo MS, Choi HS, Kwon HB, Soh J (1999) Retinoic acids up-regulate steroidogenic acute regulatory protein gene. Mol Cell Endocrinol 148:1–10
Lefevre A, Rogier E, Astraudo C, Duquenne C, Finaz C (1994) Regulation by retinoids of luteinizing hormone/chorionic gonadotropin receptor, cholesterol side-chain cleavage cytochrome P-450, 3 beta-hydroxysteroid dehydrogenase/delta (5-4)-isomerase and 17 alpha-hydroxylase/C17-20 lyase cytochrome P-450 messenger ribonucleic acid levels in the K9 mouse Leydig cell line. Mol Cell Endocrinol 106:31–39
Liu M, Kelley MH, Herson PS, Hurn PD (2010) Neuroprotection of sex steroids. Minerva Endocrinol 35:127–143
Luu-The V (2013) Assessment of steroidogenesis and steroidogenic enzyme functions. J Steroid Biochem Mol Biol 137:176–182
Macklis JD, Madison RD (1990) Progressive incorporation of propidium iodide in cultured mouse neurons correlates with declining electrophysiological status: a fluorescence scale of membrane integrity. J Neurosci Methods 31:43–46
Malagutti KS, da Silva AP, Braga HC, Mitozo PA, Soares Dos Santos AR, Dafre AL et al (2009) 17beta-estradiol decreases methylmercury-induced neurotoxicity in male mice. Environ Toxicol Pharmacol 27:293–297
McFayden MC, Melvin WT, Murray GI (1998) Regional distribution of individual forms of cytochrome P450 mRNA in normal adult human brain. Biochem Pharmacol 55:825–830
Meethal SV, Liu T, Chan HW, Ginsburg E, Wilson AC, Gray DN et al (2009) Identification of a regulatory loop for the synthesis of neurosteroids: a steroidogenic acute regulatory protein-dependent mechanism involving hypothalamic–pituitary–gonadal axis receptors. J Neurochem 110:1014–1027
Melcangi RC, Panzica G, Garcia-Segura LM (2011) Neuroactive steroids: focus on human brain. Neuroscience 191:1–5
Mellon SH (2007) Neurosteroid regulation of central nervous system development. Pharmacol Ther 116:107–124
Mellon SH, Vaudry H (2001) Biosynthesis of neurosteroids and regulation of their synthesis. Int Rev Neurobiol 46:33–78

Mora F, Segovia G, del Arco A (2007) Aging, plasticity and environmental enrichment: structural changes and neurotransmitter dynamics in several areas of the brain. Brain Res Rev 55:78–88

Munetsuna E, Hattori M, Komatsu S, Sakimoto Y, Ishida A, Sakata S et al (2009a) Social isolation stimulates hippocampal estradiol synthesis. Biochem Biophys Res Commun 379:480–484

Munetsuna E, Hojo Y, Hattori M, Ishii H, Kawato S, Ishida A et al (2009b) Retinoic acid stimulates 17beta-estradiol and testosterone synthesis in rat hippocampal slice cultures. Endocrinology 150:4260–4269

Munetsuna E, Hattori M, Sakimoto Y, Ishida A, Sakata S, Hojo Y et al (2011) Environmental enrichment alters gene expression of steroidogenic enzymes in the rat hippocampus. Gen Comp Endocrinol 171:28–32

Nakshatri H, Chambon P (1994) The directly repeated RG(G/T)TCA motifs of the rat and mouse cellular retinol-binding protein II genes are promiscuous binding sites for RAR, RXR, HNF-4, and ARP-1 homo- and heterodimers. J Biol Chem 269:890–902

Norman AW, Litwack G, Norman AW, Litwack G (1997) Hormones, 2nd edn. Academic, San Diego

Ogishima T, Mitani F, Suematsu M (2008) Cytochrome P-450(17alpha) in beta-cells of rat pancreas and its local steroidogenesis. J Steroid Biochem Mol Biol 111:80–86

Pibiri F, Nelson M, Guidotti A, Costa E, Pinna G (2008) Decreased corticolimbic allopregnanolone expression during social isolation enhances contextual fear: a model relevant for posttraumatic stress disorder. Proc Natl Acad Sci USA 105:5567–5572

Rosner JM, Macome JC, Cardinali DP (1969) In vitro biosynthesis of sterols and steroids by rat submaxillary glands. Endocrinology 85:1000–1003

Rune GM, Lohse C, Prange-Kiel J, Fester L, Frotscher M (2006) Synaptic plasticity in the hippocampus: effects of estrogen from the gonads or hippocampus? Neurochem Res 31:145–155

Sale A, Berardi N, Maffei L (2009) Enrich the environment to empower the brain. Trends Neurosci 32:233–239

Schumacher M, Liere P, Akwa Y, Rajkowski K, Griffiths W, Bodin K et al (2008) Pregnenolone sulfate in the brain: a controversial neurosteroid. Neurochem Int 52:522–540

Scott E, Zhang QG, Wang R, Vadlamudi R, Brann D (2012) Estrogen neuroprotection and the critical period hypothesis. Front Neuroendocrinol 33:85–104

Shibuya K, Takata N, Hojo Y, Furukawa A, Yasumatsu N, Kimoto T et al (2003) Hippocampal cytochrome P450s synthesize brain neurosteroids which are paracrine neuromodulators of synaptic signal transduction. Biochim Biophys Acta 1619:301–316

Shinozaki Y, Sato Y, Koizumi S, Ohno Y, Nagao T, Inoue K (2007) Retinoic acids acting through retinoid receptors protect hippocampal neurons from oxygen-glucose deprivation-mediated cell death by inhibition of c-jun-N-terminal kinase and p38 mitogen-activated protein kinase. Neuroscience 147:153–163

Singh M, Su C (2013) Progesterone and neuroprotection. Horm Behav 63:284–290

Slominski A, Zbytek B, Szczesniewski A, Semak I, Kaminski J, Sweatman T et al (2005) CRH stimulation of corticosteroids production in melanocytes is mediated by ACTH. Am J Physiol Endocrinol Metab 288:E701–E706

Takeuchi T (1982) Pathology of Minamata disease. With special reference to its pathogenesis. Acta Pathol Jpn 32(suppl 1):73–99

Taves MD, Gomez-Sanchez CE, Soma KK (2011) Extra-adrenal glucocorticoids and mineralocorticoids: evidence for local synthesis, regulation, and function. Am J Physiol Endocrinol Metab 301:E11–E24

Traish AM, Kang HP, Saad F, Guay AT (2011) Dehydroepiandrosterone (DHEA)–a precursor steroid or an active hormone in human physiology. J Sex Med 8:2960–2982, quiz 83

Tsutsui K (2006) Biosynthesis, mode of action and functional significance of neurosteroids in the developing Purkinje cell. J Steroid Biochem Mol Biol 102:187–194

Tsutsui K (2012) Neurosteroid biosynthesis and action during cerebellar development. Cerebellum 11:414–415

Tsutsui K, Ukena K, Takase M, Kohchi C, Lea RW (1999) Neurosteroid biosynthesis in vertebrate brains. Comp Biochem Physiol C Pharmacol Toxicol Endocrinol 124:121–129
Tsutsui K, Sakamoto H, Ukena K (2003) Biosynthesis and action of neurosteroids in the cerebellar Purkinje neuron. J Steroid Biochem Mol Biol 85:311–321
Tsutsui K, Haraguchi S, Fukada Y, Vaudry H (2013) Brain and pineal 7alpha-hydroxypregnenolone stimulating locomotor activity: identification, mode of action and regulation of biosynthesis. Front Neuroendocrinol 34:179–189
van Praag H, Kempermann G, Gage FH (2000) Neural consequences of environmental enrichment. Nat Rev Neurosci 1:191–198
Vitale M, Zamai L, Mazzotti G, Cataldi A, Falcieri E (1993) Differential kinetics of propidium iodide uptake in apoptotic and necrotic thymocytes. Histochemistry 100:223–229
Wickenheisser JK, Nelson-DeGrave VL, Hendricks KL, Legro RS, Strauss JF 3rd, McAllister JM (2005) Retinoids and retinol differentially regulate steroid biosynthesis in ovarian theca cells isolated from normal cycling women and women with polycystic ovary syndrome. J Clin Endocrinol Metab 90:4858–4865
Wojtal K, Trojnar MK, Czuczwar SJ (2006) Endogenous neuroprotective factors: neurosteroids. Pharmacol Rep 58:335–340
Yamazaki T, Shimodaira M, Kuwahara H, Wakatsuki H, Horiuchi H, Matsuda H et al (2005) Tributyltin disturbs bovine adrenal steroidogenesis by two modes of action. Steroids 70:913–921
Yamazaki T, Yamamoto M, Ishihara Y, Komatsu S, Munetsuna E, Onizaki M et al (2013) De novo synthesized estradiol protects against methylmercury-induced neurotoxicity in cultured rat hippocampal slices. PLoS One 8:e55559

Chapter 10
Whole Cell-Dependent Biosynthesis of Drug Metabolites Using Genetically Engineered Budding Yeast

Shinichi Ikushiro, Miyu Nishikawa, and Toshiyuki Sakaki

Abstract Xenobiotic phase I and II reactions generally render a compound more water soluble and pharmacologically inactive, thereby eliminating the need for further evaluation. However, if the metabolite forms a toxic compound such as acylglucuronide, additional safety assessment may be needed. Glucuronidation is the most common pathway for detoxification and elimination of hydrophobic xenobiotics in mammals. Thus, development of an efficient in vitro synthesis of glucuronides from parent drugs often becomes critical during studies of drug metabolism undertaken in the development of a new pharmaceutical product. To produce glucuronides as drug metabolites, we have developed coexpression systems for mammalian cytochrome P450 (P450), UDP-glucuronosyltransferase (UGT), and UDP-glucose dehydrogenase in *Saccharomyces cerevisiae* cells, and combination between each of the human P450s and UGTs was achieved. Glucuronide formation in yeast cells was performed in reaction medium containing 8 % glucose, and most of the glucuronides were readily recovered from the cell medium. In addition, we have expressed human sulfotransferase (SULT) with P450s in *S. cerevisiae* cells, and successfully obtained sulfo-conjugates from the cell medium. Coexpression of P450 electron transfer and glucuronidation or sulfo-conjugate production systems allow us to obtain the phase I metabolites and phase II metabolites from the parent compound. In conclusion, our yeast expression systems of xenobiotic-metabolizing enzymes have made it possible to produce xenobiotic phase I and phase II metabolites on the milligram to gram scale.

Keywords Budding yeast • Cytochrome P450 • Glucuronide • P450 • Phase I metabolite • Phase II metabolites • Sulfo-conjugate • Sulfotransferase • SULT • UDP-glucuronosyltransferase • UGT • Xenobiotics

---

S. Ikushiro (✉) • M. Nishikawa • T. Sakaki
Toyama Prefectural University, Toyama, Japan
e-mail: ikushiro@pu-toyama.ac.jp

H. Yamazaki (ed.), *Fifty Years of Cytochrome P450 Research*,
DOI 10.1007/978-4-431-54992-5_10, © Springer Japan 2014

10.1 Introduction

Many endogenous and exogenous compounds such as bilirubin, steroids, drugs, and environmental pollutants are biotransformed by xenobiotic phase I and phase II enzymes. The phase I enzymes such as cytochrome P450 (P450) introduce a functional group, mainly a hydroxyl group, into the compounds. Subsequently, the phase II conjugating enzymes such as UDP-glucuronosyltransferase (UGT), sulfotransferase, and glutathione-*S*-transferase can catalyze the conjugation of the hydrophilic molecules glucuronic acid, sulfuric acid, and glutathione, respectively, to the resulting functional group of the compounds (Iyanagi 2007). These reactions catalyzed by phase I and II enzymes would be expected to occur sequentially and cooperatively during the xenobiotic process. These enzymes belong to gene families containing a large number of isozymes that have different substrate specificities, resulting in variety in the xenobiotic pathway. The idea that phase I and II reactions, such as hydroxylation (oxidation) and glucuronidation (a major conjugation reaction), are coupled in the endoplasmic reticulum (ER) membranes has been proposed (von Bahr and Bertilsson 1971). Both the duration and intensity of pharmacological and toxicological drug effects are largely dependent on the functional efficiency of the biotransformation systems and their modulation by numerous endogenous and exogenous physiological and pathological factors in vivo. U.S. Food and Drug Administration (FDA) guidance in 2008 showed consideration for safety assessment of the metabolites identified only in human plasma or metabolites present at disproportionately higher levels in humans than in any of the animal test species (Anderson et al. 2010). In addition to drugs, many xenobiotic compounds including environmental pollutants, food additives, and dietary compounds are metabolized by a common biotransformation system. Therefore, detailed knowledge of the specific reactions involved in the metabolism of affected xenobiotic compounds is an essential basis for understanding the nature of xenobiotics, including drugs.

Efforts in development of whole cell-dependent production of metabolites using yeast cells have been undertaken by many research groups, including our laboratories. Sakaki et al. established yeast whole-cell biotransformation systems expressing human P450s (Sakaki 2013). They have examined the metabolism of some drugs, vitamin D, and environmental pollutants, indicating the advantage of usage of budding yeast cells for P450-dependent metabolite production. The major phase II enzymes are UGTs, which contributed to the clearance of about 10 % of the top 200 drugs presented in the USA in 2002 (Crettol et al. 2010). There are 19 UGT isoforms belonging to two major families, *UGT1* and *UGT2* (Mackenzie et al. 2005). These enzymes catalyzed the glucuronidation of xenobiotics as well as endogenous substances with a hydrophobic nature, facilitating their elimination from the body. UGT isoforms from two gene families show overlapping but sometimes very distinctive substrate specificities. UGT is an endoplasmic reticulum-localized membrane protein with the catalytically active part facing the luminal side. Their activity is dependent on the localization in the phospholipid membrane environment (Radominska-Pandya et al. 2005).

Table 10.1 Methods of preparation for drug metabolites

Methods	Yield		Disadvantage
	Standard or NMR (~mg)	Toxicity (~g)	
Chemical synthesis	○	○	Often very challenging or sometimes even impossible
Preparation from urine or bile	○	×	Cost- and time-consuming
In vitro enzymatic synthesis	○	Δ	High cost (UDP-glucuronic acid)
Whole-cell biosynthesis using recombinant yeast cells	○	○	

The formation of glucuronides in the body can lead to severe health damage by the formation of pharmacologically active and toxic metabolites such as acyl glucuronides (Spahn-Langguth and Benet 1992). Furthermore, polymorphism of the UGT gene can sometimes cause adverse effects of drug therapy by the accumulation of active or toxic metabolites (Maruo et al. 2005). These findings led to a high demand for purified glucuronides for toxicity studies and as reference standards. Table 10.1 shows the preparation methods of phase II metabolites. The first choice is the chemical synthesis method, but this is often very challenging or sometimes even impossible. Various biological techniques are the second choice: these include metabolite isolation from animal urine or bile, but these techniques are costly and time-consuming processes. Alternative methods are in vitro enzymatic synthesis using liver homogenate, or enzyme preparations from recombinant expression of enzymes in cell culture or microbial systems. However, these in vitro techniques are limited by the expense of cofactors such as UDP-glucuronic acid. To overcome these difficulties of synthesis of metabolites, we have now developed the whole-cell biotransformation system of xenobiotic conjugates (Ikushiro et al. 2010; Sakaki et al. 2011). Furthermore, coexpression of phase I and phase II metabolizing enzymes in yeast cells allows us to synthesize the xenobiotic metabolites of various compounds. This chapter describes our current advances in whole cell-dependent biosynthesis of phase I- and phase II-dependent drug and xenobiotic metabolites using genetically engineered budding yeasts.

10.2 Biosynthesis of P450-Dependent Metabolites of Xenobiotics

Among the most frequently used host organisms for the expression of P450s are *Escherichia coli* and the yeasts *Saccharomyces cerevisiae*, *Schizosaccharomyces pombe*, and *Pichia pastoris* (Yun et al. 2006; Sakaki and Inouye 2000; Peters et al. 2009; Andersen and Møller 2002). The production of large quantities of P450 metabolites required for safety testing of developing drugs is often performed using recombinant yeasts. The main advantages of yeast usage are that the yeast cell

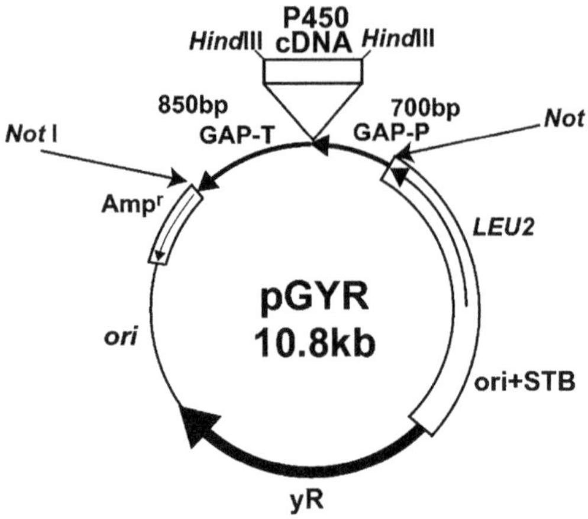

Fig. 10.1 Structure of yeast expression vector pGYR. *yR* yeast NADPH-cytochrome P450 reductase gene (from *Saccharomyces cerevisiae*), *GAP-P* glyceroaldehyde-3-phosphate dehydrogenase promotor (from *Zygosaccharomyces rouxii*), *GAP-T* glyceroaldehyde-3-phosphate dehydrogenase terminator (from *Z. rouxii*), *LEU2* isopropyl malate dehydrogenase gene (from *S. cerevisiae*), *ori + STB* replication origin in yeast and stabilized region for plasmid DNA (from *S. cerevisiae*), *Amp^r* ampicillin-resistance gene (from bacterial vector pUC19 DNA), *ori* replication origin in *Escherichia coli* (from bacterial vector pUC19 DNA)

has an endoplasmic reticulum membrane as well as protein expression and processing of maturation that resemble those of higher eukaryotes. Further advantage of the yeast expression system are the availability of the intrinsic P450 redox partner (NADPH-cytochrome P450 reductase, CPR) that functionally interacts with mammalian P450s. *S. cerevisiae* contains only three endogenous P450s that are involved in the metabolism of sterol, indicating that these are unlikely to cause the formation of unwanted side products in a mammalian P450-dependent biosynthesis in yeasts.

For optimization of P450 expression in yeast, the plasmid vector pGYR was developed (Ikezawa et al. 2003) (Fig. 10.1). pGYR contains the promoter and terminator sequences of glyceraldehyde 3-phosphate dehydrogenase from a yeast strain, *Zygosacchcaromyces rouxii*, for expression of P450 proteins. pGYR has also a single copy of the yeast CPR gene for enhancement of electron transfer to coexpressed P450. The budding yeast cell strain AH22 as host cells are widely used for heterologous expression of proteins including microsomal P450 isoforms. One of the advantage may be that the AH22 strain is a petite mutant and lacks mitochondrial DNA, resulting in developed endoplasmic reticulum membranes in the yeast cells. Finally, human major P450 isoforms were expressed in *S. cerevisiae* (Imaoka et al. 1996). In some cases, low monooxygenase activity of human P450 3A4 was attributed to the insufficient reduction of the heme iron of P450 by CPR. To enhance the efficiency of electron transfer, a fused enzyme was constructed between human P450 3A4 and yeast CPR gene, with coexpression of human cytochrome b_5 as effector (Hayashi et al. 2000).

The alternative application of whole cell-dependent P450 biotransformation allows us to achieve bioremediation for soils and sediments contaminated with industrial chemicals, such as dioxins and PCBs. Previous studies revealed that mammalian P450s are capable of the degradation of mono-, di-, and tri-chlorobenzo-*p*-dioxin (CDD) (Sakaki et al. 2013). Apart from mammalian P450, some P450 isoforms from other organisms are useful for the biotransformation of xenobiotics. Because the white-rot fungus *Phanerochaete chrysosporium* genome contains 148 isoforms of P450 genes, it is possible to assume that some *P. chrysosporium* P450s can metabolize various xenobiotic compounds. Recently, we obtained 120 clones expressing individual P450s of *P. chrysosporium* in budding yeast, and 6 of them could metabolize 2-mono-CDD (Kasai et al. 2010). Watanabe et al. reported that avian P450 isoforms were expressed in yeast and characterized these (Watanabe et al. 2013). According to numerous reports, including our results, *S. cerevisiae* is one of the most established host organisms as a platform for characterization of P450 function and the production of P450 metabolites via whole cell-dependent biosynthesis.

10.3 Biosynthesis of UGT-Dependent Metabolites of Xenobiotics

In contrast to the P450-dependent biosynthesis system in yeast, *S. cerevisiae* is incapable of producing the UDP-glucuronic acid as a cofactor of glucuronidation. Oka et al. previously reported that the functional expression of plant UDP-glucose dehydrogenase (UGDH) resulted in the production of UDP-glucuronic acid in *S. cerevisiae* (Oka and Jigami 2006). To effect biosynthesis of glucuronides in a whole-cell system, the coexpression of rat UDP-glucose dehydrogenase with UGT isoforms delivers UDP-glucuronic acid and permits self-sufficient glucuronide production. Coexpression of mammalian UGT isoforms, including human, rat, and mouse, and rat UGDH in *S. cerevisiae* was performed using the genome-integrated yeast expression vector, pAUR. pAUR is commercially available and transformants bearing the gene are selected by aureobasidin A-containing medium (Ikushiro et al. 2004). Figure 10.2 shows the construction of the expression vector. Modified pAUR contains a single copy of the rat UGDH gene with the yeast GAPDH promoter/terminator, and also the *Not*I-digested sequence of the cloning site. Insertion of a *Not*I fragment having the mammalian UGT gene with the yeast GAPDH promoter/terminator results in the coexpression of rat UGDH and various mammalian UGT isoforms. Western blot analysis of selected yeast strains from aureobasidin A-resistant colonies confirmed the protein expression of mammalian UGT and UGDH. β-glucuronidase treatment of more polar metabolites of 7-hydroxycoumarine revealed the production of glucuronides in resting yeast cells. Production of glucuronide was dependent on the concentration of glucose in the reaction medium. Most of the glucuronide of 7-hydroxycoumarine was excreted from

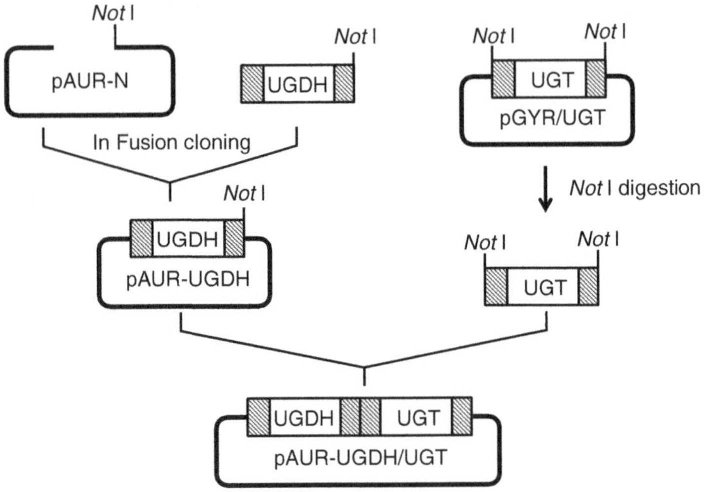

Fig. 10.2 Strategy of coexpression of UGT with UGDH using genome-integrated expression vector in yeast. *UGT* UDP-glucuronosyltransferase, *UGDH* rat UDP-glucose dehydrogenase

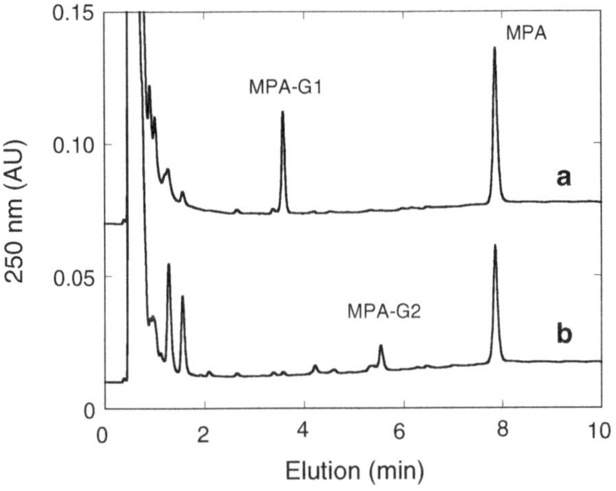

Fig. 10.3 Biosynthesis of regiospecific glucuronides of mycophenolic acid using resting yeast cells. (*A*) HPLC profile of MPA metabolite using resting yeast cell having human UGT1A9 and UGDH. (*B*) HPLC profile of MPA metabolite using resting yeast cell having rat UGT2B1 and UGDH. *MPA* mycophenolic acid, *MPA-G1* phenolic glucuronide of MPA, *MPA-G2* acyl glucuronide of MPA

the yeast cell, indicating the involvement of endogenous transporters in yeast cell membranes. Recently, Bureik's group (at PomBio Tech) successfully established a whole-cell biotransformation system using *Schizosaccharomyces pombe* (Dragan et al. 2010; Zöllner et al. 2010; Buchheit et al. 2011). Comparison of glucuronide production between *Saccharomyces cerevisiae* and *Schizosaccharomyces pombe* showed some differences in efficiency in an isoform-dependent manner.

One of the advantages of enzymatic biosynthesis is regiospecific glucuronidation of substrates with multiple conjugation sites. Mycophenolic acid used in immunosuppretant drugs has phenolic and carboxylic hydroxyl groups. Figure 10.3 shows the glucuronide production of mycophenolic acid using resting

Table 10.2 Biosynthesis of acyl glucuronides using recombinant yeast cells

Substrates	Production (mg/l)
Mefenamic acid	60
Flufenamic acid	132
Diclofenac	260
Zomepirac	47

yeast cells with the UGT-UGDH gene. Human UGT1A9 in yeast cells mainly catalyzed the phenolic glucuronide formation. In contrast, rat UGT2B1 can specifically catalyze acyl glucuronide formation. Selecting the suitable mammalian UGT isoforms in yeast, we can obtain the desired glucuronides of various compounds with multiple conjugation sites.

β-1-*O*-Acyl glucuronides have been shown to be reactive metabolites that hydrolyze, acyl migrate, and bind covalently to proteins causing potential toxicity (Spahn-Langguth and Benet 1992). This toxicity may be caused by formation of immunoreactive glucuronide protein adducts (Worrall and Dickinson 1995), but toxicity mechanisms involving modification of active sites of enzymes (Terrier et al. 1999) and interaction with structural proteins (Bailey et al. 1998) have also been suggested. Thus, So acyl glucuronide is one of the most important metabolites for safety testing of drug development. Table 10.2 shows the biosynthesis of acyl glucuronides of various drugs with carboxylic groups using resting yeast cells with the rat UGT2B1-UGDH gene. Based on our results, the genetically modified budding yeast, *Saccharomyces cerevisiae*, with the UGDH gene is one of the useful host organisms for characterization of UGT function and the production of UGT metabolites via whole cell-dependent biosynthesis.

10.4 Biosynthesis of Sulfotransferase (SULT)-Dependent Metabolites of Xenobiotics

In addition to glucuronidation, sulfonation is an important reaction in the phase II process for numerous xenobiotics, drugs, and endogenus compounds (Gamage et al. 2005). For example, biotransformation of phenolic drugs and xenobiotics yields products that are water soluble, less biologically active, and in most cases stable enough to be readily excreted from the body. These reactions are catalyzed by sulfotransferases (SULTs), which also are important phase II drug-metabolizing enzymes and have a supergene family. Of these human SULTs, 1A1, 1A3, 1B1, 1E1, and 2A1 are major cytosolic isoforms in liver, intestine, or lung (Riches et al. 2009). For biosynthesis of sulfo-conjugates in yeast, we have attempted to express these SULTs in *S. cerevisiae*, which has intrinsic enzymes to convert ATP to 3′-phosphoadenosine 5′-phosphosulfate (PAPS) as sulfonate donor. Expression of human SULT isoforms in *S. cerevisiae* was performed using yeast expression vector pGYR. Western blot analysis confirmed the protein expression of human

Table 10.3 Biosynthesis of sulfo-conjugates using recombinant yeast cells

Isoforms	Substrates	Production (mg/l)
SULT1A1	7-Hydroxycoumarine	165
SULT1A3	1-Hydroxypyrene	210
SULT1E1	Minoxidil	13
SULT2A1	Testosterone	52

SULTs. Sulfatase treatment of the more polar metabolite of 7-hydroxycoumarine revealed the production of sulfo-conjugates in resting yeast cells. Table 10.3 shows the conversion rate of sulfo-conjugated metabolites of various compounds. SULT1A1 catalyzed efficiently the sulfo-conjugation of small phenolic compounds such as 7-hydroxycoumarine. Minoxidil is an antihypertensive agent and hair growth promoter that is metabolized by sulfation to the active compound, minoxidil sulfate. In the SULT isoforms, SULT1E1 significantly catalyzed the conversion of the minoxidil sulfate. Selecting the suitable human SULT isoforms in yeast, we can obtain desirable sulfo-conjugates of various compounds including drugs, environmental pollutants, and dietary compounds.

10.5 Biosynthesis of Phase I–Phase II-Dependent Metabolites

The heterologous expression of xenobiotic-metabolizing enzymes using various host systems has been reported to characterize substrate specificity and to determine the contribution of these enzymes to drug metabolism in vivo (Guengerich et al. 1997). Most xenobiotics are thought to be biotransformed via xenobiotic phase I and II enzyme-dependent multi-step pathways, resulting in the complexity of overall drug metabolism in the body (Tirona and Pang 1996). In contrast to expression systems for individual enzymes, few studies on the coexpression of phase I and II enzymes have been performed. In the current study, a coexpression system was developed that reconstituted xenobiotic phase I and II biotransformation. P450 and the conjugating enzyme, UGT, were introduced into the yeast strain AH22 through plasmid-based, pGYR, and genome integration-based, pAUR, genetic engineering, respectively (Fig. 10.4). To achieve the coupled hydroxylation–glucuronidation reactions in the coexpression system, 7-ethoxycoumarine was used as model substrate. Oxidative deethylation of 7-ethoxycoumarine by human P450 1A1 generates 7-hydroxycoumarine, which is then susceptible to conjugation of the 7-hydroxy group by UGT1A6. Figure 10.5 shows the significant formation of 7-glucuronated coumarine in the whole-cell system. Formation of 7-hydroxycoumarine, which was catalyzed by P450, was linear versus time and then reached a plateau phase. In contrast, 7-glucuronated coumarine formation was detected after 7-hydroxycoumarine accumulation following a lag phase and then linear with time. This formation pattern of the metabolites represents a typical sequential conversion by the two enzymes. The time-course of 7-ethoxycoumarine

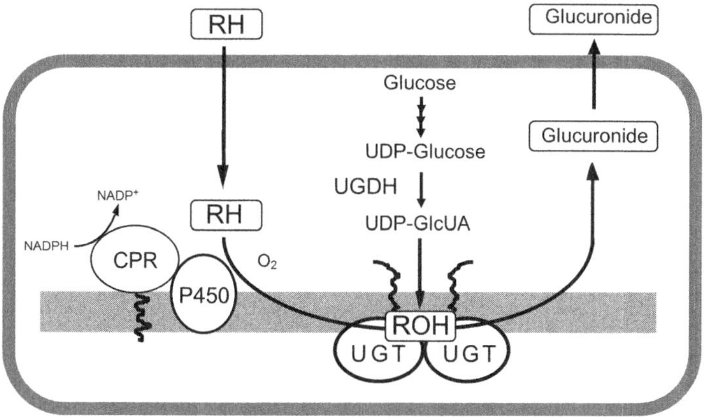

Fig. 10.4 Yeast coexpression system of P450 with conjugation enzymes for biosynthesis of xenobiotic metabolites phase I and phase II coupled reaction. *CPR* NADPH-P450 reductase, *UGDH* UDP-glucose dehydrogenase, *UDP-GlcUA* UDP-glucuronic acid, *RH* hydrophobic substrates, *ROH* hydroxylated intermediates

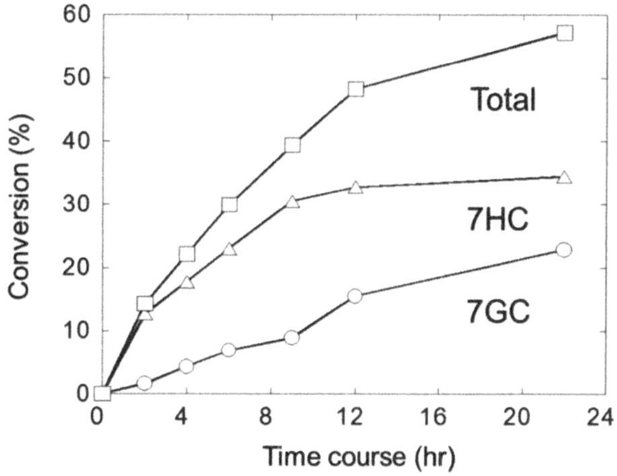

Fig. 10.5 Biosynthesis of 7-hydroxycoumarine and the glucuronide using coexpression system of human P4501A1 and UGT1A6 in yeast. This figure shows the time-course of 7-ethoxycoumarine metabolism in yeast coexpression system of human P450 1A1 and UGT1A6. *7HC* 7-hydroxycoumarine, *7GC* 7-glucuronidated coumarine

metabolism was observed in the yeast coexpression system of P450 1A1 and UGT1A6.

Another phase I- and -II-linked reaction is hydroxylation coupled with sulfo-conjugation. To mimic the reaction in yeast cells, human SULT1A1 was coexpressed with human P4501A2 using a pAUR-pGYR expression system. Figure 10.6 shows the coproduction of the hydroxylated intermediate, 3-hydroxy-7-ethoxycoumarine, and its sulfo-conjugate. This coexpression of the P450 electron-transfer and glucuronidation or sulfo-conjugate production system allows us to obtain phase I metabolites such as 7-hydroxycoumarine, and phase II metabolites, the conjugated hydroxycoumarine from the parent compound. It is an advantage that phase II

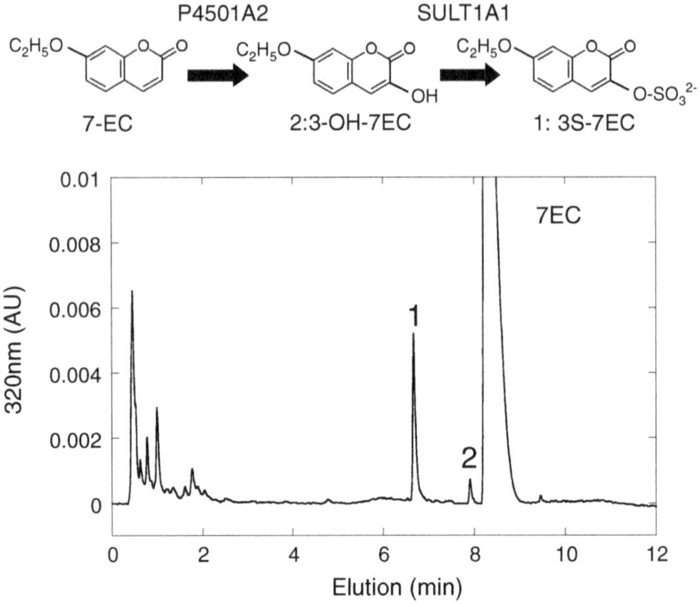

Fig. 10.6 Biosynthesis of 3-hydroxy-7-ethxycoumarine and the sulfo-conjugate using coexpression system of human P4501A2 and SULT1A1 in yeast. This figure showed HPLC profile of 7-ethoxycoumarine metabolism in yeast coexpression system of human P450 1A2 and SULT1A1. *7EC* 7-ethoxycoumarine, *3-OH-7EC* 3-hydroxy-7-ethoxycoumarine, *3S-7EC* 3-sulfated-7-ethoxycoumarine

metabolites can be obtained directly when the phase I metabolite is unknown or is barely synthesized by chemical reaction. In addition to metabolite production for safety testing, analysis of a metabolic profile using a combination of P450 and UGT isoforms can lead to a simple and high-throughput screening protocol.

10.6 Conclusion

This chapter described one of the practical applications of xenobiotic-metabolizing enzymes including P450, UGT, and SULT for biosynthesis of xenobiotic metabolites using the budding yeast cell, *Saccharomyces cerevisiae*. A coexpression system of P450, CPR, UGT, and UGDH was successfully constructed in *S. cerevisiae* cells to produce glucuronides of various xenobiotics. A coexpression system of P450 and SULT was successfully constructed to produce sulfo-conjugated metabolites of various xenobiotics. The recombinant *S. cerevisiae* cells expressing human drug-metabolizing enzymes are useful to predict drug metabolism in the human body. The recombinant *S. cerevisiae* cells appear to be useful to follow the FDA guidance in 2008 on the safety testing of drug metabolites.

References

Andersen MD, Møller BL (2002) Use of methylotropic yeast *Pichia pastoris* for expression of cytochromes P450. Methods Enzymol 357:333–342

Anderson S, Knadler MP, Luffer-Atlas D (2010) Overview of metabolite safety testing from an industry perspective. Bioanalysis 2:1249–1261

Bailey MJ, Worrall S, de Jersey J, Dickinson RG (1998) Zomepirac acyl glucuronide covalently modifies tubulin in vitro and in vivo and inhibits its assembly in an in vitro system. Chem Biol Interact 115:153–166

Buchheit D, Schmitt EI, Bischoff D, Ebner T, Bureik M (2011) S-glucuronidation of 7-mercapto-4-methylcoumarin by human UDP glycosyltransferases in genetically engineered fission yeast cells. Biol Chem 392:1089–1095

Crettol S, Petrovic N, Murray M (2010) Pharmacogenetics of phase I and phase II drug metabolism. Curr Pharm Des 16:204–219

Drăgan CA, Buchheit D, Bischoff D, Ebner T, Bureik M (2010) Glucuronide production by whole-cell biotransformation using genetically engineered fission yeast *Schizosaccharomyces pombe*. Drug Metab Dispos 38:509–515

Gamage N, Barnett A, Hempel N, Duggleby RG, Windmill KF, Martin JL, McManus ME (2005) Human sulfotransferases and their role in chemical metabolism. Toxicol Sci 90:5–22

Guengerich FP, Parikh A, Johnson EF, Richardson TH, von Wachenfeldt C, Cosme J, Jung F, Strassburg CP, Manns MP, Tukey RH, Pritchard M, Fournel-Gigleux S, Burchell B (1997) Heterologous expression of human drug-metabolizing enzymes. Drug Metab Dispos 25:1234–1241

Hayashi K, Sakaki T, Kominami S, Inouye K, Yabusaki Y (2000) Coexpression of genetically engineered fused enzyme between yeast NADPH-P450 reductase and human cytochrome P450 3A4 and human cytochrome b5 in yeast. Arch Biochem Biophys 381:164–170

Ikezawa N, Tanaka M, Nagayoshi M, Shinkyo R, Sakaki T, Inouye K, Sato F (2003) Molecular cloning and characterization of CYP719, a methylenedioxy bridge-forming enzyme that belongs to a novel P450 family, from cultured *Coptis japonica* cells. J Biol Chem 278:38557–38565

Ikushiro S, Sahara M, Emi Y, Yabusaki Y, Iyanagi T (2004) Functional co-expression of xenobiotic metabolizing enzymes, rat cytochrome P450 1A1 and UDP-glucuronosyltransferase 1A6, in yeast microsomes. Biochim Biophys Acta 1672:86–92

Ikushiro S, Masuyama Y, Yasuda K, Kamakura M, Sakaki T (2010) Development glucuronide preparation system for xenobiotic metabolites using genetically engineered budding yeast. Drug Metab Rev 42(suppl 1):S61

Imaoka S, Yamada T, Hiroi T, Hayashi K, Sakaki T, Yabusaki Y, Funae Y (1996) Multiple forms of human P450 expressed in *Saccharomyces cerevisiae*. Systematic characterization and comparison with those of the rat. Biochem Pharmacol 51:1041–1050

Iyanagi T (2007) Molecular mechanism of phase I and phase II drug-metabolizing enzymes: implications for detoxification. Int Rev Cytol 260:35–112

Kasai N, Ikushiro S, Shinkyo R, Yasuda K, Hirosue S, Arisawa A, Ichinose H, Wariishi H, Sakaki T (2010) Metabolism of mono- and dichloro-dibenzo-*p*-dioxins by *Phanerochaete chrysosporium* cytochromes P450. Appl Microbiol Biotechnol 86:773–780

Mackenzie PI, Bock KW, Burchell B, Guillemette C, Ikushiro S, Iyanagi T, Miners JO, Owens IS, Nebert DW (2005) Nomenclature update for the mammalian UDP glycosyltransferase (UGT) gene superfamily. Pharmacogenet Genomics 15:677–685

Maruo Y, Iwai M, Mori A, Sato H, Takeuchi Y (2005) Polymorphism of UDP-glucuronosyltransferase and drug metabolism. Curr Drug Metab 6:91–99

Oka T, Jigami Y (2006) Reconstruction of de novo pathway for synthesis of UDP-glucuronic acid and UDP-xylose from intrinsic UDP-glucose in *Saccharomyces cerevisiae*. FEBS J 273:2645–2657

Peters FT, Bureik M, Maurer HH (2009) Biotechnological synthesis of drug metabolites using human cytochrome P450 isozymes heterologously expressed in fission yeast. Bioanalysis 1:821–830

Radominska-Pandya A, Ouzzine M, Fournel-Gigleux S, Magdalou J (2005) Structure of UDP-glucuronosyltransferases in membranes. Methods Enzymol 400:116–147

Riches Z, Stanley EL, Bloomer JC, Coughtrie MW (2009) Quantitative evaluation of the expression and activity of five major sulfotransferases (SULTs) in human tissues: the SULT "pie". Drug Metab Dispos 37:2255–2261

Sakaki T (2013) Practical application of cytochrome P450. Biol Pharm Bull 35:844–849

Sakaki T, Inouye K (2000) Practical application of mammalian cytochrome P450. J Biosci Bioeng 90:583–590

Sakaki T, Masuyama Y, Nunome M, Takano Y, Yasuda K, Kamakura M, Ikushiro S (2011) Production of Glucurono- or Sulfo-conjugated metabolites using genetically engineered yeast. Proceedings of 17th international conference on cytochrome P450: biochemistry, biophysics and structure. Medimond SR1 (Woods Hole):57–60

Sakaki T, Yamamoto K, Ikushiro S (2013) Possibility of application of cytochrome P450 to bioremediation of dioxins. Biotechnol Appl Biochem 60:65–70

Spahn-Langguth H, Benet LZ (1992) Acyl glucuronides revisited: is the glucuronidation process a toxification as well as a detoxification mechanism? Drug Metab Rev 24:5–47

Terrier N, Benoit E, Senay C, Lapicque F, Radominska-Pandya A, Magdalou J, Fournel-Gigleux S (1999) Human and rat liver UDP-glucuronosyltransferases are targets of ketoprofen acylglucuronide. Mol Pharmacol 56:226–234

Tirona RG, Pang KS (1996) Sequestered endoplasmic reticulum space for sequential metabolism of salicylamide: coupling of hydroxylation and glucuronidation. Drug Metab Dispos 24:821–833

von Bahr C, Bertilsson L (1971) Hydroxylation and subsequent glucuronide conjugation of desmethylimipramine in rat liver microsomes. Xenobiotica 1:205–212

Watanabe KP, Kawai YK, Ikenaka Y, Kawata M, Ikushiro S, Sakaki T, Ishizuka M (2013) Avian cytochrome P450 (CYP) 1–3 family genes: isoforms, evolutionary relationships, and mRNA expression in chicken liver. PLoS One 8:e75689

Worrall S, Dickinson RG (1995) Rat serum albumin modified by diflunisal acyl glucuronide is immunogenic in rats. Life Sci 56:1921–1930

Yun CH, Yim SK, Kim DH, Ahn T (2006) Functional expression of human cytochrome P450 enzymes in *Escherichia coli*. Curr Drug Metab 7:411–429

Zöllner A, Buchheit D, Meyer MR, Maurer HH, Peters FT, Bureik M (2010) Production of human phase 1 and 2 metabolites by whole-cell biotransformation with recombinant microbes. Bioanalysis 2:1277–1290

Chapter 11
Metabolic Diversity and Cytochromes P450 of Fungi

Hirofumi Ichinose

Abstract Many fungal species exhibit unique and superior metabolic functions, including secondary metabolite production and detoxification of environmental pollutants, which are associated with cytochrome P450-dependent reactions. In the last decade, fungal genome projects have uncovered the astonishing molecular diversity of P450s in the fungal kingdom. The tremendous variation among the P450s implies that fungi have vigorously diversified P450 functions to meet novel metabolic needs. Fungal P450s discovered from genome projects are often categorized into novel families and subfamilies, suggesting that fungal P450s possess greater divergence than the animal, plant, or bacterial P450s. It is a challenging task to exploit the catalytic functions of the numerous P450s to better understand the biology of fungal metabolic systems. Comprehensive information about the functions of the P450s will also give hints about how to use their catalytic potentials in the biotechnology sector; however, experimental screening remains essential to reveal the catalytic potentials of individual P450s. In this chapter, the fungal metabolic systems in which P450s play a role are described, and the molecular and functional diversity and potential uses of fungal P450s are discussed.

Keywords Alternative electron-transfer pathway • Ascomycetous fungi • Basidiomycetous fungi • Functionomic survey • Metabolic diversity • Molecular diversity

---

H. Ichinose (✉)
Faculty of Agriculture, Kyushu University, 6-10-1 Hakozaki, Higashi-ku, Fukuoka 812-8581, Japan
e-mail: ichinose@agr.kyushu-u.ac.jp

H. Yamazaki (ed.), *Fifty Years of Cytochrome P450 Research*, DOI 10.1007/978-4-431-54992-5_11,

11.1 Metabolic Systems Associated with Cytochromes P450 in Filamentous Fungi

A fungus is a eukaryotic microorganism in a group that includes yeasts, molds, and basidiomycetes. The fungal kingdom encompasses an enormous diversity of taxa, and it has been estimated that there may be from 1.5 to 5.1 million extant fungal species (Hawksworth 2001; Blackwell 2011). Fungi inhabit a broad range of environments, and some play fundamental roles in nutrient cycling by decomposing organic matter on the Earth. As well as the ability to adapt to various nutrient sources, fungi possess unique defense systems to survive exposure to various environmental agents. Consequently, many filamentous fungi rely on P450-dependent reactions to convert environmental pollutants, as do animals (Ichinose et al. 1999; Matsuzaki and Wariishi 2004), and to produce secondary metabolites, as do plants (Ferrer-Sevillano and Fernández-Cañón 2007; Prieto and Woloshuk 1997; Tudzynski et al. 2002; Kimura et al. 2007). Thus, we can assume an evolutionary history in which molecular diversification of fungal P450s conferred survival strategies by meeting novel metabolic requirements.

11.1.1 Bioconversion of Exogenous Compounds by Basidiomycetous Fungi

Wood-rotting basidiomycetes, often categorized into white-rot and brown-rot basidiomycetes, are common inhabitants of forest litter. White-rot basidiomycetes are capable of degrading all plant cell-wall components, including cellulose, hemicellulose, and lignin. The name "white-rot" originally comes from the observation that these fungi degrade lignin, which then appears as colorless cellulose. Because lignin is the most abundant renewable aromatic polymer, and is known as one of the most recalcitrant biomaterials on earth, the biodegradation of lignin is crucial for the carbon cycle of the biosphere (Kirk and Farrell 1987; Gold et al. 1989; Eriksson et al. 1990). As important enzymes involved in lignin biodegradation, the ligninolytic extracellular enzymes, lignin and manganese peroxidase and laccase, have been identified from a series of white-rot basidiomycetes. These enzymes trigger the decomposition of polymeric lignin via nonspecific one-electron oxidation, resulting in the formation of a variety of aromatic and quinoid fragments that are further metabolized intracellularly. White-rot basidiomycetes therefore possess unique and versatile intracellular metabolic systems to support and assist extracellular ligninolytic activities. In contrast, brown-rot basidiomycetes typically decompose cellulose and hemicellulose components of plant cell walls, but leave lignin. Lignin in brown-rotted wood appears dark brown in color and is almost equal in weight to lignin in sound wood (Kirk and Farrell 1987; Blanchette 1991). However, brown-rot basidiomycetes cause some modification of lignin; for instance, brown-rotted lignins were found to be extensively demethylated and significant quantities of phenolic hydroxyl groups were introduced into brown-rotted lignin (Kirk and Adler 1970; Niemenmaa et al. 2008; Yelle et al. 2011). In general, brown-rot basidiomycetes are thought to have repeatedly evolved from white-rot basidiomycetes. Thus, the cellular mechanisms associated with their secondary metabolism may be considerably more sophisticated. Apart from biological interest in the wood-rotting process, many research studies have focused on using brown-rot basidiomycetes in biotechnological applications; however, relatively little is known about the biology of these fungi (Itoh et al. 1997; Kerem et al. 1998; Fahr et al. 1999; Wetzstein et al. 1999).

During the past two decades, a number of studies have highlighted the metabolic diversity of basidiomycetes that is associated with unique P450s. The possible involvement of a cytochrome P450-like enzyme in the metabolism of polycyclic aromatic hydrocarbon (PAH) by the white-rot basidiomycete *Phanerochaete chrysosporium* was first proposed by Sutherland and coworkers (1991). The reaction of benzo[*a*]pyrene hydroxylation was then demonstrated in vitro using the microsomal and cytosolic fractions of *P. chrysosporium* and *Pleurotus pulmonarius* (Masaphy et al. 1995, 1996). The metabolic capability of *P. chrysosporium* was investigated widely using a series of compounds that were known to be P450 substrates in other organisms, and the bioconversions of benzoic acid, camphor, 1,8-cineol, cinnamic acid, *p*-coumaric acid, coumarin, cumene, 1,12-dodecanediol, 1-dodecanol, 4-ethoxybenzoic acid, and 7-ethoxycoumarin were demonstrated (Matsuzaki and Wariishi 2004). Using these 11 molecules as substrates, 23 hydroxylation reactions and two de-ethylation reactions were determined and six novel products were observed, suggesting that *P. chrysosporium* possesses P450s with unique and novel functions. In addition to product analysis, inhibitory experiments have been conducted widely using the P450 inhibitors piperonyl butoxide (PB) and 1-aminobenzotriazole (ABT). For example, the *S*-oxidation and hydroxylation steps involved in the bioconversion of 4-methyldibenzothiophene by the white-rot basidiomycete *Coriolus versicolor* were significantly inhibited in the presence of PB (Ichinose et al. 1999). Currently, a number of reports have demonstrated that PB and ABT effectively inhibit a wide range of bioconversion steps in many white-rot basidiomycetes including *Coriolus* (*Trametes*) spp. (Marco-Urrea et al. 2009; Prieto et al. 2011), *Phanerochaete* spp. (Teramoto et al. 2004a, b; Subramanian and Yadav 2009; Wang et al. 2012), *Phlebia* spp. (Xiao et al. 2011a, b), and *Pleurotus* spp. (Golan-Rozen et al. 2011). In addition, inhibitor experiments have shown that the brown-rot basidiomycete *Tyromyces palustris* metabolizes aromatic compounds via P450-dependent hydroxylation steps (Kamada et al. 2002). However, the biochemical characterization of basidiomycetous P450s has been limited because the poor level of protein expression in fungal cells and the paucity of available gene sequences have hindered subsequent investigations.

11.1.2 Biosynthesis of Secondary Metabolites in Ascomycetous Fungi

Ascomycetes are the largest known group of fungi and important ecologically, nutritionally, and medically. Many researchers have characterized some of the biosynthetic pathways of secondary metabolites in ascomycetous fungi; for example, the metabolic pathways of mycotoxins such as the aflatoxins and trichothecenes have been well established. Mycotoxins have been implicated as causative agents of adverse health effects in humans and animals who have consumed fungus-infected agricultural products. Consequently, fungi that produce mycotoxins are potential problems from both public health and economic perspectives. Because the different enzymes involved in secondary metabolite biosynthesis pathways are often encoded by genes in gene clusters, some of the P450s found in gene clusters were investigated intensively even in the pre-genomic era.

Aflatoxins, a group of polyketide-derived furanocoumarins, are toxic and carcinogenic compounds produced by *Aspergillus* spp. (Kensler et al. 2011). The aflatoxins were first discovered in the late 1950s and, soon after, it was revealed that toxicity was associated with the presence of *Aspergillus flavus*. Thus, the name "aflatoxin" was assigned to the toxic agents from *A. flavus*. Among the 14 different types of aflatoxin known in nature, aflatoxin B1 is considered to be the most toxic of the compounds produced by both *A. flavus* and *Aspergillus parasiticus*. In these fungi, 25 genes, including six P450s, are clustered for aflatoxin production (Yu et al. 1995, 2004). The P450s involved in the aflatoxin biosynthesis pathway have been categorized into family/subfamily, namely, CYP58B (aflU; epoxidase), CYP59 (aflN; versicolorin A oxidase), CYP60A (*aflG*; averantin hydroxylase), CYP60B (*aflL*; versicolorin B desatulase), CYP62 (aflV; function unknown), and CYP64 (aflQ; *O*-methylsterigmatocystin hydroxylase) (Ehrlich et al. 2004; Yu et al. 2004).

As well as the aflatoxin mycotoxins, the trichothecene mycotoxins have a strong impact on human health because they inhibit protein synthesis. Potentially hazardous concentrations of the trichothecene mycotoxins can occur naturally in agricultural crops and commodities (Ueno 1989). Trichothecenes are a diverse family of sesquiterpenoid toxins produced by various fungi such as *Fusarium* spp., *Myrothecium* spp., *Trichothecium* spp., and *Stachybotrys* spp. In the trichothecenes biosynthetic pathway of *Fusarium* spp., the intermediate formation of trichodiene is the first committed step, and it is followed by several oxygenation steps. The subsequent C-2 hydroxylation of trichodiene is catalyzed by CYP58 (*tri4*). CYP58 has unique multifunctional properties and can catalyze four sequential reaction steps, of C-2 hydroxylation, C-12,13 epoxidation, C-11 hydroxylation, and C-3 hydroxylation, to produce isotrichotriol (Fig. 11.1) (Tokai et al. 2007). After the initial stage in the synthetic pathway, isotrichotriol is further modified by CYP65A1 (*tri11*; C-15 hydroxylase), CYP526 (*tri13*; C-4 hydroxylase), and CYP68 (*tri1*; C-8 hydroxylase) to produce the trichothecene mycotoxins (Kimura et al. 2007).

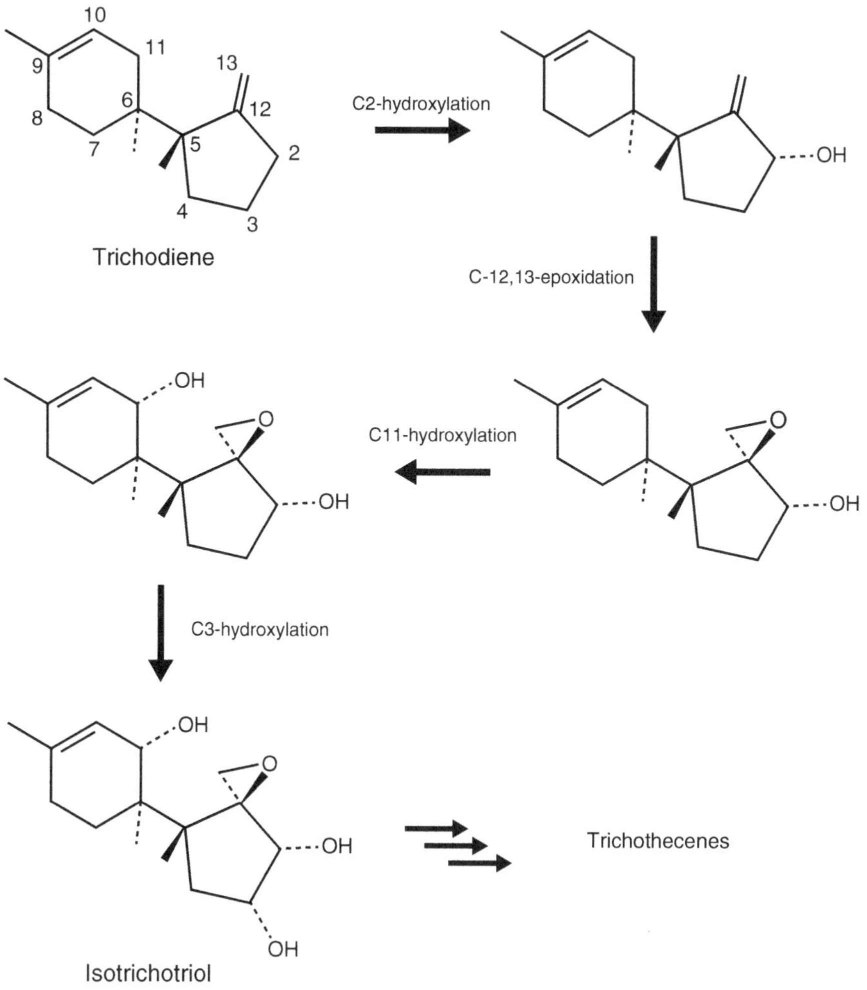

Fig. 11.1 Multi-step reaction catalyzed by CYP58 (Kimura et al. 2007)

Apart from their mycotoxic effects, ascomycetous fungi have other biological impacts by, for example, producing plant hormones such as gibberellin. The gibberellins are diterpenoid acids and a group of phytohormones that influence many developmental processes in higher plants. Although gibberellin was first identified as a secondary metabolite of the rice pathogen *Fusarium* (previously *Gibberella*) *fujikuroi*, which causes the superelongation disease of rice shoots, they are ubiquitous in higher plants as endogenous growth regulators. The similar biosynthetic pathway in plants and fungus led to the suggestion of a possible evolutionary history in which the gene cluster for gibberellin biosynthesis in *F. fujikuroi* emerged by horizontal gene transfer from the host plant (Chapman and Ragan 1980). However, gene cloning of all the pathways in plants and in

F. fujikuroi revealed dramatic differences in the character of the enzymes involved, suggesting that the gibberellin biosynthetic pathway had developed independently in plants and fungi (Tudzynski 2005). Both plants and fungi synthesize the cyclic diterpene, ent-kaurene, as an important intermediate in the initial stage of the gibberellin biosynthesis pathway, but, at the later stages, they employ different enzyme systems to produce gibberellin (Tudzynski 2005). After the intermediate formation of ent-kaurene, *F. fujikuroi* employs four different P450s (CYP68A1, CYP68B1, CYP69A1, and CYP503A1) to produce gibberellins. CYP68A1 (*P450-1*) was first cloned by differential cDNA screening. Subsequently, CYP68B1 (P450-2), CYP69A1 (P450-3), and CYP503A1 (P450-4) were identified from the gibberellin biosynthesis gene cluster. In particular, CYP68A1 and CYP503A1 are unique in their multifunctional properties; CYP68A1 catalyzes four sequential steps in the gibberellin biosynthetic pathway and CYP503A1 catalyzes three oxidation steps between ent-kaurene and ent-kaurenoic acid.

11.2 Molecular Diversity of Fungal Cytochromes P450

11.2.1 Conservative Functions of Fungal P450s

In the fungal kingdom, only the housekeeping CYP51 and CYP61 gene families are evolutionarily conserved, and their functions are mandatory for ergosterol biosynthesis (Aoyama et al. 1984; Kelly et al. 1995). CYP51 was first purified from *Saccharomyces cerevisiae* in 1984 (Yoshida and Aoyama 1984; Aoyama et al. 1984), and orthologous genes from mammals (Trzaskos et al. 1986), plants (Kahn et al. 1996), and bacteria (Aoyama et al. 1998; Bellamine et al. 1999) were then discovered. The presence of its orthologues in all biological kingdoms suggests that CYP51 is the oldest of the known eukaryotic P450s. Fungal CYP51 catalyzes the oxidative removal of the 14α-methyl group of lanosterol and eburicol (24-methylene-24,25-dihydrolanosterol) to produce Δ14,15-desaturated intermediates in the ergosterol biosynthesis pathway. The presence of a second P450 in *Saccharomyces cerevisiae* was first demonstrated by disruption of CYP51 (Kelly et al. 1993), which was then sequenced and identified as CYP61 (Kelly et al. 1995; Skaggs et al. 1996). CYP61 is responsible for sterol C-22-desaturation at the later stage of the ergosterol biosynthesis pathway, suggesting that it may have evolved after a gene duplication of CYP51.

The 14α-demethylation by CYP51 is a more complex reaction that is dissimilar from the common P450-dependent monooxygenation reaction. The 14α-demethylation involves three sequential reaction steps, each requiring one molecule of oxygen and one molecule of NADPH (Waterman and Lepesheva 2005; Lepesheva and Waterman 2007; Yoshida 1992; Shyadehi et al. 1996). The first and second reactions follow a typical P450 reaction mechanism in which the 14α-methyl group is oxidized to the related alcohol and then aldehyde.

The 14α-aldehyde group is released as formic acid, and a C14–C15 double bond is introduced in the third step. Although the catalytic properties of CYP51 are well conserved, the protein sequences are considerably diverse. The average amino acid sequence identity of CYP51 within the fungal kingdom is about 41 %, which is significantly lower than in animals (77 %), plants (76 %), and bacteria (77 %). Phylogenetic analysis revealed the presence of three clades (type A, type B, type C) of the CYP51 family in the fungal kingdom (Becher et al. 2011). In *Aspergillus fumigatus*, two different CYP51s, classified into type A and type B, are present. Neither the type-A nor type-B CYP51 variant is individually essential for fungal growth and virulence, but inactivation of both genes is lethal (Hu et al. 2007), indicating either of the proteins could probably replace the other. In contrast, only type A but not type B of CYP51 is required for conidiation of virulence of *Magnaporthe oryzae* (Yan et al. 2011).

Besides the housekeeping genes, the CYP53 family is also distributed widely in both ascomycetous and basidiomycetous fungi. CYP53 was first discovered from *Aspergillus niger* (van Gorcom et al. 1990) and the functions of this family are well characterized; they catalyze the para-hydroxylation of benzoic acid and its derivatives, which are naturally occurring antifungal plant metabolites (Amborabe et al. 2002). Because the phenolic compounds produced by CYP53 are further metabolized and degraded via the β-ketoadipate pathway (Harwood and Parales 1996), the CYP53 family was presumed to be crucial for competitive plant defenses. The numbers of CYP53 genes vary in the genomes of wood-rotting basidiomycetes, and the phylogenetic diversity of this family in *Postia placenta* is clear. Interestingly, CYP53D2 is capable of catalyzing the *O*-demethylation of stilbene derivatives, indicating its possible roles in degrading plant-based aromatic compounds, as well as benzoate derivatives, during the decomposition of woody biomass. A unique reaction mechanism might be involved in the *O*-demethylation reaction catalyzed by CYP53D2 because the CYP53 family has generally been considered to exhibit substrate specificity for benzoate; indeed, the carboxyl group in benzoate is known to be essential for enzyme–substrate binding (Matsuzaki and Wariishi 2005; Podobnik et al. 2008).

The CYP505 family is also common in ascomycetes and basidiomycetes. The first enzyme from this family, CYP505 (P450foxy), was isolated from *Fusarium oxysporum* and its catalytic activity, the subterminal hydroxylation of fatty acids, was demonstrated (Nakayama et al 1996). The characteristics of CYP505 resemble those of bacterial CYP102 in that both enzymes are self-sufficient flavocytochrome P450s composed of an amino-terminal P450 domain and a carboxy-terminal reductase domain in a single polypeptide chain. It was hypothesized that fungal CYP505 evolved from bacterial CYP102 by horizontal gene transfer (Kitazume et al. 2000), but the opposite transfer pathway (fungi to bacteria) is also possible (Moktali et al. 2012). The majority of fungal species encode multiple CYP505 genes and, currently, the white-rot basidiomycetes *Phanerochaete chrysosporium* has the largest number of isoforms (seven genes) in its genome. CYP505s are known to catalyze fatty acid hydroxylation, although the full physiological functions of this family have not yet been established. Notably, it has been demonstrated that

CYP505B1 from *Fusarium verticilloides* (*Gibberella moniliformis*) is involved in the fumonisin (a mycotoxin) synthesis pathway, where it catalyzes polyketide-amino acid hydroxylation (Proctor et al. 2003). Thus, the molecular evolution of the CYP505 family was required to meet the need for either fatty acid degradation or secondary metabolite biosynthesis.

11.2.2 Molecular Diversity of Fungal P450s

Phanerochaete chrysosporium is the first basidiomycetous fungus for which the whole-genome sequence was determined. The *P. chrysosporium* genome project provided a glimpse into the molecular diversity of basidiomycetous fungi and identified more than 100 novel P450s (Martinez et al. 2004). The discovery of more than 100 novel P450s abolished the earlier notion that fungal P450s might show limited diversity because only three isoforms had been found in the yeast (*S. cerevisiae*) genome and only a modest number of P450 families (51–69 families) were found in the pre-genomic era. With the increasing amount of information that is available as a result of the many genome projects, the current fungal cytochrome P450 database has enlarged exponentially and continues to increase (Moktali et al. 2012). To date, at least 2,784 species of fungal P450s have been identified from 53 sequenced genomes and assigned to 399 families by the P450 nomenclature committee (Nelson 2011); however, the vast majority of their functions and biological roles are still obscure. The compilation of fungal P450 sequences has highlighted a unique feature of the P450 family in the fungal kingdom: it is strikingly diverse. Consequently, CYP51–CYP69, CYP501–CYP 699, and CYP5001–CYP6999 are allocated to fungal P450s. These findings suggested that the diversification of P450s occurred more vigorously in ascomycetous and basidiomycetous fungi than in yeast (Cornell et al. 2007).

Fungal P450s show marked divergence even though the fungi are taxonomically similar. *Aspergillus oryzae*, for instance, has 155 P450 genes (including 13 pseudogenes), which is significantly more than the 111 P450 genes in *Aspergillus nidulans* (Nazir et al. 2010; Kelly et al. 2009). These differences reflect the evolution of the *Aspergillus* species, in which *A. oryzae* increased the number of genes by horizontal gene transfer and duplication (Machida et al. 2005). There are 87 and 89 P450 families in *A. oryzae* and *A. nidulans*, respectively. The ascomycetous fungi *Fusarium graminearum* and *Magnaporthe grisea* are also known to have extremely varied P450 families (Deng et al. 2007). The diversity of the P450 family in ascomycetous fungi is considerably higher than its diversity in basidiomycetous fungi; for example, only 42 P450 families in *Postia placenta* and 31 P450 families in *P. chrysosporium*. Fungal P450s appear to have diversified continuously after generic separation, and perhaps also after speciation. In fact, *A. oryzae* and *A. flavus* exhibit strikingly similar genetic features including similar P450 divergence (155 P450 genes in *A. oryzae* and 167 P450 genes in *A. flavus*), although 16 P450 genes were found in *A. oryzae* and not in *A. flavus*.

The large-scale diversity of basidiomycetous P450s was first uncovered by the white-rot *P. chrysosporium* genomic project, which identified about 150 isoforms. The importance of P450s in the decomposition of plant materials and in the detoxification of environmental pollutants is also true in the brown-rot basidiomycetes; in fact, *P. placenta* possesses a larger number of P450s than *P. chrysosporium* and at least 250 isoforms, including 60 allelic variants, were identified in the whole genome of this fungus (Ide et al. 2012). Currently, the molecular diversity of P450s in both white-rot and brown-rot basidiomycetes has become clearer because of the availability of the whole-genome sequences (Floudas et al. 2012). Comparative phylogenetic analyses of the P450s in *P. chrysosporium* and *P. placenta* have revealed interesting aspects; for example, that the large CYP5027, CYP5350, and CYP5348 families in *P. placenta* are absent in *P. chrysosporium*. In contrast, the CYP5144 family has been enlarged in *P. chrysosporium* but not in *P. placenta*. Another marked difference in the distribution of the P450 family of genes was demonstrated by a comparative analysis of the white-rot basidiomycetes, *Bjerkandera* sp., *Ganoderma* sp., and *Phlebia* sp. (Syed et al. 2013a). In basidiomycetous fungi, P450 gene expansion seemed to be achieved by tandem gene duplication, and 18–37 % of P450 genes were estimated to have appeared as a result of these events (Syed et al. 2013a; Suzuki et al. 2012). The tandem duplication events most likely occurred continuously even after speciation because 266 P450 genes were identified in the *Phanerochaete carnosa* genome, which is considerably higher than the number in the *P. chrysosporium* genome (Suzuki et al. 2012).

11.3 Genomics to Functions

11.3.1 How We Can Determine P450 Functions

A thorough understanding of the sequence–structure–function relationships of the P450s is a challenging task even in a postgenomic era. Experimental screening remains essential to determine the catalytic potentials of individual P450s. In particular, there is a paucity of literature relating to the biochemical and functional characterization of fungal P450s, making it difficult to predict their functions based on sequence homology. Under such circumstances, transcriptome studies have helped uncover novel P450 functions. For example, it has been demonstrated that basidiomycetous P450s are differently regulated in response to culture conditions, exogenous chemicals, and growth stages (Ichinose et al. 2002; Doddapaneni and Yadav 2005; Chigu et al. 2010). Syed et al. (2010) employed a two-step strategy to identify the specific P450 involved in PAH metabolism in *Phanerochaete chrysosporium*. First, P450s that were upregulated in response to exogenously added PAHs were identified, and then their catalytic functions were determined using recombinant enzymes expressed in *Pichia pastoris*. Consequently, CYP5136A2, CYP5136A3,

CYP5142A3, CYP5144A5, CYP5144A7, and CYP5145A3 were shown to be capable of oxidizing PAHs. Later, the same research group functionally expressed and characterized CYP63A2 (Syed et al. 2011, 2013b). Recently, in *Phanerochaete carnosa* it was shown that various P450s were induced when the fungus was grown on woody substrates, suggesting that some P450s play roles in the decomposition of woody biomass (MacDonald et al. 2011; Suzuki et al. 2012).

The development of a rapid functional screening system will open the door for advanced studies of fungal P450s. Recently, functionomic studies were conducted using a functional screening system in which 425 isoforms of fungal P450s from the basidiomycetous fungi *P. chrysosporium* and *P. placenta*, and from koji mold *A. oryzae*, were coexpressed with yeast NADPH-P450 reductase in *S. cerevisiae* (Hirosue et al. 2011; Nazir et al. 2011; Ide et al. 2012). A functionomic survey resulted in the discovery of novel catalytic potentials of the fungal P450s (Table 11.1), which provided new insight into their biology and demonstrated their potential for use in the biotechnology sector. CYP5145A3, CYP5136A1, CYP5136A3, CYP5141C1, and CYP5150A2, which exhibit versatile functions with broad substrate profiles against a series of aromatic compounds, have been discovered from *Phanerochaete chrysosporium*. CYP5150D1, CYP5027B1, and CYP5350B2v1 from *Postia placenta* also show multifunctional properties against PAHs such as anthracene, carbazole, phenanthrene, and pyrene (Ide et al. 2012). The multifunctional properties of a single P450 species could be physiologically linked to the xenobiotic metabolism system in basidiomycetes, similar to the drug-metabolizing P450s in mammals. A number of basidiomycetous fungi have an expanded CYP512 family in their genome, and the P450s in this family from *P. chrysosporium* and *P. placenta* exhibit catalytic activities against testosterone and progesterone. CYP512N and CYP512P convert both testosterone and dehydroabietic acid, which emphasizes the structural similarity between steroids and abietane diterpenoids. Considering the structural similarity of mammalian steroids and fungal terpenoids, the hypothesis that the CYP512 family is likely to be involved in triterpenoid biosynthesis was proposed (Ide et al. 2012; Chen et al. 2012). Although further biochemical studies are needed to elucidate the physiological functions of the CYP512 family in fungi, the provisional substrates were first demonstrated from functionomic approaches. This example illustrates how comprehensive and non-target-driven screening was able to provide interesting results.

Nazir et al. (2011) conducted functional screening to discover potential P450s for industrial applications and demonstrated that CYP57B3 from the koji mold *Aspergillus oryzae* converted genistein to 8-hydroxy-, 6-hydroxy-, and 3′-hydroxygenistein, all of which have known biological and pharmacological potential (Fig. 11.2). Although hydroxylated genistein can be isolated from natural products including fermented products, the supply of these natural compounds is limited, making it problematic for practical use. Because the synthesis of the isoflavones remains an important objective, the use of CYP57B3 for the production of these value-added rare isoflavonoids from genistein would be of great interest.

Table 11.1 Catalytic potentials of P450 from *Phanerochaete chrysosporium*

Substrate	Reaction	CYP
Steroids		
Testosterone	2β-hydroxylation	CYP5136A1, CYP5136A3, CYP5141C1, CYP5144J1
	6α-hydroxylation	CYP512E1
	6β-hydroxylation	CYP512C1, CYP512E1, CYP512F1, CYP512G2
	7β-hydroxylation	CYP5136A3
	11β-hydroxylation	CYP512C1, CYP512G2, CYP5136A1, CYP5136A3, CYP5141C1, CYP5147A3, CYP5150A2
	12β-hydroxylation	CYP512E1, CYP512G2, CYP5136A3, CYP5144J1
Progesterone	Uncharacterized	CYP512C1, CYP512E1, CYP512F1, CYP512G2, CYP5136A1, CYP5136A3, CYP5141C1, CYP5144J1
Petrochemicals		
Carbazole	3-Hydroxylation	CYP502B1, CYP5036A3, CYP5136A1, CYP5136A3, CYP5141C1, CYP5144A13, CYP5145A3, CYP5150A2
Dibenzofuran	2-Hydroxylation	CYP5136A1, CYP5136A3, CYP5145A3, CYP5150A2
Fluorene	9-Hydroxylation	CYP5136A1, CYP5136A3, CYP5150A2
Dibenzothiophene	*S*-oxidation	CYP502B1, CYP512G2, CYP5136A1, CYP5136A3, CYP5139A1, CYP5144A13, CYP5145A3, CYP5147A1, CYP5147A3, CYP5150A2
	2-hydroxylation	CYP5136A1, CYP5136A3, CYP5145A3, CYP5150A2,
Biphenyl	4-Hydroxylation	CYP5136A1, CYP5138A1, CYP5144A10, CYP5145A3, CYP5150A2
Naphthalene	1-Hydroxylation	CYP5036A3, CYP5136A1, CYP5136A3, CYP5138A1, CYP5139A1, CYP5141C1, CYP5150A2
	2-Hydroxylation	CYP5036A3, CYP5136A1, CYP5136A3, CYP5141C1, CYP5150A2
Plant-related compounds		
7-Ethoxycoumarin	*O*-Deethylation	CYP5037B2, CYP5037B3, CYP5037B4, CYP502B1, CYP5136A1, CYP5136A3, CYP5138A1, CYP5139A1, CYP5141C1, CYP5145A3, CYP5147A1, CYP5147A3, CYP5150A2
	3-Hydroxylation	CYP5136A1, CYP5139A1, CYP5141C1, CYP5147A3
4-Ethoxybenzoic acid	*O*-Deethylation	CYP5136A1, CYP5136A3, CYP5150A2,
Dehydroabietic acid	Uncharacterized	CYP512B1, CYP5035B2, CYP5136A1, CYP5141C1, CYP5150A2, CYP5150B1
Flavone	Uncharacterized	CYP505D4, CYP5035A5, CYP5136A1, CYP5141C1, CYP5142A2, CYP5147A1, CYP5150A2, CYP5150B1
Pharmacochemicals		
Diclofenac	4′-Hydroxylation	CYP5136A1, CYP5150A2, CYP5150B1
Compactin	Uncharacterized	CYP5136A3, CYP5144A9
Naproxen	Uncharacterized	CYP5035A1, CYP5035A2, CYP5035B2, CYP5136A1, CYP5139A1, CYP5141C1, CYP5150A2

Fig. 11.2 Catalytic activity of CYP57B3 against genistein

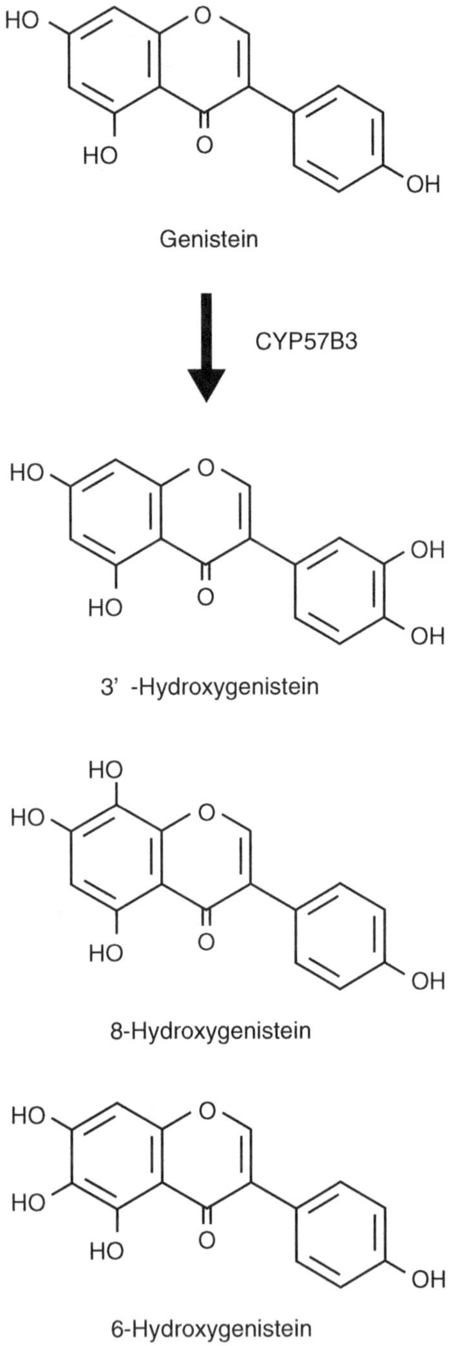

11.3.2 Mechanistic Investigations In Vitro

As well as compiling information about the catalytic potentials of P450s, it is also of great importance to thoroughly understand the biochemical features of individual P450s. Although the biochemical characterization and mechanistic understanding of basidiomycetous P450s are still in their infancy compared with the rich history of these studies in other organisms, some researchers have reported some unique aspects of the reaction mechanisms of the basidiomycetous P450s. The catalytic potentials of CYP5147A1 and CYP5136A1 were first discovered using a functionomic approach, and their kinetic properties in the 3′-hydroxylation of flavone and *O*-deethylation of 7-ethoxycoumarin were investigated intensively using a microsomal fraction of recombinant *S. cerevisiae* (Kasai et al. 2010). Based on the kinetic studies, it was suggested that CYP5147A1 and CYP5136A1 accommodate two substrates in their active sites. Molecular modeling of CYP5147A1 and a docking study of flavone to its active site supported the proposed kinetic properties (Kasai et al. 2010).

The catalytic properties of CYP5150A2 were investigated intensively using the purified enzyme and CYP5150A2 was shown to have high affinity for 4-pentylbenzoic acid (Ichinose and Wariishi 2012). Notably, CYP5150A2 is unique in exhibiting substantial activity with redox partners cytochrome b_5 (Cyt-b5) and NADH-dependent Cyt-b5 reductase (CB5R), even in the absence of cytochrome P450 oxidoreductase (CPR). These results showed that a combination of CB5R and Cyt-b5 may be capable of donating both the first and the second electrons required for the monooxygenation reaction (Fig. 11.3). An alternative electron-transfer pathway in the fungal P450 system associated with CB5R and Cyt-b5 was also demonstrated in vitro using CYP51 from *Candida albicans* and CYP63A2 from *P. chrysosporium* (Lamb et al. 1999; Syed et al. 2011). Disruption of the CPR gene in *S. cerevisiae* was not lethal, and the yeast could still synthesize ergosterol (Sutter and Loper 1989; Venkateswarlu et al. 1998), indicating that the alternative electron-transfer pathway was also active in vivo. The alternative electron-transfer pathway is most likely active in gibberellin production by *Fusarium fujikuroi* (Troncoso et al. 2008). Thus, it may be quite common that some eukaryotic P450s, including CYP5150A2, can be activated by alternative redox partners.

11.4 Conclusion

The data generated by the genomic projects have revealed the massive scale of P450 diversity in the fungal kingdom. The tremendous variations among the P450s imply that fungi have vigorously diversified P450 functions to meet novel metabolic needs, and unique aspects of this genomic feature may be related to evolutionary histories in which each fungal species has diversified differently and expanded the range of its P450 enzymes. In the postgenomic era, the numbers of

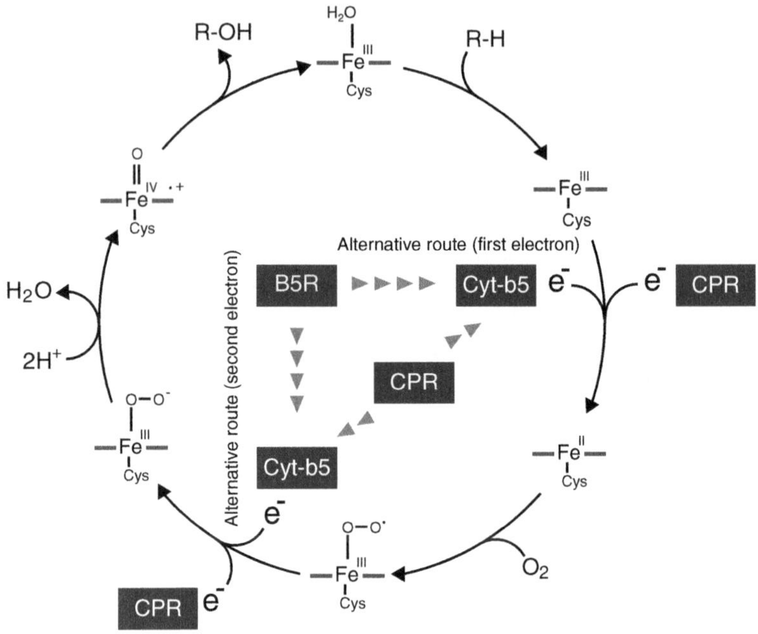

Fig. 11.3 Alternative electron-transfer pathways for P450. A combination of CB5R and Cyt-b5 is capable of donating both the first and the second electrons required for the monooxygenation reaction, as well as cytochrome P450 oxidoreductase (CPR)

"omics" studies have increased, which has contributed to the increased genome-wide understanding and advanced applications of fungal P450s. In addition to building sequence compilations, it is of great importance to explore the catalytic potentials of individual P450s. Many researchers have developed experimental approaches to investigate thoroughly the sequence–structure–function relationships of fungal P450s. Although the biochemical characterization and mechanistic understanding of basidiomycetous P450s are still in the early stages compared with the large amount of information known about P450s in other organisms, snapshots of the unique functions of the fungal P450s are already providing new insights into their fascinating biology and possible uses. Further systematic and functional compilations of fungal P450s will increase our understanding of these enzymes and give rise to new applications for them in the biotechnology sector.

References

Amborabe BE, Fleurat-Lessard P, Chollet JF, Roblin G (2002) Antifungal effects of salicylic acid and other benzoic acid derivatives towards *Eutypa lata*: structure–activity relationship. Plant Phys Biochem 40:1051–1060

Aoyama Y, Yoshida Y, Sato R (1984) Yeast cytochrome P-450 catalyzing lanosterol 14 alpha-demethylation. II. Lanosterol metabolism by purified P-450(14)DM and by intact microsomes. J Biol Chem 259:1661–1666
Aoyama Y, Horiuchi T, Gotoh O, Noshiro M, Yoshida Y (1998) CYP51-like gene of *Mycobacterium tuberculosis* actually encodes a P450 similar to eukaryotic CYP51. J Biochem (Tokyo) 124:694–696
Becher R, Weihmann F, Deising HB, Wirsel SG (2011) Development of a novel multiplex DNA microarray for *Fusarium graminearum* and analysis of azole fungicide responses. BMC Genomics 12:52
Bellamine A, Mangla AT, Nes WD, Waterman MR (1999) Characterization and catalytic properties of the sterol 14alpha-demethylase from *Mycobacterium tuberculosis*. Proc Natl Acad Sci USA 96:8937–8942
Blackwell M (2011) The Fungi: 1, 2, 3 ... 5.1 million species? Am J Bot 98:426–438
Blanchette RA (1991) Delignification by wood-decay fungi. Annu Rev Phytopathol 29:381–398
Chapman DJ, Ragan MA (1980) Evolution of biochemical pathways: evidence from comparative biochemistry. Annu Rev Plant Physiol 31:639–678
Chen S, Xu J, Liu C et al (2012) Genome sequence of the model medicinal mushroom *Ganoderma lucidum*. Nat Commun 3:913
Chigu NL, Hirosue S, Nakamura C, Teramoto H, Ichinose H, Wariishi H (2010) Cytochrome P450 monooxygenases involved in anthracene metabolism by the white-rot basidiomycete *Phanerochaete chrysosporium*. Appl Microbiol Biotechnol 87:1907–1916
Cornell MJ, Alam I, Soanes DM, Wong HM, Hedeler C, Paton NW, Rattray M, Hubbard SJ, Talbot NJ, Oliver SG (2007) Comparative genome analysis across a kingdom of eukaryotic organisms: specialization and diversification in the fungi. Genome Res 17:1809–1822
Deng J, Carbone I, Dean RA (2007) The evolutionary history of cytochrome P450 genes in four filamentous Ascomycetes. BMC Evol Biol 7:30
Doddapaneni H, Yadav JS (2005) Microarray-based global differential expression profiling of P450 monooxygenases and regulatory proteins for signal transduction pathways in the white rot fungus *Phanerochaete chrysosporium*. Mol Genet Genomics 274:454–466
Ehrlich KC, Chang PK, Yu J, Cotty PJ (2004) Aflatoxin biosynthesis cluster gene cypA is required for G aflatoxin formation. Appl Environ Microbiol 70:6518–6524
Eriksson K-EL, Blanchette RA, Ander P (1990) Biodegradation of lignin. In: Timell TE (ed) Microbial and enzymatic degradation of wood and wood components. Springer, Berlin, pp 225–333
Fahr K, Wetzstein HG, Grey R, Schlosser D (1999) Degradation of 2,4-dichlorophenol and pentachlorophenol by two brown rot fungi. FEMS Microbiol Lett 175:127–132
Ferrer-Sevillano F, Fernández-Cañón JM (2007) Novel phacB-encoded cytochrome P450 monooxygenase from *Aspergillus nidulans* with 3-hydroxyphenylacetate 6-hydroxylase and 3,4-dihydroxyphenylacetate 6-hydroxylase activities. Eukaryot Cell 6:514–520
Floudas D, Binder M, Riley R, Barry K, Blanchette RA et al (2012) The Paleozoic origin of enzymatic lignin decomposition reconstructed from 31 fungal genomes. Science 336:1715–1719
Golan-Rozen N, Chefetz B, Ben-Ari J, Geva J, Hadar Y (2011) Transformation of the recalcitrant pharmaceutical compound carbamazepine by *Pleurotus ostreatus*: role of cytochrome P450 monooxygenase and manganese peroxidase. Environ Sci Technol 45:6800–6805
Gold MH, Wariishi H, Valli K (1989) Extracellular peroxidases involved in lignin degradation by the white rot basidiomycete *Phanerochaete chrysosporium*. In: Whitaker JR, Sonnet PE (eds) Biocatalysis in agricultural biotechnology. American Chemical Society Symposium Series, vol 389. American Chemical Society, Washington, DC, pp 127–140
Harwood CS, Parales RE (1996) The beta-ketoadipate pathway and the biology of self-identity. Annu Rev Microbiol 50:553–590
Hawksworth DL (2001) The magnitude of fungal diversity: the 1.5 million species estimate revisited. Mycol Res 105:1422–1432

Hirosue S, Tazaki M, Hiratsuka N, Yanai S, Kabumoto H, Shinkyo R, Arisawa A, Sakaki T, Tsunekawa H, Johdo O, Ichinose H, Wariishi H (2011) Insight into functional diversity of cytochrome P450 in the white-rot basidiomycete *Phanerochaete chrysosporium*: involvement of versatile monooxygenase. Biochem Biophys Res Commun 407:118–123

Hu W, Sillaots S, Lemieux S, Davison J, Kauffman S, Breton A, Linteau A, Xin C, Bowman J, Becker J, Jiang B, Roemer T (2007) Essential gene identification and drug target prioritization in *Aspergillus fumigatus*. PLoS Pathog 3:e24

Ichinose H, Wariishi H (2012) Heterologous expression and mechanistic investigation of a fungal cytochrome P450 (CYP5150A2): involvement of alternative redox partners. Arch Biochem Biophys 518:8–15

Ichinose H, Wariishi H, Tanaka H (1999) Biotransformation of recalcitrant 4-methyldibenzothiophene to water-extractable products using lignin-degrading basidiomycete *Coriolus versicolor*. Biotechnol Prog 15:706–714

Ichinose H, Wariishi H, Tanaka H (2002) Identification and heterologous expression of the cytochrome P450 oxidoreductase from the white-rot basidiomycete *Coriolus versicolor*. Appl Microbiol Biotechnol 59:658–664

Ide M, Ichinose H, Wariishi H (2012) Molecular identification and functional characterization of cytochrome P450 monooxygenases from the brown-rot basidiomycete *Postia placenta*. Arch Microbiol 194:243–253

Itoh N, Yoshida M, Miyamoto T, Ichinose H, Wariishi H, Tanaka H (1997) Fungal cleavage of thioether bond found in yperite. FEBS Lett 412:281–284

Kahn RA, Bak S, Olsen CE, Svendsen I, Moller BL (1996) Isolation and reconstitution of the heme-thiolate protein obtusifoliol 14alpha-demethylase from *Sorghum bicolor* (L.) Moench. J Biol Chem 271:32944–32950

Kamada F, Abe S, Hiratsuk N, Wariish H, Tanaka H (2002) Mineralization of aromatic compounds by brown-rot basidiomycetes -mechanisms involved in initial attack on the aromatic ring. Microbiology 148:1939–1946

Kasai N, Ikushiro S, Hirosue S, Arisawa A, Ichinose H, Uchida Y, Wariishi H, Ohta M, Sakaki T (2010) Atypical kinetics of cytochromes P450 catalysing 3′-hydroxylation of flavone from the white-rot fungus *Phanerochaete chrysosporium*. J Biochem (Tokyo) 147:117–125

Kelly SL, Lamb DC, Baldwin BC, Kelly DE (1993) Benzo(*a*)pyrene hydroxylase activity in yeast is mediated by P450 other than sterol 14 alpha-demethylase. Biochem Biophys Res Commun 197:428–432

Kelly SL, Lamb DC, Corran AJ, Baldwin BC, Parks LW, Kelly DE (1995) Purification and reconstitution of activity of *Saccharomyces cerevisiae* P450 61, a sterol Δ22-desaturase. FEBS Lett 377:217–220

Kelly DE, Kraševec N, Mullins J, Nelson DR (2009) The CYPome (cytochrome P450 complement) of *Aspergillus nidulans*. Fungal Genet Biol 46:S53–S61

Kensler TW, Roebuck BD, Wogan GN, Groopman JD (2011) Aflatoxin: a 50-year odyssey of mechanistic and translational toxicology. Toxicol Sci 120(suppl 1):S28–S48

Kerem Z, Bao W, Hammel KE (1998) Rapid polyether cleavage via extracellular one-electron oxidation by a brown-rot basidiomycete. Proc Natl Acad Sci USA 95:10373–10377

Kimura M, Tokai T, Takahashi-Ando N, Ohsato S, Fujimura M (2007) Molecular and genetic studies of fusarium trichothecene biosynthesis: pathways, genes, and evolution. Biosci Biotechnol Biochem 71:2105–2123

Kirk TK, Adler E (1970) Methoxy-deficient structural elements in lignin of sweetgum decayed by brown-rot fungus. Acta Chem Scand 24:3379–3390

Kirk TK, Farrell RL (1987) Enzymatic "combustion": the microbial degradation of lignin. Annu Rev Microbiol 41:465–505

Kitazume T, Takaya N, Nakayama N, Shoun H (2000) *Fusarium oxysporum* fatty-acid subterminal hydroxylase (CYP505) is a membrane-bound eukaryotic counterpart of *Bacillus megaterium* cytochrome P450BM3. J Biol Chem 275:39734–39740

Lamb DC, Kelly DE, Manning NJ, Kaderbhai MA, Kelly SL (1999) Biodiversity of the P450 catalytic cycle: yeast cytochrome b5/NADH cytochrome b5 reductase complex efficiently drives the entire sterol 14-demethylation (CYP51) reaction. FEBS Lett 462:283–288

Lepesheva GI, Waterman MR (2007) Sterol 14alpha-demethylase cytochrome P450 (CYP51), a P450 in all biological kingdoms. Biochim Biophys Acta 1770:467–477

MacDonald J, Doering M, Canam T, Gong Y, Guttman DS, Campbell MM, Master ER (2011) Transcriptomic responses of the softwood-degrading white-rot fungus *Phanerochaete carnosa* during growth on coniferous and deciduous wood. Appl Environ Microbiol 77:3211–3218

Machida M, Asai K, Sano M et al (2005) Genome sequencing and analysis of *Aspergillus oryzae.* Nature (Lond) 438:1157–1161

Marco-Urrea E, Pérez-Trujillo M, Caminal G, Vicent T (2009) Dechlorination of 1,2,3- and 1,2,4-trichlorobenzene by the white-rot fungus *Trametes versicolor.* J Hazard Mater 166:1141–1147

Martinez D, Larrondo LF, Putnam N et al (2004) Genome sequence of the lignocellulose degrading fungus *Phanerochaete chrysosporium* strain RP78. Nat Biotechnol 22:695–700

Masaphy S, Levanon D, Henis Y, Venkateswarlu K, Kelly SL (1995) Microsomal and cytosolic cytochrome P450 mediated benzo[*a*]pyrene hydroxylation in *Pleurotus pulmonarius.* Biotechnol Lett 17:969–974

Masaphy S, Levanon D, Henis Y, Venkateswarlu K, Kelly SL (1996) Evidence for cytochrome P-450 and P-450-mediated benzo(*a*)pyrene hydroxylation in the white rot fungus *Phanerochaete chrysosporium.* FEMS Microbiol Lett 135:51–55

Matsuzaki F, Wariishi H (2004) Functional diversity of cytochrome P450s of the white-rot fungus *Phanerochaete chrysosporium.* Biochem Biophys Res Commun 324:387–393

Matsuzaki F, Wariishi H (2005) Molecular characterization of cytochrome P450 catalyzing hydroxylation of benzoates from the white-rot fungus *Phanerochaete chrysosporium.* Biochem Biophys Res Commun 334:1184–1190

Moktali V, Park J, Fedorova-Abrams ND, Park B, Choi J, Lee YH, Kang S (2012) Systematic and searchable classification of cytochrome P450 proteins encoded by fungal and oomycete genomes. BMC Genomics 13:525

Nakayama N, Takemae A, Shoun H (1996) Cytochrome P450foxy, a catalytically self-sufficient fatty acid hydroxylase of the fungus *Fusarium oxysporum.* J Biochem (Tokyo) 119:435–440

Nazir KHMNH, Ichinose H, Wariishi H (2010) Molecular characterization and isolation of cytochrome P450 genes from the filamentous fungus *Aspergillus oryzae.* Arch Microbiol 192:395–408

Nazir KHMNH, Ichinose H, Wariishi H (2011) Construction and application of a functional library of cytochrome P450 monooxygenases from the filamentous fungus *Aspergillus oryzae.* Appl Environ Microbiol 77:3147–3150

Nelson DR (2011) Progress in tracing the evolutionary paths of cytochrome P450. Biochim Biophys Acta 1814:14–18

Niemenmaa O, Uusi-Rauva A, Hatakka A (2008) Demethoxylation of [$O^{14}CH_3$]-labelled lignin model compounds by the brown-rot fungi *Gloeophyllum trabeum* and *Poria* (*Postia*) *placenta.* Biodegradation 19:555–565

Podobnik B, Stojan J, Lah L, Krasevec N, Seliskar M, Rizner TL et al (2008) CYP53A15 of *Cochliobolus lunatus*, a target for natural antifungal compounds. J Med Chem 51:3480–3486

Prieto R, Woloshuk CP (1997) ord1, an oxidoreductase gene responsible for conversion of *O*-methylsterigmatocystin to aflatoxin in *Aspergillus flavus.* Appl Environ Microbiol 63:1661–1666

Prieto A, Möder M, Rodil R, Adrian L, Marco-Urre E (2011) Degradation of the antibiotics norfloxacin and ciprofloxacin by a white-rot fungus and identification of degradation products. Bioresour Technol 102:10987–10995

Proctor RH, Brown DW, Plattner RD, Desjardins AE (2003) Co-expression of 15 contiguous genes delineates a fumonisin biosynthetic gene cluster in *Gibberella moniliformis.* Fungal Genet Biol 38:237–249

Shyadehi AZ, Lamb DC, Kelly SL, Kelly DE, Schunck WH, Wright JN, Corina D, Akhtar M (1996) The mechanism of the acyl-carbon bond cleavage reaction catalyzed by recombinant sterol 14 alpha- demethylase of *Candida albicans* (other names are lanosterol 14 alpha-demethylase, P-45014DM, and CYP51). J Biol Chem 271:12445–12450

Skaggs BA, Alexander JF, Pierson CA, Schweitzer KS, Chun KT, Koegel C, Barbuch R, Bard M (1996) Cloning and characterization of the *Saccharomyces cerevisiae* C-22 sterol desaturase gene, encoding a second cytochrome P-450 involved in ergosterol biosynthesis. Gene (Amst) 169:105–109

Subramanian V, Yadav JS (2009) Role of P450 monooxygenases in the degradation of the endocrine-disrupting chemical nonylphenol by the white rot fungus *Phanerochaete chrysosporium*. Appl Environ Microbiol 75:5570–5580

Sutherland JB, Selby AL, Freeman JP, Evans FE, Cerniglia CE (1991) Metabolism of phenanthrene by *Phanerochaete chrysosporium*. Appl Environ Microbiol 57:3310–3316

Sutter TR, Loper JC (1989) Disruption of the *Saccharomyces cerevisiae* gene for NADPH-cytochrome P450 reductase causes increased sensitivity to ketoconazole. Biochem Biophys Res Commun 160:1257–1266

Suzuki H, MacDonald J, Syed K, Salamov A, Hori C et al (2012) Comparative genomics of the white-rot fungi, *Phanerochaete carnosa* and *P. chrysosporium*, to elucidate the genetic basis of the distinct wood types they colonize. BMC Genomics 13:444

Syed K, Doddapaneni H, Subramanian V, Lam YW, Yadav JS (2010) Genome-to-function characterization of novel fungal P450 monooxygenases oxidizing polycyclic aromatic hydrocarbons (PAHs). Biochem Biophys Res Commun 399:492–497

Syed K, Kattamuri C, Thompson TB, Yadav JS (2011) Cytochrome b_5 reductase-cytochrome b_5 as an active P450 redox enzyme system in *Phanerochaete chrysosporium*: atypical properties and in vivo evidence of electron transfer capability to CYP63A2. Arch Biochem Biophys 509:26–32

Syed K, Nelson DR, Riley R, Yadav JS (2013a) Genome-wide annotation and comparative genomics of cytochrome P450 monooxygenases (P450s) in the polyporale species *Bjerkandera adusta*, *Ganoderma* sp. and *Phlebia brevispora*. Mycologia 105:1445–1455

Syed K, Porollo A, Lam YW, Grimmett PE, Yadav JS (2013b) CYP63A2, a catalytically versatile fungal P450 monooxygenase capable of oxidizing higher-molecular-weight polycyclic aromatic hydrocarbons, alkylphenols, and alkanes. Appl Environ Microbiol 79:2692–2702

Teramoto H, Tanaka H, Wariishi H (2004a) Degradation of 4-nitrophenol by the lignin-degrading basidiomycete *Phanerochaete chrysosporium*. Appl Microbiol Biotechnol 66:312–317

Teramoto H, Tanaka H, Wariishi H (2004b) Fungal cytochrome P450s catalyzing hydroxylation of substituted toluenes to form their hydroxymethyl derivatives. FEMS Microbiol Lett 234:255–260

Tokai T, Koshino H, Takahashi-Ando N, Sato M, Fujimura M, Kimura M (2007) Fusarium Tri4 encodes a key multifunctional cytochrome P450 monooxygenase for four consecutive oxygenation steps in trichothecene biosynthesis. Biochem Biophys Res Commun 353:412–417

Troncoso C, Cárcamo J, Hedden P, Tudzynski B, Rojas MC (2008) Influence of electron transport proteins on the reactions catalyzed by *Fusarium fujikuroi* gibberellin monooxygenases. Phytochemistry 69:672–683

Trzaskos J, Kawata S, Gaylor JL (1986) Microsomal enzymes of cholesterol biosynthesis. Purification of lanosterol 14 alpha-methyl demethylase cytochrome P-450 from hepatic microsomes. J Biol Chem 261:14651–14657

Tudzynski B (2005) Gibberellin biosynthesis in fungi: genes, enzymes, evolution, and impact on biotechnology. Appl Microbiol Biotechnol 66:597–611

Tudzynski B, Rojas MC, Gaskin P, Hedden P (2002) The gibberellin 20-oxidase of *Gibberella fujikuroi* is a multifunctional monooxygenase. J Biol Chem 277:21246–21253

Ueno Y (1989) Trichothecene mycotoxins: mycology, chemistry, and toxicology. Adv Nutr Res 3:301–353 van Gorcom RF, Boschloo JG, Kuijvenhoven A, Lange J, van Vark AJ, Bos CJ, van Balken JA, Pouwels PH, van den Hondel CA (1990) Isolation and molecular characterisation of the benzoate-para-hydroxylase gene (bphA) of *Aspergillus niger*: a member of a new gene family of the cytochrome P450 superfamily. Mol Gen Genet 223:192–197

Venkateswarlu K, Lamb DC, Kelly DE, Manning NJ, Kelly SL (1998) The N-terminal membrane domain of yeast NADPH-cytochrome P450 (CYP) oxidoreductase is not required for catalytic activity in sterol biosynthesis or in reconstitution of CYP activity. J Biol Chem 273:4492–4496

Wang J, Hirai H, Kawagishi H (2012) Biotransformation of acetamiprid by the white-rot fungus *Phanerochaete sordida* YK-624. Appl Microbiol Biotechnol 93:831–835

Waterman MR, Lepesheva GI (2005) Sterol 14 alpha-demethylase, an abundant and essential mixed-function oxidase. Biochem Biophys Res Commun 338:418–422

Wetzstein HG, Stadler M, Tichy HV, Dalhoff A, Karl W (1999) Degradation of ciprofloxacin by basidiomycetes and identification of metabolites generated by the brown rot fungus *Gloeophyllum striatum*. Appl Environ Microbiol 65:1556–1563

Xiao P, Mori T, Kondo R (2011a) Biotransformation of the organochlorine pesticide trans-chlordane by wood-rot fungi. N Biotechnol 29:107–115

Xiao P, Mori T, Kamei I, Kondo R (2011b) A novel metabolic pathway for biodegradation of DDT by the white rot fungi, *Phlebia lindtneri* and *Phlebia brevispora*. Biodegradation 22:859–867

Yan X, Ma WB, Li Y, Wang H, Que YW, Ma ZH, Talbot NJ, Wang ZY (2011) A sterol 14α-demethylase is required for conidiation, virulence and for mediating sensitivity to sterol demethylation inhibitors by the rice blast fungus *Magnaporthe oryzae*. Fungal Genet Biol 48:144–153

Yelle DJ, Wei D, Ralph J, Hammel KE (2011) Multidimensional NMR analysis reveals truncated lignin structures in wood decayed by the brown rot basidiomycete *Postia placenta*. Environ Microbiol 13:1091–1100

Yoshida Y (1992) Sterol biosynthesis. In: Omura T, Oshimura Y, Fujii-Kuriyama Y (eds) Cytochrome P450, 2nd edn. Kodansha, Tokyo, pp 93–101

Yoshida Y, Aoyama Y (1984) Yeast cytochrome P-450 catalyzing lanosterol 14 alpha-demethylation. I. Purification and spectral properties. J Biol Chem 259:1655–1660

Yu J, Chang PK, Cary JW, Wright M, Bhatnagar D, Cleveland TE, Payne GA, Linz JE (1995) Comparative mapping of aflatoxin pathway gene clusters in *Aspergillus parasiticus* and *Aspergillus flavus*. Appl Environ Microbiol 61:2365–2371

Yu J, Chang PK, Ehrlich KC, Cary JW, Bhatnagar D, Cleveland TE, Payne GA, Linz JE, Woloshuk CP, Bennett JW (2004) Clustered pathway genes in aflatoxin biosynthesis. Appl Environ Microbiol 70:1253–1262

Chapter 12
Metabolic Engineering of Flower Color Pathways Using Cytochromes P450

Yoshikazu Tanaka and Filippa Brugliera

Abstract A single plant species can contain more than 250 cytochromes P450. Cytochromes P450 catalyze various reactions in plant biosynthetic pathways and play critical and diversified roles in the biosynthesis of primary and specialized (secondary) compounds, including flavonoids. Flavonoids and their colored derivatives, anthocyanins, are major constituents of flower color. Functional combinations of the cytochromes P450, flavonoid 3′-hydroxylase (F3′H, CYP75B), flavonoid 3′, 5′-hydroxylase (F3′5′H, mostly CYP75A), and flavone synthase II (FNSII, CYP93B) determine flavonoid structure and flower color. The number of hydroxyl groups on the B-ring of anthocyanins catalyzed by F3′H and F3′5′H have an impact on the color: the more the bluer. Wild-type or traditionally bred carnations, roses, and chrysanthemums lack blue/violet flower colors because of the deficiency of a F3′5′H and therefore lack the trihydroxylated (B-ring) anthocyanins based upon delphinidin. Metabolic engineering of the anthocyanin pathway and specifically manipulation of *F3′5′H* and *F3′H* genes has produced an array of flower color modifications in many species. By optimizing transgene expression and suppressing endogenous genes in a species-specific manner, transgenic carnations, roses, and chrysanthemums producing novel violet/blue-hued flowers were engineered. Downregulation of *F3′5′H* and *F3′H* genes, often with expression of a dihydroflavonol 4-reductase gene, yielded redder flowers in petunia, tobacco, and torenia. Modulation of *FNSII* gene expression also impacts flower color as flavones act as co-pigments to induce bluing of anthocyanins and flavones and anthocyanins also share the same precursors.

Y. Tanaka (✉)
Research Institute, Suntory Global Innovation Center Ltd., Shimamoto, Osaka 618-8503, Japan
e-mail: Yoshikazu_Tanaka@suntory.co.jp

F. Brugliera
Florigene Pty Ltd., Melbourne, VIC, Australia

LaTrobe Institute for Molecular Science, La Trobe University, Bundoora, VIC 3086, Australia

H. Yamazaki (ed.), *Fifty Years of Cytochrome P450 Research*,
DOI 10.1007/978-4-431-54992-5_12, © Springer Japan 2014

Keywords Anthocyanidin • Anthocyanin • Flavone synthase • Flavonoid • Flavonoid 3′, 5′-hydroxylase • Flavonoid 3′-hydroxylase

12.1 Diversity of Plant Cytochromes P450

In this chapter, we focus on the recent biotechnological achievements relevant to cytochromes P450 (P450s) determining flower color, as we recently published a review on the same topic (Tanaka 2006; Tanaka and Brugliera 2013).

When the authors started isolation of a flavonoid 3, 5′-hydroxylase (*F3′5′H*) gene in 1987, only a few reactions in plants were known to be catalyzed by P450s and no plant P450 genes had been isolated. Using a polymerase chain reaction (PCR) approach and degenerate primers designed to the conserved P450 heme-binding domain, a number of PCR products were generated using petunia petal cDNA as template where the predicted amino acid sequence suggested the occurrence of a large number of P450s in plant petals. Today many plant P450 sequences have been identified in plants. A land plant species typically contains 250–350 predicted cytochrome P450 (P450) amino acid sequences in the genome, and P450s occupy about 1 % of plant proteins encoded by a genome. The P450 family is the third largest gene family after F-box proteins and receptor-like kinases, although the functions of a majority of plant P450s are yet to be determined (Nelson and Werck-Reichhart 2011).

It has been suggested that plant P450s have been continuously recruited for specialized compounds, either by divergence from general metabolism after gene duplication, or from preexisting specialized compound pathways (Hamberger and Bak 2013).

According to Mizutani, the role of plant P450s can be classified into four categories (Mizutani 2012; Mizutani and Ohta 2010). Category I includes three families (CYP51, CYP710, and CYP97). CYP51G and CYP710A are involved in sterol biosynthesis and the CYP97 family is involved in xanthophyll (a carotenoid) biosynthesis.

Category II members are involved in core metabolisms that are conserved in all land plants, including the biosynthesis of biopolymers such as phenylpropanoids and fatty acids covering the plant surface. Members involved in the phenylpropanoid pathway include CYP73A (cinnamic acid 4-hydroxylase, C4H), CYP98A (3′-hydroxylase of coumaroyl esters), and CYP84A (coniferaldehyde/coniferyl alcohol 5′-hydroxylase). P450s that catalyze modification of fatty acids in *Arabidopsis* are CYP86 (ω-hydroxylation), CYP94 (ω-hydroxylation), and CYP703 (in-chain hydroxylation). CYP77, CYP78, CYP96, CYP709, and CYP726 catalyze hydroxylation of fatty acids in various plants.

Category III includes P450s for essential metabolisms that emerged during the evolution of flowering plants, such as plant hormone homeostasis: CYP735 in the metabolism of cytokinin, CYP707 in the abscidic acid pathway, CYP88A, CYP701A, CYP714A, and CYP714D in biosynthesis and degradation of gibberellins, CYP85A, CYP90A, CYP90B, CYP90C, CYP90D, and CYP724B involved in brassinolide metabolism, CYP74A in jasmonic acid, and CYP711A1 in the biosynthesis of strigolactone plant hormones.

Category IV members contribute to a diversity of specialized compounds. CYP75, CYP93, and CYP81 are involved in flavonoid biosynthesis and are reviewed in this section. P450s involved in terpenoid biosynthesis have been reviewed recently (Hamberger and Bak 2013), as further discussed in another chapter of this book (Fukushima et al. 2014). Benzylisoquinoline alkaloid biosynthesis includes CYP80 and CYP719 members. The dhurrin (a cyanogenic glucoside) pathway includes CYP79A1 and CYP71E1 (Nielsen et al. 2008). Members of CYP79, CYP81, and CYP83 are included in the biosynthesis of versatile glucosinolates in *Arabidopsis* (Sonderby et al. 2010). These P450s involved in specialized metabolites sometimes catalyze atypical reactions for P450s such as methylenedioxy bridge formation, intermolecular C–O phenol coupling, and intramolecular C–C phenol (reviewed by Mizutani and Sato 2011).

12.2 Flower Color

Flower color can be primarily attributed to three classes of pigment: flavonoids, carotenoids, and betalains (Tanaka et al. 2008). Among them, a colored class of flavonoids, the anthocyanins, confers a diverse range of color from orange to red to violet and blue. Carotenoids and betalains confer yellow or red colors (Tanaka et al. 2008).

The color of a flower containing anthocyanins is determined by a combination of various factors: anthocyanin structures, the pH of the vacuole where anthocyanins are stored, coexisting flavonoids (co-pigments), and metal ions (Yoshida et al. 2009). It is interesting that of all the flower color hues present and the large numbers of anthocyanins identified, they are primarily derivatives of only six major classes of anthocyanidins (chromophores of anthocyanins): pelargonidin, cyanidin, peonidin, delphinidin, petunidin, and malvidin. Peonidin is derived from cyanidin, and petunidin and malvidin are methylated derivatives of delphinidin (Fig. 12.1). The number of hydroxyl groups on the B-ring of the anthocyanin molecule is a major determinant of color, with more hydroxyl groups shifting the color to blue. Blue/violet flowers tend to contain delphinidin-based anthocyanins, magenta/red flowers contain predominantly cyanidin-based anthocyanins, and orange/brick-red flowers contain pelargonidin-based anthocyanins (Yoshida et al. 2009). Anthocyanins modified with plural aromatic acyl groups (polyacylanthocyanins) tend to be bluer, as are gentian and butterfly pea. Vacuolar pH greatly affects anthocyanin color, with anthocyanins being redder and stable in lower pH and bluer but more unstable in higher (neutral) pH. The vacuolar pH of morning glory petals increases from pH 6.6 to 7.7 as the flowers open, resulting in intense blue colors in fully open flowers. Co-pigments, typically flavones and flavonols, cause bathochromic shifts (bluer and darker colors) of anthocyanins, as observed in iris flowers (Yabuya et al. 1997).

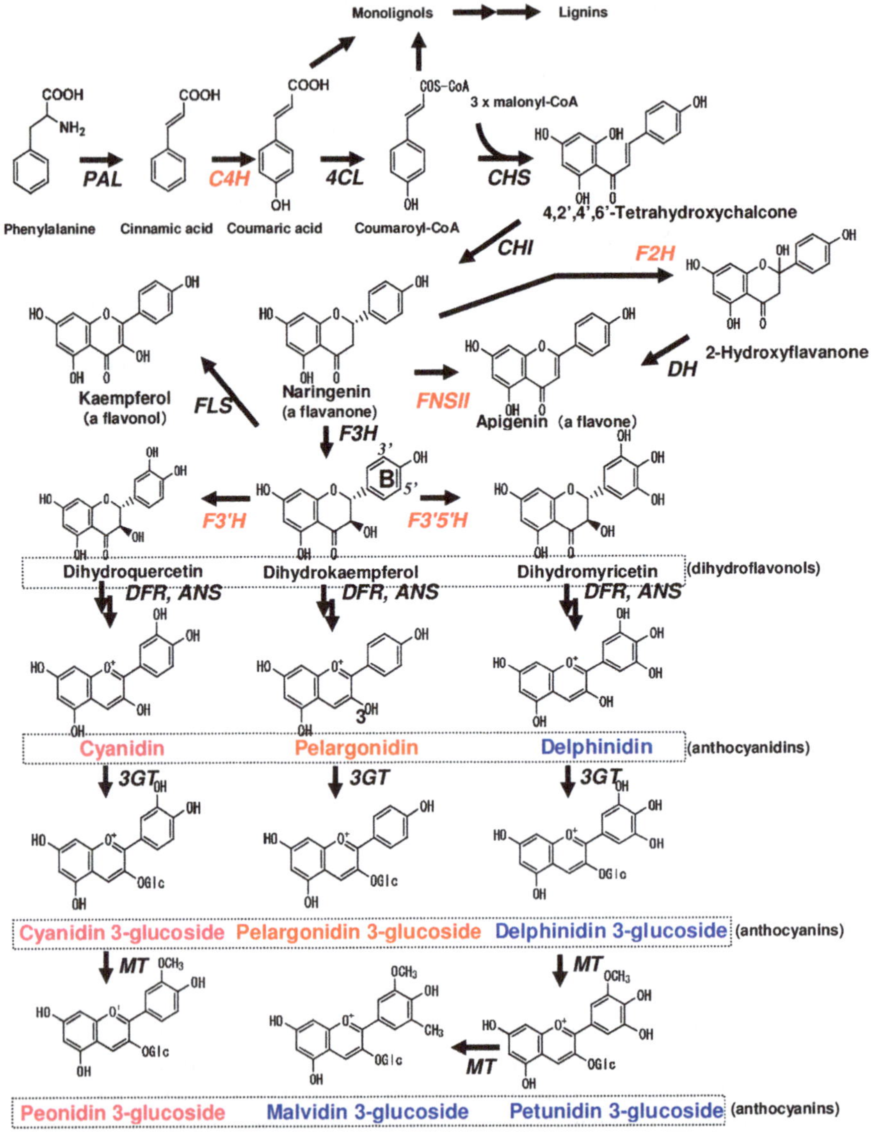

Fig. 12.1 The flavonoid biosynthetic pathway relevant to flower color. The pathways leading to anthocyanidin 3-glucosides are conserved in seed plants. Anthocyanidin 3-glucosides are further modified by glycosylation, acylation, and methylation in a species-specific manner. P450 enzymes are shown in *red letters*. In this review, anthocyanins based on peonidin are included as cyanidin-based anthocyanins whereas those based on petunidin and malvidin are included as delphinidin-based anthocyanins. *PAL* phenylalanine ammonia lyase, *C4H* cinnamic acid 4-hydroxylase, *CL* 4-coumarate Co-A ligase, *CHS* chalcone synthase, *CHI* chalcone isomerase, *F3H* flavanone 3-hydroxylase, *F3′H* flavonoid 3′-hydroxylase, F3′5′H, flavonoid 3′,5′-hydroxylase; *DFR* dihydroflavonol 4-reductase, *ANS* anthocyanidin synthase, *FNS* flavone synthase, *F2H* flavanones 2-hydroxylase, *DH* dehydratase, *FLS* flavonol synthase; 3GT, UDP-glucose: anthocyanidin 3-*O*-glucosyltransferase; MT, *S*-adenosylmethionine: anthocyanin methyltransferase

Metal ions such as ferrous and aluminium ions cause bluing of color by chelating the hydroxyl groups on the B-ring of anthocyanins, as is the case in the blue sections of tulips and blue hydrangea. Super-complexes consisting of anthocyanins, flavonoids, and metal ions are found in cornflower and *Commelina communis* and result in blue flower colors. Plants achieve their own color, especially blue, by a combination of these factors.

12.3 Cytochromes P450 in Pigment Formation

12.3.1 Hydroxylation of the B-Ring

The hydroxylation of the B-ring, a critical step for flower color, is catalyzed by two related P450 enzymes, flavonoid 3′-hydroxylase (F3′H) and F3′5′H. These P450 enzymes have broad substrate specificity and catalyze hydroxylation of flavanones, dihydroflavonols, flavonols, and flavones, but have different substrate preferences depending on the molecular species. Because flavanones and dihydroflavonols are precursors of anthocyanidins and anthocyanins, F3′H and F3′5′H determine the hydroxylation pattern of the B-ring and thus flower color. F3′5′H is essential for the biosynthesis of delphinidin, which is the major pigment of most violet/blue flowers. *Rosa hybrida* (rose), *Chrysanthemum morifolium* (chrysanthemum), *Dianthus caryophyllus* (carnation), and *Lilium* spp. (lily), which occupy more than 50 % of the cut-flower market, lack delphinidin-based anthocyanins because of the deficiency of a *F3′5′H* gene, and so heterologous expression of a *F3′5′H* gene in these species was expected to yield novel blue flower colors.

F3′H and F3′5′H activities were detected in the microsomal fractions of petals of plant species using radioisotope-labeled flavonoids (usually naringenin) and NADPH as an electron donor (Stotz and Forkmann 1981). The genes encoding F3′H (Brugliera et al. 1999) and F3′5′H (Holton et al. 1993) were isolated first from petunia, followed by many other plant species. More and more genes encoding these enzymes are being identified thanks to the determination of various plant genomes. The isolation from many species revealed that F3′H and F3′5′H were classified to different subfamilies in the same family in P450 nomenclature (CYP75B and CYP75A, respectively), except for F3′5′Hs of the Compositae family belonging to CYP75B. CYP75 is found in gymnosperm and angiosperm plants and appeared as early as the ferns but is not found in mosses or algae (Nelson and Werck-Reichhart 2011). Although the genes encoding CYP75A and CYP75B diverged before the speciation of higher plants (Tanaka and Brugliera 2013; Ueyama et al. 2002), many plants including *Arabidopsis*, carnation, and rose do not synthesize delphinidin consequent to the F3′5′H deficiency. The genome sequences of *Arabidopsis*, carnation, and Rosaceae plants (apple, strawberry, peach) revealed that CYP75A-like sequences are not found, indicating these plants lost the CYP75A sequence and thus F3′5′H activity during their evolution.

Convergent evolution of the *F3′5′H* gene in Compositae plants is intriguing. The Compositae plants have a F3′5′H belonging to the CYP75B family: *Callistephus chinensis* (China aster) (F3′5′H, CYP75B5; F3′H, CYP75B6), *Osteospermum hybrida* (African daisy) (F3′5′H, CYP75B17; F3′H, CYP75B14), and *Senecio cruentus* (cineraria) (F3′5′H, CYP75B18; F3′H, CYP75B58). This observation suggests that ancestors of these species had lost a CYP75A-type *F3′5′H* gene and then reacquired the *F3′5′H* gene by duplication and neo-functionalization of a CYP75B gene (Seitz et al. 2006).

It was shown that F3′H and F3′5′H activities were exchangeable by substitution of a small number of amino acid residues. Substitution of Thr-487 to a serine residue of the gerbera F3′H resulted in clear F3′5′H activity, and reverse substitution in an *Osteospermum* F3′5′H reduced F3′5′H activity without sacrificing their specific activity (Seitz et al. 2007).

Dahlia variabilis petals contain 3-hydroxylated chalcones. Its two allelic variants of F3′H differ by only 3 of 508 amino acid residues, and their recombinant enzymes exhibited differences in chalcone 3-hydroxylase (CH3H) activity. Substitution of Ala-425 with a valine residue conferred CH3H activity, and the reciprocal substitution resulted in a loss of CH3H activity. The dahlia F3′H acquired CH3H activity with a single amino acid substitution (Schlangen et al. 2010). These reports suggest plasticity of P450 substrate specificity, which contributes to the diversity of plant compounds.

12.3.2 Flavone Biosynthesis

Flavones are biosynthesized from corresponding flavanones by the catalysis of flavone synthase (FNS). Two types of FNS are known: FNSI belongs to the 2-oxoglutarate-dependent dioxygenase family and FNSII belongs to the P450 family. FNSI from parsley has been well characterized (Martens et al. 2003). Flavone biosynthesis by P450s was initially considered to consist of two enzymes [flavanones 2-hydroxylase (F2H) and dehydratase] via 2-hydroxyl flavanones. The first F2H identified was CYP93B1 from licorice (*Glycyrrhiza echinata*) (Akashi et al. 1998b), followed by *Medicago truncatula* CYP93B10 and -B11 (Zhang et al. 2007). FNSII genes have been isolated from gerbera (CYP93B2) (Martens and Forkmann 1999), snapdragon (CYP93B3), torenia (CYP93B4) (Akashi et al. 1999b), and some other species, including soybean (CYP93B16) (Fleigmann et al. 2010). Legume plants have both F2H and FNSII.

Flavone *C*-glucoside is a strong co-pigment inducing bluing of anthocyanins as observed in *Iris ensata* (Yabuya et al. 1997) and a transgenic blue carnation (Fukui et al. 2003). Recently, the biosynthetic pathway of flavone *C*-glucoside has been elucidated in rice. F2H (Du et al. 2010) catalyzes the 2-hydroxylation of flavanones to result in 2-hydroxyflavanones. A rice glucosyltransferase preferentially *C*-glucosylates the dibenzoylmethane tautomers formed in equilibrium with 2-hydroxyflavanones (Brazier-Hicks et al. 2009). The genes encoding these enzymes could be useful molecular tools to modify flower color and perhaps enable the production of bluer colors of genetically modified roses and carnations.

Other P450s acting in flavonoid biosynthetic pathways include 2-hydroxyisoflavanone synthase (commonly known as isoflavone synthase, CYP93Cs) (Akashi et al. 1999a; Jung et al. 2000), isoflavone 2′-hydroxylase (CYP81E1) (Akashi et al. 1998a), and flavone 6-hydroxylase (CYP82D33, D62) (Berim and Gang 2013). However, they are not involved in floral pigment biosynthesis or have not been used to modify flower color.

12.3.3 Flower Color and P450s: Recent Reports

Alterations in expression of F3′H, F3′5′H, and FNSII or their mutations result in color changes. Because we have recently reviewed this subject (Tanaka and Brugliera 2013), only recent reports that were not covered previously are described here.

A mutational change from purple (accumulation of cyanidin) to deep pink (accumulation of pelargonidin) in carnation petals was attributed to insertion of the Tdic110 (an active hAT-type transposable element) in the second intron of the *F3′H* gene. Excision of Tdic110 restored the color to purple (Momose et al. 2013). Excision of a miniature inverted-repeat transposable element from a red-skinned potato (accumulating pelargonidin) from the first exon of the *F3′5′H* gene restored F3′5′H activity and resulted in a purple-skinned line containing petunidin (Momose et al. 2010). Purple (*B*) and pink (*b*) flowers of pea (*Pisum sativum*) contain delphinidin- and cyanidin/pelargonidin-based anthocyanins, respectively. Characterization of mutants generated by fast neutron bombardment revealed that the allelic pink-flowered *b* mutant lines carried a variety of lesions in the *F3′5′H* gene, including complete gene deletions. The *F3′5′H* gene cosegregated with the *B* locus. These results showed locus *B* encodes F3′5′H (Moreau et al. 2012).

Fragaria × ananassa (garden strawberry) and *Fragaria vesca* (wild strawberry) contain pelargonidin- and cyanidin-based anthocyanins, respectively, in their ripe fruit. Expression of the *F3′H* genes (99 % identity at the amino acid level) from the two species in yeast confirmed that both have functional *F3′H* genes and both F3′Hs can catalyze dihydroflavonols, flavanones, flavonols, and flavones. Expression of the *F. ananassa F3′H* gene declines drastically whereas that of the *F. vesca F3′H* gene was highly expressed during fruit ripening. Such temporal expression patterns help to explain the differences in accumulating anthocyanins in these fruits (Thill et al. 2013).

Dahlia (*Dahlia variabilis*) mainly contains pelargonidin- or cyanidin-based anthocyanins and flavones in the petals. Its black petal color is derived from a high level of anthocyanins. The majority of the black cultivars have relatively low levels of expression of *FNSII* and small amounts of flavones. The low expression level of the *FNSII* gene was presumed to allow for the provision of increased flavanones for anthocyanin biosynthesis (Thill et al. 2012).

12.3.4 P450s in Carotenoid and Betalain Biosynthesis

Carotenoids contribute to the yellow or red color of flowers (such as chrysanthemums, roses, and sunflowers) and fruits (including tomatoes and pumpkins) (Tanaka et al. 2008). Lutein, the most abundant carotenoid in plant photosynthetic tissues contributing to light harvesting, is derived from α-carotene by the catalysis of β-hydroxylases and ε-hydroxylases. Although initially identified β-hydroxylases were non-heme di-iron monooxygenases, β-hydroxylation is also catalyzed by CYP97C1 in *Arabidopsis* (Tian et al. 2004). *Arabidopsis* CYP97A3 encodes another carotenoid hydroxylase catalyzing mainly β-hydroxylation of α-carotene (Kim and DellaPenna 2006).

Betalains also provide yellow or red color to flowers and other organs of species belonging to the families of Caryophyllales such as cactus, portulaca, and beet, except for Caryophyllaceae and Molluginaceae (Tanaka et al. 2008). P450s also play an important role in betalain biosynthesis. CYP76AD1 encodes a novel cytochrome P450 and is required to produce the red betacyanin pigments in beets (*Beta vulgaris*). Silencing of CYP76AD1 results in only yellow betaxanthin pigment, and yellow betalain mutants were complemented by transgenic expression of CYP76AD1. CY76AD1 catalyzes a biosynthetic step from L-DOPA (L-3,4-dihydroxyphenylalanine) to *cyclo*-DOPA. CYP76AD2 and CYP76AD3 from *Amaranthus cruentus* and *Mirabilis jalapa*, respectively, which accumulate betalains, may encode the same activity (Hatlestad et al. 2012).

12.3.5 Nonenzymatic Role of P450s in Biosynthesis of Specialized Compounds

Most flavonoid biosynthetic enzymes are soluble cytosolic proteins. These enzymes have been proposed to form a multi-enzymatic complex (metabolon) by protein–protein interaction, and the complex is anchored to endoplasmic reticulum (ER) membranes via P450s enzymes in the pathway such as C4H and F3′H. Such complexes facilitate metabolic channeling for efficient biosynthesis of flavonoids (Winkel 2004).

Synthesis of dhurrin (the tyrosine-derived cyanogenic glucoside) in *Sorghum bicolor* is catalyzed by membrane-bound CYP79A1 and CYP71E1 and a soluble UDP-glucose-dependent glucosyltransferase, UGT85B1. In the absence of CYP79A1 and CYP71E1, UGT85B1 is localized in the cytosol; however, in the presence of CYP79A1 and CYP71E1, the localization of UGT85B1 shifted toward the surface of the ER. It was suggested that the formation of a complex (metabolon) of these proteins was necessary for efficient biosynthesis of dhurrin (Nielsen et al. 2008). In these cases, the key role of P450s is as an anchor of the metabolons to the membranes rather than the specific activity.

12.4 Engineering of the Flavonoid Biosynthetic Pathway by Modulation of P450 Gene Expression

12.4.1 Tactics to Generate Transgenic Plants

An advantage of plant breeding via genetic engineering over hybridization breeding is that practically any gene from any organism can be utilized to improve or modify the target species while hybridization breeding suffers from genetic constraints of the species. Breeding by genetic engineering consists of four key steps: gene isolation, selection of proper host cultivars, development of transformation and regeneration protocols, and optimization of transgene expression. For release and commercialization of genetically modified plants, it is also necessary to obtain regulatory clearance from government bodies.

Tactics that have been used to engineer blue flower color in a target include attempts to replicate the vacuolar environments found in the species just mentioned. However, changing vacuolar pH or incorporation of metal ions can be detrimental to a plant, and genes regulating these processes are not yet fully understood. On the other hand, modifying the structures of anthocyanins is a proven path to alter flower color because subtle differences (B-ring hydroxyl group number) cause visible effects on color. The anthocyanin biosynthetic pathway has been well characterized with most key genes isolated.

The top-selling cut flowers (roses, chrysanthemums, and carnations) lack delphinidin because of the deficiency of a *F3′5′H* gene. It was predicted that therefore expression of a *F3′5′H* gene in these species would modify the anthocyanin complementation and thus alter the flower color toward violet/blue. However, to achieve this, development of species- or sometimes cultivar-specific transformation protocols and optimization of transgene expression were essential. Because flower color not only depends on anthocyanin structure but also on cell physiology, such as vacuolar pH and coexisting compounds and metal ions, selection of proper host cultivars was important to achieve a novel flower color (blue-violet in this case).

12.4.2 Color Modification of Petunia

Petunia hybrida has been a favorite choice in flower color modification because it is well characterized in terms of flower color genetics and it is easy to transform. Petunia petals contain delphinidin- or cyanidin-based anthocyanins and flavonols but not flavones. The first achievement of flower color modification using a P450 gene was also carried out in petunia: expression of petunia *F3′5′H* cDNAs (CYP75A1 and CYP75A3; petunia has two *F3′5′H* genes) in a pale pink petunia (a low vacuolar pH line) that was deficient in F3′5′H and F3′H activities resulted in reddish-purple transgenic plants that had elevated levels of delphinidin-based anthocyanins in the petals (Holton et al. 1993). Expression of CYP75A1 in a higher vacuolar pH pink flowering line (deficient in F3′5′H and F3′H activities) resulted in violet-colored flowers accumulating delphinidin-based anthocyanins (unpublished results).

Similar flower color and anthocyanidin changes have been reported by using *F3′5′H* genes from *Eustoma grandiflorum* (lisianthus, CYP75A7) (Shimada et al. 1999), *Vinca major* (periwinkle, CYP75A8) (Mori et al. 2004), *Senecio cruentus* (cineraria, CYP58B18v4) (Tanaka and Brugliera 2013), *Antirrhinum kelloggii* (climbing snapdragon, CYP75A48 and CYP75A59) (Ishiguro et al. 2012), and *Phalaenopsis* (Qi et al. 2013).

Tobacco (*Nicotiana tabacum*) petals contain cyanidin-based anthocyanins and flavonols but not delphinidin or flavones. It is noteworthy that expression of a *Campanula medium* (Canterbury bells, CYP75A6) *F3′5′H* gene in tobacco yielded a higher percentage of delphinidin (up to 99 %) than the petunia and lisianthus *F3′5′H* genes, probably because of a more efficient enzymatic character of the Canterbury bells F3′5′H (Okinaka et al. 2003). The high efficiency of the *Campanula F3′5′H* gene has been also shown in chrysanthemum, as described next. Expression of butterfly pea *F3′5′H* (CYP75A24) and a *Viola wittrockiana* (pansy) *F3′5′H* genes resulted in higher delphinidin production and clearer color changes than that of a *Verbena hybrida* (verbena) *F3′5′H* (CYP75A23) gene in transgenic verbena plants (Togami et al. 2006). These results suggest that the choice of the gene source is important for successful diversion of the flavonoid biosynthetic pathway. Although yeast is a suitable host to identify the function of a P450, including F3′5′H, F3′H and FNSII, it is not still possible to select a gene that functions well in an objective plant with yeast, and it is necessary to express candidate genes in the plant, which is time consuming and laborious.

Expression of *F3′H* gene sequences isolated from petunia (Brugliera et al. 1999), cineraria (Tanaka and Brugliera 2013), rose (Fig. 12.2), carnation, and snapdragon (Brugliera et al. 1997), and *Antirrhinum kelloggii* (Ishiguro et al. 2012) in a petunia line that is deficient in *F3′H* gene (*ht1*) resulted in an elevated amount of cyanidin-based anthocyanins and flower color changes from pale pink to intense pink. Tobacco petals expressing a gentian *F3′H* gene showed a slight increase in anthocyanin content and flower color intensity, with conversion of the flavonol quercetin from kaempferol (Nakatsuka et al. 2006).

Petunia does not accumulate pelargonidin-based anthocyanins because the petunia DFR does not catalyze DHK (Forkman and Ruhnau 1987). Expression of maize (Meyer et al. 1987), rose (Tanaka et al. 1995), and gerbera (Helariutta et al. 1993) *DFR* genes in petunia, which accumulate DHK from the deficiency of *F3′5′H*, *F3′H*, and flavonol synthase (FLS) activities, resulted in pelargonidin-based anthocyanins and thus brick-red or orange colors. Downregulation of the endogenous *F3′H* gene and overexpression of a rose *DFR* gene in a red petunia line accumulating cyanidin-based anthocyanins in its petals yielded orange-colored flowers accumulating pelargonidin-based anthocyanins (Tsuda et al. 2004). Moreover, the levels of flavonols also increased, suggesting that the endogenous FLS competed

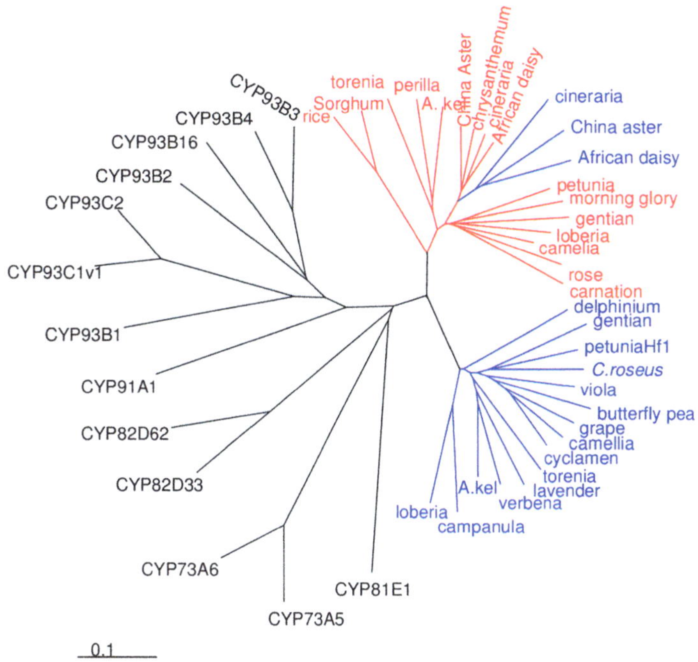

Fig. 12.2 Phylogenetic tree consisting of CYP75A, CYP75B, and P450s members in flavonoid biosynthetic pathway from various plant species. F3′5′H and F3′H are shown in *blue* and *red letters*, respectively. CYP75A subfamily: *Delphinium* (delphinium) (AAX51796), *Viola* (BAF93855), petunia (CYP75A1, P48418), butterfly pea (CYP75A24, BAE72870), grape (CYP75A38v4, ABH06585), *Camellia* (AAY23287), cyclamen (ACX37698), *C. roseus* (*Catharanthus*, CYP75A6, CAA09850), gentian (CYP75A4, BAA12735), lisianthus (CYP75A5, Q96418), torenia (CYP75A10, BAB20076), lavender (CYP75A50, ADA34527), verbena (CYP75A19v2, BAE72871), A. kel. (*Antirrhinum kelloggii*, CYP75A48, BAJ16329), campanula (CYP75A6, O04773), *Lobelia* (BAF49321). CYP75B subfamily: carnation (CYP75B28, AAW01411), torenia (CYP75B10, BAB87838), *Perilla* (CYP75B4, BAB59005), A. kel (*A. kelloggii*, CYP75B50, BAB87838), gerbera (CYP75B15, ABA64468), chrysanthemum (AAW01419), China Aster F3′H(CYP75B6, AAG49298), cineraria F3′H (CYP75B58), African daisy F3′H (CYP75B14, ABB29899), cineraria F3′5′H (CYP75B18v4), *Aster* F3′5′H (CYP75B5, AAG49299), African daisy F3′5′H (CYP75B17, ABB43031), camellia (ADZ28515), rose (AAW01418), *Lobelia* (BAF49324), morning glory (CYP75B19, BAD00190), gentian (CYP75B9, BAD91808). petunia (CYP75B2, Q9SBQ9). CYP73A5 (Arabidopsis cinnamic acid 4-hydroxylase), CYP73A6 (*Zea maize* cinnamic acid 4-hydroxylase), CYP82D33 (basil flavone 6-hydroxylase, AGF30364), CYP82D62 (peppermint flavone 6-hydroxylase, AGF30366). CYP81E1 (*Glycyrrhiza echinata* isoflavone 2-hydroxylase, BAA22422), CYP93A1 (soybean dihydroxypterocarpan 6α-hydroxylase, BAA12159), CYP93B1 (*G. echinata* flavanone 2-hydroxylase, BAA22423), CYP93B2 (gerbera flavone synthase II, AAD39549), CYP93B16 (soybean flavone synthase II, AAF34520), CYP93C1v1 (soybean isoflavone synthase, AAF34520), CYP93C2 (*G. echinata* isoflavone synthase, BAA76380) numbers (http://drnelson.uthsc.edu/Nomenclature.html) and/or Protein database accession numbers are shown in parentheses when available. *Bar* 0.1 substitutions/site

with the introduced rose DFR for substrate. By also downregulating the endogenous *FLS* gene, a deeper orange color may have been obtained. Suppression of the endogenous *F3′5′H* and *FLS* genes and overexpression of a rose *DFR* gene in *Nierembergia*, a species closely related to petunia, resulted in novel pink-colored petals accumulating pelargonidin-based anthocyanins (Tanaka et al. 2010). Similarly, red tobacco flowers with pelargonidin-based anthocyanins and decreased amounts of flavonols were generated by downregulation of the endogenous *F3′H* and *FLS* genes and overexpression of a gerbera *DFR* gene (Nakatsuka et al. 2007).

Expression of a torenia *FNSII* gene in violet petunia yielded paler flower colors resulting from a decrease of anthocyanins and generation of flavones, which the native petunia does not synthesize (Tsuda et al. 2004). Similarly, the expression of a gentian *FNSII* gene in tobacco flowers resulted in flavone accumulation, dramatic decrease of anthocyanins, and paler flower color (Nakatsuka et al. 2006). These results are not surprising because flavones and anthocyanins are derived from the same precursors. These results indicate it is possible to add a nonexistent flavonoid biosynthetic pathway to a plant, but also that attempts to engineer the co-pigments such as flavones to enhance bluing of the anthocyanins may not be easily achieved.

12.4.3 Color Modification of Torenia

Torenia is another model plant used to study flower color and other floral characters (Aida 2008; Nishihara et al. 2013). Blue- or violet-colored *Torenia hybrida* (torenia) petals mainly contain delphinidin-based anthocyanins and some cyanidin-based anthocyanins. Sense suppression of the *F3′5′H* gene led to an increase of cyanidin-based anthocyanins, sacrificing delphinidin-based anthocyanins, with the resulting flower color being pale pink (Suzuki et al. 2000). Additional expression of the torenia *F3′H* gene increased the amount of cyanidin and resulted in darker pink flowers (Ueyama et al. 2002). Suppression of the endogenous *F3′5′H* and *F3′H* genes by RNAi yielded low levels of pelargonidin-based anthocyanins (pale pink). The level was increased by additional expression of a rose or geranium *DFR* gene. Expression of the *DFR* gene coding region from geranium resulted in higher levels of pelargonidin-based anthocyanins (dark pink) (Nakamura et al. 2010). Selected transgenic pink torenia lines were vegetatively propagated and subjected to a field trial. They were found to grow as well as the host control, and the modified color was stable over the course of the trial (Tanaka et al. 2010).

In an attempt to produce darker torenia flowers by increasing the level of anthocyanins, the expression of FNS was downregulated; however, this instead resulted in decreased amounts of both flavones and anthocyanins and an accumulation of flavanones (Ueyama et al. 2002). In this case, the reason for the decrease in anthocyanins was not elucidated.

12.4.4 Generation of a Novel Violet Carnation

Carnation (*Dianthus caryophyllus* L.) originates from the Mediterranean and is one of the major floricultural crops in the world. The yellow flower color of carnations is derived from tetrahydroxychalcone 2′-glucoside whereas the pink, red, and magenta flower colors result from macrocyclic malyl pelargonidin or cyanidin 3,5-diglucosides or their derivatives. They also accumulate flavonols but not usually flavones. The biosynthetic pathway to the production of anthocyanidin 3-glucoside of carnation is common to other plant species. Malylation and 5-glucosylation of anthocyanidin 3-glucoside are catalyzed by malyl-glucose-dependent acyltransferase (Abe et al. 2008) and acyl-glucose dependent 5-*O*-glucosyltransferase (Matsuba et al. 2010), which are localized in the vacuoles. These modification steps are unique to carnation (Sasaki et al. 2013). Classically bred carnations do not produce delphinidin-based anthocyanins because of a deficiency of F3′5′H. Moreover, *F3′5′H* gene sequences have not been found in the recently determined carnation draft genome (Yagi et al. 2013).

A number of strategies have been employed to engineer the production of anthocyanins derived from delphinidin using *F3′5′H* (CYP75A) genes in carnations as described previously (Tanaka and Brugliera 2013), and the transgenic carnations have been marketed since 1996 (Fig. 12.3b, c).

There are multiple paths to achieving exclusive accumulation of delphinidin such as (1) use of host cultivars lacking a competing pathway against F3′5′H, (2) enhancement of the F3′5′H pathway, (3) downregulation of a competing pathway, and (4) a combination of the three strategies.

The first carnation accumulating delphinidin exclusively and having violet flowers was generated by expressing a petunia or pansy *F3′5′H* gene and a petunia *DFR* gene in a white cultivar that lacked *DFR* and *F3′H* genes. Using this strategy a total of six varieties of carnation with novel violet hues have been successfully commercialized (Tanaka and Brugliera 2013).

A second tactic employed was coexpression of a petunia *F3′5′H* gene along with the gene of petunia cytochrome b_5 of which the protein specifically transfers electrons to the petunia F3′5′H. The coexpression in a carnation cultivar producing pelargonidin-based anthocyanins resulted in efficient production of delphinidin-based anthocyanins and subsequent petal color change to produce the dark purple-colored FLORIGENE Moonvelvet (Tanaka and Brugliera 2013). Alternatively, expression of two *F3′5′H* genes from different species such as pansy and *Salvia guaranitica* (a blue salvia) with the petunia *DFR* gene was a successful strategy employed in a cyanidin-producing carnation line. The resulting product was a dark purple-colored carnation FLORIGENE Moonique (Tanaka and Brugliera 2013).

Third, endogenous *DFR* gene expression can be downregulated by transcribing its double-strand RNA (RNAi) (Wang and Waterhouse 2001; Wesley et al. 2001) by which more host varieties can be used. RNAi-induced silencing of carnation *DFR* with expression of a pansy *F3′5′H* and a petunia *DFR* gene in the pelargonidin

Fig. 12.3 Color-modified transgenic flowers by expressing flavonoid 3′-hydroxylase (F3′H) or flavonoid 3′,5′-hydroxylase (F3′5′H) genes. (**a**) Petunia expressing a rose *F3′H* gene (*right*) and its host petunia (*left*) deficient in *F3′H* and *F3′5′H* genes. (**b**) Eight transgenic carnation cultivars expressing *F3′5′H* genes with novel violet flower color are currently sold. Photograph is a vase life test at a carnation grower. (**c**) A flower arrangement containing the transgenic rose and carnations expressing *F3′5′H* genes. (**d**) Expression of the *F3′5′H* gene in chrysanthemums alters flower color from pink to violet (*right*, hosts; *left*, transgenic flowers)

variety successfully resulted in two products, the purple-colored FLORIGENE Moonberry and the light purple-colored FLORIGENE Moonpearl (Tanaka and Brugliera 2013). About 25 million stems of the "Moon" series carnations are sold per year, mainly in the USA, EU, Japan, and Australia.

Because the novel-colored carnations were genetically modified plants, it was necessary to obtain appropriate regulatory approvals from government statutory bodies to sell or grow these flowers. After showing that the release of these carnations was most unlikely to affect biodiversity, permission was granted in several jurisdictions including USA, Japan, Europe, and Australia. The regulatory systems and procedures that apply vary extensively among countries and regions (Tanaka et al. 2009). Data of regulatory clearance that are required for (cut) flowers to be released commercially have been published (Chandler et al. 2013).

12.4.5 Generation of a Novel Violet Rose

Cultivated roses (*Rosa hybrida*) were produced by extensive interspecies hybridization of about eight wild species over hundreds of years (*Rosa gigantean*, *R. gallica*, *R. moschata*, *R. chinensis*, *R. multiflora*, *R. lutea*, *R. foetida*). Rose petals contain cyanidin- and pelargonidin-based anthocyanins and flavonols but not delphinidin or flavones (Mikanagi et al. 1995; Mikanagi et al. 2000). They tend to have lower vacuolar pH. In cultivated rose petals, 5-glucosylation of anthocyanidin precedes 3-glucosylation, differing from other plants. The reaction is catalyzed by a single anthocyanidin 5,3-glucosytransferase (Ogata et al. 2004). Common anthocyanidin 3-glucosyltransferase and its related flavonol 3-glucosyltransferases are also found, and they contribute to anthocyanidin 3-glucosylation in petals and other organs (Fukuchi-Mizutani et al. 2011).

The introduction of pansy-derived *F3′5′H* genes (two *F3′5′H* genes were isolated from pansy) under the control of a constitutive cauliflower mosaic virus 35S (CaMV35S) promoter into rose resulted in a significant amount (up to 60 %) of delphinidin-based anthocyanins in petals and a subsequent color change from red to dark red (Tanaka and Brugliera 2013). The color shift toward blue was not observed, presumably because of the relatively low vacuolar pH (around pH 4–5) of the host rose.

The pansy *F3′5′H* genes under the control of the CaMV35S promoter were subsequently introduced into rose cultivars having higher petal vacuolar pH and lower F3′H activity. The resultant flower color and delphinidin content were variable. Transgenic roses accumulating greater than 90 % delphinidin were obtained for some cultivars with delphinidin 3,5-diglucoside as the dominant anthocyanin (Katsumoto et al. 2007). The two transgenic lines showing most significant and similar color change derived from the same host were selected for commercialization as cut flowers.

After proving that cultivated roses and the transgenic roses do not hybridize with Japanese wild roses under natural conditions, the growth characters of transgenic roses are comparable to their host and that the transgenic plants do not produce substances more harmful than those of the host, general release permission (commercial production and sales permissions) was granted for transgenic roses in Japan (Nakamura et al. 2011a, b). One was first commercialized in 2009 in Japan as the Suntory blue rose Applause (Fig. 12.3c).

To avoid competition with the endogenous *DFR* gene, a genetic construct designed to downregulate the endogenous rose *DFR* gene via RNAi-mediated silencing and overexpress the pansy *F3′5′H* and *Iris hollandica* (iris) *DFR* genes was introduced into rose cultivars. The resultant transgenic events produced flowers that exclusively accumulated anthocyanins derived from delphinidin (more than 99 %) with significant color changes (Katsumoto et al. 2007). It was also shown that the higher the delphinidin content, the bluer the flower color. One of the transgenic roses was crossed with a deep red rose cultivar. The progenies carrying the transgene produce flowers exclusively accumulating delphinidin although flower colors were variable (Katsumoto et al. 2007). Although the results revealed the efficacy of this genetic construct to engineer the rose anthocyanin biosynthetic pathway to delphinidin biosynthesis, the transgenic roses obtained exhibited growth retardation for unknown reasons and were not suitable for commercialization.

12.4.6 Generation of Novel Violet-Blue Chrysanthemum

Cultivated chrysanthemums (*Chrysanthemum × morifolium* Ramat.) presumably originated in China and are one of the important horticultural crops worldwide. All cultivated chrysanthemums are allohexaploid ($2n = 6\times = 54$) and their ancestry includes ten or more primarily hexaploid species (Anderson 2007). In Japan, they are the top-selling cut flowers and occupy about 40 % of the cut-flower market, partly because they are used for decoration in Buddhist rites. A chrysanthemum's inflorescence with 16 ray florets is used as the armorial bearing of the Japanese Emperor's family.

Pink, red, and magenta flower color in chrysanthemums are derived from flavonoids whereas the yellow color is from carotenoids. Flavonoids in chrysanthemum are rather simple: they contain cyanidin 3-(malonyl) glucoside or cyanidin 3-(dimalonyl)-glucoside as anthocyanins (Nakayama et al. 1997) and apigenin, acacetin (4′-methoxyapigenin), luteolin, and diosmetin (4′-methoxyluteolin) glucosides as flavonoids (Schwinn et al. 1994). They do not produce pelargonidin-based anthocyanins because of the presence of F3′H activity (Schwinn et al. 1994). Treatment of the petals with tetcyclasis, a P450 inhibitor, resulted in pelargonidin production. They do not produce delphinidin-based anthocyanin because of the F3′5′H activity deficiency.

Transformation of the chrysanthemum has been difficult for a number of reasons. Chrysanthemum transformation efficiency was low. In addition to development of transformation protocols of chrysanthemums, achieving stable expression of the transgene has been a challenge. In transgenic chrysanthemum plants containing β-glucuronase (GUS) driven by the CaMV35S promoter, GUS activity was reduced to zero level in most of the transgenic lines after 12 months (Takatsu et al. 2000). Modification of petal color in the chrysanthemum was unsuccessful even when using transgenes that had previously been effective in other species such as carnations and rose (unpublished data). Recently, two research programs conquered the unstable transgene expression in chrysanthemum by using petal-specific promoters and generated the violet/blue color (Brugliera et al. 2013; Noda et al. 2013).

Noda et al. introduced *F3′5′H* genes from various plants under the control of various promoters in line 94–765 and cultivar Taihei. Transgenic plants harboring pansy *F3′5′H* genes under the control of various promoters [flavanone 3-hydroxylase (F3H) promoter from chrysanthemum, CaMV35S, rose chalcone synthase (CHS), pansy F3′5′H, *Rosa rugosa* DFR, *R. rugosa* F3H, Gerbera CHS] were generated to assess the ability of promoters. Among them, a chrysanthemum F3H promoter fragment performed best and produced up to 27 % of delphinidin of total anthocyanidins. The 5′-untranslated sequences from alcohol dehydrogenase (Adh), which had been suggested to enhance transgene expression (Satoh et al. 2004), were assessed in chrysanthemum. Campanula, cineraria, verbena, pansy, lisianthus, *A. keloggii*, gentian, two lobelia *F3′5′H* sequences under the control of the chrysanthemum F3H promoter, and the tobacco Adh 5′-untranslated sequence were assessed in line 94–765 and the campanula F3′5′H performed best (maximum, 88 % and 95.1 %; average, 37.5 % and 50.0 %; in line 94–765 and Taihei, respectively). The result may reflect possible high efficiency of *Campanula* F3′5′H.

The higher the delphinidin content, the bluer the chrysanthemum petals became as reported in the transgenic rose accumulating delphinidin (Katsumoto et al. 2007). The flower colors of transgenic lines producing delphinidin-based anthocyanins changed from a red-purple to a purple-violet hue in the Royal Horticultural Society Colour Charts. Such color has not been achieved by hybridization breeding.

Brugliera et al. (2013) analyzed flavonoids of 75 cultivars and selected 53 cultivars (mainly pink) that have a high flavone/cyanidin ratio. Judging from the color of the petal section after feeding with dihydromyricetin (a precursor of delphinidin), 16 cultivars were selected for transformation. Cultivar Improved Reagan, which gave high transformation frequency, was utilized to identify an effective *F3′5′H* transgene that would lead to delphinidin pigment production in chrysanthemum petals. The *F3′5′H* transgenes evaluated included carnation anthocyanidin synthase (ANS) promoter: pansy *F3′5′H*: carnation *ANS* terminator, CaMV35S promoter: pansy *F3′5′H*: CaMV35S terminator, cineraria *F3′5′H* genomic clone, and a rose *CHS* promoter: pansy *F3′5′H*: nopaline synthase terminator (*tnos*). Among them, transgenic Improved Reagan harboring the rose *CHS* promoter: pansy *F3′5′H*: *tnos* transgene resulted in the highest delphinidin content (up to 37 % of total anthocyanidin) and the most significant color changes toward violet. To downregulate cyanidin biosynthesis and further increase delphinidin production, the endogenous *F3′H* gene was suppressed by RNAi-mediated silencing. The authors were careful not to include the conserved sequences around the heme-binding domain of P450 sequences to avoid downregulation of the introduced F3′5′H or perhaps other P450 genes. The highest delphinidin level detected was 80 % in an Improved Reagan line containing the rose *CHS* promoter: pansy *F3′5′H*: tnos transgene along with the rose *CHS* promoter: ds*F3′H* : tnos transgene containing partial sense and antisense fragments of the chrysanthemum F3′H cDNA with an intron from a cineraria *F3′5′H* gene. Downregulation of the endogenous *F3′H* gene was confirmed with Northern blot analysis.

The T-DNA from these *F3′5′H* constructs was then introduced into a further seven chrysanthemum cultivars (Dark Splendid Reagan, Sei Spire, Sei Figaro, Sei Titan, Sei Titan406, Sei Florea, and Sei050-0382). Significant novel bluish petal color changes and high levels of delphinidin-based anthocyanins were achieved (Fig. 12.3d). Such novel petal colors have not been achieved by traditional breeding. A puzzling result is that efficient delphinidin production and color change in chrysanthemum using a campanula *F3′5′H* gene under the control of the chrysanthemum *F3H* promoter attached to a tobacco alcohol dehydrogenase translation enhancer (Noda et al. 2013) was not successfully repeated in Improved Reagan (a daisy-type cultivar) (Brugliera et al. 2013). This result might be caused by differences in host cultivars or transformation protocols.

12.4.7 Flower Color Modification of Other Plants

Expression of a *Commelina F3′5′H* gene in dahlia and phalaenopsis yielded blue/violet color from accumulation of delphinidin-derived anthocyanins (Mii 2012). The details are yet to be published. The color of the transgenic dahlia and phalaenopsis looks bluer than that of the transgenic roses and carnations accumulating delphinidin, probably because dahlia and phalaenopsis have vacuolar contents that are more appropriate to blue color in the presence of delphinidin-based anthocyanins, such as higher vacuolar pH or presence of co-pigments.

Suppression of *F3′H* or *F3′5′H* genes to alter flower color has been carried in a few important ornamental species. By suppressing the endogenous *F3′5′H* gene in *Cyclamen persicum* Mill. (cyclamen), the levels of cyanidin-derived anthocyanins increased and the color changed from purple to red/pink (Boase et al. 2010). Suppression of the *F3′5′H* gene in blue gentian resulted in a decrease of delphinidin and an increase of cyanidin levels with a color change to magenta (Nakatsuka et al. 2008). Accumulation of pelargonidin-based anthocyanins in these species by additional suppression of the *F3′H* genes and expression of a proper *DFR* gene would perhaps redden the flower color.

Flavonoids attract attention as compounds beneficial to health. Flavonoid biosynthetic genes are expressed in rice to accumulate flavonoids in its endosperm where flavonoids are hardly found. Expression of parsley *FNSI*, soybean *FNSII*, rice *PAL*, and rice *CHS* genes resulted in apigenin, chrysoeriol (3′-methoxy apigenin), and tricin (3′,5′-dimethoxy apigenin). Additional expression of pansy *F3′5′H* and rice methyltransferase further elevated the amount of tricetin in endosperm (about 100 times more than wild rice) (Ogo et al. 2013).

12.5 Concluding Remarks

The successful modification of flower color by genetic engineering was possible with the isolation and subsequent manipulation of the plant cytochromes P450 F3′5′H and F3′H, which are key determinants of flower color. In the past 20 years the modification of flower color utilizing P450 sequences has successfully contributed to a better understanding of the flavonoid pathway and metabolic engineering methods. It was the use of a heterologous P450 sequence that led to the introduction of a new pathway and novel products in carnation and rose, and to the world's first commercial launch of a genetically modified cut flower (the Florigene Moondust carnation) in Melbourne, Australia, in 1996. Since then, 11 novel-colored flowers (carnations and rose) have been launched, and more novel-colored products of chrysanthemums and other species will follow.

Acknowledgments The authors are grateful to all colleagues in Suntory Holdings Ltd. (Japan) and Florigene Pty Ltd (Australia), Florigene Europe (Netherlands), and Paso de Luna (Colombia). Dr. David Nelson is acknowledged for his contribution to P450 nomenclature for many years.

References

Abe Y, Tera M, Sasaki N, Okamura M, Umemoto N, Momose M, Kawahara N, Kamakura H, Goda Y, Nagasawa K, Ozeki Y (2008) Detection of 1-*O*-malylglucose: pelargonidin 3-*O*-glucose-6″-*O*-malyltransferase activity in carnation (*Dianthus caryophyllus*). Biochem Biophys Res Commun 373:473–477

Aida R (2008) *Torenia fournieri* (torenia) as a model plant for transgenic studies. Plant Biotechnol 25:541–545

Akashi T, Aoki T, Ayabe S (1998a) CYP81E1, a cytochrome P450 cDNA of licorice (*Glycyrrhiza echinata* L.), encodes isoflavone 2′-hydroxylase. Biochem Biophys Res Commun 251:67–70

Akashi T, Aoki T, Ayabe S (1998b) Identification of a cytochrome P450 cDNA encoding (2*S*)-flavanone 2-hydroxylase of licorice (*Glycyrrhiza echinata* L.; Fabaceae) which represents licodione synthase and flavone synthase II. FEBS Lett 431:287–290

Akashi T, Aoki T, Ayabe S (1999a) Cloning and functional expression of a cytochrome P450 cDNA encoding 2-hydroxyisoflavanone synthase involved in biosynthesis of the isoflavonoid skeleton in licorice. Plant Physiol 121:821–828

Akashi T, Fukuchi-Mizutani M, Aoki T, Ueyama Y, Yonekura-Sakakibara K, Tanaka Y, Kusumi T, Ayabe S (1999b) Molecular cloning and biochemical characterization of a novel cytochrome P450, flavone synthase II, that catalyzes direct conversion of flavanones to flavones. Plant Cell Physiol 40:1182–1186

Anderson NO (2007) Chrysanthemum (*Dendrathema* × *grandiflora* Tzelv.). In: Anderson NO (ed) Flower breeding and genetics. Springer, Amsterdam, pp 389–437

Berim A, Gang DR (2013) The roles of a flavone-6-hydroxylase and 7-*O*-demethylation in the flavone biosynthetic network of sweet basil. J Biol Chem 288:1795–1805

Boase MR, Lewis DH, Davies KM, Marshall GB, Patel D, Schwinn KE, Deroles SC (2010) Isolation and antisense suppression of flavonoid 3′,5′-hydroxylase modifies flower pigments and colour in cyclamen. BMC Plant Biol 10:107

Brazier-Hicks M, Evans KE, Gershater MC, Puschmann H, Steel PG, Edwards R (2009) The *C*-glycosylation of flavonoids in cereals. J Biol Chem 284:17926–17934

Brugliera F, Holton TA, Michael MW (1997) Genetic sequences encoding flavonoid pathway enzymes and uses therefore. International Patent Publication Number WO97/32023

Brugliera F, Barri-Rewell G, Holton TA, Mason JG (1999) Isolation and characterization of a flavonoid 3′-hydroxylase cDNA clone corresponding to the *Ht1* locus of *Petunia hybrida*. Plant J 19:441–451

Brugliera F, Tao GQ, Tems U, Kalc G, Mouradova E, Price K, Stevenson K, Nakamura N, Stacey I, Katsumoto Y, Tanaka Y, Mason JG (2013) Violet/blue chrysanthemums–metabolic engineering of the anthocyanin biosynthetic pathway results in novel petal colors. Plant Cell Physiol 54:1696–1710

Chandler SF, Senior M, Nakamura N, Tsuda S, Tanaka Y (2013) Expression of flavonoid 3′,5′-hydroxylase and acetolactate synthase genes in transgenic carnation: assessing the safety of a nonfood plant. J Agric Food Chem 61:11711–11720

Du Y, Chu H, Chu IK, Lo C (2010) CYP93G2 is a flavanone 2-hydroxylase required for *C*-glycosylflavone biosynthesis in rice. Plant Physiol 154:324–333

Fleigmann J, Furtwangler R, Malterer G, Cantarello C, Schuler G, Ebel J, Mithofer A (2010) Flavone synthase II (CYP93B16) from soybean (*Glycine max* L.). Phytochem 71:508–514

Forkman G, Ruhnau B (1987) Distinct substrate specificity of dihydroflavonol 4-reductase from flowers of *Petunia hybrida*. Z Naturforsch 42c:1146–1148

Fukuchi-Mizutani M, Akagi M, Ishiguro K, Katsumoto Y, Fukui Y, Togami J, Nakamura N, Tanaka Y (2011) Biochemical and molecular characterization of anthocyanidin/flavonol 3-glucosylation pathways in *Rosa* × *hybrida*. Plant Biotechnol 28:239–244

Fukui Y, Tanaka Y, Kusumi T, Iwashita T, Nomoto K (2003) A rationale for the shift in colour towards blue in transgenic carnation flowers expressing the flavonoid 3′,5′-hydroxylase gene. Phytochemistry 63:15–23

Fukushima EO, Seki H, Muranaka T (2014) Plant cytochrome P450s in triterpenoid biosynthesis: diversity and application to combinatorial biosynthesis

Hamberger B, Bak S (2013) Plant P450s as versatile drivers for evolution of species-specific chemical diversity. Philos Trans R Soc Lond B Biol Sci 368:20120426

Hatlestad GJ, Sunnadeniya RM, Akhavan NA, Gonzalez A, Goldman IL, McGrath JM, Lloyd AM (2012) The beet *R* locus encodes a new cytochrome P450 required for red betalain production. Nat Genet 44:816–820

Helariutta Y, Elomaa P, Kotilainen H, Seppanen P, Teeri TH (1993) Cloning of cDNA for dihydroflavonol 4-reductase (DFR) and charaterization of dfr expression in the corollas of *Gerbera hybrida* var. Regina (Compositae). Plant Mol Biol 22:183–193

Holton TA, Brugliera F, Lester DR, Tanaka Y, Hyland CD, Menting JG, Lu CY, Farcy E, Stevenson TW, Cornish EC (1993) Cloning and expression of cytochrome P450 genes controlling flower colour. Nature (Lond) 366:276–279

Ishiguro K, Masumi T, Tanaka Y (2012) Functional analysis of *Antirrhinum kelloggii* flavonoid 3′-hydroxylase and flavonoid 3′,5′-hydroxylase genes; critical role in flower color and evolution in the genus *Antirrhinum*. J Plant Res 124:451–456

Jung W, Yu O, Lau SM, O'Keefe DP, Odell J, Fader G, McGonigle B (2000) Identification and expression of isoflavone synthase, the key enzyme for biosynthesis of isoflavones in legumes. Nat Biotechnol 18:208–212

Katsumoto Y, Mizutani M, Fukui Y, Brugliera F, Holton T, Karan M, Nakamura N, Yonekura-Sakakibara K, Togami J, Pigeaire A, Tao G-Q, Nehra N, Lu C-Y, Dyson B, Tsuda S, Ashikari T, Kusumi T, Mason J, Tanaka Y (2007) Engineering of the rose flavonoid biosynthetic pathway successfully generated blue-hued flowers accumulating delphinidin. Plant Cell Physiol 48:1589–1600

Kim J, DellaPenna D (2006) Defining the primary route for lutein synthesis in plants: the role of *Arabidopsis* carotenoid beta-ring hydroxylase CYP97A3. Proc Natl Acad Sci USA 103:3474–3479

Martens S, Forkmann G (1999) Cloning and expression of flavone synthase II from *Gerbera* hybrids. Plant J 20:611–618

Martens S, Forkmann G, Britsch L, Wellmann F, Matern U, Lukacin R (2003) Divergent evolution of flavonoid 2-oxoglutarate-dependent dioxygenases in parsley. FEBS Lett 544:93–98

Matsuba Y, Sasaki N, Tera M, Okamura M, Abe Y, Okamoto E, Nakamura H, Funabashi H, Takatsu M, Saito M, Matsuoka H, Nagasawa K, Ozeki Y (2010) A novel glucosylation reaction on anthocyanins catalyzed by acyl-glucose-dependent glucosyltransferase in the petals of carnation and delphinium. Plant Cell 22:3374–3389

Meyer P, Heidemann I, Forkmann G, Saedler H (1987) A new petunia flower colour generated by transformation of a mutant with a maze gene. Nature (Lond) 330:677–678

Mii M (2012) Interspecific hybridization, somatic hybridization and genetic transformation. Acta Hortic 953:43–54

Mikanagi Y, Yokoi M, Ueda Y, Saito N (1995) Flower flavonol and anthocyanin distribution in subgenus *Rosa*. Biochem Syst Ecol 23:183–200

Mikanagi Y, Saito N, Yokoi M, Tatsuzawa F (2000) Anthocyanins in flowers of genus *Rosa*, sections *Cinnamomeae* (=*Rosa*), *Chinenses*, *Gallicanae* and some modern garden roses. Biochem Syst Ecol 28:887–902

Mizutani M (2012) Impacts of diversification of cytochrome P450 on plant metabolism. Biol Pharm Bull 35:824–832

Mizutani M, Ohta D (2010) Diversification of P450 genes during land plant evolution. Annu Rev Plant Biol 61:291–315

Mizutani M, Sato F (2011) Unusual P450 reactions in plant secondary metabolism. Arch Biochem Biophys 507:194–203

Momose M, Abe Y, Ozeki Y (2010) Miniature inverted-repeat transposable elements of Stowaway are active in potato. Genetics 186:59–66

Momose M, Nakayama M, Itoh Y, Umemoto N, Toguri T, Ozeki Y (2013) An active hAT transposable element causing bud mutation of carnation by insertion into the flavonoid 3′-hydroxylase gene. Mol Genet Genomics 288:175–184

Moreau C, Ambrose MJ, Turner L, Hill L, Ellis TH, Hofer JM (2012) The B gene of pea encodes a defective flavonoid 3′,5′-hydroxylase, and confers pink flower color. Plant Physiol 159:759–768

Mori S, Kobayashi H, Hoshi Y, Kondo M, Nakano M (2004) Heterologous expression of the flavonoid 3′,5′-hydroxylase gene of *Vinca major* alters flower color in transgenic *Petunia hybrida*. Plant Cell Rep 22:415–421

Nakamura N, Fukuchi-Mizutani M, Fukui Y, Ishiguro K, Suzuki K, Tanaka Y (2010) Generation of red flower varieties from blue *Torenia hybrida* by redirection of the flavonoid pathway from delphinidin to pelargonidin. Plant Biotechnol 27:375–383

Nakamura N, Fukuchi-Mizutani M, Katsumoto Y, Togami J, Senior M, Matsuda Y, Furuichi K, Yoshimoto M, Matsunaga A, Ishiguro K, Aida M, Tasaka M, Fukui H, Tsuda S, Chandler S, Tanaka Y (2011a) Environmental risk assessment and field performance of rose (*Rosa* × *hybrida*) genetically modified for delphinidin production. Plant Biotechnol 28:251–261

Nakamura N, Tems U, Fukuchi-Mizutani M, Chandler S, Matsuda Y, Takeuchi S, Matsumoto S, Tanaka Y (2011b) Molecular based evidence for a lack of gene-flow between *Rosa* × *hybrida* and wild *Rosa* species in Japan. Plant Biotechnol 28:245–250

Nakatsuka T, Nishihara M, Mishiba K, Yamamura S (2006) Heterologous expression of two gentian cytochrome P450 genes can modulate the intensity of flower pigmentation in transgenic tobacco plants. Mol Breed 17:91–99

Nakatsuka T, Abe Y, Kakizaki Y, Yamamura S, Nishihara M (2007) Production of red-flowered plants by genetic engineering of multiple flavonoid biosynthetic genes. Plant Cell Rep 26:1951–1959

Nakatsuka T, Mishiba K, Abe Y, Kubota A, Kakizaki Y, Yamamura S, Nishihara M (2008) Flower color modification of gentian plants by RNAi-mediated gene silencing. Plant Biotechnol 25:61–68

Nakayama M, Koshioka M, Shibata M, Hiradate S, Sugie H, Yamaguchi M (1997) Identification of cyanidin 3-*O*-(3,6-*O*-dimalonyl-β-glucopyranoside) as a flower pigment of chrysanthemum (*Dendranthema grandiflorum*). Biosci Biotechnol Biochem 61:1607–1608

Nelson D, Werck-Reichhart D (2011) A P450-centric view of plant evolution. Plant J 66:194–211

Nielsen KA, Tattersall DB, Jones PR, Moller BL (2008) Metabolon formation in dhurrin biosynthesis. Phytochemistry 69:88–98

Nishihara M, Shimoda T, Nakatsuka T, Arimura G (2013) Frontiers of torenia research: innovative ornamental traits and study of ecological interaction networks through genetic engineering. Plant Methods 9:23

Noda N, Aida R, Kishimoto S, Ishiguro K, Fukuchi-Mizutani M, Tanaka Y, Ohmiya A (2013) Genetic engineering of novel bluer-colored chrysanthemums produced by accumulation of delphinidin-based anthocyanins. Plant Cell Physiol 54:1684–1695

Ogata J, Itoh Y, Ishida M, Yoshida H, Ozeki Y (2004) Cloning and heterologous expression of a cDNA encoding flavonoid glucosyltransferases from *Dianthus caryophyllus*. Plant Biotechnol 2003

Ogo Y, Ozawa K, Ishimaru T, Murayama T, Takaiwa F (2013) Transgenic rice seed synthesizing diverse flavonoids at high levels: a new platform for flavonoid production with associated health benefits. Plant Biotechnol J 11:734–746

Okinaka Y, Shimada Y, Nakano-Shimada R, Ohbayashi M, Kiyokawa S, Kikuchi Y (2003) Selective accumulation of delphinidin derivatives in tobacco using a putative flavonoid 3′,5′-hydroxylase cDNA from *Campanula medium*. Biosci Biotechnol Biochem 67:161–165

Qi Y, Lou Q, Quan Y, Liu Y, Wang Y (2013) Flower-specific expression of the Phalaenopsis flavonoid 3′,5′-hydoxylase modifies flower color pigmentation in *Petunia* and *Lilium*. Plant Cell Tissue Organ Cult 115:263–273

Sasaki N, Matsuba Y, Abea Y, Okamura M, Momose M, Umemoto N, Nakayama M, Itoh Y, Ozekia Y (2013) Recent advances in understanding the anthocyanin modification steps in carnation flowers. Sci Hortic 163:37–45

Satoh J, Kato K, Shinmyo A (2004) The 5′-untranslated region of the tobacco alcohol dehydrogenase gene functions as an effective translational enhancer in plant. J Biosci Bioeng 98:1–8

Schlangen K, Miosic S, Halbwirth H (2010) Allelic variants from *Dahlia variabilis* encode flavonoid 3′-hydroxylases with functional differences in chalcone 3-hydroxylase activity. Arch Biochem Biophys 494:40–45

Schwinn KE, Markham KR, Giveno N (1994) Floral flavonoids and the potential for pelargonidin biosynthesis in commercial chrysanthemum cultivars. Phytochemistry 35:145–150

Seitz C, Eder C, Deiml B, Kellner S, Martens S, Forkmann G (2006) Cloning, functional identification and sequence analysis of flavonoid 3′-hydroxylase and flavonoid 3′,5′-hydroxylase cDNAs reveals independent evolution of flavonoid 3′,5′-hydroxylase in the Asteraceae family. Plant Mol Biol 61:365–381

Seitz C, Ameres S, Forkmann G (2007) Identification of the molecular basis for the functional difference between flavonoid 3′-hydroxylase and flavonoid 3′,5′-hydroxylase. FEBS Lett 581:3429–3434

Shimada Y, Nakano-Shimada R, Ohbayashi M, Okinaka Y, Kiyokawa S, Kikuchi Y (1999) Expression of chimeric P450 genes encoding flavonoid-3′, 5′-hydroxylase in transgenic tobacco and petunia plants. FEBS Lett 461:241–245

Sonderby IE, Geu-Flores F, Halkier BA (2010) Biosynthesis of glucosinolates: gene discovery and beyond. Trends Plant Sci 15:283–290

Stotz G, Forkmann G (1981) Hydroxylation of the B-ring of flavonoids in the 3- and 5-position with enzyme extracts from flowers of *Verbena hybrida*. Z Naturforsch 37c:19–23

Suzuki K, Xue H, Tanaka Y, Fukui Y, Fukuchi-Mizutani M, Murakami Y, Katsumoto Y, Tsuda S, Kusumi T (2000) Flower color modifications of *Torenia hybrida* by cosuppression of anthocyanin biosynthesis genes. Mol Breed 6:239–246

Takatsu Y, Hayashi M, Sakumao F (2000) Transgene inactivation in *Agrobacterium*-mediated chrysanthemum (*Dendranthema grandiflorum* (Ramat.) Kitamura) transformants. Plant Biotechnol 17:241–245

Tanaka Y (2006) Flower colour and cytochromes P450. Phyochem Rev 5:283–291

Tanaka Y, Brugliera F (2013) Flower colour and cytochromes P450. Philos Trans R Soc Lond B Biol Sci 368:20120432

Tanaka Y, Fukui Y, Fukuchi-Mizutani M, Holton TA, Higgins E, Kusumi T (1995) Molecular cloning and characterization of *Rosa hybrida* dihydroflavonol 4-reductase. Plant Cell Physiol 36:1023–1031

Tanaka Y, Sasaki N, Ohmiya A (2008) Plant pigments for coloration: anthocyanins, betalains and carotenoids. Plant J 54:733–749

Tanaka Y, Brugliera F, Chandler S (2009) Recent progress of flower colour modification by biotechnology. Int J Mol Sci 10:5350–5369

Tanaka Y, Brugliera F, Kalc G, Senior M, Dyson B, Nakamura N, Katsumoto Y, Chandler S (2010) Flower color modification by engineering of the flavonoid biosynthetic pathway: practical perspectives. Biosci Biotechnol Biochem 74:1760–1769

Thill J, Miosic S, Ahmed R, Schlangen K, Muster G, Stich K, Halbwirth H (2012) Le Rouge et le Noir: a decline in flavone formation correlates with the rare color of black dahlia (*Dahlia variabilis* hort.) flowers. BMC Plant Biol 12:225

Thill J, Miosic S, Gotame TP, Mikulic-Petkovsek M, Gosch C, Veberic R, Preuss A, Schwab W, Stampar F, Stich K, Halbwirth H (2013) Differential expression of flavonoid 3′-hydroxylase during fruit development establishes the different B-ring hydroxylation patterns of flavonoids in *Fragaria × ananassa* and *Fragaria vesca*. Plant Physiol Biochem 72:72–78

Tian L, Musetti V, Kim J, Magallanes-Lundback M, DellaPenna D (2004) The Arabidopsis *LUT1* locus encodes a member of the cytochrome p450 family that is required for carotenoid epsilon-ring hydroxylation activity. Proc Natl Acad Sci USA 101:402–407

Togami J, Tamura M, Ishiguro K, Hirose C, Okuhara H, Ueyama Y, Nakamura N, Yonekura-Sakakibara K, Fukuchi-Mizutani M, Suzuki K, Fukui Y, Kusumi T, Tanaka Y (2006) Molecular characterization of the flavonoid biosynthesis of *Verbena hybrida* and the functional analysis of verbena and *Clitoria ternatea* F3′5′H genes in transgenic verbena. Plant Biotechnol 23:5–11

Tsuda S, Fukui Y, Nakamura N, Katsumoto Y, Yonekura-Sakakibara K, Fukuchi-Mizutani M, Ohira K, Ueyama Y, Ohkawa H, Holton TA, Kusumi T, Tanaka Y (2004) Flower color modification of *Petunia hybrida* commercial varieties by metabolic engineering. Plant Biotechnol 21:377–386

Ueyama U, Suzuki K, Fukuchi-Mizutani M, Fukui Y, Miyazaki K, Ohkawa H, Kusumi T, Tanaka Y (2002) Molecular and biochemical characterization of torenia flavonoid 3′-hydroxylase and flavone synthase II and modification of flower color by modulating the expression of these genes. Plant Sci 163:253–263

Wang M-B, Waterhouse PM (2001) Application of gene silencing in plants. Curr Opin Plant Biol 5:146–150

Wesley SV, Helliwell CA, Smith NA (2001) Construct design for efficient, effective and high-throughput gene silencing in plants. Plant J 27:581–590

Winkel BSJ (2004) Metabolic channeling in plants. Annu Rev Plant Biol 55:85–107

Yabuya T, Nakamura M, Iwashina T, Yamaguchi M, Takehara T (1997) Anthocyanin-flavone copigmentaion in bluish purple flowers of Japanese garden iris (*Iris ensata* Thunb.). Euphytica 98:163–167

Yagi M, Kosugi S, Hirakawa H, Ohmiya A, Tanase K, Harada T, Kishimoto K, Nakayama M, Ichimura K, Onozaki T, Yamaguchi H, Sasaki N, Miyahara T, Nishizaki Y, Ozeki Y, Nakamura N, Suzuki S, Tanaka Y, Sato S, Shirasawa K, Isobe S, Miyamura Y, Watanabe A, Nakayama S, Kishida Y, Kohara M, Tabata S (2013) Sequence analysis of the genome of carnation (*Dianthus caryophyllus* L.). DNA Res. doi:10.1093/dnares/dst053

Yoshida K, Mori M, Kondo T (2009) Blue flower color development by anthocyanins: from chemical structure to cell physiology. Nat Prod Rep 26:884–915

Zhang J, Subramanian S, Zhang Y, Yu O (2007) Flavone synthases from *Medicago truncatula* are flavanone-2-hydroxylases and are important for nodulation. Plant Physiol 144:741–751

Part III
Gene Regulation of P450

Chapter 13
Aryl Hydrocarbon Receptor Suppresses Cecal Carcinogenesis

Togo Ikuta, Yasuhito Kobayashi, Yoshiaki Fujii-Kuriyama, and Kaname Kawajiri

Abstract The aryl hydrocarbon receptor (AhR) has dual roles in regulating intracellular protein levels as both a ligand-activated transcription factor and a ligand-dependent E3 ubiquitin ligase. We showed that the AhR E3 ubiquitin ligase has a role in the suppression of intestinal carcinogenesis by a previously undescribed ligand-dependent β-catenin degradation pathway. This function of AhR is activated by both xenobiotics and natural AhR ligands, such as indole derivatives that are converted from dietary tryptophan and glucosinolates by intestinal microbes, and which suppresses intestinal tumor development in $Apc^{\mathrm{Min}/+}$ mice. To elucidate whether the tumors develop autonomously in $AhR^{-/-}$ mice as a result of impaired β-catenin degradation or in association with accelerated inflammation, we performed two kinds of experiments using germ-free (GF) $AhR^{-/-}$ mice or compound mutant mice lacking genes for AhR and adaptor apoptosis-associated speck-like protein containing a caspase recruitment domain (ASC), which play an essential role in caspase-1 activation in inflammasomes. Both GF $AhR^{-/-}$ and $AhR^{-/-}$•$ASC^{-/-}$ mice showed considerably reduced tumor development compared with that in $AhR^{-/-}$ mice, albeit in a "cancer-prone" state with aberrant β-catenin

T. Ikuta (✉) • K. Kawajiri
Research Institute for Clinical Oncology, Saitama Cancer Center,
818 Komuro, Ina, Saitama 362-0806, Japan
e-mail: togo@cancer-c.pref.saitama.jp; k.kawajiri@email.plala.or.jp

Y. Kobayashi
Pathology Division, Department of Clinical Laboratory, Saitama Cardiovascular and Respiratory Center, 1696 Itai, Kumagaya 360-0105, Japan
e-mail: kobayashi.yasuhito@pref.saitama.lg.jp

Y. Fujii-Kuriyama
Department of Cell and Molecular Biology, Karolinska Institute,
171 77, Stockholm, Sweden
e-mail: y.k_fujii@nifty.com

H. Yamazaki (ed.), *Fifty Years of Cytochrome P450 Research*,
DOI 10.1007/978-4-431-54992-5_13, © Springer Japan 2014

accumulation. These results revealed an important role of the bacteria-triggered or ASC-mediated inflammation signaling pathway in mouse intestinal tumorigenesis. Furthermore, they also suggest a possible chemical therapeutic intervention that involves AhR ligands and inhibitors of the inflammation pathway.

Keywords Ah receptor • ASC • Caspase-1 • Colon cancer • Inflammasome • Inflammation • Interleukin-1β • STAT3 • β-Catenin

13.1 Introduction

The aryl hydrocarbon receptor (AhR), also known as the dioxin receptor, is a member of the basic helix-loop-helix/Per-AhR nuclear translocator-Sim homology (bHLH-PAS) superfamily and mediates a wide variety of pharmacological and toxicological effects, such as the induction of drug-metabolizing enzymes, tumor promotion, teratogenesis, immunosuppression, and wasting syndrome (Gu et al. 2000; Kawajiri and Fujii-Kuriyama 2007). It has recently been revealed that AhR is also involved in the normal development and homeostasis of multiple physiological processes such as reproduction, innate immunity, and tumor suppression (Fujii-Kuriyama and Kawajiri 2010). In addition to ligand-dependent transcriptional regulation, AhR regulates intracellular protein levels as a ligand-dependent E3 ubiquitin ligase of nuclear receptors such as estrogen and androgen receptors (Ohtake et al. 2007). In this study, we show that $AhR^{-/-}$ mice spontaneously develop colonic tubular adenocarcinomas with abnormal accumulation of β-catenin, particularly near the ileocecal junction of the cecum (Kawajiri et al. 2009). The tumor suppression mechanism involves AhR-mediated ubiquitylation and proteasomal degradation of β-catenin.

Although it is currently unknown why $AhR^{-/-}$ mice develop cancers specifically in the cecum, the host genetic predisposition to these cancers may be potentiated by microbial interaction or subsequent inflammation (Maggio-Price et al. 2006). Severe inflammation in the intestines of $AhR^{-/-}$ mice has been observed (Fernandez-Salguero et al. 1997) together with high levels of inflammatory cytokine expression (Sekine et al. 2009). $AhR^{-/-}$ mice become hypersensitive to lipopolysaccharide (LPS)-induced septic shock (Sekine et al. 2009; Kimura et al. 2009), and the increased susceptibility to endotoxin is associated with elevated plasma concentrations of interleukin (IL)-1β, which is assumed to be a master mediator or initiator of inflammation (Mills and Dunne 2009).

We performed carcinogenesis experiments using germ-free (GF) $AhR^{-/-}$ mice and compound mutant mice lacking genes for both AhR and ASC, which play a critical role in the inflammasome-dependent activation of caspase-1 (Ikuta et al. 2013). Both GF $AhR^{-/-}$ and the compound $AhR^{-/-}$•$ASC^{-/-}$ (*DKO*) mutant mice showed markedly reduced tumor development in the cecum compared with that in $AhR^{-/-}$ mice, albeit in a "cancer-prone" state with aberrant β-catenin accumulation, suggesting critical roles for microflora and the resulting inflammation in cancer development.

13.2 Cecal Tumor Development in $AhR^{-/-}$ Mice

After thoroughly examining the digestive tracts of $AhR^{-/-}$ mice, we observed that $AhR^{-/-}$ mice, but not heterozygous $AhR^{+/-}$ or wild-type $AhR^{+/+}$ mice, frequently developed colonic tumors, mainly in the cecum near the ileocecal junction (Fig. 13.1a). $AhR^{-/-}$ mice bred at two independent animal houses displayed a similar time-course of macroscopic tumor incidence, with tumor size increasing gradually with age before reaching a plateau at 30–40 weeks (Fig. 13.1b).

Randomly selected mice were examined histologically for atypia classified according to the standard methods. Although $AhR^{+/+}$ and $AhR^{+/-}$ mice of all ages had normal to mild hyperplasia at worst, $AhR^{-/-}$ mice older than 11 weeks had abnormal histology with atypia ranging from mild malignancy of polyps to severe carcinomas that were exacerbated with age. Close microscopic examination revealed that the $AhR^{-/-}$ mice bore cecal lesions with a moderate or a high grade of atypia, adenoma, and adenocarcinoma. Among the 17 diagnosed adenocarcinomas, 12 tumors (71 %) invaded the submucosal region or beyond, with the remainder being located within the intramucosal region. Overall survival rates estimated by the Kaplan–Meyer method revealed that $AhR^{-/-}$ mice had a significantly shorter lifespan than wild-type or heterozygous mice, although this reduced longevity may not be solely because of the presence of cecal tumors (Fernandez-Salguero et al. 1997). The detected cecal cancers were predominantly tubular adenocarcinomas with various degrees of malignancy. Moderately differentiated

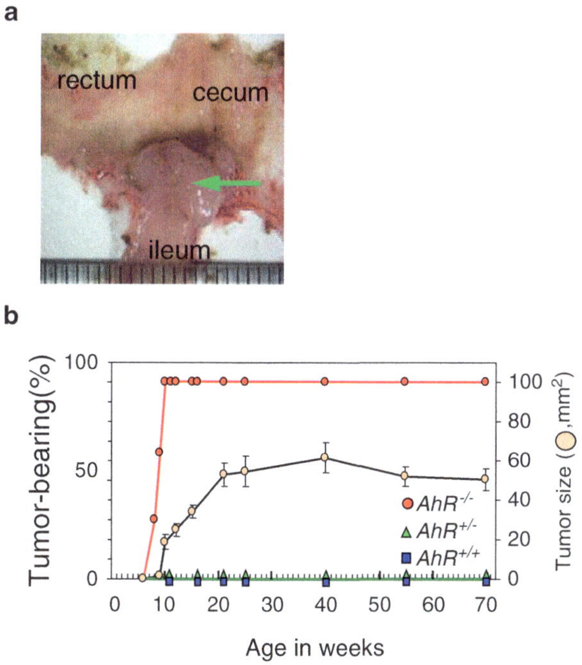

Fig. 13.1 Cecal tumor development in $AhR^{-/-}$ mice. (**a**) Representative profiles of colon tumors at the cecum in $AhR^{-/-}$ mice. (**b**) Relationship between the time-course of macroscopic tumor incidence and tumor growth by age. Tumor size was estimated based on NIH images as shown by *beige circles*. *Ahr*, aryl hydrocarbon receptor. *Error bars*, means ± SD

adenocarcinomas with irregularly shaped and fused tubular structures that sometimes invaded the submucosal regions were observed. In these cells, immunohistochemical staining showed concomitant overexpression of β-catenin and c-*myc*, a target gene of β-catenin/TCF4 (He et al. 1998).

13.3 β-Catenin Accumulation in $AhR^{-/-}$ Mice

To examine the molecular mechanism underlying tumor development in $AhR^{-/-}$ mice, we analyzed the expression of β-catenin in the intestines of 6-week-old $AhR^{+/+}$ and $AhR^{-/-}$ mice, which had morphologically normal epithelia. β-Catenin expression was abnormally high in epithelial cells of the ileum, colon, and cecum of $AhR^{-/-}$ mice, suggesting that the intestines of $AhR^{-/-}$ mice may be in a "cancer-prone" or "precancerous" state (van de Wetering et al. 2002). In particular, these elevated levels of β-catenin were observed in the nuclei of Paneth cells when compared with the corresponding regions in wild-type mice. Using Western blotting, we confirmed that $AhR^{-/-}$ mice had significantly higher levels of β-catenin in the cecum than did wild-type mice, whereas β-catenin mRNA expression levels were unchanged, suggesting that the stabilization of β-catenin protein, but not enhanced synthesis of the β-catenin protein in the $AhR^{-/-}$ intestine, leads to β-catenin accumulation. Consistent with the abnormal accumulation of β-catenin, expression of its downstream target, c-*myc*, showed an approximately twofold induction.

Furthermore, we examined whether the AhR E3 ubiquitin ligase participates in the degradation of β-catenin as previously reported (Ohtake et al. 2007). Experimental data suggested that the ligand-dependent E3 ubiquitin ligase activity of AhR participates in β-catenin degradation (Fig. 13.2) (Kawajiri et al. 2009). We were interested to investigate whether β-catenin protein is reduced in the intestines of

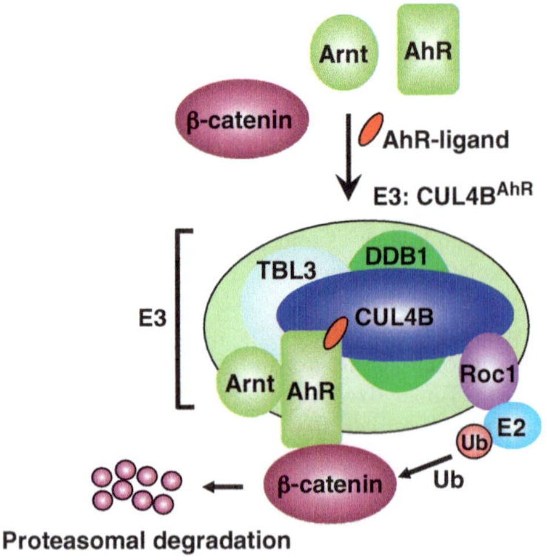

Fig. 13.2 Novel ligand-dependent ubiquitylation and proteasomal degradation of β-catenin

mice in vivo after AhR ligand treatment. In vivo degradation of β-catenin was observed after intraperitoneal injection of the natural AhR ligands such as indole-acetic acid and indole-3-carbinol (I3C). These in vivo observations are highly consistent with the in vitro experiments and provide a basis for possible chemoprevention against intestinal carcinogenesis by using natural AhR ligands.

13.4 Tumor Suppression by AhR Natural Ligands

We were interested to study whether natural AhR ligands actually suppress carcinogenesis in the cecum or small intestine in $Apc^{Min/+}$ mice. This chemoprevention (Wattenberg 1985) study was designed so that $Apc^{Min/+}$ or $Apc^{Min/+}$•$AhR^{+/-}$ mice were fed diets containing natural AhR ligands such as I3C (Xu et al. 1996) and diindolylmethane (DIM) (Chen et al. 1998) immediately after weaning at 3–4 weeks of age. When fed the control diet, $Apc^{Min/+}$ mice started to develop small intestinal polyps at 7 weeks of age, with the number of tumors containing polyps plateauing (approximately 30 tumors per mouse) at approximately 10–15 weeks (Fig. 13.3). However, when fed an I3C- or DIM-containing diet (0.1 % and 0.01 %, respectively), $Apc^{Min/+}$ mice showed a cecal tumor incidence of approximately 50 % of the total at 25 weeks of age as well as a markedly reduced number of tumors in the small intestine. Similar chemopreventive effects were also clearly observed with the compound $Apc^{Min/+}$•$AhR^{+/-}$ mutant mice. However, no suppressive effect was observed in $AhR^{-/-}$ mice, suggesting that AhR ligand-dependent chemoprevention requires the presence of AhR. Using immunohistochemical analysis, we showed a marked reduction of β-catenin, except for the molecules associated with adherence junctions in the intestines of $Apc^{Min/+}$ and $Apc^{Min/+}$•$AhR^{+/-}$ mice fed AhR ligand-containing diets compared with those fed a control diet. These results clearly demonstrate that chemoprevention of intestinal carcinogenesis by AhR ligands in $Apc^{Min/+}$ and $Apc^{Min/+}$•$AhR^{+/-}$ mice is the result of β-catenin degradation mediated by the natural ligand-activated AhR E3 ubiquitin ligase.

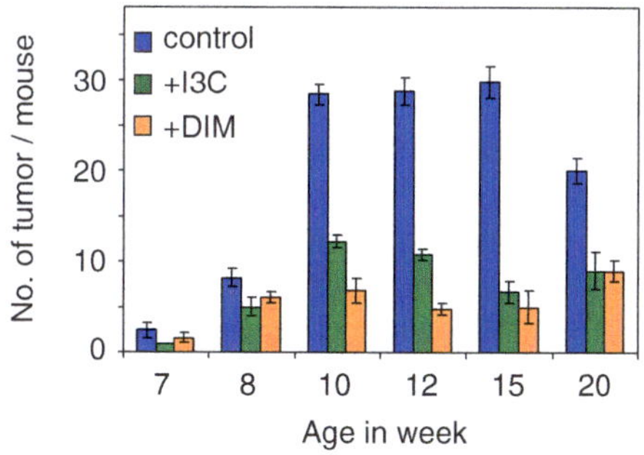

Fig. 13.3 Natural AhR ligands suppress intestinal carcinogenesis. Four to five mice were used in each group. Number of small intestinal polyps in $Apc^{Min/+}$ mice fed a control diet (*blue squares*). Number of polyps in the small intestines of mice fed an I3C-containing (*green squares*) or DIM-containing (*beige squares*) diet. Data are presented as means ± SD

13.5 No Tumor Development in GF *AhR*$^{-/-}$ Mice

To elucidate whether the tumors develop autonomously in *AhR*$^{-/-}$ mice because of impaired β-catenin degradation or in association with accelerated inflammation, we performed two kinds of experiments using germ-free (GF) *AhR*$^{-/-}$ mice and compound mutant mice lacking the AhR and ASC genes, the latter of which plays an essential role in caspase-1 activation in inflammasomes. We examined whether GF *AhR*$^{-/-}$ and GF *AhR*$^{+/-}$ mice develop tumors in the cecum. No tumors developed in GF mice until they were 25 weeks of age, regardless of their genotypes as determined by macroscopic observation and histopathological examination (Ikuta et al. 2013). In contrast, all *AhR*$^{-/-}$ mice maintained under specific pathogen-free (SPF) or conventional conditions developed adenomas or adenocarcinomas by that age (Kawajiri et al. 2009). In addition, close microscopic examination revealed a markedly severe inflammation in the intestines of SPF *AhR*$^{-/-}$ mice, consistent with previous results (Fernandez-Salguero et al. 1997; Sekine et al. 2009), but not in those of GF *AhR*$^{-/-}$ mice. These results suggest that microbial interaction or subsequent inflammation is required for cecal tumor development in *AhR*$^{-/-}$ mice.

13.6 Functional Cross-Talk Between AhR and ASC

It is well known that ASC plays an essential role in inflammatory responses during host defense against infectious diseases, and that it is involved in the release of IL-1β/IL-18 through caspase-1 activation in inflammasomes (Taniguchi and Sagara 2007; Yamamoto et al. 2004). In contrast, AhR enhances Pai-2 expression to inhibit caspase-1 activity (Sekine et al. 2009), resulting in an antiinflammatory function. To reveal the functional cross-talk between ASC-mediated activation and AhR-mediated suppression of caspase-1 in vivo and to investigate the involvement of ASC-mediated inflammation in cecal carcinogenesis of *AhR*$^{-/-}$ mice (Kawajiri et al. 2009), we generated *AhR*$^{-/-}$•*ASC*$^{-/-}$ (*DKO*) mice with the same genetic background. We analyzed the in vitro activation of caspase-1 and secretion of IL-1β in macrophages isolated from wild-type (WT), *AhR*$^{-/-}$, and *DKO* mice. In the LPS-primed macrophages of WT mice, procaspase-1 (p45) was evidently processed to a mature form of caspase-1 (p10). The activated form of caspase-1 was markedly increased in *AhR*$^{-/-}$ macrophages, whereas it was hardly detectable in the *DKO* macrophages (Fig. 13.4a) (Ikuta et al. 2013). Consistent with these observations, essentially no mature form of IL-1β (p17) was secreted from the *DKO* macrophages, whereas its secretion from *AhR*$^{-/-}$ macrophages treated with LPS was increased compared with that from WT macrophages. These results clearly indicate that AhR suppressed ASC-mediated processing of IL-1β by inhibiting caspase-1 activity. Plasma concentrations of the inflammatory cytokines interleukin (IL)-1β, IL-6 (Fig. 13.4b), and IL-18 were increased in *AhR*$^{-/-}$ mice, whereas DKO mice

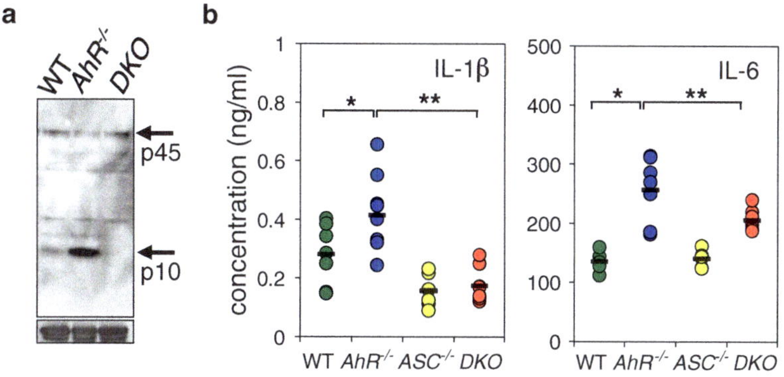

Fig. 13.4 Functional cross-talk between AhR and ASC. (**a**) Induction of caspase-1 activation in peritoneal macrophages of $AhR^{-/-}$ mice. Activated caspase-1 (p10) was detected by Western blot. Procaspase-1 (p45) and p10 are indicated by *arrows*. (**b**) Plasma concentrations of interleukin (*IL*)-1β (*$P < 0.05$, **$P < 0.001$) and IL-6 (*$P < 0.001$, **$P < 0.05$). Each *symbol* represents the cytokine concentrations of an individual animal. Mean values are indicated by *bars*

showed low levels of IL-6 secretion and even lower levels of IL-1β than WT mice. These results indicate a functional association between ASC- and AhR-directed pathways, which regulate the IL-1β-dependent inflammatory response in vivo.

13.7 Reduced Cecal Carcinogenesis in $AhR^{-/-}$•$ASC^{-/-}$ Mice

To investigate whether the cecal carcinogenesis observed in $AhR^{-/-}$ mice developed autonomously because of a lack of ligand-dependent β-catenin degradation by AhR or whether it also involves inflammation caused by a lack of antiinflammatory function, we performed experimental carcinogenesis using *DKO* mice. No cecal tumor development was observed in WT or $ASC^{-/-}$ mice (Fig. 13.5a). When the tumor incidence in *DKO* mice was compared with that of $AhR^{-/-}$ mice, *DKO* mice presented no tumor development until 30 weeks of age, which was markedly delayed in *DKO* mice in comparison with $AhR^{-/-}$ mice (Ikuta et al. 2013). As observed in $AhR^{-/-}$ mice, cancers that developed later in *DKO* mice were tubular adenocarcinomas, (Kawajiri et al. 2009), with varying degrees of malignancy. Immunohistochemical staining of these cells showed concomitant overexpression of β-catenin and c-*myc*, a target gene of the β-catenin/TCF4 pathway.

To elucidate why *DKO* mice were more resistant to cecal carcinogenesis than the $AhR^{-/-}$ mice, β-catenin levels were monitored in the cecum of *DKO* mice at 14 weeks of age, when tumors were already observed in the cecum of $AhR^{-/-}$ mice. Although tumors and morphologically aberrant epithelia had not developed, abnormal β-catenin accumulation was observed in the cecum of 14-week-old *DKO* mice using immunohistochemical staining and Western blot analysis, which

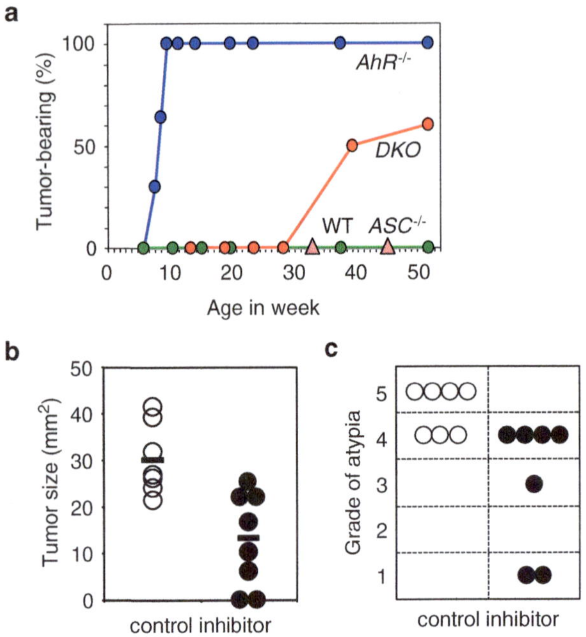

Fig. 13.5 Reduced cecal carcinogenesis in compound DKO mice. (**a**) Time-course of cecal tumorigenesis in *AhR*$^{-/-}$ (*blue circles*), DKO (*red circles*), and WT (*green circles*) mice. No cecal tumor development was observed in *ASC*$^{-/-}$ (*pink triangles*) mice. (**b**, **c**) The caspase-1 inhibitor YVAD suppressed tumor development. (**b**) Tumor size was estimated based on ImageJ. Mean values are indicated by *bars*. $P = 0.0041$, control versus YVAD groups. (**c**) Histological grades of atypia are indicated. $P = 0.04$, control versus YVAD groups

revealed an approximately threefold increase compared to WT cecum. These results suggest that the intestinal epithelia of *DKO* mice are in a "cancer-prone" state with aberrant β-catenin accumulation, similar to the intestinal epithelia of *AhR*$^{-/-}$ (Kawajiri et al. 2009) and GF *AhR*$^{-/-}$ mice.

Furthermore, we used mice with different genotypes to investigate tissue-specific IL-1β expression in the cecum near the ileocecal region. The expression levels of IL-1β protein in *AhR*$^{-/-}$ mice were significantly higher than those expressed in WT mice, whereas its expression levels were lower in *ASC*$^{-/-}$ and *DKO* mice. IL-18 levels were also elevated in *AhR*$^{-/-}$ mice. Furthermore, IL-6 expression, which was upregulated in response to IL-1β, changed in a manner similar to the IL-1β expression, depending on *AhR* and *ASC* genotypes. Antitumor effects of IL-18 have been reported in experimental tumor models (Micallef et al. 1997; Osaki et al. 1998). Recent studies also showed that IL-18 production is critically involved in protection against colorectal tumorigenesis (Zaki et al. 2010). Based on these results, it has been suggested that the enhanced expression of inflammatory cytokines such as IL-1β and IL-6 is involved in the cecal carcinogenesis in *AhR*$^{-/-}$ mice, in addition to aberrant β-catenin accumulation. In support of these observations, intraperitoneal injections of YVAD, an IL-1β-converting enzyme inhibitor II, administered to $AhR^{-/-}$ mice once a week from 5 to 18 weeks of age exerted significant inhibitory effects on both the growth (Fig. 13.5b) and the histological grade (Fig. 13.5c) of tumor atypia compared with the sham-treated $AhR^{-/-}$ mice.

13.8 $AhR^{-/-}$-Dependent, Cecal-Specific STAT3 Activation

Members of the signal transducers and activators of transcription (STAT) family of proteins are central in determining whether immune responses in the tumor microenvironment promote or inhibit cancer development (Yu et al. 2009), and since levels of IL-6, which are known to activate STAT3, were elevated in $AhR^{-/-}$ mice, we investigated the activation of these proteins in the intestines of the $AhR^{-/-}$ animals (Ikuta et al. 2013). Among the five STAT proteins examined, STAT3 hyperactivation (phosphorylated Y705) was particularly observed in the normal cecal tissues of 8-week-old $AhR^{-/-}$ mice (Fig. 13.6a, left) and later observed in 23-week-old *DKO* mice as well (Fig. 13.6a, right). Furthermore, STAT3 was activated most prominently in the cecum, which is consistent with the site of tumor development in $AhR^{-/-}$ mice (Kawajiri et al. 2009), and weak STAT3 activation was observed in the upper part of the small intestine (duodenum) and in the lower part of the colon (rectum) (Fig. 13.6b). Immunohistochemical analyses also revealed specific staining of activated STAT3 in the cecum of $AhR^{-/-}$ mice. STAT3 activation was observed only in stromal cells in 8-week-old mice, whereas it was detectable in stroma and epithelial cell nuclei at 19 weeks of age (Fig. 13.7). Further immunofluorescent analysis revealed that STAT3 was activated in a subset

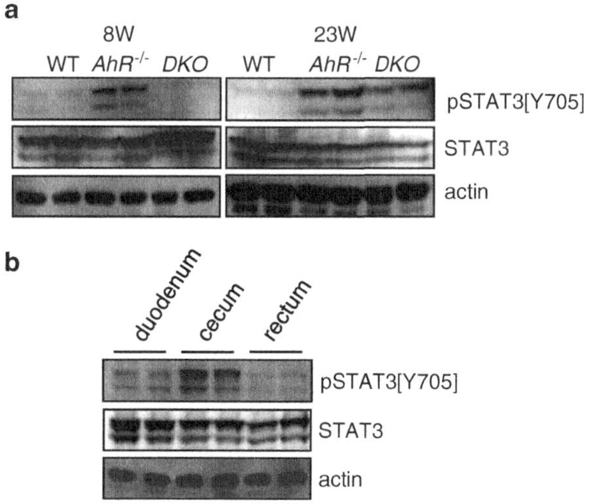

Fig. 13.6 Cecal-specific activation of STAT3 in $AhR^{-/-}$ mice. (**a**) Activation of STAT3 in the cecum of mice at 8 weeks (*left*) and 23 weeks (*right*) of age. (**b**) Tissue lysates were prepared from $AhR^{-/-}$ mice at 8 weeks of age and analyzed for STAT3 activation. *WT*, wild type

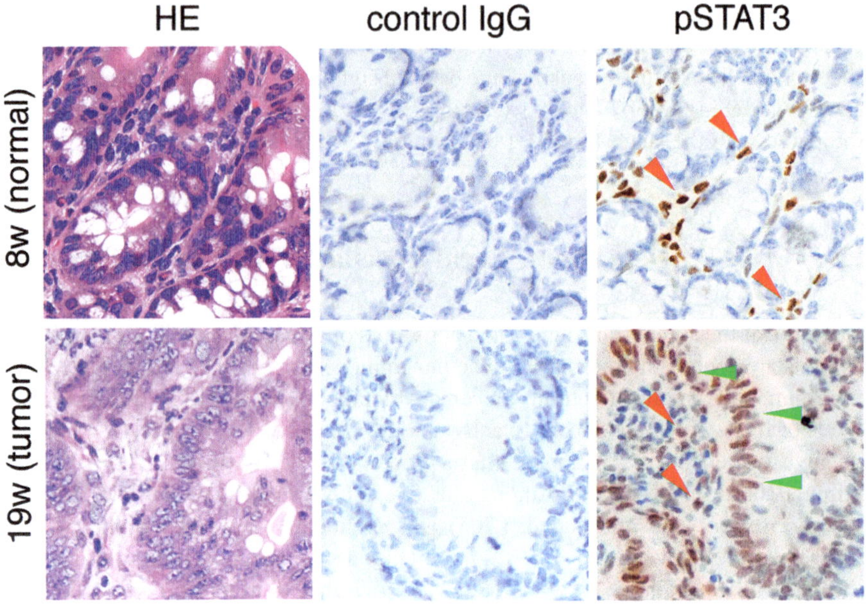

Fig. 13.7 Hematoxylin and eosin (H&E, *HE*) and immunohistochemical staining of phosphoSTAT3 (Y705) in the cecum of 8- or 19-week-old $AhR^{-/-}$ mice. Normal rabbit IgG was used as a negative control. Positive cells in the stroma are indicated by *red arrowheads*; those in the epithelium are indicated by *green arrowheads*

of CD45-positive cells. Significantly higher levels of IL-6 in the $AhR^{-/-}$ cecum may result in the specific activation of STAT3 in this tissue.

Subsequently, we investigated the effect of AG490, an inhibitor of JAK which phosphorylates STAT3, on cecal tumor development in $AhR^{-/-}$ mice. These mice were intraperitoneally injected with AG490 at an interval of 5 days from 5 to 18 weeks of age. In agreement with the results of successive injections of YVAD, AG490 significantly inhibited tumor growth and improved the histological grade of atypia in $AhR^{-/-}$ mice compared with the in sham-treated $AhR^{-/-}$ mice. As expected, intraperitoneal injections of AG490 in $AhR^{-/-}$ mice decreased STAT3 activation in the cecum.

13.9 Conclusion

In this study, we examined whether tumors develop autonomously in $AhR^{-/-}$ mice as a result of impaired β-catenin degradation or in association with accelerated inflammation. Our experimental evidence showed that GF $AhR^{-/-}$ mice did not develop tumors until 25 weeks of age. Furthermore, *DKO* mice that were generated to investigate how intestinal carcinogenesis was associated with the pro-inflammatory state in $AhR^{-/-}$ mice did not develop macroscopically detectable

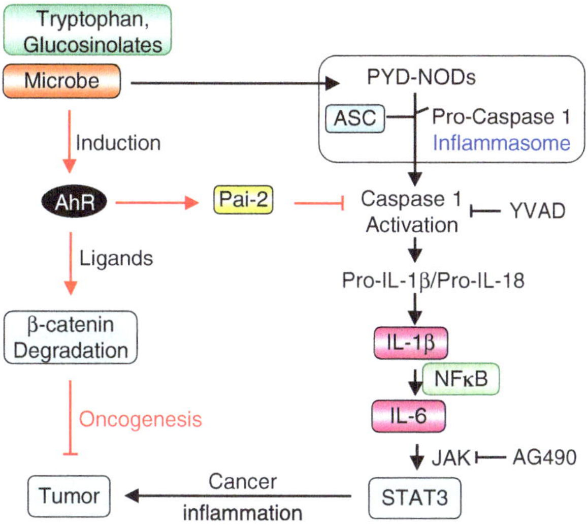

Fig. 13.8 A model of AhR-dependent cecal tumor suppression. AhR plays a dual role as a tumor suppressor of β-catenin E3 ubiquitylation and in intestinal antiinflammatory activity

tumors until 30 weeks of age. These mice are proposed to be in a "cancer-prone" state with accumulated β-catenin as well as additional cues involving the presence of intestinal microbes that may cause inflammation leading to carcinogenesis (Fig. 13.8).

We presented evidence that natural AhR ligands converted from dietary Trp and glucosinolates in the intestine are as efficient as exogenous AhR ligands in promoting the degradation of endogenous β-catenin. These results provide a molecular basis for the chemopreventive mechanisms against intestinal carcinogenesis that were observed in $Apc^{Min/+}$ mice fed diets containing the AhR ligands I3C and DIM. In addition, our findings lend credence to previous reports on the chemopreventive effects on human colorectal cancers resulting from cruciferous vegetables containing a high content of glucosinolates (Kim and Milner 2005; Bonnesen et al. 2001; Potter and Steinmetz 1996) and suggest that AhR ligands define a potent strategy for dietary chemoprevention of intestinal cancer.

Thus, together with abnormal β-catenin accumulation, we have shown that the loss of AhR-associated anti-inflammatory functions is a crucial determinant for the promotion of cecal carcinogenesis in $AhR^{-/-}$ mice. We conclude that the $AhR^{-/-}$ mouse is a useful model for elucidating the molecular mechanisms of inflammation-associated intestinal carcinogenesis. Further investigation into the interactions between the host and microbes will be needed to provide a new strategy for chemoprevention and chemotherapy of intestinal cancer using AhR ligands.

References

Bonnesen C, Eggleston IM, Hayes JD (2001) Dietary indoles and isothiocyanates that are generated from cruciferous vegetables can both stimulate apoptosis and confer protection against DNA damage in human colon cell lines. Cancer Res 61:6120–6130

Chen I, McDougal A, Wang F, Safe S (1998) Aryl hydrocarbon receptor-mediated antiestrogenic and antitumorigenic activity of diindolylmethane. Carcinogenesis (Oxf) 19:1631–1639

Fernandez-Salguero PM, Ward JM, Sundberg JP, Gonzalez FJ (1997) Lesions of aryl hydrocarbon receptor-deficient mice. Vet Pathol 34:605–614

Fujii-Kuriyama Y, Kawajiri K (2010) Molecular mechanisms of the physiological functions of the aryl hydrocarbon (dioxin) receptor, a multifunctional regulator that senses and responds to environmental stimuli. Proc Jpn Acad Ser B Phys Biol Sci 86:40–53

Gu YZ, Hogenesch J, Bradfield CA (2000) The PAS superfamily: sensors of environmental and developmental signals. Annu Rev Pharmacol 40:519–561

He TC, Sparks AB, Rago C, Hermeking H, Zawel L, da Costa LT, Morin PJ, Vogelstein B, Kinzler KW (1998) Identification of c-MYC as a target of the APC pathway. Science 281:1509–1512

Ikuta T, Kobayashi Y, Kitazawa M, Shiizaki K, Itano N, Noda T, Pettersson S, Poellinger L, Fujii-Kuriyama Y, Taniguchi S, Kawajiri K (2013) ASC-associated inflammation promotes cecal tumorigenesis in aryl hydrocarbon receptor-deficient mice. Carcinogenesis (Oxf) 34:1620–1627

Kawajiri K, Fujii-Kuriyama Y (2007) Cytochrome P450 gene regulation and physiological functions mediated by the aryl hydrocarbon receptor. Arch Biochem Biophys 464:207–212

Kawajiri K, Kobayashi Y, Ohtake F, Ikuta T, Matsushima Y, Mimura J, Pettersson S, Pollenz RS, Sakaki T, Hirokawa T, Akiyama T, Kurosumi M, Poellinger L, Kato S, Fujii-Kuriyama Y (2009) Aryl hydrocarbon receptor suppresses intestinal carcinogenesis in ApcMin/+ mice with natural ligands. Proc Natl Acad Sci USA 106:13481–13486

Kim YS, Milner JA (2005) Targets for indole-3-carbinol in cancer prevention. J Nutr Biochem 16:65–73

Kimura A, Naka T, Nakahama T, Chinen I, Masuda K, Nohara K, Fujii-Kuriyama Y, Kishimoto T (2009) Aryl hydrocarbon receptor in combination with Stat1 regulates LPS-induced inflammatory responses. J Exp Med 206:2027–2035

Maggio-Price L, Treuting P, Zeng W, Tsang M, Bielefeldt-Ohmann H, Iritani BM (2006) *Helicobacter* infection is required for inflammation and colon cancer in Smad3-deficient mice. Cancer Res 66:828–838

Micallef MJ, Yoshida K, Kawai S, Hanaya T, Kohno K, Arai S, Tanimoto T, Torigoe K, Fujii M, Ikeda M, Kurimoto M (1997) In vivo antitumor effects of murine interferon gamma-inducing factor/interleukin-18 in mice bearing syngeneic Meth A sarcoma malignant ascites. Cancer Immunol Immunother 43:361–367

Mills KH, Dunne A (2009) Immune modulation: IL-1, master mediator or initiator of inflammation. Nat Med 15:1363–1364

Ohtake F, Baba A, Takada I, Okada M, Iwasaki K, Miki H, Takahashi S, Kouzmenko A, Nohara K, Chiba T, Fujii-Kuriyama Y, Kato S (2007) Dioxin receptor is a ligand-dependent E3 ubiquitin ligase. Nature (Lond) 446:562–566

Osaki T, Péron JM, Cai Q, Okamura H, Robbins PD, Kurimoto M, Lotze MT, Tahara H (1998) IFN-γ-inducing factor/IL-18 administration mediates IFN-γ- and IL-12-independent antitumor effects. J Immunol 160:1742–1749

Potter JD, Steinmetz K (1996) Vegetables, fruit, and phytoestrogens as preventive agents. IARC Sci Publ 139:61–90

Sekine H, Mimura J, Oshima M, Okawa H, Kanno J, Igarashi K, Gonzalez FJ, Ikuta T, Kawajiri K, Fujii-Kuriyama Y (2009) Hypersensitivity of aryl hydrocarbon receptor-deficient mice to lipopolysaccharide-induced septic shock. Mol Cell Biol 29:6391–6400

Taniguchi S, Sagara J (2007) Regulatory molecules involved in inflammasome formation with special reference to a key mediator protein, ASC. Semin Immunopathol 29:231–238

van de Wetering M, Sancho E, Verweij C, de Lau W, Oving I, Hurlstone A, van der Horn K, Batlle E, Coudreuse D, Haramis AP, Tjon-Pon-Fong M, Moerer P, van den Born M, Soete G, Pals S, Eilers M, Medema R, Clevers H (2002) The β-catenin/TCF-4 complex imposes a crypt progenitor phenotype on colorectal cancer cells. Cell 111:241–250

Wattenberg LW (1985) Chemoprevention of cancer. Cancer Res 45:1–8

Xu M, Bailey AC, Hernaez JF, Taoka CR, Schut HA, Dashwood RH (1996) Protection by green tea, black tea, and indole-3-carbinol against 2-amino-3-methylimidazo[4,5-f]quinoline-induced DNA adducts and colonic aberrant crypts in the F344 rat. Carcinogenesis (Oxf) 17:1429–1434

Yamamoto M, Yaginuma K, Tsutsui H, Sagara J, Guan X, Seki E, Yasuda K, Yamamoto M, Akira S, Nakanishi K, Noda T, Taniguchi S (2004) ASC is essential for LPS-induced activation of procaspase-1 independently of TLR-associated signal adaptor molecules. Genes Cells 9:1055–1067

Yu H, Pardoll D, Jove R (2009) STATs in cancer inflammation and immunity: a leading role for STAT3. Nat Rev Cancer 9:798–809

Zaki MH, Vogel P, Body-Malapel M, Lamkanfi M, Kanneganti TD (2010) IL-18 production downstream of the Nlrp3 inflammasome confers protection against colorectal tumor formation. J Immunol 185:4912–4920

Chapter 14
Epidermal Growth Factor Receptor: The Phenobarbital Receptor that Elicits CAR Activation Signal for P450 Induction

Shingo Mutoh and Masahiko Negishi

Abstract Phenobarbital is the classic inducer of cytochrome P450 and drug metabolism. Here we describe the cell signaling mechanism by which phenobarbital binds to antagonize epidermal growth factor receptor to indirectly activate the nuclear constitutive androstane receptor (CAR). In response to this signaling, residue threonine 38 of CAR is dephosphorylated in the cytoplasm. Nonphosphorylated CAR translocates into the nucleus, thereby activating genes that encode enzymes and transporters to induce drug metabolism and excretion.

Keywords Cell signaling • Dephosphorylation • Drug metabolism • EGF receptor • Gene regulation • Nuclear receptor CAR • P450 • Phenobarbital • Phosphorylation • RACK1

14.1 Introduction

Phenobarbital has widely been used to treat epilepsy for nearly a century. A half century ago, barbiturates such as phenobarbital were first reported to induce enzymes in the liver endoplasmic reticulum and increase their own metabolism (Remmer and Merker 1963). With this finding, drug induction of drug metabolism was linked with the mechanism associated with drug tolerance. Subsequently, cytochrome P450 (P450), the key enzyme that metabolizes drugs, was characterized, and P450 induction was conceptualized as the key regulatory system affecting the pharmacological as well as toxicological consequences of drug treatments

S. Mutoh • M. Negishi (✉)
Pharmacogenetics Section, Laboratory of Reproductive and Developmental Toxicology, National Institute of Environmental Health Sciences, National Institutes of Health, Research Triangle Park, NC 27709, USA
e-mail: negishi@niehs.nih.gov

H. Yamazaki (ed.), *Fifty Years of Cytochrome P450 Research*,
DOI 10.1007/978-4-431-54992-5_14,

(Omura and Sato 1962). Soon after, research into this induction phenomenon was dramatically widened by the expansion of known inducers from drugs to include numerous xenobiotics and known metabolizing enzymes from P450s to comprise various transferases (e.g., UDP-glucronosytransferases and sulfotransferases) as well as drug transporters. Phenobarbital (PB) and 3-methylcholanthrene (3-MC) emerged as two major inducers for which the mechanisms have been intensively studied: PB and 3MC represent distinct groups capable of inducing different P450 forms (Alvares et al. 1967). 3-MC-type inducers are typically characterized by their similar structural features of polycyclic aromatic hydrocarbons with 2,3,7,8-tetrachlorodibenzodioxin (TCDD) being the most potent inducer. The aryl hydrocarbon receptor AHR was identified as the receptor for this type of inducers to activate the *P450* gene in the late 1970s (Greenlee and Poland 1979; Okey et al. 1979). Subsequently, the AHR-mediated activation mechanism was quickly determined toward the end of the 1980s (Okey 2007). In contrast, the PB-type inducers exhibit extremely diverse structures and are not capable of activating their target *CYP2B* gene in transformed cell lines, which long prevented us from discovering their induction mechanism. The 1998s finding that established the nuclear receptor CAR as the PB-activated transcription factor provided us with the molecular basis to decipher the mechanism of PB induction. CAR is only the PB-activated nuclear receptor in both mice and humans (Scheer et al. 2008). This chapter focuses on our recent work that led to the discovery of the epidermal growth factor receptor (EGFR) as the phenobarbital receptor which elicits cell signaling for CAR activation (Mutoh et al. 2013). Figure 14.1 shows the activation mechanism of CAR by PB based on our findings. This discovery has not only solved a long-standing question regarding the initial site of PB action to induce drug metabolism but also opened a new era for investigations into drug–cell signal interactions via cell membrane receptors as well as their biological, pharmacological, and toxicological implications. For more details that summarized research efforts before discovery, we suggest readers refer to previous review articles (Honkakoski et al. 1998b; Sueyoshi and Negishi 2001).

14.2 Phenobarbital Response Enhancer Module (PBREM)

The first cDNA and gene encoding rat CYP2B mRNA were cloned and sequenced in the early 1980s (Fujii-Kuriyama et al. 1982; Mizukami et al. 1983), followed by subsequent cloning of the mouse and human *CYP2B* genes (Lakso et al. 1991). By the mid-1990s, primary hepatocytes became available for the investigation of PB-induced promoter activity. The first PB-induced enhancing activity was delineated to a 163-bp sequence from −2,318 bp to −2,155 bp of the rat *CYP2B2* gene, which was dubbed "phenobarbital response element" or PBRE. The PBRE sequence was further minimized to the 51-bp phenobarbital response enhancer module (PBREM) of the mouse *Cyp2b10* gene (Honkakoski et al. 1998a). PBREM contains two DR4 motifs (NR1 and NR2) separated by an NF1-binding

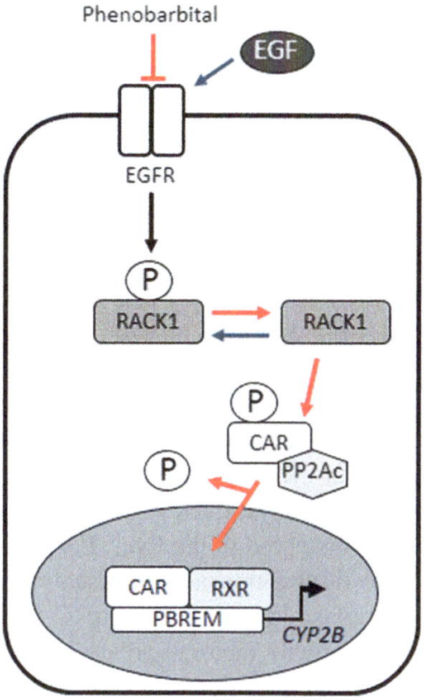

Fig. 14.1 Phenobarbital-induced signaling mechanism that indirectly activates CAR. Phenobarbital directly binds to the epidermal growth factor (EGF) receptor, antagonizes EGF activation, and represses EGF receptor signaling. Because EGF receptor signaling phosphorylates tyrosine 51 of RACK1, repression by phenobarbital results in dephosphorylation. Nonphosphorylated RACK1 binds to a phosphorylated CAR-PP2Ac complex and stimulates PP2Ac to dephosphorylate CAR. Dephosphorylated CAR translocates from the cytoplasm into the nucleus, forms a heterodimer with RXR, and binds to PBREM, thereby activating the *CYP2B* gene. This mechanism is depicted based on the findings reported in our previous works (Mutoh et al. 2009, 2013)

site and can be activated in response to various PB-type inducers. The PBREM characterized in the mouse *Cyp2b10* gene is conserved within rat and human *CYP2B* genes (Sueyoshi et al. 1999). PBREM provided us with the basis for determining the transcription factors that respond to PB induction and their mechanism of regulation.

14.3 Nuclear Receptor CAR

Utilizing an NR1 oligonucleotide as an affinity ligand, binding proteins were purified from liver nuclear extracts prepared from PBS- or PB-treated mice. Western blot analysis of these purified fractions revealed both nuclear receptors RXR and CAR increased their binding to NR1 after PB treatment (Honkakoski and Negishi 1998). Subsequent gel-shift and reporter analyses confirmed that a RXR-CAR heterodimer bound to NR1 and transactivated PBREM. Subsequent to our finding that CAR transactivated PBREM in transient transfection assays, CAR KO mice were generated and provided the direct evidence that CAR was essential for the induction of the *Cyp2b10* gene (Wei et al. 2000). CAR (NR1I3; first given name MB67), short for nuclear constitutive active/androstane receptor, was originally cloned as a nuclear receptor that activates an empirical set of retinoic acid response elements in the absence of ligand (Baes et al. 1994). Thus, from the beginning CAR has been a nuclear receptor with high constitutive activity, and this activity quickly became a key aspect in the investigation of CAR activation by drugs and xenobiotics. CAR can be activated by the two types of P450 inducers in liver in vivo, either directly by ligands such as 1,4-bis[2-(3,5-dicholoropyriyloxy)] benzene (TCPOBOP) (Sueyoshi et al. 1999; Tzameli et al. 2000) and 6-(4-chlorophenyl) imidazo [2,1-*b*] [1,3]-thiazole-5-carbaldehyde *O*-(3,4-dichlorobenzyl) oxime (CITCO) (Maglich et al. 2003) or indirectly by therapeutic drugs such as phenobarbital. However, CAR activates PBREM without the presence of either a ligand or phenobarbital in cell-based transient transfection assays; CAR must suppress its constitutive activity to acquire the responsiveness observed in the liver. For that, CAR is retained in the cytoplasm to suppress its constitutive transactivation activity. Three different factors were found to regulate cytoplasmic localization of CAR in hepatocytes: protein phosphatase 2A, epidermal growth factor, and co-chaperone cytoplasmic CAR retention protein (CCRP/DNAJ7) (Kawamoto et al. 1999; Bauer et al. 2004; Koike et al. 2007; Kobayashi et al. 2003; Yoshinari et al. 2003).

14.4 Phosphorylation of Threonine 38

The role of protein phosphatase 2A was first suggested by the fact that okadaic acid prevented PB-induced nuclear accumulation of CAR as well as CYP2B induction (Kawamoto et al. 1999). Later, the target of PP2A dephosphorylation was identified to be threonine 38 in the DNA-binding domain of CAR. Phosphorylation of threonine 38 sequesters CAR in the cytoplasm and phenobarbital indirectly dephosphorylates CAR to activate it. However, the molecular mechanism of how phenobarbital stimulates dephosphorylation still remained elusive at that time. Epidermal growth factor (EGF) was first reported to repress phenobarbital activation of CAR in transient transfection assays using rat primary hepatocytes by the laboratory of George Kahl (Bauer et al. 2004). In this report, growth hormone also repressed CAR activation, which supported an old finding that growth hormone represses phenobarbital induction of CYP2B in rat liver (Yamazoe et al. 1987). In 2007, Koike et al. utilized mouse primary hepatocytes and MEK inhibitor and demonstrated that EGF repressed nuclear CAR accumulation through activation of the MEK-ERK1/2 pathway. Thus, CAR is in principle an EGF-regulated nuclear receptor, and whether EGF regulates CAR phosphorylation at threonine 38 has emerged as a key question in deciphering the CAR activation mechanism. CCRP was characterized as the co-chaperone that retains CAR in the cytoplasm of HepG2 cells. CCRP forms a complex with CAR to retain it in the cytoplasm of HepG2 cells

(Kobayashi et al. 2003). CCRP regulates interactions between CAR and Hsp70, and TCPOBOP dissociates CCRP from this CAR complex to translocate CAR into the nucleus (Timsit and Negishi, in preparation). Whether CCRP is also involved in phosphorylation of threonine 38 remains unexplained at the present time. Recently, CCRP KO mice were generated to investigate the in vivo roles of CCRP in CAR regulation.

14.5 EGFR as the PB Receptor

PB treatment is known to cause hepatocytes to develop hypertrophy as well as hyperplasia and to promote development of hepatocellular carcinoma. From the standpoint of trying to grasp the mechanism of these phenobarbital actions, whether PB modulates the function of growth factor receptors (e.g., epidermal growth factor receptor and insulin receptor) was previously examined (Hwang et al. 1986; Meyer et al. 1989; Lindros and Michalopoulos 1993). These studies consistently concluded that rat liver attenuates growth factor receptors after phenobarbital treatment. Whether phenobarbital directly bound to the receptors remained elusive, although Meyer et al. did suggest that phenobarbital indirectly repressed EGF receptor activity (1989). The availability of advanced technology such as isothermal titration calorimetry (ITC) allowed us to revisit the question. First, phenobarbital binding to EGFR was indirectly examined utilizing in vitro competitive binding assays in which EGF binding to recombinant EGFR was inhibited by increasing concentrations of PB with the ED_{50} for phenobarbital being around 10 μM (Mutoh et al. 2013). Subsequently, PB binding was directly confirmed using recombinant EGFR protein and ITC. ITC analysis revealed the presence of five binding sites and a K_d value of 10 μM. This K_d value corroborated the ED_{50} value. Having the results that PB directly binds to EGFR, molecular dynamic simulation was employed to dock phenobarbital molecules into a three-dimensional (3D) structure of EGFR. Multiple phenobarbital binding sites were revealed at locations that were overlapped with EGF binding sites or in regions which constrain the structural changes required for EGFR activation. An apparent K_d value was calculated from this dynamic simulation, which also turned out to be 10 μM. Results obtained from three different types of binding assays all corroborate the conclusion that phenobarbital directly binds to EGFR. Moreover, the structural features of PB binding suggested that phenobarbital may interfere with EGFR activation by EGF.

Upon EGF binding, EGFR is phosphorylated at various residues such as tyrosine 845 and tyrosine 1179 to elicit cell signaling. These tyrosine residues were in fact phosphorylated in mouse primary hepatocytes after EGF treatment in a dose- and time-dependent manner. Once phosphorylated, these residues were dephosphorylated in response to phenobarbital treatment. On the other hand, pretreatment with phenobarbital prevented EGF from phosphorylating these tyrosine residues. In agreement with the physiologically effective concentration of EGF to activate EGFR and the K_d value of phenobarbital binding to EGFR, 10 μM of phenobarbital could repress EGFR activation by EGF at 10 nM in mouse primary hepatocytes. Thus, phenobarbital was found to directly bind to EGFR and to antagonize EGF to activate EGFR, thereby repressing EGFR signaling. Although EGFR appeared to be the phenobarbital receptor, a critical question raised then was whether and how phenobarbital converts the EGFR pathway to CAR activation signal.

14.6 RACK1 Enables PP2c to Dephosphorylate CAR

As had been described, PB activates CAR through dephosphorylation of threonine 38 by PP2A (Kawamoto et al. 1999; Mutoh et al. 2009). Before it could be possible to perform experiments to connect EGFR signaling with CAR activation, the mechanism of this dephosphorylation first needed to be deciphered. PP2A consists of three subunits: a, b, and c. PP2Ac is the catalytic subunit and forms an inactive complex with the scaffold PP2Aa subunit, the so-called PP2A core enzyme. The regulatory subunit PP2Ab binds to and activates the core enzyme. There are more than 20 known b-subunits, from which the core enzyme selects to confer substrate specificity. Our yeast two-hybrid screening identified the receptor for activated kinase C 1 (RACK1) as an associated protein with the CAR T38D mutant. Because of its binding to phosphorylated CAR and as the RACK1 3D structure resembles that of known PP2Ab subunit 52, the possibility that RACK1 might act as the PP2Ab subunit to activate the core enzyme and facilitate dephosphorylation of threonine 38 of CAR was tested. Subsequent in vitro dephosphorylation assays using phosphorylated CAR as a substrate confirmed that the core enzyme requires RACK1 to dephosphorylate threonine 38. RACK1 is also known to mediate signal transduction by interacting with various protein kinases and phosphatases as well as to be regulated by stimuli such as EGF. Therefore, RACK1 provided us with the opportunity to investigate whether EGFR transduces the signal to dephosphorylate threonine 38 of CAR through RACK1.

14.7 RACK1 Transduces EGFR Signal to PP2Ac

Tyrosine 52 of RACK1 was first found to be phosphorylated in response to EGF in Huh 7 cells. RACK1 activates PP2Ac to dephosphorylate threonine 38 for CAR activation. Recombinant RACK1 Y52F and RACK1Y52E proteins were utilized as the regulatory b-subunit to examine whether phosphorylation determines the ability of RACK1 to activate PP2Ac. As a result, only the nonphosphorylated Y52F form enabled PP2Ac to dephosphorylate threonine 38. Thus, EGFR transduces a phosphorylation signal to RACK1 at tyrosine 52, thereby repressing RACK1 activation of PP2Ac. In the mouse liver, this tyrosine 52 is phosphorylated and dephosphorylated after phenobarbital treatment. Therefore, a scenario that has evolved is the following: EGF-EGFR signaling inhibits CAR activation by phosphorylating RACK1, whereas phenobarbital directly binds to EGFR and antagonizes the EGFR signal to dephosphorylate RACK1. Then, nonphosphorylated RACK1 functions as the regulatory b-subunit to activate PP2Ac, which dephosphorylates threonine 38 to activate CAR.

14.8 From a General EGFR to a CAR-Specific Signal

Phenobarbital action to induce P450 and drug metabolism begins by directly binding to EGFR and antagonizing its signaling. Subsequently, RACK1 mediates this antagonistic signaling to PP2Ac, which dephosphorylates CAR for activation (Fig. 14.1). Then, a critical question arises as to the mechanism by which CAR specifies a general EGFR signaling to utilize for its own activation. The site (threonine 38) that is dephosphorylated is located in the DBD and the binding site for PP2C and RACK1 reside near the N-terminal region of the LBD, whereas these two sites are in the same space within the CAR structure (Mutoh et al. 2009). However, dephosphorylation was regulated by the C-terminal region of CAR, which is located on the opposite surface of the LBD structure (Osabe and Negishi 2011). Our hypothesis is that CAR possesses a peptide within its molecule that acts as an intramolecular signal and converts EGF signaling to the CAR-specific signal, facilitating dephosphorylation. Given the caveat that this hypothesis is still in its embryonic stage, the previously defined peptide designated xenochemical response signal (XRS) near the C-terminus of CAR may be this intramolecular signal. XRS was first characterized as a peptide motif (LXXLXXL) that regulates drug-induced nuclear translocation of CAR in mouse liver; CAR is unable to translocate into the nucleus in the absence of XRS (Zelko et al. 2001). Recently, XRS was revealed to be the active ERK1/2-binding site (Osabe and Negishi 2011). In response to EGF treatment, XRS binds active ERK1/2 to prevent dephosphorylation of CAR, thereby restraining CAR in an inactive form and sequestering it in the cytoplasm. Inactivation of ERK1/2 by MEK inhibitor U0126 results in ERK1/2 dissociation from CAR for dephosphorylation and activation. Now, our work in progress has demonstrated that XRS–ERK1/2 interactions in fact regulate RACK1 binding to phosphorylated CAR in response to EGFR signaling. Phenobarbital appears to antagonize multiple EGFR signals, among which MEK-ERK1/2 and RACK1 signaling converge on XRS to regulate CAR activation (Fig. 14.2). The RACK1 signaling is to activate PP2A for dephosphorylation and the MEK-ERK1/2 signaling is for XRS to regulate the RACK1 signal to phosphorylated CAR. Thus, CAR cannot be properly activated in the absence of XRS as the intramolecular peptide signal. Deciphering the molecular mechanism of this XRS-mediated intramolecular peptide signal is the next critical step to fully comprehend phenobarbital induction. A transgenic mouse expressing CAR without XRS is in production, which should help us with this endeavor.

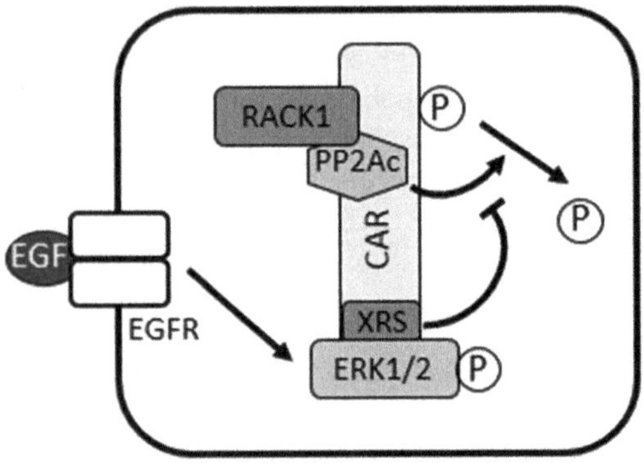

Fig. 14.2 XRS converts EGFR signaling into the specific CAR activation mechanism. XRS is an internal peptide signal near the C-terminus of CAR and the binding site for active (phosphorylated) ERK1/2. When EFGR signaling is stimulated, XRS binds to active ERK1/2 and prevents CAR from being dephosphorylated by a PP2Ac-RACK1 complex. When EGFR signaling is attenuated after phenobarbital treatment, XRS dissociates from ERK1/2, allowing PP2Ac-RACK1 to facilitate PP2Ac to dephosphorylate CAR

14.9 Perspectives

Phenobarbital is the archetype that has been intensively investigated for induction of drug metabolism. It is a difficult one: we have been struggling to decipher the mechanism of its actions for a long time. EGFR is now characterized as a direct target of phenobarbital action (i.e., a phenobarbital receptor), which has opened up a whole new field of both basic and clinical research of phenobarbital actions (Fig. 14.3). Phenobarbital antagonizes its signaling to elicit CAR activation and induce drug metabolism; the phrase “antagonistic activation” describes the key feature of this CAR activation mechanism. Phenobarbital may also antagonize other cell membrane receptors such as growth hormone receptor, hepatocyte growth factor receptor, insulin receptor, and insulin-like growth factor receptor. As is true with EGF, activators of these membrane receptors are known to repress phenobarbital induction. Similar to phenobarbital, numerous therapeutic drugs and xenobiotics are known to activate CAR indirectly. It can now be tested whether they also antagonize EGFR or membrane receptors to activate CAR. Therefore, the concept of antagonistic activation (i.e., indirect CAR activation via membrane receptors) should open a new era for the study of drug metabolism and induction. CAR can be activated by direct ligand binding, typified by TCPOBOP and CITCO. However, ligand activation of CAR is no longer understood by a simple binding mechanism because CITCO cannot activate CAR so long as it is phosphorylated at threonine 38 (Mutoh et al. 2009). CAR is, in principle, a cell signal-regulated nuclear receptor, and its underlying mechanism is dephosphorylation of threonine 38. The ligand activation mechanism is an important subject of future

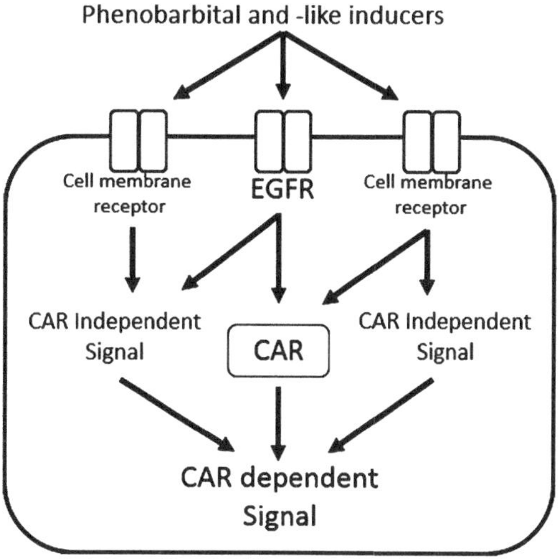

Fig. 14.3 Phenobarbital elicits diverse signaling via binding to membrane receptors. Phenobarbital antagonizes growth factor receptors such as EGFR and elicits multiple signaling, both CAR-mediated and non-CAR-mediated signaling. Both types of signaling can be independent of each other but may also be coordinators to regulate liver functions

investigations. Nevertheless, CAR activation not only regulates drug metabolism but also other liver functions such as energy metabolism and development of disease such as hepatocellular carcinoma (Konno et al. 2008; Yamamoto et al. 2004; Huang et al. 2005).

Acknowledgments This research was supported by the Intramural Research Program of National Institute of Environmental Health Sciences (Grants Z01ES1005-01).

References

Alvares AP, Schilling G, Levin W, Kuntzman R (1967) Studies on the induction of CO-binding pigments in liver microsomes by phenobarbital and 3-methylcholanthrene. Biochem Biophys Res Commun 9:521–526

Baes M, Gulick T, Choi HS, Martinoli MG, Simha D, Moore DD (1994) A new orphan member of the nuclear hormone receptor superfamily that interacts with a subset of retinoic acid response elements. Mol Cell Biol 14:1544–1552

Bauer D, Wolfran N, Kahl GF, Hirsch-Ernst KI (2004) Transcriptional regulation of CYP2b1 induction in primary rat hepatocyte cultures: repression by epidermal growth factor is mediated via a distal enhancer region. Mol Pharmacol 65:172–180

Fujii-Kuriyama Y, Mizukami Y, Kawajiri K, Sogawa K, Muramatsu M (1982) Primary structure of a cytochrome P-450: coding nucleotide sequence of phenobarbital-inducible cytochrome P-450 cDNA from rat liver. Proc Natl Acad Sci USA 79:2793–2797

Greenlee WF, Poland A (1979) Nuclear uptake of 2,3,7,8-tetrachlorodibenzo-*p*-dioxin in C57BL/6J and DBA/2J mice. Role of the hepatic cytosol receptor protein. J Biol Chem 254:9814–9821

Honkakoski P, Negishi M (1998) Regulatory DNA elements of phenobarbital-responsive cytochrome P450 CYPΔ genes. J Biochem Mol Toxicol 12:3–9

Honkakoski P, Moore R, Washburn KA, Negishi M (1998a) Activation by diverse xenochemicals of the 51-base pair phenobarbital-responsive enhancer module in the CYP2B10 gene. Mol Pharmacol 53:597–601

Honkakoski P, Zelko I, Sueyoshi T, Negishi M (1998b) The nuclear orphan receptor CAR-retinoid X receptor heterodimer activates the phenobarbital-responsive enhancer module of the *CYP2B* gene. Mol Cell Biol 18:5652–5658

Huang W, Zhang J, Washington M, Liu J, Parant JM, Lozano G, Moore DD (2005) Xenobiotic stress induces hepatomegaly and liver tumors via the nuclear receptor constitutive androstane receptor. Mol Endocrinol 19:1646–1653

Hwang DL, Roitman A, Lev-Ran A, Carr BI (1986) Chronic treatment with phenobarbital decreases the expression of rat EGF and insulin receptors. Biochem Biophys Res Commun 135:501–506

Kawamoto T, Sueyoshi T, Zelko I, Moore R, Washburn K, Negishi M (1999) Phenobarbital responsive nuclear translocation of the receptor CAR in induction of the CYP2B gene. Mol Cell Biol 19:6318–6322

Kobayashi K, Sueyoshi T, Inoue K, Moore R, Negishi M (2003) Cytoplasmic accumulation of the nuclear receptor CAR by a tetratricopeptide repeat protein in HepG2 cells. Mol Pharmacol 64:1069–1075

Koike C, Moore R, Negishi M (2007) Extracellular signal-regulated kinase is an endogenous signal retaining the nuclear constitutive active/androstane receptor (CAR) in the cytoplasm of mouse primary hepatocytes. Mol Pharmacol 71:1217–1221

Konno Y, Negishi M, Kodama S (2008) The roles of nuclear receptors CAR and PXR in hepatic energy metabolism. Drug Methab Pharmacokinet 23:8–13

Lakso M, Masaki R, Noshiro M, Negishi M (1991) Structures and characterization of sex-specific cytochrome-P450 genes as members within a large family: duplication and evolution. Eur J Biochem 195:477–486

Lindros P, Michalopoulos GK (1993) Response of phenobarbital- and ciprofibrate-exposed hepatocytes to growth factors in type I collagen gels. Carcinogenesis (Oxf) 14:731–735

Maglich JM, Parks DJ, Moore LB, Collins JL, Goodwin B, Billin AN, Stoltz CA, Kliewer SA, Lambert MH, Willson TM, Moore JT (2003) Identification of a novel human constitutive androstane receptor (CAR) agonist and its use in the identification of CAR target genes. J Biol Chem 278:17277–17283

Meyer SA, Gibbs TA, Jirtle RL (1989) Independent mechanisms for tumor promoters phenobarbital and 12-*O*-tetradcanoylphorbol-13-acetate in reduction of epidermal growth factor binding by rat hepatocytes. Cancer Res 49:5907–5912

Mizukami Y, Sogawa K, Suwa Y, Muramatsu M, Fuji-Kuriyama Y (1983) Gene structure of a phenobarbital-inducible cytochrome P-450 in rat liver. Proc Natl Acad Sci USA 80:3958–3962

Mutoh S, Osabe M, Inoue K, Moore R, Pedersen L, Perera L, Rebolloso Y, Sueyoshi T, Negishi M (2009) Dephosphorylation of threonine 38 is required for nuclear translocation and activation of human xenobiotic receptor CAR (NR1I3). J Biol Chem 284:34785–34792

Mutoh S, Sobhany M, Moore R, Perera L, Pedersen L, Sueyoshi T, Negishi M (2013) Phenobarbital indirectly activates the constitutive active androstane receptor (CAR) by inhibition of epidermal growth factor receptor signaling. Sci Signal 6:ra31

Okey AB (2007) An aryl hydrocarbon receptor odyssey to the shores of toxicology: The Deichmann Lecture, International Congress of Toxicology-XI. Toxicol Sci 98:5–38

Okey AB, Bondy GP, Mason ME, Kahl GF, Eisen HJ, Guenthner TM, Nebert DW (1979) Regulatory gene product of the Ah locus. Characterization of the cytosolic inducer-receptor complex and evidence for its nuclear translocation. J Biol Chem 254:11636–11648

Omura T, Sato R (1962) A new cytochrome in liver microsomes. J Biol Chem 237:1375–1376

Osabe M, Negishi M (2011) Active ERK1/2 protein interacts with the phosphorylated nuclear constitutive active/androstane receptor (CAR; NR1I3), repressing dephosphorylation and sequestering CAR in the cytoplasm. J Biol Chem 286:35763–35779

Remmer H, Merker HJ (1963) Drug-induced changes in the liver endoplasmic reticulum: association with drug-metabolizing enzymes. Science 142:1657–1658

Scheer N, Ross J, Rode A, Zevnik B, Niehaves S, Faust N, Wolf CR (2008) A novel panel of mouse models to evaluate the role of human pregnane X receptor and constitutive androstane receptor in drug response. J Clin Invest 118:3228–3239

Sueyoshi T, Negishi M (2001) Phenobarbital response elements of cytochrome P450 genes and nuclear receptors. Annu Rev Pharmacol Toxicol 41:123–143

Sueyoshi T, Kawamoto T, Zelko I, Honkakoski P, Negishi M (1999) The repressed nuclear receptor CAR responds to phenobarbital in activating the human CYP2B6 gene. J Biol Chem 274:6043–6046

Tzameli I, Pissios P, Schuetz EG, Moore DD (2000) The xenobiotic compound 1,4-bis [2-(3,5-dichloropyridyloxy)]benzene is an agonist ligand for the nuclear receptor CAR. Mol Cell Biol 20:2951–2958

Wei P, Zhang J, Egan-Hafley M, Liang S, Moore DD (2000) The nuclear receptor CAR mediates specific xenobiotic induction of drug metabolism. Nature (Lond) 407:920–923

Yamamoto Y, Moore R, Goldsworthy TL, Negishi M, Maronpot RR (2004) The orphan nuclear receptor constitutive active/androstane receptor is essential for liver tumor promotion by phenobarbital in mice. Cancer Res 64:7197–7200

Yamazoe Y, Shimada M, Murayama N, Kato R (1987) Suppression of levels of phenobarbital-inducible rat liver cytochrome P-450 by pituitary hormone. J Biol Chem 262:7423–7428

Yoshinari K, Kobayashi K, Moore R, Kawamoto T, Negishi M (2003) Identification of the nuclear receptor:HSP90 complex in mouse liver and recruitment of protein phosphatase 2A in response to phenobarbital. FEBS Lett 548:17–20

Zelko I, Sueyoshi T, Kawamoto T, Moore R, Negishi M (2001) The peptide near the C terminus regulates receptor CAR nuclear translocation induced by xenochemicals in mouse liver. Mol Cell Biol 21:2838–2846

Chapter 15
Steroidogenic Cytochrome P450 Gene *CYP11A1*: Functions and Regulation

Monica Meng-Chun Shih, Hwei-Jan Hsu, Hsin-Chieh Lan, Jui-Hsia Weng, Yu Chien, Meng-Chun Hu, and Bon-chu Chung

Abstract Steroids belong to the class of endogenous substrates that are metabolized by cytochromes P450. These steroids are produced in our body, mainly from the adrenals and gonads, and circulate throughout the body in minute amounts to exert their physiological functions. The first and rate-limiting step for the production of all steroids is the conversion of cholesterol into pregnenolone, catalyzed by P450scc, or CYP11A1. Deficiency of CYP11A1 results in adrenal insufficiency that can be lethal when untreated. Hypomorphic expression of *Cyp11a1* is less detrimental, but still resulted in decreased stress response in mice. *CYP11A1* is expressed mainly in the adrenals and gonads, and to a lesser extent brain, skin, and intestine. Transcription factor NR5A1 controls the expression of *Cyp11a1* in the adrenals and testis, whereas NR5A2 controls the expression

M.M.-C. Shih
Institute of Molecular Biology, Academia Sinica, Taipei 115, Taiwan

Department of International Cooperation, National Science Council, Taipei, Taiwan

H.-J. Hsu
Institute of Molecular Biology, Academia Sinica, Taipei 115, Taiwan

Institute of Cellular and Organismic Biology, Academia Sinica, Taipei, Taiwan

H.-C. Lan
Institute of Molecular Biology, Academia Sinica, Taipei 115, Taiwan

Graduate Institute of Biology and Anatomy, National Defense Medical Center, Taipei, Taiwan

J.-H. Weng • Y. Chien • B.-c. Chung (✉)
Institute of Molecular Biology, Academia Sinica, Taipei 115, Taiwan
e-mail: mbchung@sinica.edu.tw

M.-C. Hu
Institute of Molecular Biology, Academia Sinica, Taipei 115, Taiwan

Graduate Institute of Physiology, National Taiwan University, Taipei, Taiwan

H. Yamazaki (ed.), *Fifty Years of Cytochrome P450 Research*,
DOI 10.1007/978-4-431-54992-5_15, © Springer Japan 2014

of *Cyp11a1* in the intestine. Both NR5A proteins cooperate with other general factors such as Sp1 and AP1 to activate *Cyp11a1* expression in response to the stimulation of cAMP. The *cyp11a1* mRNA in zebrafish is also deposited as a maternal transcript for the production of pregnenolone in early embryos. Pregnenolone stabilizes microtubules and promotes embryonic cell migration in zebrafish. Thus, use of the zebrafish model reveals a new function of CYP11A1 during early embryogenesis.

Keywords Adrenal • Gonad • NR5A • P450scc • Steroidogenesis • Zebrafish

15.1 Introduction

15.1.1 Classification of Steroids

Steroids are a group of compounds composed of a core cyclopentanophenanthrene structure and various attached functional groups. In addition to a role in tissue development, steroids exert profound physiological effects. Steroids are hormones because they are secreted from the endocrine glands in small quantities and circulate in the whole body via the bloodstream. Steroids are therefore often classified by their physiological functions. For example, progesterone is important for the maintenance of pregnancy, and sex steroids control sexual differentiation and sexual characteristics. Mineralocorticoids control salt balance, and glucocorticoids control sugar balance. In addition, glucocorticoids have additional roles in modulating immune response and inflammation. This antiinflammatory action is perhaps more widely recognized and used in medicine, and many common drugs are synthetic glucocorticoids.

Steroids function by binding to their receptors in target organs. Steroid receptors belong to the nuclear receptor superfamily (Nuclear Receptors Nomenclature 1999): these are transcription factors composed of a ligand-binding domain, a DNA-binding domain, and transcriptional activation domains. These receptors are activated by their steroid ligands to stimulate transcription after binding to the promoters of their target genes. Steroids have important roles in development by activating genes that control the differentiation of their target organs such as adrenals and gonads.

15.1.2 Tissue-Specific Secretion of Steroids

Two major sites of steroid secretion are the adrenal glands and the gonads (Hu et al. 2001). In the adrenal, steroids are secreted from the cortex, and therefore the adrenal steroids are also termed corticoids. The steroids secreted from the adrenal cortex are mineralocorticoids, glucocorticoids, and sex steroid precursors.

The adrenal cortex is divided in three layers: zonae glomerulosa, fasciculata, and reticularis. Mineralocorticoids are secreted from zona glomerulosa, and glucocorticoids are mainly secreted from zona fasciculata. The zona reticularis mainly secretes sex steroid precursors. As is discussed next, mineralocorticoids and glucocorticoids control salt and sugar balance, which is important for life. The absence of mineralocorticoid or glucocorticoid leads to dehydration or hypoglycemia, either of which is life threatening.

Gonads secrete sex steroids. The main female sex steroid, estrogen, is secreted from the ovary. The major male hormone, testosterone, is secreted from the testis. In addition to these two major sites of steroid secretion, the placenta also secretes a large quantity of progesterone during pregnancy for the maintenance of pregnancy. Other than these major sites of steroid secretion, tissues such as the brain, intestine, and skin also secrete a small amount of steroids. The functions of these extra-adrenal steroids is discussed in the following section.

15.1.3 HPA/HPG/RAA Axes of Steroid Regulation

The secretion of sex steroids is controlled by the hypothalamus–pituitary–gonad axis. This axis invokes the hypothalamus in the brain, which secrets gonadotropin-releasing hormone (GnRH) that signals the release of follicle-stimulating hormone (FSH) and leuteinizing hormone (LH) from the pituitary (Marieb and Hoehn 2007). FSH and LH then stimulate the gonad, which responds by secreting sex steroids. Sex steroids exert their actions in the circulation, and at high enough concentration they reach the brain to prevent the hypothalamus from further stimulation.

Similar to gonadal steroid production, the production of adrenal glucocorticoids is also subject to the regulation of the hypothalamus and pituitary forming an HPA axis (Papadimitriou and Priftis 2009). In this axis, the brain is the major sensor of physiological changes such as stress that require the production of extra steroids. The hypothalamus in the brain will respond to stress by secreting corticotropin-releasing hormone (CRH), which signals the pituitary to secrete adrenocorticotropic hormone (ACTH). ACTH travels to and stimulates the adrenal to secrete corticosteroids, which circulate in the bloodstream to reach all body organs and to stimulate transcription of target genes to cope with the stress. High levels of corticosteroids will then cause feedback to inhibit the secretion of CRH and ACTH in the brain to dampen this stress response.

The regulation of mineralocorticoids is under the renin-angiotensin-aldosterone (RAA) system in response to salt concentration (Williams and Dluhy 1972). In this system, the kidney is the main organ that senses salt concentration in the blood. When sodium concentration is low, the kidney responds by secreting more renin, which triggers the maturation of angiotensin II. Angiotensin II then stimulates the secretion of mineralocorticoids from the adrenal for the regulation of salt balance.

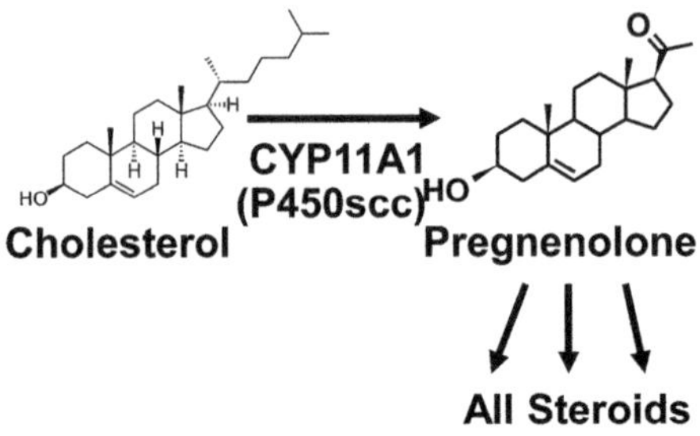

Fig. 15.1 Catalytic reaction of P450scc (CYP11A1). CYP11A1 catalyzes the conversion of cholesterol into pregnenolone, which is then converted by other enzymes to form all steroids

15.1.4 Developmental Regulation of Steroid Secretion

Steroids have a strong developmental role, as can be revealed by the ontogeny of steroid secretion. Steroids are made in large quantities during embryogenesis; they decline in childhood, and steroid levels then soar again during adolescence. Steroid levels gradually decline with aging (Schumacher et al. 2003; Purifoy et al. 1980). This developmental profile indicates that steroids are most important for embryogenesis and also in adult life.

15.1.5 CYP11A1 in the First Step of Steroid Synthesis

Steroids are compounds derived from cholesterol; they have structures composed of four carbon rings similar to those of cholesterol, but they do not have the long side chain of cholesterol (Fig. 15.1). The side chain of cholesterol is first cleaved by the cholesterol side-chain cleavage enzyme, or in short, SCC, to produce the first steroid product, pregnenolone. This first enzyme in the steroidogenic pathway has been classified as CYP11A1 (P450scc).

Cytochromes P450 catalyze mono-oxygenation reactions of substrates derived from either an exogenous or endogenous source, and steroids are an important class of endogenous substrates for cytochromes P450. The nomenclature of cytochromes P450 has been unified to become that of CYPs (Nelson 2006). In this chapter, P450scc protein is termed CYP11A1. For the gene nomenclature, the human gene is termed *CYP11A1*, the mouse gene *Cyp11a1*, and the zebrafish gene *cyp11a1* for clarity. In this chapter, if a statement is applicable to all species, the term *CYP11A1* will be used to represent all species.

15.2 Regulation of CYP11A1

Steroids circulate in our body in minute amounts: the amounts of steroids are tightly regulated to precisely exert their physiological functions. This control is exerted both acutely and chronically. The acute response is usually at the enzymatic levels to secrete more steroids to ensure immediate reactions to physiological changes to stress. Steroidogenic acute regulatory protein (StAR) has a key role in this acute response (Clark et al. 1994). It is translated in response to stimulation and simultaneously triggers the transfer of cholesterol into the inner mitochondrial membrane as a supply of substrate for the catalysis of CYP11A1 in the first and rate-limiting step of steroid synthesis (Bose et al. 1996). This acute regulation occurs within minutes of stimulation and can quickly accommodate the immediate needs for steroids.

The chronic changes usually involve changes of gene expression; it is achieved to ascertain the maintenance of the cells at the new physiological step. This tight control is achieved at the temporal, developmental, physiological, and hormonal levels. Although all steroidogenic genes are coordinately regulated, in this chapter we focus on the regulation of *CYP11A1* in the first and rate-limiting step of steroid synthesis.

15.2.1 Tissue-Specific Expression of CYP11A1

Similar to steroid secretion, *CYP11A1* is mainly expressed in steroidogenic tissues such as the adrenals and gonads, and to a lesser extent in brain, intestine, and skin. Nuclear receptor 5A proteins are the major transcription factors that control the expression of *CYP11A1*. In the adrenal, the main transcription factor for *Cyp11a1* is NR5A1 (steroidogenic factor 1, or SF1/Ad4BP) (Parker et al. 2002; Honda et al. 1993). However, in the ovaries and in the intestine, the major transcription factor that controls *Cyp11a1* becomes NR5A2 (liver receptor homolog-1, LRH-1) (Mueller et al. 2006). Both NR5A1 and NR5A2 are orphan nuclear receptors that bind to the TCAAGGTCA sequence as a monomer. They activate the transcription of *Cyp11a1* and other steroidogenic genes to maintain the differentiated state of the steroidogenic cells. Both NR5A1 and NR5A2 are classified as orphan nuclear receptors, and their activities are not regulated by ligand binding, despite numerous attempts to look for their ligands (Mellon and Bair 1998; Desclozeaux et al. 2002). They are instead regulated mainly by translational modifications such as sumoylation (Chen et al. 2004; Lee et al. 2011), acetylation (Chen et al. 2005), phosphorylation (Hammer et al. 1999), and ubiquitination (Chen et al. 2007).

15.2.2 Developmental Regulation of CYP11A1 *Expression*

CYP11A1 is expressed in a developmentally regulated fashion. It first appears in the steroidogenic cells during embryogenesis when adrenocortical primordial cells are being formed. The factors that control the developmental expression of *Cyp11a1* in the adrenals and gonads are NR5A1 and NR5A2, respectively (Parker et al. 2002; Mueller et al. 2006). These two factors may serve as the master regulator that determines the differentiation of their progenitor cells into the steroid-secreting cells by turning on the transcription of steroidogenic genes. The continued expression of NR5As in these steroidogenic cells ensures the continued expression of steroidogenic genes as characteristics of this cell type.

Steroidogenic genes are upregulated during puberty, mainly because of the increased secretion of pituitary gonadotropins that stimulate the expression of *CYP11A1*. Therefore, this second wave of steroid increase is controlled by hormones at the level of transcription.

15.2.3 In Vivo and In Vitro Study of Cyp11a1 *Promoter*

The regulatory elements that govern the tissue-specific, developmentally and hormonally controlled transcription of *CYP11A1* have been extensively investigated (Shih et al. 2011). The basal promoter of *CYP11A1* is mainly composed of the *TATA* sequence (Guo and Chung 1999), the NR5A-binding site, and the binding site for transcription factor Sp1 (Guo et al. 1994). More distal to this basal promoter are the binding sites for many ubiquitous transcription factors that interact with NR5A and thus lead to maximal gene expression. The upstream region of the basal promoter controls hormonal regulation of *CYP11A1* (Guo et al. 1994); it includes the binding sites for NR5A1 and AP1 family members that respond to hormonal stimulation for maximal gene activation. Generally speaking, the activation of both the basal and the upstream regions of *CYP11A1* promoter is stimulated by the interaction of tissue-specific NR5A and general transcriptional co-activators, such as AP1 and Sp1 (Guo et al. 2007; Li et al. 1998). Factors that contribute to the expression of *CYP11A1* promoter has been listed in a recent review and therefore are not discussed in detail here (Shih et al. 2011).

The importance of a few *CYP11A1* promoter elements has been verified in transgenic mice (Hu et al. 1999). When the basal NR5A-binding site is mutated, the resulting reporter gene expression is greatly reduced (Guo et al. 2007), demonstrating the absolute requirement of this basal element for *CYP11A1* transcription. The upstream NR5A-binding site, however, is important for hormone-stimulated transcription. The reporter gene fails to respond to LH when this upstream NR5A site is mutated.

15.2.4 Signal Pathway Regulating Cyp11a1 *Transcription in the Adrenal*

CYP11A1 expression is stimulated in the adrenal when the level of circulating ACTH is increased. ACTH binds to the cell-surface G protein-coupled MC2R receptor and stimulates the activity of adenylyl cyclase, leading to the increase of intracellular cAMP levels (Chida et al. 2007). The transcriptional activation of

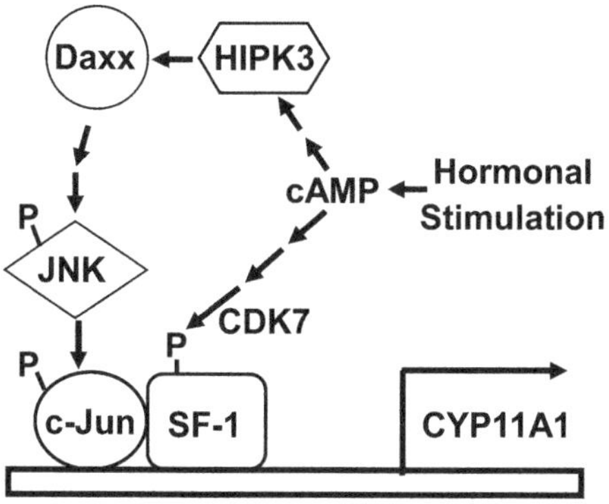

Fig. 15.2 Diagram of signaling pathways that regulate Cyp11a1 transcription in response to cAMP. Hormonal stimulation leads to an increase in the intracellular cAMP levels. This signal is transmitted via many proteins. One signaling pathway goes through HIPK3, Daxx, and JNK, leading to c-Jun phosphorylation. In the meantime, the activity of transcription factor SF-1 is increased after its Ser-203 is phosphorylated by CDK7. SF-1 and c-Jun form a complex that binds to *CYP11A1* promoter and enhances *CYP11A1* transcription in response to hormonal stimulation

CYP11A1 by cAMP takes many hours to complete and requires many intermediate steps, including the translation of labile intermediate protein factors such as salt-inducible kinase (SIK1) (Lin et al. 2001). The signaling pathways that respond to cAMP stimulation to activate *CYP11A1* transcription are very complex, and there are many cross-talks among them. One of these activation pathways is the HIPK3/Daxx/JNK/c-Jun protein cascade (Fig. 15.2) (Lan et al. 2007, 2012). This pathway usually operates in other cell types in response to the death signal, but in steroidogenic cells it is used for the activation of *CYP11A1* transcription. cAMP can also trigger the activation of CDK7 kinase, leading to the phosphorylation of NR5A1 (SF-1) at Ser-209 (Winnay and Hammer 2006; Lewis et al. 2008). Phosphorylation of SF-1 at Ser-209 is important for co-activator recruitment and full transcriptional activity (Hammer et al. 1999). By the actions of concurrent activation pathways and co-activator interaction, ultimately *CYP11A1* promoter is activated in response to cAMP stimulation.

15.3 Functions of CYP11A1

CYP11A1 is known to convert cholesterol into pregnenolone biochemically. The physiological roles of this enzymatic reaction have also been characterized in *CYP11A1*-deficient patients and in mice mutated in *Cyp11a1*. The results from human patients and from mice manifest the importance of steroids in maintaining vitality, metabolism, and the response to stress.

Table 15.1 Phenotypes of mice with disturbances of Cyp11a1 expression

Genotype (reference)	Adrenal CYP11A1 level	Lifespan	Steroid levels	Phenotype
WT	100 %	Normal	Normal	Normal
Null (Hu et al. 2002)	0 %	0–7 days[a]	Nil	Steroid insufficiency, abnormal sexual differentiation
L/L Promoter Mutation (Shih et al. 2008)	15 %	Normal	Insufficient during stress	Decreased stress response
Transgenic *Tg8* (Chien et al. 2013)	~2–3 fold	Normal	Insufficient progesterone at pregnancy	Impaired pregnancy

The phenotypes of *Cyp11a1* wild type (WT), transgenic, and mutant mice are listed. The null mice have an insertional disruption of the *Cyp11a1* gene leading to no protein production. The *Cyp11a1*$^{L/L}$ mice have a 2-bp mutation at the basal promoter causing reduction of gene expression. The transgenic line *Tg8* has extra *Cyp11a1* genes inserted as a bacterial artificial chromosome
[a]Lifespan varies from a few hours in a pure inbred background to around 7 days in a mixed genetic background

15.3.1 Human CYP11A1 *Deficiency*

Because the production of steroids catalyzed by CYP11A1 is essential for life, CYP11A1 deficiency is usually associated with death; there are very few cases of survival in the clinic. Most of these cases are associated with partial or complete adrenal insufficiency diagnosed very early in the childhood (Sahakitrungruang et al. 2011; Kim et al. 2008; Hiort et al. 2005). Patients present with adrenal insufficiency and XY sex reversal and have to be treated early. Most patients have mutations in both alleles of the *CYP11A1* gene, but there has been one patient with a loss-of-function mutation in one allele and normal coding sequence in the other allele (Sahakitrungruang et al. 2011). *CYP11A1* deficiency is a recessive disease caused by mutations in both alleles, and heterozygous parents of these patients are normal, which infers that this patient with normal coding sequence may have a yet unidentified mutation (Sahakitrungruang et al. 2011), most likely in the regulatory region. This hypothesis remains to be tested if the entire genome of this patient can be sequenced and other mutation sites searched.

15.3.2 Phenotypes of Cyp11a1 *Null Mice*

Cyp11a1 null mice have been generated (Hu et al. 2002). They share the same phenotypes as other mouse models of steroid deficiency such as *StAR* knockout mice. *Cyp11a1* null mice die neonatally of the complete loss of adrenal steroids (Table 15.1). Their life can be rescued by the daily injection of mineralocorticoids and glucocorticoids. Furthermore, these mice also suffer from incomplete development of accessory sexual organs. XY mice are feminized with female external genitalia. Internally, these knockout mice have underdeveloped testis, epididymis, and vas deferens. The seminal vesicles, prostate, bulbourethral glands, and penis are not present, revealing the functions of male sex steroids in sexual differentiation.

Although fetal *Cyp11a1* null mice develop to term because of supplementation from the maternal steroid supply, these knockout mice still suffer from insufficient development of the adrenal medulla and the hypothalamus–pituitary–adrenal axis (Huang et al. 2012). This observation indicates that de novo synthesis of steroids is required for proper development of the adrenal and HPA axis.

15.3.3 Phenotypes of Cyp11a1 *Hypomorphic Mice*

A hypomorphic *Cyp11a1* allele has been created using the gene knock-in strategy. This allele, termed $Cyp11a1^{L/L}$, has two base-pair mutations in the proximal NR5A-binding site of the promoter (Shih et al. 2008). This mutation reduces *Cyp11a1* expression to 15 % of the wild-type level in the adrenal (Table 15.1). The mutation verifies the importance of this site in the adrenal. The expression of $Cyp11a1^{L/L}$ in the ovary, however, is the same as that of the wild type, indicating a tissue-specific regulation.

One interesting question is the steroid output of these hypomorphic mice when their CYP11A1 levels are about 15 % of the wild-type level. It turns out that these hypomorphs secrete the same low levels of corticosterone, the major glucocorticoid in the rodent, at rest. Their circadian rhythm of steroid secretion, however, is dampened during the active hours. The secretion of corticosteroids in response to stress is also reduced. The decreased secretion of corticosteroids is reflected by the super-high levels of inflammatory cytokines when mice are stressed.

15.3.4 CYP11A1 *Polymorphism and* Cyp11a1 *Transgenic Mice*

In addition to creating mice with *Cyp11a1* deficiency, we have generated transgenic mice that overexpress *Cyp11a1* (Table 15.1). The transgenes in these mice are fully functional, as they can rescue the defect of *Cyp11a1* null mice when crossed together (Chien et al. 2013). However, female transgenic mice suffer from pregnancy failure, in contrast to the expectation (Chien et al. 2013). The male transgenic mice, however, are fertile. Female transgenic mice produce insufficient amounts of progesterone during early pregnancy; placental development is also defective. The defective placental development, the loss of fetuses, and pregnancy impairment can all be accounted for by the reduction of progesterone secretion, revealing the importance of progesterone in the maintenance of pregnancy.

The reduction of progesterone from transgenic female ovaries is an unexpected result that indicates that steroid secretion is tightly regulated and can go the opposite way when the balance is disturbed. This result is also different from that of the human studies when *CYP11A1* polymorphisms are detected. Increased *CYP11A1* mRNA levels are associated with higher risks to polycystic ovary syndrome (Strauss 2003; Pusalkar et al. 2009). *CYP11A1* polymorphism may also modify the risks of breast cancer and endometrial hyperplasia (Yaspan et al. 2007; Terry et al. 2010). One similarity between the studies of human beings and transgenic mice is that both suffer from reduced pregnancy. Because only young female mice were the subject of the study, perhaps these transgenic mice will develop other problems in the endometrium or ovary when they become older. This possibility remains to be examined.

15.4 Roles of Cyp11a1 in Zebrafish

In addition to mice, zebrafish has been a popular animal model for the studies of diseases, drug discovery, and development. Zebrafish culture is much less expensive than mouse husbandry; they lay eggs weekly in large quantities with ready access of experimental materials. Zebrafish embryos are transparent and develop quickly, enabling easy manipulation. Furthermore, many genomic tools are available, rendering zebrafish an easy and useful tool for the study of human diseases (Hsu et al. 2006a). The following description exemplifies the advantage of using zebrafish for the study of steroid function and regulation.

15.4.1 Expression of Two cyp11a1 *Genes*

One major difference between zebrafish and rodent genomes is that zebrafish genomes are partially duplicated, which has resulted in the presence of two sets of genes in many cases. For example, there are two functional *cyp11a1* genes in the genome (Parajes et al. 2013). These two genes are arranged in tandem on the same region of chromosome 25. The first gene, termed *cyp11a1*, is expressed in early embryos as a maternal transcript (Hsu et al. 2002). Later on, it is expressed in the yolk syncytial layer, a site that simulates the placental syncytial trophoblast in mammals. The maternal deposit and expression of *cyp11a1* mRNA in the yolk syncytial trophoblast probably ensure adequate steroid supply during embryogenesis. Because zebrafish embryos develop outside maternal protection, it is important that enough nutrients are supplied maternally and in the yolk for proper development of the embryo. This novel function of *cyp11a1* is probably the result of evolution as a consequence of zebrafish genome duplication to ensure development of the oviparous embryos.

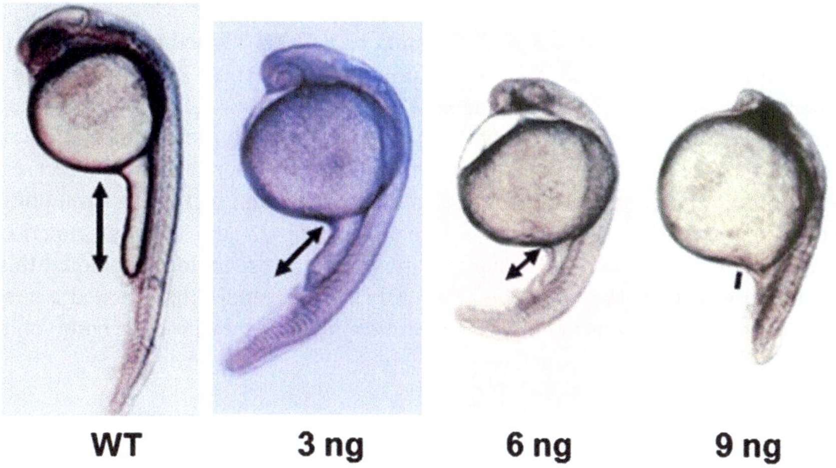

Fig. 15.3 Gross appearance of zebrafish devoid of *cyp11a1* expression at 24 h post fertilization. Zebrafish embryos have a reduced anterior–posterior axis and become fatter after the injection of morpholinos that deplete *cyp11a1* expression. *Double arrows* indicate the length of yolk extension, which is also reduced. Amounts of morpholino are shown below each fish. *WT*, wild-type fish

The second zebrafish *cyp11a1* gene is termed *cyp11a2* (Parajes et al. 2013). It is expressed in the inter-renal gland, the zebrafish equivalent of adrenal glands. When this gene is knocked down, zebrafish display defects of steroidogenesis. Thus, this gene is mainly functional in steroidogenesis during adult life; it is the orthologue of mammalian *CYP11A1*.

15.4.2 Embryonic cyp11a1 *Functions in Cell Migration*

The role of embryonic *cyp11a1* has been investigated in zebrafish. When *cyp11a1* morpholinos were injected into fish embryos, the level of pregnenolone was reduced. Furthermore, the anterior–posterior axis became shorter with increasing concentrations of morpholino (Fig. 15.3). The shorter axis usually represents defects in cell migration. Indeed, these *cyp11a1* "morphants" are delayed in their embryonic epiboly migration, and this migration delayed can be rescued by the supplementation of pregnenolone (Hsu et al. 2006b). This observation indicates that embryonic *cyp11a1* and its product pregnenolone function in supporting cell migration during zebrafish early embryogenesis.

The mechanism by which pregnenolone promotes zebrafish embryonic cell migration has been further investigated. Using photoaffinity labeling, we have identified the pregnenolone receptor as a microtubule-plus-end-tracking protein, CLIP-1 (CLIP-170) (Weng et al. 2013). CLIP-170 has an N-terminal microtubule-binding domain, the middle coiled-coil domain, and the C-terminal zinc knuckles that bind to the dynein–dynactin motor complex (Mishima et al. 2007). When phosphorylated, CLIP-170 exhibits a folded inactive conformation (Lee et al. 2010). Binding to pregnenolone relieves the conformational constraint, and CLIP-170 becomes extended capable of binding to microtubules and the dynein–dynactin motor complex (Weng et al. 2013), resulting in increased microtubule polymerization and increased cell motility. Therefore, CLIP-170 mediates pregnenolone action in promoting zebrafish embryonic cell migration. In zebrafish, *clip-1a* and *cyp11a1* function synergistically to promote embryonic cell migration; pregnenolone promoted this cell migration via CLIP-170 (Weng et al. 2013). These studies have added a new dimension to the functions of embryonic steroids in which they promote embryonic cell migration and development.

15.5 Extra-Adrenal CYP11A1 Expression

The secretion of steroids from tissues other than adrenals and gonads has long been detected. These extra-adrenal steroids, although present in small amounts, have their unique functions that cannot be ignored. The synthesis of these extra-adrenal steroids goes through pathways similar to that in the adrenal glands. The synthesis of extra-adrenal steroids also involves the catalysis of CYP11A1 as the first step of steroidogenic pathway.

15.5.1 Brain

CYP11A1 protein has been detected in many parts of the brain by immunohistochemical staining, including the white matter (Le Goascogne et al. 1987), hippocampus (Kimoto et al. 2001), Purkinjie neurons (Ukena et al. 1998), and neural crest derivatives in the central and peripheral nervous systems during embryogenesis (Compagnone et al. 1995). Many of these data, however, cannot be repeated by other laboratories, probably because the anti-CYP11A1 antibodies that these authors used may have detected cross-reacting materials as the abundance of CYP11A1 is very low in the brain. The validity of these experiments, therefore, is questionable.

The detection of *Cyp11a1* mRNA may be more convincing than that of CYP11A1 protein. *Cyp11a1* mRNA has been detected in the glia and in the peripheral nervous system but not in the central nervous system by RNase protection assay (Mellon and Deschepper 1993). By the more sensitive RT-PCR analysis, however, *Cyp11a1* mRNA was also detected in many brain regions (Chiang et al. 2011). Its level in the brain, however, is very low as estimated by quantitative RT-PCR at about 1/3,000 of that in the adrenal (Chiang et al. 2011). Transgenic mice harboring DNA fragments containing different lengths of the *CYP11A1* promoter can drive the expression of the reporter gene in various brain regions, indicating that the *CYP11A1* promoter is active in these regions (Wu et al. 2007).

In addition to CYP11A1 protein and mRNA, the level of its steroid product, pregnenolone, has also been measured in the brain. A study determined pregnenolone concentration in the cerebellum to be about 8 pmol/mg tissue (Ukena et al. 1998). The concentration of pregnenolone and pregnenolone sulfate in the brain in another study was about 10 ng/g tissue, much higher than their plasma level at about 2 ng/ml (Le Goascogne et al. 1987). This result indicates that de novo synthesis of pregnenolone may be more important than the supply from the circulation. Furthermore, pregnenolone synthesis in the brain can be enhanced by calcium and NMDA neuronal activity (Kimoto et al. 2001), implying its regulation by neuronal activities.

The steroids that are synthesized de novo in the brain are termed neurosteroids. The functions of neurosteroid pregnenolone have been extensively studied. It is the precursor of many neurosteroids such as pregnenolone sulfate, progesterone, estradiol, and dehydro-epiandrosterone. In addition, pregnenolone by itself also functions as a neurosteroid. It enhances memory (Flood et al. 1992), promotes spinal cord recovery after injury (Guth et al. 1994), improves cognition (Vallee et al. 2001; Mayo et al. 2001), and reduces depression in mouse behavior assays (Bianchi and Baulieu 2012). It also has therapeutic effects against schizophrenia (Marx et al. 2011).

15.5.2 Intestine

Intestines also synthesize glucocorticoids to guard against intestinal inflammation. Steroids are synthesized in the crypt cells of the intestine, and steroidogenic genes such as *Cyp11a1* and *Cyp11b1* are also expressed in the crypt cells. The transcription of *Cyp11a1* in the intestinal crypt cells is controlled by NR5A2 (Mueller et al. 2006), just as that in the ovary. *Cyp11a1* is usually present at a very low level and is elevated when T lymphocytes are activated by stressors such as lipopolysaccharide (Noti et al. 2010). Tumor necrosis factor (TNF) plays an important role in the activation of *Cyp11a1* expression for the synthesis of glucocorticoids that suppress inflammation.

15.5.3 Skin

Skin is another tissue that contains CYP11A1. The presence of *Cyp11a1* mRNA in the skin has been detected by RT-PCR and the presence of CYP11A1 protein by Western blot analysis (Slominski et al. 2004). Skin CYP11A1 catalyzes the conversion of 7-dehydro-cholesterol into 7-dehydropregnenolone, which can be further metabolized into vitamin D3-like derivatives (Guryev et al. 2003). It also catalyzes the metabolism of vitamin D3, converting it into 20-hydroxyvitamin D3 and 20,22-dihydroxyvitamin D3. 20,22-Dihydroxyvitamin D3 stimulates the differentiation and inhibits proliferation of human keratinocytes (Janjetovic et al. 2010). Thus, the activity of CYP11A1 may overlap or augment that of CYP27B1 in vitamin D metabolism and keratinocyte differentiation.

Acknowledgments Part of the work described in this review was supported by grants NSC 102-2311-B-001-013-MY3 and 102-2923-B-001-003-MY3 from National Science Council, NHRI- EX102-10210SI from National Health Research Institutes, and AS-101-TP-B05 from Academia Sinica, Taiwan.

References

Bianchi M, Baulieu EE (2012) 3beta-Methoxy-pregnenolone (MAP4343) as an innovative therapeutic approach for depressive disorders. Proc Natl Acad Sci USA 109(5):1713–1718. doi:10.1073/pnas.1121485109

Bose HS, Sugawara T, Strauss JF 3rd, Miller WL, International Congenital Lipoid Adrenal Hyperplasia Consortium (1996) The pathophysiology and genetics of congenital lipoid adrenal hyperplasia. N Engl J Med 335(25):1870–1878. doi:10.1056/NEJM199612193352503

Chen WY, Lee WC, Hsu NC, Huang F, Chung BC (2004) SUMO modification of repression domains modulates function of nuclear receptor 5A1 (steroidogenic factor-1). J Biol Chem 279 (37):38730–38735. doi:10.1074/jbc.M405006200

Chen WY, Juan LJ, Chung BC (2005) SF-1 (nuclear receptor 5A1) activity is activated by cyclic AMP via p300-mediated recruitment to active foci, acetylation, and increased DNA binding. Mol Cell Biol 25(23):10442–10453. doi:10.1128/MCB.25.23.10442-10453.2005

Chen WY, Weng JH, Huang CC, Chung BC (2007) Histone deacetylase inhibitors reduce steroidogenesis through SCF-mediated ubiquitination and degradation of steroidogenic factor 1 (NR5A1). Mol Cell Biol 27(20):7284–7290. doi:10.1128/MCB.00476-07

Chiang YF, Lin HT, Hu JW, Tai YC, Lin YC, Hu MC (2011) Differential regulation of the human CYP11A1 promoter in mouse brain and adrenals. J Cell Physiol 226(8):1998–2005. doi:10.1002/jcp.22534

Chida D, Nakagawa S, Nagai S, Sagara H, Katsumata H, Imaki T, Suzuki H, Mitani F, Ogishima T, Shimizu C, Kotaki H, Kakuta S, Sudo K, Koike T, Kubo M, Iwakura Y (2007) Melanocortin 2 receptor is required for adrenal gland development, steroidogenesis, and neonatal gluconeogenesis. Proc Natl Acad Sci USA 104(46):18205–18210. doi:10.1073/pnas.0706953104

Chien Y, Cheng WC, Wu MR, Jiang ST, Shen CK, Chung BC (2013) Misregulated progesterone secretion and impaired pregnancy in CYP11A1 transgenic mice. Biol Reprod 89(4):91. doi:10.1095/biolreprod.113.110833

Clark BJ, Wells J, King SR, Stocco DM (1994) The purification, cloning, and expression of a novel luteinizing hormone-induced mitochondrial protein in MA-10 mouse Leydig tumor cells. Characterization of the steroidogenic acute regulatory protein (StAR). J Biol Chem 269(45):28314–28322

Compagnone NA, Bulfone A, Rubenstein JL, Mellon SH (1995) Expression of the steroidogenic enzyme P450scc in the central and peripheral nervous systems during rodent embryogenesis. Endocrinology 136(6):2689–2696. doi:10.1210/endo.136.6.7750493

Desclozeaux M, Krylova IN, Horn F, Fletterick RJ, Ingraham HA (2002) Phosphorylation and intramolecular stabilization of the ligand binding domain in the nuclear receptor steroidogenic factor 1. Mol Cell Biol 22(20):7193–7203

Flood JF, Morley JE, Roberts E (1992) Memory-enhancing effects in male mice of pregnenolone and steroids metabolically derived from it. Proc Natl Acad Sci USA 89(5):1567–1571

Guo IC, Tsai HM, Chung BC (1994) Actions of two different cAMP-responsive sequences and an enhancer of the human CYP11A1 (P450scc) gene in adrenal Y1 and placental JEG-3 cells. J Biol Chem 269(9):6362–6369

Guo IC, Chung BC (1999) Cell-type specificity of human CYP11A1 TATA box. J Steroid Biochem Mol Biol 69(1–6):329–334

Guo IC, Huang CY, Wang CK, Chung BC (2007) Activating protein-1 cooperates with steroidogenic factor-1 to regulate 3',5'-cyclic adenosine 5'-monophosphate-dependent human CYP11A1 transcription in vitro and in vivo. Endocrinology 148(4):1804–1812. doi:10.1210/en.2006-0938

Guryev O, Carvalho RA, Usanov S, Gilep A, Estabrook RW (2003) A pathway for the metabolism of vitamin D3: unique hydroxylated metabolites formed during catalysis with cytochrome P450scc (CYP11A1). Proc Natl Acad Sci USA 100(25):14754–14759. doi:10.1073/pnas.2336107100

Guth L, Zhang Z, Roberts E (1994) Key role for pregnenolone in combination therapy that promotes recovery after spinal cord injury. Proc Natl Acad Sci USA 91(25):12308–12312

Hammer GD, Krylova I, Zhang Y, Darimont BD, Simpson K, Weigel NL, Ingraham HA (1999) Phosphorylation of the nuclear receptor SF-1 modulates cofactor recruitment: integration of hormone signaling in reproduction and stress. Mol Cell 3(4):521–526

Hiort O, Holterhus PM, Werner R, Marschke C, Hoppe U, Partsch CJ, Riepe FG, Achermann JC, Struve D (2005) Homozygous disruption of P450 side-chain cleavage (CYP11A1) is associated with prematurity, complete 46, XY sex reversal, and severe adrenal failure. J Clin Endocrinol Metab 90(1):538–541. doi:10.1210/jc.2004-1059

Honda S, Morohashi K, Nomura M, Takeya H, Kitajima M, Omura T (1993) Ad4BP regulating steroidogenic P-450 gene is a member of steroid hormone receptor superfamily. J Biol Chem 268(10):7494–7502

Hsu HJ, Hsiao P, Kuo MW, Chung BC (2002) Expression of zebrafish cyp11a1 as a maternal transcript and in yolk syncytial layer. Gene Expr Patterns 2(3-4):219–222

Hsu HJ, Hsu NC, Hu MC, Chung BC (2006a) Steroidogenesis in zebrafish and mouse models. Mol Cell Endocrinol 248(1-2):160–163. doi:10.1016/j.mce.2005.10.011

Hsu HJ, Liang MR, Chen CT, Chung BC (2006b) Pregnenolone stabilizes microtubules and promotes zebrafish embryonic cell movement. Nature (Lond) 439(7075):480–483. doi:10.1038/nature04436

Hu MC, Chou SJ, Huang YY, Hsu NC, Li H, Chung BC (1999) Tissue-specific, hormonal, and developmental regulation of SCC-LacZ expression in transgenic mice leads to adrenocortical zone characterization. Endocrinology 140(12):5609–5618

Hu MC, Chiang EF, Tong SK, Lai W, Hsu NC, Wang LC, Chung BC (2001) Regulation of steroidogenesis in transgenic mice and zebrafish. Mol Cell Endocrinol 171(1-2):9–14

Hu MC, Hsu NC, El Hadj NB, Pai CI, Chu HP, Wang CK, Chung BC (2002) Steroid deficiency syndromes in mice with targeted disruption of Cyp11a1. Mol Endocrinol 16(8):1943–1950

Huang CC, Shih MC, Hsu NC, Chien Y, Chung BC (2012) Fetal glucocorticoid synthesis is required for development of fetal adrenal medulla and hypothalamus feedback suppression. Endocrinology 153(10):4749–4756. doi:10.1210/en.2012-1258

Janjetovic Z, Tuckey RC, Nguyen MN, Thorpe EM Jr, Slominski AT (2010) 20,23-Dihydroxyvitamin D3, novel P450scc product, stimulates differentiation and inhibits proliferation and NF-kappaB activity in human keratinocytes. J Cell Physiol 223(1):36–48. doi:10.1002/jcp.21992

Kim CJ, Lin L, Huang N, Quigley CA, AvRuskin TW, Achermann JC, Miller WL (2008) Severe combined adrenal and gonadal deficiency caused by novel mutations in the cholesterol side chain cleavage enzyme, P450scc. J Clin Endocrinol Metab 93(3):696–702. doi:10.1210/jc.2007-2330

Kimoto T, Tsurugizawa T, Ohta Y, Makino J, Tamura H, Hojo Y, Takata N, Kawato S (2001) Neurosteroid synthesis by cytochrome p450-containing systems localized in the rat brain hippocampal neurons: *N*-methyl-D-aspartate and calcium-dependent synthesis. Endocrinology 142(8):3578–3589. doi:10.1210/endo.142.8.8327

Lan HC, Li HJ, Lin G, Lai PY, Chung BC (2007) Cyclic AMP stimulates SF-1-dependent CYP11A1 expression through homeodomain-interacting protein kinase 3-mediated Jun N-terminal kinase and c-Jun phosphorylation. Mol Cell Biol 27(6):2027–2036. doi:10.1128/MCB.02253-06

Lan HC, Wu CF, Shih HM, Chung BC (2012) Death-associated protein 6 (Daxx) mediates cAMP-dependent stimulation of Cyp11a1 (P450scc) transcription. J Biol Chem 287(8):5910–5916. doi:10.1074/jbc.M111.307603

Le Goascogne C, Robel P, Gouezou M, Sananes N, Baulieu EE, Waterman M (1987) Neurosteroids: cytochrome P-450scc in rat brain. Science 237(4819):1212–1215

Lee HS, Komarova YA, Nadezhdina ES, Anjum R, Peloquin JG, Schober JM, Danciu O, van Haren J, Galjart N, Gygi SP, Akhmanova A, Borisy GG (2010) Phosphorylation controls autoinhibition of cytoplasmic linker protein-170. Mol Biol Cell 21(15):2661–2673. doi:10.1091/mbc.E09-12-1036

Lee FY, Faivre EJ, Suzawa M, Lontok E, Ebert D, Cai F, Belsham DD, Ingraham HA (2011) Eliminating SF-1 (NR5A1) sumoylation in vivo results in ectopic hedgehog signaling and disruption of endocrine development. Dev Cell 21(2):315–327. doi:10.1016/j.devcel.2011.06.028

Lewis AE, Rusten M, Hoivik EA, Vikse EL, Hansson ML, Wallberg AE, Bakke M (2008) Phosphorylation of steroidogenic factor 1 is mediated by cyclin-dependent kinase 7. Mol Endocrinol 22(1):91–104. doi:10.1210/me.2006-0478

Li LA, Lala D, Chung BC (1998) Function of steroidogenic factor 1 (SF1) ligand-binding domain in gene activation and interaction with AP1. Biochem Biophys Res Commun 250(2):318–320. doi:10.1006/bbrc.1998.9305

Lin X, Takemori H, Katoh Y, Doi J, Horike N, Makino A, Nonaka Y, Okamoto M (2001) Salt-inducible kinase is involved in the ACTH/cAMP-dependent protein kinase signaling in Y1 mouse adrenocortical tumor cells. Mol Endocrinol 15(8):1264–1276. doi:10.1210/mend.15.8.0675

Marieb EN, Hoehn K (2007) Human anatomy and physiology, 7th edn. Pearson Benjamin Cummings, San Francisco

Marx CE, Bradford DW, Hamer RM, Naylor JC, Allen TB, Lieberman JA, Strauss JL, Kilts JD (2011) Pregnenolone as a novel therapeutic candidate in schizophrenia: emerging preclinical and clinical evidence. Neuroscience 191:78–90. doi:10.1016/j.neuroscience.2011.06.076

Mayo W, Le Moal M, Abrous DN (2001) Pregnenolone sulfate and aging of cognitive functions: behavioral, neurochemical, and morphological investigations. Horm Behav 40(2):215–217. doi:10.1006/hbeh.2001.1677

Mellon SH, Bair SR (1998) 25-Hydroxycholesterol is not a ligand for the orphan nuclear receptor steroidogenic factor-1 (SF-1). Endocrinology 139(6):3026–3029. doi:10.1210/endo.139.6.6129

Mellon SH, Deschepper CF (1993) Neurosteroid biosynthesis: genes for adrenal steroidogenic enzymes are expressed in the brain. Brain Res 629(2):283–292

Mishima M, Maesaki R, Kasa M, Watanabe T, Fukata M, Kaibuchi K, Hakoshima T (2007) Structural basis for tubulin recognition by cytoplasmic linker protein 170 and its autoinhibition. Proc Natl Acad Sci USA 104(25):10346–10351. doi:10.1073/pnas.0703876104

Mueller M, Cima I, Noti M, Fuhrer A, Jakob S, Dubuquoy L, Schoonjans K, Brunner T (2006) The nuclear receptor LRH-1 critically regulates extra-adrenal glucocorticoid synthesis in the intestine. J Exp Med 203(9):2057–2062. doi:10.1084/jem.20060357

Nelson DR (2006) Cytochrome P450 nomenclature, 2004. Methods Mol Biol 320:1–10. doi:10.1385/1-59259-998-2:1

Noti M, Corazza N, Tuffin G, Schoonjans K, Brunner T (2010) Lipopolysaccharide induces intestinal glucocorticoid synthesis in a TNFalpha-dependent manner. FASEB J 24(5):1340–1346. doi:10.1096/fj.09-140913

Nuclear Receptors Nomenclature Committee (1999) A unified nomenclature system for the nuclear receptor superfamily. Cell 97(2):161–163

Papadimitriou A, Priftis KN (2009) Regulation of the hypothalamic-pituitary-adrenal axis. Neuroimmunomodulation 16(5):265–271. doi:10.1159/000216184

Parajes S, Griffin A, Taylor AE, Rose IT, Miguel-Escalada I, Hadzhiev Y, Arlt W, Shackleton C, Muller F, Krone N (2013) Redefining the initiation and maintenance of zebrafish interrenal

steroidogenesis by characterizing the key enzyme cyp11a2. Endocrinology 154(8):2702–2711. doi:10.1210/en.2013-1145
Parker KL, Rice DA, Lala DS, Ikeda Y, Luo X, Wong M, Bakke M, Zhao L, Frigeri C, Hanley NA, Stallings N, Schimmer BP (2002) Steroidogenic factor 1: an essential mediator of endocrine development. Recent Prog Horm Res 57:19–36
Purifoy FE, Koopmans LH, Tatum RW (1980) Steroid hormones and aging: free testosterone, testosterone and androstenedione in normal females aged 20–87 years. Hum Biol 52(2):181–191
Pusalkar M, Meherji P, Gokral J, Chinnaraj S, Maitra A (2009) CYP11A1 and CYP17 promoter polymorphisms associate with hyperandrogenemia in polycystic ovary syndrome. Fertil Steril 92(2):653–659. doi:10.1016/j.fertnstert.2008.07.016
Sahakitrungruang T, Tee MK, Blackett PR, Miller WL (2011) Partial defect in the cholesterol side-chain cleavage enzyme P450scc (CYP11A1) resembling nonclassic congenital lipoid adrenal hyperplasia. J Clin Endocrinol Metab 96(3):792–798. doi:10.1210/jc.2010-1828
Schumacher M, Weill-Engerer S, Liere P, Robert F, Franklin RJ, Garcia-Segura LM, Lambert JJ, Mayo W, Melcangi RC, Parducz A, Suter U, Carelli C, Baulieu EE, Akwa Y (2003) Steroid hormones and neurosteroids in normal and pathological aging of the nervous system. Prog Neurobiol 71(1):3–29
Shih MC, Hsu NC, Huang CC, Wu TS, Lai PY, Chung BC (2008) Mutation of mouse Cyp11a1 promoter caused tissue-specific reduction of gene expression and blunted stress response without affecting reproduction. Mol Endocrinol 22(4):915–923. doi:10.1210/me.2007-0222
Shih MC, Chiu YN, Hu MC, Guo IC, Chung BC (2011) Regulation of steroid production: analysis of Cyp11a1 promoter. Mol Cell Endocrinol 336(1–2):80–84. doi:10.1016/j.mce.2010.12.017
Slominski A, Zjawiony J, Wortsman J, Semak I, Stewart J, Pisarchik A, Sweatman T, Marcos J, Dunbar C, CT R (2004) A novel pathway for sequential transformation of 7-dehydrocholesterol and expression of the P450scc system in mammalian skin. Eur J Biochem/FEBS 271(21):4178–4188. doi:10.1111/j.1432-1033.2004.04356.x
Strauss JF 3rd (2003) Some new thoughts on the pathophysiology and genetics of polycystic ovary syndrome. Ann N Y Acad Sci 997:42–48
Terry K, McGrath M, Lee IM, Buring J, De Vivo I (2010) Genetic variation in CYP11A1 and StAR in relation to endometrial cancer risk. Gynecol Oncol 117(2):255–259. doi:10.1016/j.ygyno.2010.02.002
Ukena K, Usui M, Kohchi C, Tsutsui K (1998) Cytochrome P450 side-chain cleavage enzyme in the cerebellar Purkinje neuron and its neonatal change in rats. Endocrinology 139(1):137–147. doi:10.1210/endo.139.1.5672
Vallee M, Mayo W, Le Moal M (2001) Role of pregnenolone, dehydroepiandrosterone and their sulfate esters on learning and memory in cognitive aging. Brain Res Rev 37(1–3):301–312
Weng JH, Liang MR, Chen CH, Tong SK, Huang TC, Lee SP, Chen YR, Chen CT, Chung BC (2013) Pregnenolone activates CLIP-170 to promote microtubule growth and cell migration. Nat Chem Biol 9(10):636–642. doi:10.1038/nchembio.1321
Williams GH, Dluhy RG (1972) Aldosterone biosynthesis. Interrelationship of regulatory factors. Am J Med 53(5):595–605
Winnay JN, Hammer GD (2006) Adrenocorticotropic hormone-mediated signaling cascades coordinate a cyclic pattern of steroidogenic factor 1-dependent transcriptional activation. Mol Endocrinol 20(1):147–166. doi:10.1210/me.2005-0215
Wu HS, Lin HT, Wang CK, Chiang YF, Chu HP, Hu MC (2007) Human CYP11A1 promoter drives Cre recombinase expression in the brain in addition to adrenals and gonads. Genesis 45(2):59–65. doi:10.1002/dvg.20266
Yaspan BL, Breyer JP, Cai Q, Dai Q, Elmore JB, Amundson I, Bradley KM, Shu XO, Gao YT, Dupont WD, Zheng W, Smith JR (2007) Haplotype analysis of CYP11A1 identifies promoter variants associated with breast cancer risk. Cancer Res 67(12):5673–5682. doi:10.1158/0008-5472.CAN-07-0467

Chapter 16
Cooperative Regulation of Expression of Cytochrome P450 Enzymes by Aryl Hydrocarbon Receptor and Vitamin D Receptor

Makoto Makishima

Abstract Benzo[*a*]pyrene (BaP), a polycyclic aromatic hydrocarbon produced by cigarette combustion, is implicated as a causative agent of smoking-related diseases such as cancer and atherosclerosis. BaP activates the aryl hydrocarbon receptor (AHR) and induces the expression of genes involved in xenobiotic metabolism, including cytochrome P450 1A1 (CYP1A1). CYP1A1 is involved not only in BaP detoxification but also in its metabolic activation, which results in DNA adduct formation and reactive oxygen species production. The vitamin D receptor (VDR) mediates vitamin D signaling in the regulation of calcium metabolism, cellular growth and differentiation, inflammation, immunity, and cardiovascular function. VDR belongs to the NR1I subfamily of the nuclear receptor superfamily along with other nuclear receptors involved in xenobiotic metabolism. The active form of vitamin D_3, 1,25-dihydroxyvitamin D_3 (1,25$(OH)_2D_3$), acts as a physiological VDR ligand and is catalyzed by CYP24A1, a VDR target gene. We have investigated cross-talk between the AHR and VDR signaling pathways. BaP effectively enhances 1,25$(OH)_2D_3$-dependent induction of *CYP24A1* and inactivation of 1,25 $(OH)_2D_3$ by CYP24A1 in human monocyte/macrophage-derived cells. The effect of BaP on *CYP24A1* induction is mediated by AHR activation and de novo protein synthesis. On the other hand, 1,25$(OH)_2D_3$ enhances BaP-induced transcription of *CYP1A1* in these cells. VDR and AHR directly bind to specific elements in the human *CYP1A1* promoter. Induction of *CYP24A1* and *CYP1A1* by the activation of VDR and AHR may contribute to BaP-mediated toxicity.

Keywords Aryl hydrocarbon receptor • Benzo[*a*]pyrene • CYP1A1 • CYP24A1 • Cytochrome P450 • Metabolic activation • Nuclear receptor • Vitamin D • Vitamin D receptor • Xenobiotic metabolism

---

M. Makishima (✉)
Division of Biochemistry, Department of Biomedical Sciences, Nihon University School of Medicine, 30-1 Oyaguchi-kamicho, Itabashi-ku, Tokyo 173-8610, Japan
e-mail: makishima.makoto@nihon-u.ac.jp

H. Yamazaki (ed.), *Fifty Years of Cytochrome P450 Research*,
DOI 10.1007/978-4-431-54992-5_16, © Springer Japan 2014

16.1 Vitamin D and Vitamin D Receptor

Vitamin D is a lipid-soluble vitamin that is involved in numerous physiological processes, including bone and calcium metabolism, cellular growth and differentiation, inflammation and immunity, and cardiovascular function (Nagpal et al. 2005; Makishima and Yamada 2005). Vitamin D is a secosteroid, in which the B-ring of the canonical steroid structure is ruptured, and is synthesized from 7-dehydrocholesterol, an intermediate metabolite in cholesterol synthesis, or is derived from dietary sources (Jones et al. 1998). Ultraviolet irradiation of the skin, most commonly by sunlight exposure, induces a photochemical reaction of 7-dehydrocholesterol to produce the secosteroid vitamin D_3 (cholecalciferol). Vitamin D_3 is hydroxylated at the 25-position by a mitochondrial P450 enzyme, sterol 27-hydroxylase [cytochrome P450 (CYP) 27A1 (CYP27A1)], and a microsomal P450 enzyme, vitamin D 25-hydroxylase (CYP2R1), to yield 25-hydroxyvitamin D_3 (25(OH)D_3; 25-hydroxycholecalciferol) in the liver (Choi and Makishima 2009). 25(OH)D_3 is the major circulating form of vitamin D, is bound by vitamin D-binding protein (Holick 2006), and is further hydroxylated at the 1α-position by a P450 enzyme, 25-hydroxyvitamin D 1α-hydroxylase (CYP27B1). The 1α-hydroxylation reaction is tightly regulated and occurs mainly in the kidney to yield the active metabolite 1α,25-dihydroxyvitamin D_3 (1,25 $(OH)_2D_3$; 1α,25-dihydroxycholecalciferol or calcitriol). 1,25$(OH)_2D_3$ is also synthesized by CYP27B1 expressed in extrarenal tissues, such as macrophages, and has been suggested to exert noncalcemic effects (Hewison 2010). Dietary vitamin D_2 (ergocalciferol) and vitamin D_3 are metabolized through the same activation process, involving 25-hydroxylation in the liver and subsequent 1α-hydroxylation in the kidney (Holick 2008). 1,25$(OH)_2D_2$ has a different catabolic pathway than 1,25$(OH)_2D_3$ (Horst et al. 2003), and vitamin D_2 is less effective than vitamin D_3 in maintaining serum 25(OH)D levels (Armas et al. 2004). Vitamin D deficiency causes rickets and osteomalacia and is also associated with an increased risk of osteoporosis, cancer, autoimmune diseases, infections, cardiovascular diseases, obesity, and diabetes (Rosen et al. 2012).

1,25$(OH)_2D_3$, 1,25$(OH)_2D_2$, and their synthetic derivatives exert physiological and pharmacological effects by binding to the vitamin D receptor (VDR), a member of the nuclear receptor superfamily of transcription factors (Haussler et al. 1998; Makishima and Yamada 2005; Choi and Makishima 2009) (Fig. 16.1). Forty-eight human nuclear receptors have been identified, including the endocrine receptors for steroid and thyroid hormones, metabolic sensors for fatty acids, bile acids, oxysterols and xenobiotics, and orphan receptors whose natural ligands are unknown (Makishima 2005; Shulman and Mangelsdorf 2005). Similar to other nuclear receptors, VDR is activated in a ligand-dependent manner. Upon ligand binding, VDR undergoes a conformational change in the cofactor binding site and activation function 2 domain, structural rearrangements that result in dynamic interaction with the heterodimer partner retinoid X receptor (RXR) and exchange of cofactor complexes (Makishima and Yamada 2005; Choi et al. 2011).

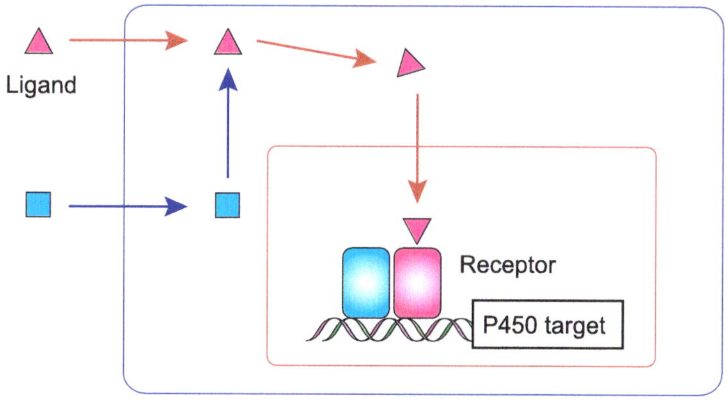

Receptor	VDR	AHR
Ligand	$1,25(OH)_2D_3$ Lithocholic acid	Benzo[*a*]pyrene (BaP) TCDD Indole derivatives
P450 target	CYP24A1, CYP3A4 CYP27B1 (negative) CYP1A1	CYP1A1, CYP1A2, CYP1B1
Function of P450 induction	Vitamin D metabolism Xenobiotic metabolism	Xenobiotic metabolism (detoxification, metabolic activation)

Fig. 16.1 Regulation of P450 enzymes by vitamin D receptor (VDR) and aryl hydrocarbon receptor (AHR). VDR is a nuclear receptor for $1,25(OH)_2D_3$ and lithocholic acid. VDR induces expression of many target genes including the P450 enzymes, *CYP24A1* and *CYP3A4*. VDR negatively regulates *CYP27B1* expression through an unknown mechanism. We have identified that VDR is involved in *CYP1A1* transcription (Matsunawa et al. 2012). CYP24A1 and CYP3A4 inactivate $1,25(OH)_2D_3$, and CYP27B1 produces $1,25(OH)_2D_3$ from the precursor $25(OH)D_3$. CYP3A4 and CYP1A1 are involved in xenobiotic metabolism. AHR is activated by BaP, 2,3,7,8-tetrachlorodibenzo-*p*-dioxin (TCDD), and indole derivatives. AHR does not belong to the nuclear receptor superfamily. AHR activates transcription of *CYP1A1*, *CYP1A2*, and *CYP1B1*. These P450 enzymes are involved in both detoxification and metabolic activation of xenobiotics, including BaP

Although corepressors bind to the VDR-RXR heterodimer in the absence of ligand, ligand binding reduces the affinity of corepressors and increases the affinity for coactivators, a structural transition that induces transcription of specific genes. The VDR-RXR heterodimer binds preferentially to a DNA response element that consists of a two-hexanucleotide (AGGTCA or a related sequence) direct repeat motif separated by three nucleotides (direct repeat 3, DR3) (Haussler et al. 1998). The DR3 VDR-binding element has been identified in the regulatory regions of many target genes, including 1,25$(OH)_2D_3$ 24-hydroxylase (CYP24A1; gene symbol, *CYP24A1*), cathelicidin antimicrobial peptide (*CAMP*), and osteopontin (Haussler et al. 1998; Makishima and Yamada 2005) (Fig. 16.1). Inverted palindromes of the hexanucleotide motif, called everted repeat elements, separated by six, seven, eight, or nine nucleotides (ER6, ER7, ER8, or ER9) have been identified as vitamin D response elements in genes including *CYP3A4* (Thummel et al. 2001; Choi and Makishima 2009) (Fig. 16.1). Two VDR-inducible P450 enzymes, CYP24A1 and CYP3A4, are involved in inactivation of 1,25$(OH)_2D_3$ (Sawada et al. 2004; Xu et al. 2006). VDR activation by 1,25$(OH)_2D_3$ induces negative feedback regulation of vitamin D signaling by inhibiting the expression of CYP27B1 and parathyroid hormone, and also negatively regulates genes involved in inflammation and cell growth, such as tumor necrosis factor and *MYC*, through a poorly characterized mechanism (Haussler et al. 1998; Van Cromphaut et al. 2001; Nagpal et al. 2005) (Fig. 16.1). Lithocholic acid, a secondary bile acid, acts as an additional physiological VDR ligand (Makishima et al. 2002) (Fig. 16.1). CYP3A4 metabolizes lithocholic acid into 3-ketocholanic acid by 3-oxidation, hyodeoxycholic acid by 6α-hydroxylation, and 1β,3α-dihydroxy-5β-cholanoic acid by 1β-hydroxylation (Bodin et al. 2005).

VDR activation induces the expression of genes involved in xenobiotic metabolism, such as *CYP3A4*, multidrug resistance 1, multidrug resistance-associated protein (MRP) 2, MRP3, MRP4, and sulfotransferase 2A1 (Saeki et al. 2008; McCarthy et al. 2005; Fan et al. 2009; Echchgadda et al. 2004; Ogura et al. 2009; Nishida et al. 2009; Ishizawa et al. 2012). VDR belongs to the NR1I nuclear receptor subfamily along with pregnane X receptor (PXR) and constitutive androstane receptor (CAR), both of which play a role in the regulation of xenobiotic metabolism (Gao and Xie 2010). Structural analysis reveals that PXR is the most closely related nuclear receptor to VDR (Adachi et al. 2004). Compared with PXR and CAR, however, our understanding of the role of VDR in xenobiotic metabolism remains limited.

16.2 Benzo[*a*]Pyrene and Aryl Hydrocarbon Receptor

The environmental pollutant benzo[*a*]pyrene (BaP) is a polycyclic aromatic hydrocarbon found in industrial incineration products, cigarette smoke, and charcoal-grilled food and is implicated as a causative agent in malignancies, such as lung cancer, head and neck cancers, and atherosclerosis (Alexandrov et al. 2010; Miller and Ramos 2001). BaP activates the aryl hydrocarbon receptor (AHR), a transcription factor containing basic helix-loop-helix and Per-Arnt-Sim domains (Nebert et al. 2000) (Fig. 16.1). AHR forms an active transcription factor heterodimer with the AHR nuclear translocator and induces expression of a group of genes called the [*Ah*] gene battery, including phase I enzymes (*CYP1A1*, *CYP1A2*, *CYP1B1*, and

NAD(P)H:quinone oxidoreductase I) and the phase II enzymes (glutathione transferase A1 and UDP glucuronosyltransferase 1A6) (Nebert et al. 2000) (Fig. 16.1). BaP has been implicated to mediate carcinogenic, mutagenic, and cytotoxic effects after conversion to toxic metabolites through an AHR-dependent mechanism of metabolic activation (Miller and Ramos 2001; Shimada and Fujii-Kuriyama 2004) (Fig. 16.1). Studies of BaP-resistant mouse hepatoma cells have shown that resistance to BaP toxicity is associated with mutations in the *Cyp1a1* gene or dysfunction of the AHR transcription factor (Hankinson et al. 1991). Expression of exogenous CYP1A1 in CYP1A1-deficient cells restores the formation of BaP-induced DNA adducts (Maier et al. 2002). These findings indicate that the AHR-CYP1A1 cascade is involved in the metabolic activation of BaP. BaP is first oxidized by CYP1A1 and CYP1B1 to epoxides that nonenzymatically rearrange rapidly to form phenols, such as 3-hydroxy-BaP and 9-hydroxy-BaP, as well as to the more stable BaP-7,8-epoxide (Shimada 2006; Uno and Makishima 2009). BaP-7,8-epoxide is then metabolized by epoxide hydrolase to BaP-7,8-diol, which serves as a substrate for a second P450-dependent oxidation reaction, generating the highly reactive BaP-7,8-diol-9,10-epoxide (Shimada 2006). BaP-7,8-diol-9,10-epoxide effectively forms DNA adducts and serves as a putative carcinogen (Miller and Ramos 2001). BaP-7,8-diol-9,10-expoxide forms adducts with guanosine residues in the DNA of epithelial cells in the lungs of smokers, and characteristic base substitutions in mutant p53 alleles suggest a role for smoking-induced lung carcinogenesis (Denissenko et al. 1996; Hainaut and Pfeifer 2001). BaP-7,8-diol is also metabolized to BaP-7,8-dione by aldoketoreductase, and the BaP-1,3-, 1,6-, and 3,6-diones are also formed through metabolism of BaP to phenols by P450 enzymes (Shimada 2006). Although BaP-7,8-diol does not activate AHR, BaP-7,8-dione is an effective AHR activator (Burczynski and Penning 2000). The BaP quinones are involved in DNA adduct formation and are especially important in generating reactive oxygen species.

Studies using mice deficient in CYP1 family members have shown that these enzymes are responsible for both metabolic activation and detoxification of BaP (Nebert et al. 2004). When given intraperitoneal BaP injections (500 mg/kg/day), *Cyp1a1*$^{-/-}$ mice have enhanced survival compared with *Cyp1a1*$^{+/-}$ mice, a finding consistent with CYP1A1-mediated BaP metabolic activation (Uno et al. 2001). Interestingly, BaP-induced DNA adducts are increased in the liver and BaP is cleared from the blood more slowly in *Cyp1a1*$^{-/-}$ mice, indicating that CYP1A1 metabolizes BaP in vivo and leads to decreased survival while inhibiting DNA adduct formation. When administered a lower dose of oral BaP (125 mg/kg/day), *Cyp1a1*$^{-/-}$ mice show decreased survival and increased DNA adducts in the intestine, liver, spleen, and bone marrow, indicating that CYP1A1 is necessary for detoxification of BaP (Uno et al. 2004). In mice deficient for both CYP1A1 and CYP1B1, CYP1B1 deficiency ameliorates BaP-induced toxicity, such as body weight loss, increased serum transaminases, and increased DNA adducts in intestine, spleen, and bone marrow when compared to *Cyp1a1*$^{-/-}$ mice, indicating that CYP1B1 is involved in BaP-induced DNA adduct formation (Uno et al. 2006). Interestingly, when *Cyp1b1*$^{-/-}$ mice are administered oral BaP (125 mg/kg/day),

DNA adducts decrease in spleen and bone marrow but increase in intestine (Uno et al. 2006). Overexpression of CYP1A1 and CYP1A2, but not CYP1B1, inhibits DNA adduct formation in hepatocytes (Endo et al. 2008). CYP1A1 and CYP1B1, but not CYP1A2, convert BaP to metabolites with diminished AHR transactivation activity (Endo et al. 2008). Thus, involvement of the AHR-CYP1 cascade in the metabolic activation or detoxification of BaP is likely determined by the expression patterns of AHR, CYP1 enzymes, and other phase I and II enzymes (Fig. 16.1).

The high-affinity AHR ligand 2,3,7,8-tetrachlorodibenzo-*p*-dioxin (TCDD) causes a wide spectrum of toxic effects including hepatotoxicity, immunosuppression, wasting, and carcinogenesis (Bock and Kohle 2006; Safe et al. 2013) (Fig. 16.1). Industrial and military exposures to TCDD have been linked to detrimental health effects. Because TCDD has virtually no cellular metabolism, abnormal AHR activation has been implicated in the carcinogenesis and teratogenesis of this compound by modulating expression of genes involved in cell proliferation and differentiation. AHR activation also exerts nongenomic actions, such as activation of mitogen-activated protein kinase pathways and formation of the ubiquitin ligase complex (Henklova et al. 2008; Ohtake et al. 2007). AHR is activated not only by ligands but also through phosphorylation in a ligand-independent manner (Fujii-Kuriyama and Kawajiri 2010). Studies using AHR-null mice have demonstrated that AHR is essential for the induction of [*Ah*] gene battery members and in mediating most TCDD toxicities (Fernandez-Salguero et al. 1996; Nebert et al. 2000). Expression of a constitutively active AHR in transgenic mice induces stomach tumors and enhances hepatocarcinogenesis (Andersson et al. 2002; Moennikes et al. 2004), consistent with a carcinogenic effect of AHR activation. In contrast, AHR deletion results in enhanced formation of colonic tumors including adenomas and adenocarcinomas (Kawajiri et al. 2009). AHR activation by xenobiotic and natural ligands, such as indole derivatives that are converted from dietary tryptophan and glucosinolates by intestinal microbes, induces E3 ubiquitin ligase-dependent degradation of β-catenin and suppresses intestinal tumor development in $Apc^{Min/+}$ mice (Kawajiri et al. 2009) (Fig. 16.1). AHR deletion leads to increased tumor formation in the liver of diethylnitrosamine-treated mice (Fan et al. 2010). The role of AHR in carcinogenesis requires further investigation. Recently, emerging data suggest that AHR also plays a role in additional physiological processes, such as reproduction and innate immunity (Fujii-Kuriyama and Kawajiri 2010).

16.3 CYP24A1

The epidemiology of vitamin D deficiency-related diseases, such as cancer, cardiovascular disease, autoimmune disease, infection, and osteoporosis, reveals an association with cigarette smoking (Irigaray et al. 2007). This relationship suggests that AHR activation by environmental pollutants such as BaP modulates vitamin D signaling. We have investigated the effect of BaP on the vitamin D signaling

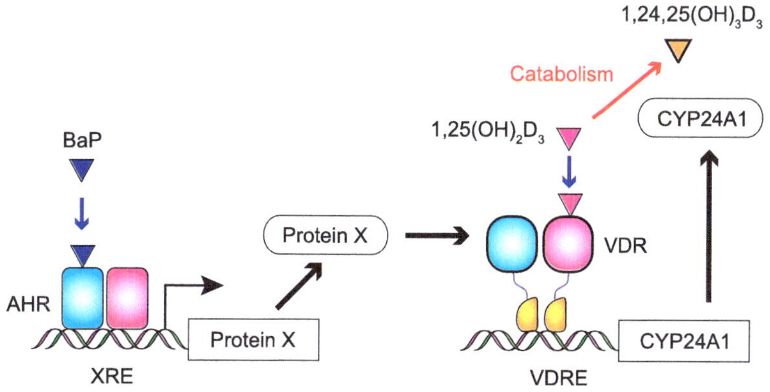

Fig. 16.2 AHR activation enhances CYP24A1 expression induced by VDR. AHR activation by BaP or TCDD enhances *CYP24A1* transcription induced by $1,25(OH)_2D_3$-activated VDR via de novo synthesis of a unknown protein (shown as *protein X*), leading to stimulation of $1,25(OH)_2D_3$ catabolism to $1,24,25(OH)_3D_3$ in monocyte/macrophage-derived cells

pathway and shown that AHR activation by BaP enhances catabolism of 1,25 $(OH)_2D_3$ (Matsunawa et al. 2009). $1,25(OH)_2D_3$ treatment increases expression of VDR target genes, *CYP24A1*, *CD14*, arachidonate 5-lipoxygenase (gene symbol, *ALOX5*), and *CAMP* in monocyte/macrophage-derived THP-1 cells. BaP effectively enhances the expression of *CYP24A1* induced by $1,25(OH)_2D_3$ in these cells (Fig. 16.2). BaP co-treatment less effectively enhances the $1,25(OH)_2D_3$-induced expression of *CD14* and *ALOX5* and is not effective in induction of *CAMP*. Thus, the potentiating effects of BaP differ among VDR target genes and are most apparent for *CYP24A1*. An AHR antagonist, α-naphthoflavone, inhibits the enhancing effect of BaP on *CYP24A1* induction, and other AHR activators, such as TCDD, also increase the expression of *CYP24A1* induced by $1,25(OH)_2D_3$. The combined effect of $1,25(OH)_2D_3$ and BaP is also observed in breast cancer MCF-7 cells. Treatment of cells with a protein synthesis inhibitor inhibits the enhancing effect of BaP on *CYP24A1* induction. These findings indicate that the BaP effect is mediated by AHR activation and de novo protein synthesis (Fig. 16.2). BaP pretreatment increases $1,25(OH)_2D_3$-dependent recruitment of VDR and RXR to the *CYP24A1* promoter (Fig. 16.2). It remains unknown how AHR activation modifies the transactivation activity of the VDR-RXR heterodimer or induces a promoter-selective mechanism. Mutations in the XPD subunit of transcription/repair factor TFIIH cause xeroderma pigmentosum, and VDR activation of *CYP24A1* is selectively inhibited in XPD-deficient cells (Drane et al. 2004). Mutant TFIIH is unable to phosphorylate Ets-1 in XPD mutant cells, leading to deficient binding of Ets-1 and VDR to the *CYP24A1* promoter. The effect of AHR activation on *CYP24A1* induction may be mediated by Ets-1 or related proteins (Cui et al. 2009).

CYP24A1 is a mitochondrial cytochrome P450 enzyme that catalyzes the 24-hydroxylation of $1,25(OH)_2D_3$ and its precursor $25(OH)D_3$ through the C-23*S* and C-24*R* hydroxylation reactions (Sawada et al. 2004). Human CYP24A1 has preference for C-24*R* over C-23*S* hydroxylation (Sawada et al. 2004; Hamamoto et al. 2006). Because 24-hydroxylation of vitamin D_3 metabolites decreases VDR agonist activity, the induction of CYP24A1 by VDR provides negative feedback regulation of vitamin D signaling. Pretreatment of THP-1 cells with BaP alone does not affect 1,25$(OH)_2D_3$ metabolism, but 1,25$(OH)_2D_3$ pretreatment enhances 1,25 $(OH)_2D_3$ catabolism to 1,24*R*,25-trihydroxyvitamin D3 (1,24R,25$(OH)_3D_3$) (Matsunawa et al. 2009) (Fig. 16.2). Pretreatment of cells with 1,25$(OH)_2D_3$ plus BaP further increases 1,24R,25$(OH)_3D_3$ production. CYP3A4 has also been reported to catalyze 1,25$(OH)_2D_3$ through 23- and 24-hydroxylation, leading to production of 1,23R,25$(OH)_3D_3$, 1,24S,25$(OH)_3D_3$, and 1,23S,25$(OH)_3D_3$ (Xu et al. 2006). The combination of 1,25$(OH)_2D_3$ and BaP induces expression of *CYP24A1* but not *CYP3A4* in THP-1 cells, consistent with enhanced 1,25 $(OH)_2D_3$ metabolism to 1,24R,25$(OH)_3D_3$ (Matsunawa et al. 2009), indicating that 1,25$(OH)_2D_3$ catabolism in THP-1 cells is mediated by CYP24A1. Activation of Toll-like receptors upregulates expression of *CYP27B1* and *VDR* genes in macrophages and enhances vitamin D-mediated innate immunity (Liu et al. 2006). Hypercalcemia and hypercalciuria associated with sarcoidosis are caused by deregulated production of 1,25$(OH)_2D_3$ by activated macrophages (Sharma 1996). Additionally, macrophages may have a role in vitamin D metabolism in other contexts. The BaP effect on *CYP24A1* expression is also observed in breast cancer MCF-7 cells (Matsunawa et al. 2009). Serum levels of 25(OH)D_3 have been demonstrated to be decreased in individuals who smoke more than 20 cigarettes per day compared with those who had never smoked (Black and Scragg 2005). Although further studies are needed to elucidate whether BaP affects vitamin D metabolism in vivo, modulation of vitamin D signaling by AHR may be one of the underlying mechanisms of cigarette smoking-related diseases.

16.4 CYP1A1

VDR belongs to the NR1I nuclear receptor subfamily along with the xenobiotic receptors PXR and CAR. Recently, CAR has been found to induce *CYP1A1* and *CYP1A2* expression by binding to a common regulatory element in the human *CYP1A1* and *CYP1A2* genes in hepatocytes (Yoshinari et al. 2010). We have investigated the effect of VDR signaling on *CYP1A1* transcription and found that 1,25$(OH)_2D_3$ enhances BaP-induced CYP1A1 expression and activity (Matsunawa et al. 2012) (Fig. 16.3). 1,25$(OH)_2D_3$ treatment enhances BaP-induced transcription of *CYP1A1* in human monocytic U937 cells and THP-1 cells, breast cancer MCF-7 cells, and embryonic kidney epithelium-derived cells. 1,25$(OH)_2D_3$ alone is not effective in inducing *CYP1A1*, and a combination of 1,25$(OH)_2D_3$ and BaP does not induce *CYP1A2* and *CYP1B1* mRNA expression in U937 cells. Combined treatment with 1,25$(OH)_2D_3$ and BaP increases CYP1A1 protein levels, BaP hydroxylation activity, and BaP-DNA adduct formation in U937 cells and THP-1 cells more effectively than BaP alone (Fig. 16.3). The effect of 1,25$(OH)_2D_3$ plus BaP on

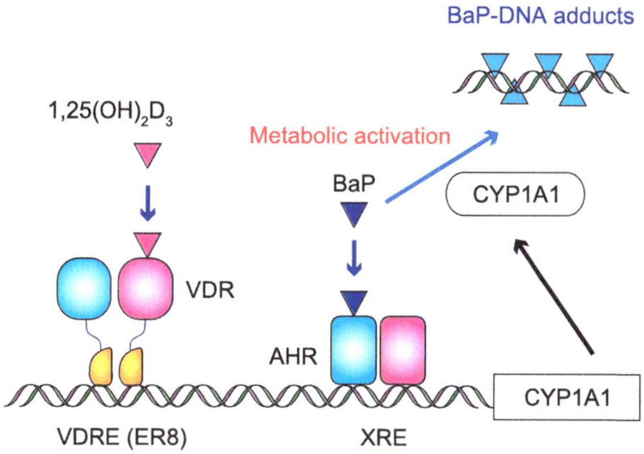

Fig. 16.3 VDR and AHR cooperatively induce CYP1A1 expression. VDR and AHR bind the ER8 VDR response element (VDRE) and xenobiotic response element (XRE), respectively, in the *CYP1A1* promoter in a ligand-dependent manner and induce *CYP1A1* transcription. Increased CYP1A1 enzyme levels enhance BaP-induced DNA adduct formation in monocyte/macrophage-derived cells

CYP1A1 mRNA expression in these cells is inhibited by VDR knockdown, VDR antagonist, and α-naphthoflavone. The CAR-RXR heterodimer binds to an ER8 element in the human *CYP1A1* promoter (Yoshinari et al. 2010). VDR-RXR also binds to this ER8 element in the *CYP1A1* promoter (Matsunawa et al. 2012). BaP and 1,25$(OH)_2D_3$ recruit AHR to a xenobiotic responsive element and VDR to ER8, respectively, in the *CYP1A1* promoter in U937 cells and THP-1 cells (Fig. 16.3). Thus, VDR regulates *CYP1A1* transcription by direct binding to the promoter in cells. The human *CYP1A1* and *CYP1A2* genes are located in a head-to-head orientation on chromosome 15 and share a common regulatory region (Ueda et al. 2006). Although CAR activation and AHR activation induce expression of both *CYP1A1* and *CYP1A2* in hepatocytes (Xu et al. 2000; Yoshinari et al. 2010), combined treatment of BaP and 1,25$(OH)_2D_3$ does not induce *CYP1A2* expression in U937 cells (Matsunawa et al. 2012). Although administration of BaP to mice induces *Cyp1a1* mRNA levels in liver, small intestine, spleen, and bone marrow, it induces *Cyp1a2* mRNA expression in liver, small intestine, and to a lesser degree in spleen, but not in bone marrow (Uno et al. 2006). Bone marrow-derived cells may be less responsible for *CYP1A2* induction.

Increased expression of CYP1A1 by BaP plus 1,25$(OH)_2D_3$ results in enhanced BaP-DNA adduct formation (Matsunawa et al. 2012) (Fig. 16.3). Exogenous expression of the oncogenic AML-ETO fusion protein in U937 cells upregulates *CYP1A1* expression and increases BaP-DNA adduct formation (Xu et al. 2007). By contrast, DNA adduct formation is increased in CYP1A1-deficient bone marrow (Uno et al. 2004), and overexpression of CYP1A1 in hepatocytes suppresses BaP-induced DNA adduct formation (Endo et al. 2008). Monocytes and monocyte-derived leukemia cells produce reactive oxygen species in response to 1,25$(OH)_2D_3$ (Levy and Malech 1991; Sly et al. 2001). Increased reactive oxygen species production is associated with monocyte/macrophage differentiation and may modify BaP metabolism, leading to increased DNA adduct formation. Recently, eicosanoid metabolism has been reported to change in *Cyp1* triple-knockout mice (Divanovic et al. 2013). Increased differentiation of myeloid leukemia cells by 1,25$(OH)_2D_3$ is associated with increased expression of enzymes involved in eicosanoid metabolism, such as *ALOX5* and prostaglandin-endoperoxidase synthase 1 (Amano et al. 2009; Matsunawa et al. 2009). Regulation of *CYP1A1* gene by VDR may be related to eicosanoid production and other physiological functions of leukocytes.

16.5 Conclusion

We have demonstrated cross-talk between AHR signaling and VDR signaling in transcriptional regulation of P450 enzymes. First, AHR activation enhances VDR-dependent CYP24A1 induction by a mechanism mediated by de novo protein synthesis (Matsunawa et al. 2009) (Fig. 16.2). Increased CYP24A1 expression enhances 1,25$(OH)_2D_3$ inactivation, a possible mechanism of BaP toxicity. Identification of the AHR target protein that enhances *CYP24A1* transcription will be required to elucidate the underlying molecular mechanism. Second, VDR and AHR regulate *CYP1A1* transcription by direct binding to the promoter (Matsunawa et al. 2012) (Fig. 16.3). Cells treated with 1,25$(OH)_2D_3$ plus BaP have increased BaP-DNA adduct formation. Metabolic activation of BaP may be a consequence of increased monocyte/macrophage function, and physiological roles of CYP1A1 should be further investigated. NF-E2-related factor 2 (NRF2) is another transcription factor regulating xenobiotic metabolism (Maher and Yamamoto 2010). Coordination of AHR, NR1I nuclear receptors, and NRF2 appears to be important to metabolic homeostasis, and disturbance of these regulatory systems may be a cause of diseases such as cancer, inflammation, and atherosclerosis. Further investigation in this area has the potential to lead to novel strategies in the prevention and treatment of these conditions.

Acknowledgments The author thanks Dr. Andrew I. Shulman for editorial assistance. This work was supported by MEXT KAKENHI Grant number 18077995, Nihon University Multidisciplinary Research Grant for 2005 and 2006, Uehara Memorial Foundation, and Ono Medical Research Foundation.

References

Adachi R, Shulman AI, Yamamoto K, Shimomura I, Yamada S, Mangelsdorf DJ, Makishima M (2004) Structural determinants for vitamin D receptor response to endocrine and xenobiotic signals. Mol Endocrinol 18(1):43–52

Alexandrov K, Rojas M, Satarug S (2010) The critical DNA damage by benzo(*a*)pyrene in lung tissues of smokers and approaches to preventing its formation. Toxicol Lett 198(1):63–68

Amano Y, Cho Y, Matsunawa M, Komiyama K, Makishima M (2009) Increased nuclear expression and transactivation of vitamin D receptor by the cardiotonic steroid bufalin in human myeloid leukemia cells. J Steroid Biochem Mol Biol 114(3-5):144–151

Andersson P, McGuire J, Rubio C, Gradin K, Whitelaw ML, Pettersson S, Hanberg A, Poellinger L (2002) A constitutively active dioxin/aryl hydrocarbon receptor induces stomach tumors. Proc Natl Acad Sci USA 99(15):9990–9995

Armas LA, Hollis BW, Heaney RP (2004) Vitamin D_2 is much less effective than vitamin D_3 in humans. J Clin Endocrinol Metab 89(11):5387–5391

Black PN, Scragg R (2005) Relationship between serum 25-hydroxyvitamin D and pulmonary function in the Third National Health and Nutrition Examination Survey. Chest 128(6):3792–3798

Bock KW, Kohle C (2006) Ah receptor: dioxin-mediated toxic responses as hints to deregulated physiologic functions. Biochem Pharmacol 72(4):393–404

Bodin K, Lindbom U, Diczfalusy U (2005) Novel pathways of bile acid metabolism involving CYP3A4. Biochim Biophys Acta 1687(1-3):84–93

Burczynski ME, Penning TM (2000) Genotoxic polycyclic aromatic hydrocarbon ortho-quinones generated by aldo-keto reductases induce CYP1A1 via nuclear translocation of the aryl hydrocarbon receptor. Cancer Res 60(4):908–915

Choi M, Makishima M (2009) Therapeutic applications for novel non-hypercalcemic vitamin D receptor ligands. Expert Opin Ther Pat 19(5):593–606

Choi M, Yamada S, Makishima M (2011) Dynamic and ligand-selective interactions of vitamin D receptor with retinoid X receptor and cofactors in living cells. Mol Pharmacol 80(6):1147–1155

Cui M, Zhao Y, Hance KW, Shao A, Wood RJ, Fleet JC (2009) Effects of MAPK signaling on 1,25-dihydroxyvitamin D-mediated CYP24 gene expression in the enterocyte-like cell line, Caco-2. J Cell Physiol 219(1):132–142

Denissenko MF, Pao A, Tang M-S, Pfeifer GP (1996) Preferential formation of benzo[*a*]pyrene adducts at lung cancer mutational hotspots in p53. Science 274(5286):430–432

Divanovic S, Dalli J, Jorge-Nebert LF, Flick LM, Galvez-Peralta M, Boespflug ND, Stankiewicz TE, Fitzgerald JM, Somarathna M, Karp CL, Serhan CN, Nebert DW (2013) Contributions of the three CYP1 monooxygenases to pro-inflammatory and inflammation-resolution lipid mediator pathways. J Immunol 191(6):3347–3357

Drane P, Compe E, Catez P, Chymkowitch P, Egly JM (2004) Selective regulation of vitamin D receptor-responsive genes by TFIIH. Mol Cell 16(2):187–197

Echchgadda I, Song CS, Roy AK, Chatterjee B (2004) Dehydroepiandrosterone sulfotransferase is a target for transcriptional induction by the vitamin D receptor. Mol Pharmacol 65(3):720–729

Endo K, Uno S, Seki T, Ariga T, Kusumi Y, Mitsumata M, Yamada S, Makishima M (2008) Inhibition of aryl hydrocarbon receptor transactivation and DNA adduct formation by CYP1 isoform-selective metabolic deactivation of benzo[*a*]pyrene. Toxicol Appl Pharmacol 230(2):135–143

Fan J, Liu S, Du Y, Morrison J, Shipman R, Pang KS (2009) Up-regulation of transporters and enzymes by the vitamin D receptor ligands, 1α,25-dihydroxyvitamin D_3 and vitamin D analogs, in the Caco-2 cell monolayer. J Pharmacol Exp Ther 330(2):389–402

Fan Y, Boivin GP, Knudsen ES, Nebert DW, Xia Y, Puga A (2010) The aryl hydrocarbon receptor functions as a tumor suppressor of liver carcinogenesis. Cancer Res 70(1):212–220

Fernandez-Salguero PM, Hilbert DM, Rudikoff S, Ward JM, Gonzalez FJ (1996) Aryl-hydrocarbon receptor-deficient mice are resistant to 2,3,7,8-tetrachlorodibenzo-*p*-dioxin-induced toxicity. Toxicol Appl Pharmacol 140(1):173–179

Fujii-Kuriyama Y, Kawajiri K (2010) Molecular mechanisms of the physiological functions of the aryl hydrocarbon (dioxin) receptor, a multifunctional regulator that senses and responds to environmental stimuli. Proc Jpn Acad Ser B Phys Biol Sci 86(1):40–53

Gao J, Xie W (2010) Pregnane X receptor and constitutive androstane receptor at the crossroads of drug metabolism and energy metabolism. Drug Metab Dispos 38(12):2091–2095

Hainaut P, Pfeifer GP (2001) Patterns of p53G→T transversions in lung cancers reflect the primary mutagenic signature of DNA-damage by tobacco smoke. Carcinogenesis (Oxf) 22(3):367–374
Hamamoto H, Kusudo T, Urushino N, Masuno H, Yamamoto K, Yamada S, Kamakura M, Ohta M, Inouye K, Sakaki T (2006) Structure-function analysis of vitamin D 24-hydroxylase (CYP24A1) by site-directed mutagenesis: amino acid residues responsible for species-based difference of CYP24A1 between humans and rats. Mol Pharmacol 70(1):120–128
Hankinson O, Brooks BA, Weir-Brown KI, Hoffman EC, Johnson BS, Nanthur J, Reyes H, Watson AJ (1991) Genetic and molecular analysis of the Ah receptor and of Cyp1a1 gene expression. Biochimie 73(1):61–66
Haussler MR, Whitfield GK, Haussler CA, Hsieh JC, Thompson PD, Selznick SH, Dominguez CE, Jurutka PW (1998) The nuclear vitamin D receptor: biological and molecular regulatory properties revealed. J Bone Miner Res 13(3):325–349
Henklova P, Vrzal R, Ulrichova J, Dvorak Z (2008) Role of mitogen-activated protein kinases in aryl hydrocarbon receptor signaling. Chem Biol Interact 172(2):93–104
Hewison M (2010) Vitamin D and the intracrinology of innate immunity. Mol Cell Endocrinol 321(2):103–111
Holick MF (2006) Resurrection of vitamin D deficiency and rickets. J Clin Invest 116(8):2062–2072
Holick MF (2008) The vitamin D deficiency pandemic and consequences for nonskeletal health: mechanisms of action. Mol Aspects Med 29(6):361–368
Horst RL, Omdahl JA, Reddy S (2003) Rat cytochrome P450C24 (CYP24) does not metabolize 1,25-dihydroxyvitamin D2 to calcitroic acid. J Cell Biochem 88(2):282–285
Irigaray P, Newby JA, Clapp R, Hardell L, Howard V, Montagnier L, Epstein S, Belpomme D (2007) Lifestyle-related factors and environmental agents causing cancer: an overview. Biomed Pharmacother 61(10):640–658
Ishizawa M, Ogura M, Kato S, Makishima M (2012) Impairment of bilirubin clearance and intestinal interleukin-6 expression in bile duct-ligated vitamin D receptor null mice. PLoS One 7(12):e51664
Jones G, Strugnell SA, DeLuca HF (1998) Current understanding of the molecular actions of vitamin D. Physiol Rev 78(4):1193–1231
Kawajiri K, Kobayashi Y, Ohtake F, Ikuta T, Matsushima Y, Mimura J, Pettersson S, Pollenz RS, Sakaki T, Hirokawa T, Akiyama T, Kurosumi M, Poellinger L, Kato S, Fujii-Kuriyama Y (2009) Aryl hydrocarbon receptor suppresses intestinal carcinogenesis in ApcMin/+ mice with natural ligands. Proc Natl Acad Sci USA 106(32):13481–13486
Levy R, Malech HL (1991) Effect of 1,25-dihydroxyvitamin D3, lipopolysaccharide, or lipoteichoic acid on the expression of NADPH oxidase components in cultured human monocytes. J Immunol 147(9):3066–3071
Liu PT, Stenger S, Li H, Wenzel L, Tan BH, Krutzik SR, Ochoa MT, Schauber J, Wu K, Meinken C, Kamen DL, Wagner M, Bals R, Steinmeyer A, Zugel U, Gallo RL, Eisenberg D, Hewison M, Hollis BW, Adams JS, Bloom BR, Modlin RL (2006) Toll-like receptor triggering of a vitamin D-mediated human antimicrobial response. Science 311(5768):1770–1773
Maher J, Yamamoto M (2010) The rise of antioxidant signaling: the evolution and hormetic actions of Nrf2. Toxicol Appl Pharmacol 244(1):4–15
Maier A, Schumann BL, Chang X, Talaska G, Puga A (2002) Arsenic co-exposure potentiates benzo[*a*]pyrene genotoxicity. Mutat Res 517(1–2):101–111
Makishima M (2005) Nuclear receptors as targets for drug development: regulation of cholesterol and bile acid metabolism by nuclear receptors. J Pharmacol Sci 97(2):177–183
Makishima M, Yamada S (2005) Targeting the vitamin D receptor: advances in drug discovery. Expert Opin Ther Pat 15(9):1133–1145
Makishima M, Lu TT, Xie W, Whitfield GK, Domoto H, Evans RM, Haussler MR, Mangelsdorf DJ (2002) Vitamin D receptor as an intestinal bile acid sensor. Science 296(5571):1313–1316

Matsunawa M, Amano Y, Endo K, Uno S, Sakaki T, Yamada S, Makishima M (2009) The aryl hydrocarbon receptor activator benzo[*a*]pyrene enhances vitamin D3 catabolism in macrophages. Toxicol Sci 109(1):50–58

Matsunawa M, Akagi D, Uno S, Endo-Umeda K, Yamada S, Ikeda K, Makishima M (2012) Vitamin D receptor activation enhances benzo[*a*]pyrene metabolism via CYP1A1 expression in macrophages. Drug Metab Dispos 40(11):2059–2066

McCarthy TC, Li X, Sinal CJ (2005) Vitamin D receptor-dependent regulation of colon multidrug resistance-associated protein 3 gene expression by bile acids. J Biol Chem 280(24):23232–23242

Miller KP, Ramos KS (2001) Impact of cellular metabolism on the biological effects of benzo[*a*]pyrene and related hydrocarbons. Drug Metab Rev 33(1):1–35

Moennikes O, Loeppen S, Buchmann A, Andersson P, Ittrich C, Poellinger L, Schwarz M (2004) A constitutively active dioxin/aryl hydrocarbon receptor promotes hepatocarcinogenesis in mice. Cancer Res 64(14):4707–4710

Nagpal S, Na S, Rathnachalam R (2005) Noncalcemic actions of vitamin D receptor ligands. Endocr Rev 26(5):662–687

Nebert DW, Roe AL, Dieter MZ, Solis WA, Yang Y, Dalton TP (2000) Role of the aromatic hydrocarbon receptor and [Ah] gene battery in the oxidative stress response, cell cycle control, and apoptosis. Biochem Pharmacol 59(1):65–85

Nebert DW, Dalton TP, Okey AB, Gonzalez FJ (2004) Role of aryl hydrocarbon receptor-mediated induction of the CYP1 enzymes in environmental toxicity and cancer. J Biol Chem 279(23):23847–23850

Nishida S, Ozeki J, Makishima M (2009) Modulation of bile acid metabolism by 1α-hydroxyvitamin D3 administration in mice. Drug Metab Dispos 37(10):2037–2044

Ogura M, Nishida S, Ishizawa M, Sakurai K, Shimizu M, Matsuo S, Amano S, Uno S, Makishima M (2009) Vitamin D3 modulates the expression of bile acid regulatory genes and represses inflammation in bile duct-ligated mice. J Pharmacol Exp Ther 328(2):564–570

Ohtake F, Baba A, Takada I, Okada M, Iwasaki K, Miki H, Takahashi S, Kouzmenko A, Nohara K, Chiba T, Fujii-Kuriyama Y, Kato S (2007) Dioxin receptor is a ligand-dependent E3 ubiquitin ligase. Nature (Lond) 446(7135):562–566

Rosen CJ, Adams JS, Bikle DD, Black DM, Demay MB, Manson JE, Murad MH, Kovacs CS (2012) The nonskeletal effects of vitamin D: an Endocrine Society scientific statement. Endocr Rev 33(3):456–492

Saeki M, Kurose K, Tohkin M, Hasegawa R (2008) Identification of the functional vitamin D response elements in the human MDR1 gene. Biochem Pharmacol 76(4):531–542

Safe S, Lee SO, Jin UH (2013) Role of the aryl hydrocarbon receptor in carcinogenesis and potential as a drug target. Toxicol Sci 135(1):1–16

Sawada N, Kusudo T, Sakaki T, Hatakeyama S, Hanada M, Abe D, Kamao M, Okano T, Ohta M, Inouye K (2004) Novel metabolism of 1α,25-dihydroxyvitamin D3 with C24-C25 bond cleavage catalyzed by human CYP24A1. Biochemistry 43(15):4530–4537

Sharma OP (1996) Vitamin D, calcium, and sarcoidosis. Chest 109(2):535–539

Shimada T (2006) Xenobiotic-metabolizing enzymes involved in activation and detoxification of carcinogenic polycyclic aromatic hydrocarbons. Drug Metab Pharmacokinet 21(4):257–276

Shimada T, Fujii-Kuriyama Y (2004) Metabolic activation of polycyclic aromatic hydrocarbons to carcinogens by cytochromes P450 1A1 and 1B1. Cancer Sci 95(1):1–6

Shulman AI, Mangelsdorf DJ (2005) Retinoid X receptor heterodimers in the metabolic syndrome. N Engl J Med 353(6):604–615

Sly LM, Lopez M, Nauseef WM, Reiner NE (2001) 1α,25-Dihydroxyvitamin D3-induced monocyte antimycobacterial activity is regulated by phosphatidylinositol 3-kinase and mediated by the NADPH-dependent phagocyte oxidase. J Biol Chem 276(38):35482–35493

Thummel KE, Brimer C, Yasuda K, Thottassery J, Senn T, Lin Y, Ishizuka H, Kharasch E, Schuetz J, Schuetz E (2001) Transcriptional control of intestinal cytochrome P-4503A by 1α,25-dihydroxy vitamin D3. Mol Pharmacol 60(6):1399–1406

Ueda R, Iketaki H, Nagata K, Kimura S, Gonzalez FJ, Kusano K, Yoshimura T, Yamazoe Y (2006) A common regulatory region functions bidirectionally in transcriptional activation of the human CYP1A1 and CYP1A2 genes. Mol Pharmacol 69(6):1924–1930

Uno S, Makishima M (2009) Benzo[*a*]pyrene toxicity and inflammatory disease. Curr Rheumatol Rev 5:266–271

Uno S, Dalton TP, Shertzer HG, Genter MB, Warshawsky D, Talaska G, Nebert DW (2001) Benzo [a]pyrene-induced toxicity: paradoxical protection in Cyp1a1(−/−) knockout mice having increased hepatic BaP-DNA adduct levels. Biochem Biophys Res Commun 289(5):1049–1056

Uno S, Dalton TP, Derkenne S, Curran CP, Miller ML, Shertzer HG, Nebert DW (2004) Oral exposure to benzo[*a*]pyrene in the mouse: detoxication by inducible cytochrome P450 is more important than metabolic activation. Mol Pharmacol 65(5):1225–1237

Uno S, Dalton TP, Dragin N, Curran CP, Derkenne S, Miller ML, Shertzer HG, Gonzalez FJ, Nebert DW (2006) Oral benzo[*a*]pyrene in Cyp1 knockout mouse lines: CYP1A1 important in detoxication, CYP1B1 metabolism required for immune damage independent of total-body burden and clearance rate. Mol Pharmacol 69(4):1103–1114

Van Cromphaut SJ, Dewerchin M, Hoenderop JG, Stockmans I, Van Herck E, Kato S, Bindels RJ, Collen D, Carmeliet P, Bouillon R, Carmeliet G (2001) Duodenal calcium absorption in vitamin D receptor-knockout mice: functional and molecular aspects. Proc Natl Acad Sci USA 98(23):13324–13329

Xu L, Li AP, Kaminski DL, Ruh MF (2000) 2,3,7,8-Tetrachlorodibenzo-*p*-dioxin induction of cytochrome P4501A in cultured rat and human hepatocytes. Chem Biol Interact 124(3):173–189

Xu Y, Hashizume T, Shuhart MC, Davis CL, Nelson WL, Sakaki T, Kalhorn TF, Watkins PB, Schuetz EG, Thummel KE (2006) Intestinal and hepatic CYP3A4 catalyze hydroxylation of 1α,25-dihydroxyvitamin D3: implications for drug-induced osteomalacia. Mol Pharmacol 69(1):56–65

Xu M, Li D, Lu Y, Chen GQ (2007) Leukemogenic AML1-ETO fusion protein increases carcinogen-DNA adduct formation with upregulated expression of cytochrome P450-1A1 gene. Exp Hematol 35(8):1249–1255

Yoshinari K, Yoda N, Toriyabe T, Yamazoe Y (2010) Constitutive androstane receptor transcriptionally activates human CYP1A1 and CYP1A2 genes through a common regulatory element in the 5′-flanking region. Biochem Pharmacol 79(2):261–269

Part IV
Drug Metabolism

Chapter 17
Species, Ethnic, and Individual Differences in Human Drug-Metabolizing Cytochrome P450 Enzymes

Hiroshi Yamazaki

Abstract Safety testing of drug metabolites resulting from P450 enzyme activity in humans is important for developing drug candidates and investigating the potential for idiosyncratic drug reactions. Drug-induced hepatotoxicity may be caused by active intermediates formed by animal or, more commonly, human cytochrome P450 enzymes from common toxicants or drugs. Safety assessments of drug metabolites are often carried out in minipigs, monkeys, and genetically engineered mouse models that express human enzymes. Consequently, it is important to understand and take into account species differences in the roles of P450 enzymes in drug metabolism. During the past 30 years, a vast amount of research has focused on the substrates of human P450 enzymes and the metabolites formed; however, the issue of human-specific metabolites is complex and much remains to be learned.

Keywords Cytochrome P450 • Drug metabolism • Human • Monkey • Pig

17.1 Introduction

In the past 30 years, human cytochrome P450 (P450) enzymes have been characterized with respect to drug metabolism. Research has elucidated liver microsomal P450 isoform contents (Shimada et al. 1994), their relatively broad but selective substrate specificities (Guengerich and Rendic 2010; Rendic and Guengerich 2010), P450 induction and inhibition (Niwa et al. 2011), P450-mediated drug interactions (Hisaka et al. 2010), and genetic polymorphisms of

H. Yamazaki (✉)
Laboratory of Drug Metabolism and Pharmacokinetics, Showa Pharmaceutical University, 3-3165 Higashi-tamagawa Gakuen, Machida, Tokyo 194-8543, Japan
e-mail: hyamazak@ac.shoyaku.ac.jp

H. Yamazaki (ed.), *Fifty Years of Cytochrome P450 Research*,
DOI 10.1007/978-4-431-54992-5_17,

Table 17.1 Numbers of articles mentioning drug interactions in PubMed or Medline

	17 March 1995	17 June 2007	17 April 2013
Drug interactions[a]	5,690	53,849	101,369
+ Metabolism	2,570	30,438	62,418
+ P450 or CYP	185	2,868	3,780
+ Transporter	–	1,611	5,705

[a]Drug interactions and (pharmacology or pharmacokinetics or toxicology or toxicokinetics) and (human or adult) were used for keyword search (Yamazaki 2007a and updated)

P450 enzymes. In addition, a nomenclature system for polymorphic human P450 alleles (http://www.cypalleles.ki.se/) has been established. The human *P450* gene superfamily is made up of 57 functional genes and 58 pseudogenes (Nelson et al. 2004). Research has focused on the P450s involved in the oxidation of compounds associated with pharmacological and/or toxicological actions. The knowledge gained so far has been impressive, but many challenges remain.

Systematic research into drug interactions has become more comprehensive and complex. Metabolic factors are involved in approximately half of all reported drug interactions. Although the number of articles citing P450s as the source of metabolic drug interactions has increased since the mid-1980s, these comprise only a small proportion of overall reports on drug interactions. By 2007 and 2013, the number of reports of drug interactions had increased 10- and 20 fold compared with the number in 1995 (Table 17.1). Metabolic drug interactions are clearly still relevant in the twenty-first century.

Reactive metabolites are among the most important research topics for academic, industrial, and clinical toxicologists (Yamazaki 2007b). It is not surprising that many drug regulatory authorities require evaluation of the formation and detoxification of reactive metabolites before market approval as well as during the postmarketing period. Both in vitro and in vivo studies are carried out to develop and use medicines effectively. Information on species differences and epidemiological and mechanistic studies will further reveal the importance of genetic polymorphisms of P450s and the deactivation by glutathione *S*-transferases (GSTs, another enzyme superfamily) to the toxicology of drugs and drug candidates. Troglitazone is a good example of a drug (Yamamoto et al. 2002) that was marketed after approval by the U.S. Food and Drug Administration and the Ministry of Health, Labour and Welfare of Japan but was subsequently withdrawn as a result of idiosyncratic liver toxicity associated with impaired GST activity (Watanabe et al. 2003). The current increasing trend for the mechanistic evaluation of reactive metabolites and their potential for predictions of individual risk indicates that this area of investigation will likely remain a hot topic challenging the P450 research community for many years to come.

17.2 P450 1A-Mediated Caffeine Demethylation and 5-*n*-Butyl-Pyrazolo[1,5-*a*]Pyrimidine (M-5) Activation

Caffeine (1,3,7-trimethylxanthine) is a phenotyping substrate for human P450 1A2. The predominant reaction in caffeine metabolism in humans is the 3-demethylation of caffeine to generate 1,7-dimethylxanthine (Berthou et al. 1991; Gu et al. 1992; Grant et al. 1987). In humans, the 3-*N*-demethylation of caffeine is the major metabolic pathway (81 % of total dimethylxanthines); in monkeys, however, 7-*N*-demethylation to form pharmacological active theophylline is reportedly predominant (89 %) (Berthou et al. 1992). Interspecies differences regarding caffeine oxidation exist, but the predominant 8-hydroxylation of caffeine in mice, rabbits, and rats has been shown to be broadly similar (Berthou et al. 1992). In an analytical approach to investigate species differences in P450 activities, rat P450 1A2 (75 % homologous with human 1A2) was aligned with the crystal structure of human 1A2 (PDB 2HI4) using MOE software (Chemical Computing Group, Montreal, Canada) to model the three-dimensional (3-D) structure. Before running docking simulations, the energies of the P450 structures were minimized using the CHARM22 force field for caffeine binding to P450 enzymes using the MMFF94x force field distributed in the MOE Dock software. A docking simulation, shown in Fig. 17.1, was able to explain a typical species difference in caffeine metabolism by the orthologous enzyme forms with possibly different substrate pockets.

During drug development, similar species differences to those described for caffeine metabolism by the orthologous enzyme forms have been encountered. Treatment of rats or dogs with 1,000 mg/kg OT-7100 (5-*n*-butyl-7-(3,4,5-trimethoxy-benzoylamino)-pyrazolo[1,5-*a*]pyrimidine) (Fig. 17.2), an experimental analgesic

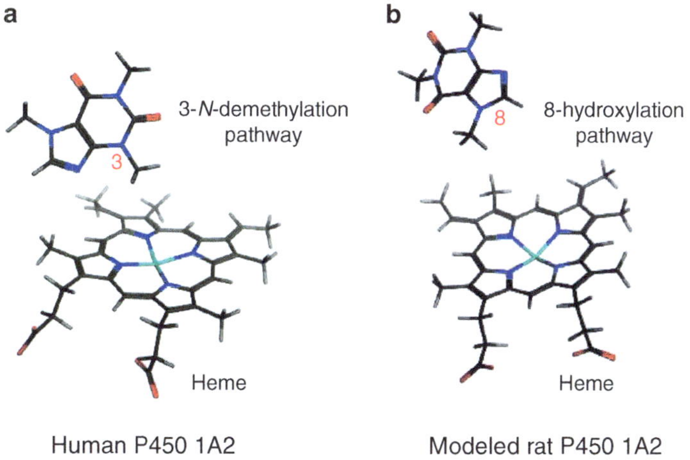

Fig. 17.1 Docking of caffeine into human (**a**) and rat (**b**) P450 1A2 to produce different main metabolites. The heme group of the P450 is shown in the *lower part* of each part. Oxygen, nitrogen, and iron are colored *red*, *blue*, and *light blue*, respectively

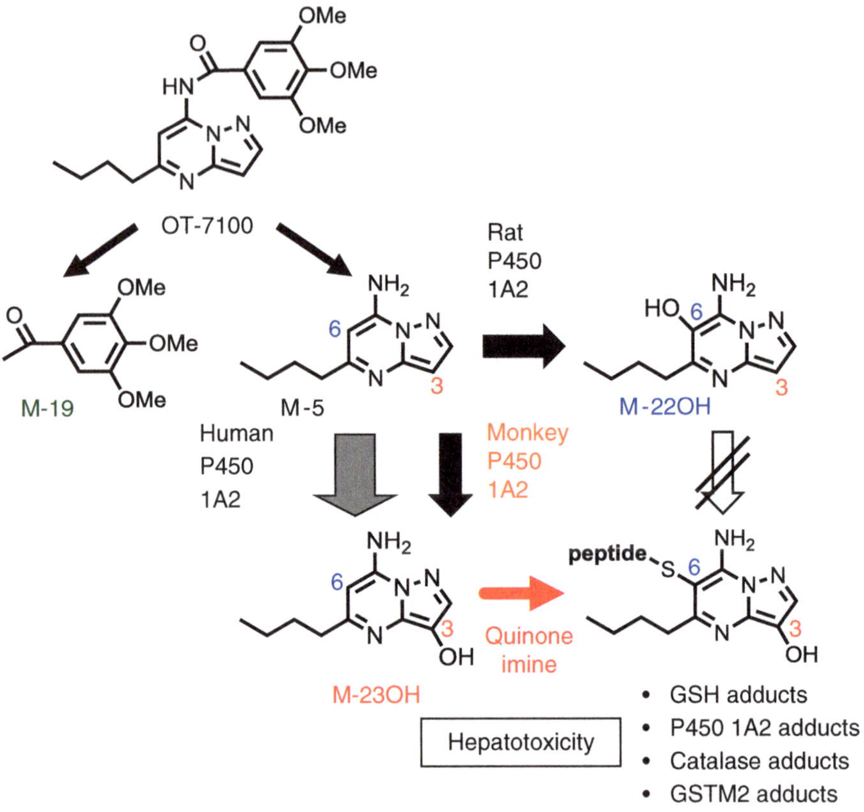

Fig. 17.2 Analgesic drug OT-7100 is metabolized by P450 1A2 from humans, monkeys and rats. Metabolic differences result in liver toxicity in humans but not in rats

drug, or with its primary metabolite, M-5, produced no apparent hepatotoxicity (Kuribayashi et al. 2006). However, limited and reversible elevations in serum levels of aspartate and alanine aminotransferases (markers for liver damage) were occasionally observed. Most of these cases (7 of 11) occurred with more than 1,400 mg/day OT-7100 for more than 14 days after 28-day treatments in a dose-escalation study (Kuribayashi et al. 2009). M-5, the primary metabolite of OT-7100, is metabolized by human P450 1A2 to form M-23OH, which conjugates with a peptide to form a covalent adduct. In rats, the same enzyme forms some M-23OH, but predominantly yields M-22OH, which exhibits no toxicity (Fig. 17.2). These metabolic differences highlight some of the perils of testing drug candidates for metabolite toxicity in animals (Kuribayashi et al. 2012). Relatively nonspecific binding may play a role in the reversible effects of high doses of the hepatotoxic former drug candidate OT-7100 (Yamazaki et al. 2010). As in this example, toxicologists pool their knowledge to advance safety testing of drug candidates based on information on P450 orthologues and their activities.

17.3 P450 2C-Mediated Warfarin Oxidation

Drug-metabolizing P450 isoforms play important roles in the metabolism of a variety of endogenous and exogenous compounds in humans. Metabolically active metabolites may be formed, some of which can be toxic (Rendic and Guengerich 2012). The cynomolgus monkey (*Macaca fascicularis*) is one of the most widely used primate species in preclinical studies because of its evolutionary closeness to humans (Iwasaki and Uno 2009). More than 20 P450 enzymes in the cynomolgus monkey have been identified and were found to be highly identical to orthologous human P450s (Uno et al. 2011a). Although cynomolgus monkey P450 2C76 is not orthologous to any human P450 (Uno et al. 2011b), monkey P450 2C8 (previously named P450 2C20) is orthologous to human P450 2C8. In addition, monkey P450 2C9 (previously P450 2C43) and 2C19 (previously P450 2C75) are 91–93 % identical to human P450 2C9 and 2C19, respectively (Uno et al. 2011a). Cynomolgus monkey P450 2C76, 2C8, 2C9, and 2C19 (Table 17.2) are expressed

Table 17.2 Autologous P450 isoforms in humans, macaques, and pigs

Human	Cynomolgus monkey	Minipig
1A1	1A1	1A1
1A2	1A2	1A2
1B1	1B1	
[1D1P]	1D1	
2A6	2A23	2A19
[2A7]	2A24	
2A13	2A26	
2B6	2B6	2B22
2C8	2C8(20)	2C33
[2C18]	2C18	2C42
2C9	2C9(43)	2C49
2C19	2C19(75)	
	2C76	
	2C93	
2D6	2D17	2D25
	2D44	
2E1	2E1	2E1
2G2P	2G2	
2J2	2J2	2J34
3A4	3A4(8)	3A22
3A5	3A5	3A29
3A7	(3A7)	3A39
3A43	3A43	3A46
4A11	4A11	4A21
4A22		4A24

Rendic and Guengerich (Rendic and Guengerich 2010), Uno et al. (Uno et al. 2011a), Achour et al. (Achour et al. 2011), and updated

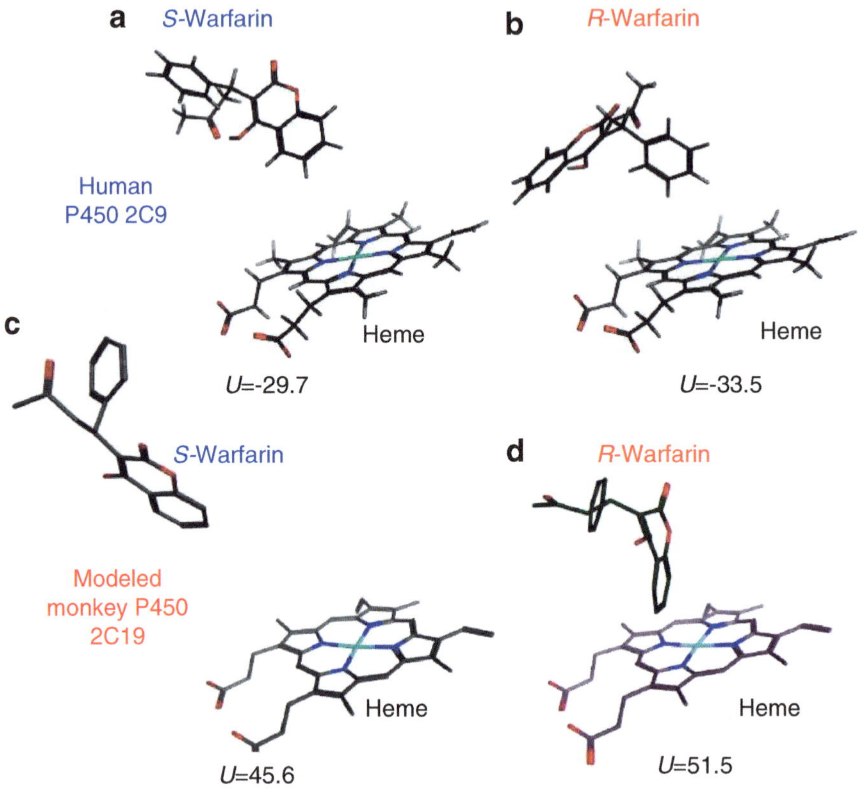

Fig. 17.3 Docking of *S*- (**a**, **c**) and *R*- (**b**, **d**) warfarin with reported human P450 2C9 (**a**, **b**: 1OG5) and modeled monkey 2C19 (**c**, **d**). The heme group of the P450 is shown in the *lower part* of each part. *U* values indicate the interaction energy. Oxygen and iron are colored *red* and *light blue*, respectively. The docking simulations explain species activity differences

as functional drug-metabolizing enzymes in monkey livers, just as the latter three are in the human liver (Uno et al. 2011c).

The important, well-established medicine *S*-warfarin is a typical human P450 2C9 substrate (Yamazaki and Shimada 1997). High activity of monkey P450 2C19 toward *R*-warfarin 7-hydroxylation has been observed (Hosoi et al. 2012), whereas in humans, *R*-warfarin 7-hydroxylation is much slower than *S*-warfarin 7-hydroxylation. A high capacity of monkey liver P450 2C19 with respect to *R*- and *S*-warfarin 6- and 7-hydroxylation was supported by molecular docking simulations (Fig. 17.3). The activities of orthologous P450 enzymes from humans and monkeys (Table 17.2) are generally similar, but some differences were recently recognized, such as those for warfarin hydroxylation as just described. An additional example of such species differences is predominant caffeine 7-demethylation mediated by monkey P450 2C9 in contrast to predominant caffeine 3-demethylation mediated by human P450 1A2 (Utoh et al. 2013).

17.4 P450 2D-Mediated Dextromethorphan Demethylation

A new combination of five P450 probe drugs for humans has been reported (Turpault et al. 2009). This combination of caffeine (human P450 1A2 probe), warfarin (human P450 2C9 probe), omeprazole (human P450 2C19 probe), metoprolol (human P450 2D6 probe), and midazolam (human P450 3A probe) was designed to overcome potential shortcomings of other reported drug combinations. When multiple P450 probe drugs were administered concurrently in minipigs in vivo, minipig P450 2D enzymes activities were revealed to be fast (Mogi et al. 2012). In in vitro analyses, minipig P450 2D25 and monkey P450 2D17 in liver microsomes showed similar substrate selectivity (such as bufuralol and dextromethorphan) and had high drug oxidation activities compared with human P450 2D6 (Fig. 17.4). Monkey and minipig P450 3A enzymes also showed similar substrate selectivity to human P450 3A4 and 3A5 enzymes, but monkey P450 3A

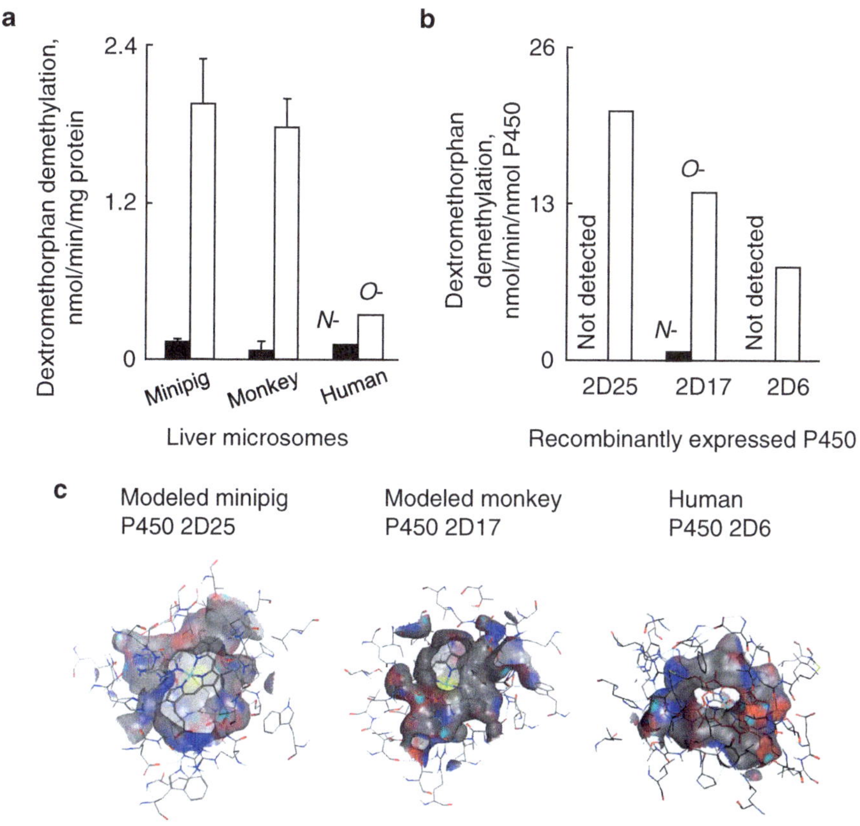

Fig. 17.4 Dextromethorphan *O*- (open) and *N*- (closed) demethylation by liver microsomes (**a**) and recombinantly expressed P450 2D enzymes (**b**) in minipigs, monkeys, and humans. (**c**) Modeled P450 2D enzyme structures

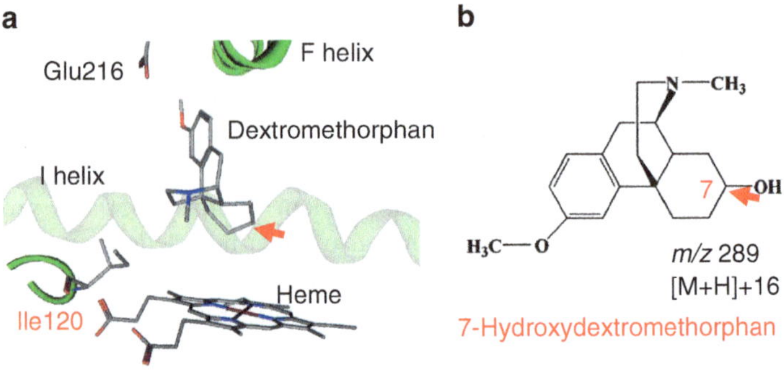

Fig. 17.5 Docking of dextromethorphan into P450 2D6.49 in human liver microsomes (**a**) resulting in a unique human metabolite of dextromethorphan, 7-hydroxydextromethorphan (**b**)

enzymes exhibited wider substrate selectivity toward human P450 2D substrates (Iwasaki et al. 2010), resulting in high P450 2D-related drug clearance in monkeys. Modeled minipig P450 2D25 (human 2D6 homology, 77 %) seemed to have a large substrate pocket (Fig. 17.4c). Minipig P450 2D-mediated drug oxidation activities were high in vitro and in vivo compared with humans, probably because of high P450 2D concentrations in minipig livers as determined by liquid chromatography/mass spectrometry (Achour et al. 2011) in comparison with mean P450 isoform contents determined immunochemically in liver microsomes from humans (Shimada et al. 1994) and monkeys (Uehara et al. 2011).

During the course of sequencing the human *P450 2D6* gene, a novel F120I variant (*2D6*49*) was found (Matsunaga et al. 2009), which was previously identified as a *2D6*10*-associated allele in a Japanese population. With dextromethorphan as a substrate, P450 2D6.49 formed a novel 7-hydroxydextromethorphan (Fig. 17.5), with a V_{max}/K_m value roughly similar to that of *O*-demethylation (Matsunaga et al. 2009). The *P450 2D6*10* haplotype is a possible cause of Asian interindividual variations in the activities and the substrate specificity of P450 2D6. Variability in P450 2D substrate pockets and content in the liver may result in species and individual differences.

17.5 P450 3A-Mediated Activation of, and Drug Interactions with, Thalidomide

The sedative drug thalidomide [α-(*N*-phthalimido)glutarimide] was withdrawn in the early 1960s because of its potent teratogenic effects (Speirs 1962), but it was subsequently approved for the treatment of multiple myeloma (Palumbo et al. 2008; Nakamura et al. 2013). Because of the recent emergence of thalidomide as a drug with clinical potential, there is renewed interest in both its toxicity and pharmacological mechanisms. Various hypotheses have been proposed to explain teratogenic

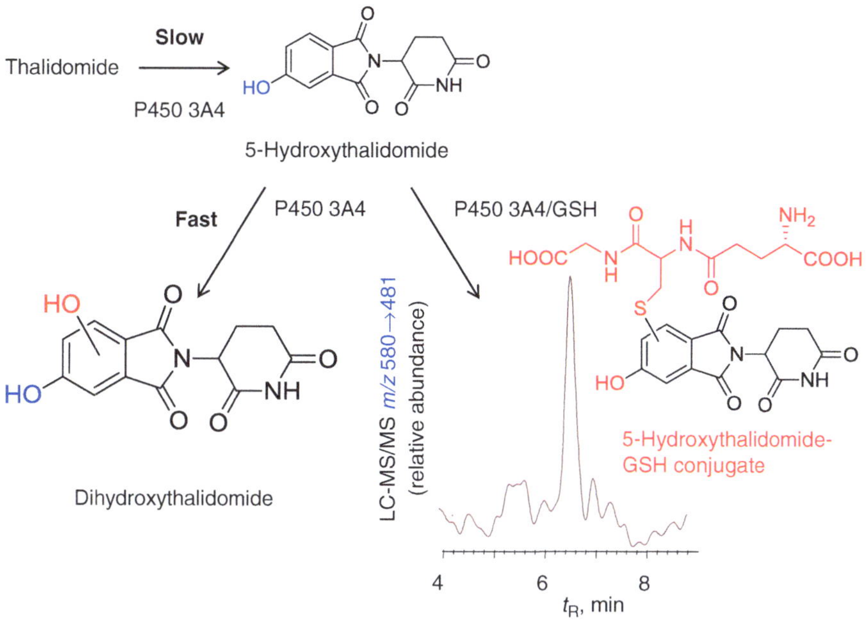

Fig. 17.6 Secondary hydroxylation of 5-hydroxythalidomide by P450 3A enzymes in human livers. *GSH* glutathione, *LC-MS/MS* liquid chromatography-tandem mass spectrometry

effects, including the generation of reactive oxygen species (Parman et al. 1999), the generation of reactive acylating compounds (Fabro et al. 1965) and arene oxide intermediates (Gordon et al. 1981), inhibition of angiogenesis (D'Amato et al. 1994), and inhibition of the protein cereblon (Ito et al. 2010).

We reported the cooperativity of human P450 subfamily 3A enzymes with respect to thalidomide (Okada et al. 2009) and 5-hydroxy and dihydroxy product formation from thalidomide (Fig. 17.6) mediated by human P450 3A4 (Yamazaki et al. 2011, 2012). Also, the secondary hydroxylation of 5-hydroxythalidomide involves a reactive intermediate, possible an arene oxide, that can be trapped by glutathione (GSH) to give GSH adducts (Chowdhury et al. 2010). Because this unique human metabolite(s) was identified as a conjugate of 5-hydroxythalidomide formed on the phenyl ring, reactive metabolite(s) from aromatic 5-hydroxythalidomide mediated by human P450 3A enzymes were identified. Two aspects of in vivo drug interaction involving thalidomide have been reported (Yamazaki et al. 2013): an enhanced clearance of midazolam and a higher area under the curve of 4-hydroxymidazolam following pre-treatment with thalidomide in humanized-liver mice, presumably caused by human P450 3A induction. In vivo cooperativity of human P450 3A enzymes has also been reported, with a higher area under the curve for 1′-hydroxymidazolam following co-treatment with thalidomide in humanized mice (Yamazaki et al. 2013). Metabolic activation of thalidomide occurs, but the details are not yet understood. In our current study, we are investigating the induction of P450 3A enzymes to fully understand the metabolic activation of thalidomide. Interestingly, there was little effect of 5-hydroxy- or 5′-hydroxythalidomide on in vitro induction. Auto-induction by thalidomide of human P450 3A is likely associated with the pregnane X receptor (PXR), a member of the steroid hormone nuclear receptor family, as revealed by a novel microarray assay for real-time analysis of co-regulator–nuclear receptor interaction (Murayama et al. 2014). Taken together, thalidomide was capable of binding to PXR directly and induced transcriptional regulation of the human *P450 3A* gene. Drug interactions involving thalidomide should be evaluated with the knowledge that the drug and its human metabolite are P450 inducers.

17.6 Epilogue

The world of P450 research has likely been a source of excitement for many readers of this book. The success of P450 research in drug discovery and development has been remarkable, and the applications of P450 research continue to grow. We have demonstrated that genetic polymorphisms in *P450 2A6*, including copy number polymorphisms, are determinants of smoking behavior and tobacco-related lung cancer risk, particularly for squamous cell and small cell carcinoma, which are known to be associated with cigarette smoking in male Japanese heavy smokers (Fujieda et al. 2004; Shimizu et al. 2011, 2013). Many P450 researchers, including myself (Yamazaki and Shimizu 2013), are also interested in another monooxygenase, flavin-containing monooxygenase 3 (FMO3), because drug oxygenations are linked to their pharmacological and toxicological actions.

The International Conference on Cytochrome P450 is an international series of biennial meetings held to discuss the latest findings on P450 enzymes (http://www.p450meetings.com). This P450 series was established in 1976 and has developed into an important international event, with meetings rotating throughout the world. The next meeting will be held in Tokyo in 2015 and will be the third held in Japan (the others were held in Europe and North America). As P450 has become a more mature field, the issues addressed at these P450 meetings have grown to include biochemistry, biophysics, and gene regulation as well as new applications of P450 science in today's practical problems.

On behalf of the organizers, I cordially invite you to Tokyo, Japan, for the 19th International Conference on Cytochrome P450 (19th ICCP450 Tokyo) at the Olympics Memorial Youth Center, Shibuya-ku, Tokyo 151-0052 from June 12 to 15, 2015. The Olympics Memorial Youth Center was used 50 years ago as part of the Tokyo Olympic Village site (1964). I look forward to seeing you in Tokyo in June 2015.

Acknowledgments I thank Drs. Tsutomu Shimada and Fred Guengerich for supporting my three decades of human P450 research, and I also thank my many collaborators and laboratory members for all their hard work. This work was partly supported by MEXT-Supported Program for the Strategic Research Foundation at Private Universities (2013–2017).

References

Achour B, Barber J, Rostami-Hodjegan A (2011) Cytochrome P450 pig liver pie: determination of individual cytochrome P450 isoform contents in microsomes from two pig livers using liquid chromatography in conjunction with mass spectrometry. Drug Metab Dispos 39:2130–2134

Berthou F, Flinois JP, Ratanasavanh D, Beaune P, Riche C (1991) Evidence for the involvement of several cytochromes P-450 in the first steps of caffeine metabolism by human liver microsomes. Drug Metab Dispos 19:561–567

Berthou F, Guillois B, Riche C, Dreano Y, Jacqz-Aigrain E, Beaune PH (1992) Interspecies variations in caffeine metabolism related to cytochrome P4501A enzymes. Xenobiotica 22:671–680

Chowdhury G, Murayama N, Okada Y, Uno Y, Shimizu M, Shibata N, Guengerich FP, Yamazaki H (2010) Human liver microsomal cytochrome P450 3A enzymes involved in thalidomide 5-hydroxylation and formation of a glutathione conjugate. Chem Res Toxicol 23:1018–1024

D'Amato RJ, Loughnan MS, Flynn E, Folkman J (1994) Thalidomide is an inhibitor of angiogenesis. Proc Natl Acad Sci USA 91:4082–4085

Fabro S, Smith RL, Williams RT (1965) Thalidomide as a possible biological acylating agent. Nature (Lond) 208:1208–1209

Fujieda M, Yamazaki H, Saito T, Kiyotani K, Gyamfi MA, Sakurai M, Dosaka-Akita H, Sawamura Y, Yokota J, Kunitoh H, Kamataki T (2004) Evaluation of *CYP2A6* genetic polymorphisms as determinants of smoking behavior and tobacco-related lung cancer risk in male Japanese smokers. Carcinogenesis (Oxf) 25:2451–2458

Gordon GB, Spielberg SP, Blake DA, Balasubramanian V (1981) Thalidomide teratogenesis: evidence for a toxic arene oxide metabolite. Proc Natl Acad Sci USA 78:2545–2548

Grant DM, Campbell ME, Tang BK, Kalow W (1987) Biotransformation of caffeine by microsomes from human liver. Kinetics and inhibition studies. Biochem Pharmacol 36:1251–1260

Gu L, Gonzalez FJ, Kalow W, Tang BK (1992) Biotransformation of caffeine, paraxanthine, theobromine and theophylline by cDNA-expressed human CYP1A2 and CYP2E1. Pharmacogenetics 2:73–77

Guengerich FP, Rendic S (2010) Update information on drug metabolism systems 2009: part I. Curr Drug Metab 11:1–3

Hisaka A, Ohno Y, Yamamoto T, Suzuki H (2010) Prediction of pharmacokinetic drug–drug interaction caused by changes in cytochrome P450 activity using in vivo information. Pharmacol Ther 125:230–248

Hosoi Y, Uno Y, Murayama N, Fujino H, Shukuya M, Iwasaki K, Shimizu M, Utoh M, Yamazaki H (2012) Monkey liver cytochrome P450 2C19 is involved in *R*- and *S*-warfarin 7-hydroxylation. Biochem Pharmacol 84:1691–1695

Ito T, Ando H, Suzuki T, Ogura T, Hotta K, Imamura Y, Yamaguchi Y, Handa H (2010) Identification of a primary target of thalidomide teratogenicity. Science 327:1345–1350

Iwasaki K, Uno Y (2009) Cynomolgus monkey CYPs: a comparison with human CYPs. Xenobiotica 39:578–581

Iwasaki K, Murayama N, Koizumi R, Uno Y, Yamazaki H (2010) Comparison of cytochrome P450 3A enzymes in cynomolgus monkeys and humans. Drug Metab Pharmacokinet 25:388–391

Kuribayashi S, Ueda N, Naito S, Yamazaki H, Kamataki T (2006) Species differences in hydrolase activities toward OT-7100 responsible for different bioavailability in rats, dogs, monkeys and humans. Xenobiotica 36:301–314

Kuribayashi S, Goto K, Naito S, Kamataki T, Yamazaki H (2009) Human cytochrome P450 1A2 involvement in the formation of reactive metabolites from a species-specific hepatotoxic pyrazolopyrimidine derivative, 5-*n*-butyl-7-(3,4,5-trimethoxybenzoylamino)pyrazolo[1,5-*a*] pyrimidine. Chem Res Toxicol 22:323–331

Kuribayashi S, Uno Y, Naito S, Yamazaki H (2012) Different metabolites of human hepatotoxic pyrazolopyrimidine derivative 5-*n*-butyl-pyrazolo[1,5-*a*]pyrimidine produced by human, rat, and monkey cytochrome P450 1A2 and liver microsomes. Basic Clin Pharmacol Toxicol 110:405–408

Matsunaga M, Yamazaki H, Kiyotani K, Iwano S, Saruwatari J, Nakagawa K, Soyama A, Ozawa S, Sawada J, Kashiyama E, Kinoshita M, Kamataki T (2009) Two novel *CYP2D6*10* haplotypes as possible causes of a poor metabolic phenotype in Japanese. Drug Metab Dispos 37:699–701

Mogi M, Toda A, Iwasaki K, Kusumoto S, Takahara H, Shimizu M, Murayama N, Izumi H, Utoh M, Yamazaki H (2012) Simultaneous pharmacokinetics assessment of caffeine, warfarin, omeprazole, metoprolol, and midazolam intravenously or orally administered to microminipigs. J Toxicol Sci 37:1157–1164

Murayama N, van Beuningen R, Suemizu H, Guguen-Guillouzo C, Shibata N, Yajima K, Utoh M, Shimizu M, Chesne C, Nakamura M, Guengerich FP, Houtman R, Yamazaki H (2014) Thalidomide increases human hepatic cytochrome P450 3A enzymes by direct activation of pregnane X receptor. Chem Res Toxicol 27:304–308

Nakamura K, Matsuzawa N, Ohmori S, Ando Y, Yamazaki H, Matsunaga T (2013) Clinical evidence of the pharmacokinetics change in thalidomide therapy. Drug Metab Pharmacokinet 28:38–43

Nelson DR, Zeldin DC, Hoffman SMG, Maltais LJ, Wain HM, Nebert DW (2004) Comparison of cytochrome P450 (*CYP*) genes from the mouse and human genomes, including nomenclature recommendations for genes, pseudogenes and alternative-splice variants. Pharmacogenetics 14:1–18

Niwa T, Murayama N, Yamazaki H (2011) Stereoselectivity of human cytochrome p450 in metabolic and inhibitory activities. Curr Drug Metab 12:549–569

Okada Y, Murayama N, Yanagida C, Shimizu M, Guengerich FP, Yamazaki H (2009) Drug interactions of thalidomide with midazolam and cyclosporine A: heterotropic cooperativity of human cytochrome P450 3A5. Drug Metab Dispos 37:18–23

Palumbo A, Facon T, Sonneveld P, Blade J, Offidani M, Gay F, Moreau P, Waage A, Spencer A, Ludwig H, Boccadoro M, Harousseau JL (2008) Thalidomide for treatment of multiple myeloma: 10 years later. Blood 111:3968–3977

Parman T, Wiley MJ, Wells PG (1999) Free radical-mediated oxidative DNA damage in the mechanism of thalidomide teratogenicity. Nat Med 5:582–585

Rendic S, Guengerich FP (2010) Update information on drug metabolism systems 2009, part II. Summary of information on the effects of diseases and environmental factors on human cytochrome P450 (CYP) enzymes and transporters. Curr Drug Metab 11:4–84

Rendic S, Guengerich FP (2012) Contributions of human enzymes in carcinogen metabolism. Chem Res Toxicol 25:1316–1383

Shimada T, Yamazaki H, Mimura M, Inui Y, Guengerich FP (1994) Interindividual variations in human liver cytochrome P-450 enzymes involved in the oxidation of drugs, carcinogens and toxic chemicals: studies with liver microsomes of 30 Japanese and 30 Caucasians. J Pharmacol Exp Ther 270:414–423

Shimizu M, Kiyotani K, Kunitoh H, Kamataki T, Yamazaki H (2011) Different effects of *TERT*, *TP63*, and *CYP2A6* polymorphism on individual risk of tobacco-related lung cancer in male Japanese smokers. J Cancer Ther 2:690–696

Shimizu M, Ishii Y, Okubo M, Kunitoh H, Kamataki T, Yamazaki H (2013) Effects of *ADH1C*, *ALDH2*, and *CYP2A6* polymorphisms on individual risk of tobacco-related lung cancer in male Japanese smokers. J Cancer Ther 4:29–35

Speirs AL (1962) Thalidomide and congenital abnormalities. Lancet 1:303–305

Turpault S, Brian W, Van HR, Santoni A, Poitiers F, Donazzolo Y, Boulenc X (2009) Pharmacokinetic assessment of a five-probe cocktail for CYPs 1A2, 2C9, 2C19, 2D6 and 3A. Br J Clin Pharmacol 68:928–935

Uehara S, Murayama N, Nakanishi Y, Zeldin DC, Yamazaki H, Uno Y (2011) Immunochemical detection of cytochrome P450 enzymes in liver microsomes of 27 cynomolgus monkeys. J Pharmacol Exp Ther 339:654–661

Uno Y, Iwasaki K, Yamazaki H, Nelson DR (2011a) Macaque cytochromes P450: nomenclature, transcript, gene, genomic structure, and function. Drug Metab Rev 43:346–361

Uno Y, Matsuno K, Murayama N, Nakamura C, Yamazaki H (2011b) Metabolism of P450 probe substrates by cynomolgus monkey CYP2C76. Basic Clin Pharmacol Toxicol 109:315–318

Uno Y, Uehara S, Kohara S, Iwasaki K, Nagata R, Fukuzaki K, Utoh M, Murayama N, Yamazaki H (2011c) Newly identified CYP2C93 is a functional enzyme in rhesus monkey, but not in cynomolgus monkey. PLoS One 6:e16923

Utoh M, Murayama N, Uno Y, Onose Y, Hosaka S, Fujino H, Shimizu M, Iwasaki K, Yamazaki H (2013) Monkey liver cytochrome P450 2C9 is involved in caffeine 7-*N*-demethylation to form theophylline. Xenobiotica 43:1037–1042

Watanabe I, Tomita A, Shimizu M, Sugawara M, Yasumo H, Koishi R, Takahashi T, Miyoshi K, Nakamura K, Izumi T, Matsushita Y, Furukawa H, Haruyama H, Koga T (2003) A study to survey susceptible genetic factors responsible for troglitazone-associated hepatotoxicity in Japanese patients with type 2 diabetes mellitus. Clin Pharmacol Ther 73:435–455

Yamamoto Y, Yamazaki H, Ikeda T, Watanabe T, Iwabuchi H, Nakajima M, Yokoi T (2002) Formation of a novel quinone epoxide metabolite of troglitazone with cytotoxicity to HepG2 cells. Drug Metab Dispos 30:155–160

Yamazaki H (2007a) Drug interaction studies in the 21st century: research into cytochrome P450s, transporters, and simulations informing their role in drug–drug interactions. Drug Metab Pharmacokinet 22:223–224

Yamazaki H (2007b) Individual differences in toxicological response caused by a diversity of chemicals: observations in Japan. Chem Res Toxicol 21:3–4

Yamazaki H, Shimada T (1997) Human liver cytochrome P450 enzymes involved in the 7-hydroxylation of *R*- and *S*-warfarin enantiomers. Biochem Pharmacol 54:1195–1203

Yamazaki H, Shimizu M (2013) Survey of variants of human flavin-containing monooxygenase 3 (FMO3) and their drug oxidation activities. Biochem Pharmacol 85:1588–1593

Yamazaki H, Kuribayashi S, Inoue T, Tateno C, Nishikura Y, Oofusa K, Harada D, Naito S, Horie T, Ohta S (2010) Approach for in vivo protein bindings of 5-*n*-butyl-pyrazolo[1,5-*a*] pyrimidine bioactivated in chimeric mice with humanized liver by two-dimensional electrophoresis with accelerator mass spectrometry. Chem Res Toxicol 23:152–158

Yamazaki H, Suemizu H, Igaya S, Shimizu M, Shibata M, Nakamura M, Chowdhury G, Guengerich FP (2011) In vivo formation of a glutathione conjugate derived from thalidomide in humanized uPA-NOG mice. Chem Res Toxicol 24:287–289

Yamazaki H, Suemizu H, Shimizu M, Igaya S, Shibata N, Nakamura N, Chowdhury G, Guengerich FP (2012) In vivo formation of dihydroxylated and glutathione conjugate metabolites derived from thalidomide and 5-hydroxythalidomide in humanized TK-NOG mice. Chem Res Toxicol 25:274–276

Yamazaki H, Suemizu H, Murayama N, Utoh M, Shibata N, Nakamura M, Guengerich FP (2013) In vivo drug interactions of the teratogen thalidomide with midazolam: heterotropic cooperativity of human cytochrome P450 in humanized TK-NOG mice. Chem Res Toxicol 26:486–489

Chapter 18
Cytochrome P450-Dependent Change in UDP-Glucuronosyltransferase Function and Its Reverse Regulation

Yuji Ishii, Yuu Miyauchi, and Hideyuki Yamada

Abstract Cytochrome P450 (P450, CYP) and UDP-glucuronosyltransferase (UGT) are important enzymes involved in phase I and II drug metabolism, respectively. It has long been believed that these enzymes work separately, because their topology with regard to the endoplasmic reticulum membrane is very different and the location of the two enzymes is separated by the membrane. However, cumulative evidence suggests that these enzymes interact with each other to modify their respective functions. This review mainly focuses on a P450-dependent alteration in UGT function and discusses the relevance of this modification to the polymorphic nature of drug metabolism mediated by UGTs. Also, we describe the selectivity of P450/UGT isoforms in terms of their pairing and possible P450/UGT domains serving the interactions. The reverse modulation, that is, the alteration of P450 function by UGT, is also described.

Keywords Cytochrome P450 • Heterologous expression • P450 • Protein–protein interaction • UDP-glucuronosyltransferase • UGT

18.1 General Introduction

There are large interindividual differences in drug metabolism catalyzed by cytochrome P450 (P450, CYP). Thus, it is very important to know all the factors affecting the function of P450. To date, it has been shown that genetic variation caused by single nucleotide polymorphisms (SNPs) is important for the interindividual differences (see Chap. 17, 20 and 21 in this book). Furthermore, accumulating evidence suggests the importance of interindividual variations in microRNAs in the posttranscriptional regulation of P450s (Yokoi and Nakajima 2013; also see

Y. Ishii (✉) • Y. Miyauchi • H. Yamada
Graduate School of Pharmaceutical Sciences, Kyushu University, Fukuoka, Japan
e-mail: ishii@phar.kyushu-u.ac.jp

H. Yamazaki (ed.), *Fifty Years of Cytochrome P450 Research*,
DOI 10.1007/978-4-431-54992-5_18, © Springer Japan 2014

Chap. 19). In addition, transcriptional regulation of P450 genes is an important mechanism determining the level of P450 expression (Timsit and Negishi 2007; Gonzalez 2008; see also Chap. 14). However, although the role of modifications such as phosphorylation has been discussed (Oesch-Bartlomowicz and Oesch 2008), the posttranslational regulation of P450 has not been fully investigated.

It is widely known that microsomal P450 needs an electron donor protein, NADPH-P450 reductase (POR) (Horecker 1950), and cytochrome b_5 (Strittmatter and Ball 1951) also has a role in P450-catalyzed reactions (Guengerich 1999). Accordingly, it is possible that differences in the expression level or function of the aforementioned redox partners is linked to interindividual differences in drug metabolism. However, little attention has been paid to the modulator protein(s) of P450 other than POR and cytochrome b_5. The term metabolosome, which first appeared in the literature in 1963 (Roizin 1963), is now attracting attention again with regard to the better understanding of complicated enzyme complexes (Parsons et al. 2010). Generally speaking, a metabolosome represents a complex that consists of multiple proteins working cooperatively. Because the endoplasmic reticulum (ER) where P450 is located also expresses other membrane-bound drug-metabolizing enzymes, such as microsomal epoxide hydrolase (mEH) (Fretland and Omiecinski 2000) and uridine diphosphate (UDP)-glucuronosyltransferase (UGT) (Dutton 1980), we proposed a hypothesis that these proteins associate with each other and modulate their functions. Indeed, we applied the concept of metabolosome to the microsomal drug-metabolizing enzymes and examined whether a complex exists. During the past decade, we have focused on the protein–protein interaction between P450 and UGT and a change in their functions produced by this mechanism. Cumulative evidence suggests that this hypothesis is true, and CYP3A4 modulates the catalytic function of multiple UGT isoforms. In this chapter, we consider the effect of CYP3A4 on UGT function, mainly based on our own research findings. The reverse effect, that is, UGT-produced modulation of CYP3A4 function, is also discussed.

18.1.1 Structure and Function of UGT

UGT catalyzes glucuronidation, which accounts for approximately 35 % of phase II drug metabolism (Evans and Relling 1999). This enzyme transfers the glucuronic acid moiety of a cofactor, UDP-glucuronic acid (UDPGA), to the functional group of a substrate to form β-glucuronide (Dutton 1980; Radominska-Pandya et al. 1999). There are a huge number of UGT substrates. In many cases of drug metabolism, oxidation and conjugation take place sequentially to eliminate xenobiotics as highly polarized metabolites. Indeed, many drug metabolites formed by P450 are metabolized by UGT. In mammals, UGT is categorized as a member of the UGT1∼3 families (Mackenzie et al. 1997, 2005, 2008). Of these UGTs, the isoforms belonging to the UGT1A and -2B subfamilies are mainly involved in drug metabolism. There are at least 19 UGT isoforms in humans (Mackenzie et al. 1997, 2005). The main body of the UGT, including its catalytic site, is located at the luminal side of the ER, and UGT is a type I membrane-bound enzyme (Shepherd et al. 1989; Vanstapel and Blanckaert 1988; Meech and Mackenzie 1998). The obligate cofactor of UGT, UDPGA, is biosynthesized in the cytosol and delivered to the lumen of the ER by UDPGA transporter(s) (Hauser et al. 1988). A membrane-bound transporter capable of transporting UDP-*N*-acetylglucosamine and UDPGA has been identified (Muraoka et al. 2001, 2007). The domain-spanning Trp 356 and Gln 399 of UGT2B7 are thought to serve as a binding site with sugar nucleotides, based on an X-ray crystallographic analysis with 1.8 Å resolution (Miley et al. 2007). Furthermore, from site-directed mutagenesis studies, His 35 and Asp 151 or Asp150 are suggested to have a role in catalysis involving UGT2B7 and -1A6, respectively (Miley et al. 2007; Li et al. 2007). The sites where phosphorylation and disulfide bonding occur have also been identified by similar approaches (Basu et al. 2004b; Ikushiro et al. 2002). However, the complete structure of UGT and the fine mechanism by which it recognizes its substrate and cofactor remain largely unknown.

18.1.2 Posttranslational Modulation of UGT Function

The function of UGT is changed by phosphorylation and disulfide bonding (Basu et al. 2004b; Ikushiro et al. 2002). Furthermore, UGTs form oligomers to alter their substrate specificity (Ishii et al. 2001; Kurkela et al. 2004; see also review Ishii et al. 2010b; Finel and Kurkela 2008). In UGT isoforms belonging to the 1A subfamily, their mRNAs are formed by alternative splicing from variable exon 1 and common exons 2 to 5 (Iyanagi 1991; Ritter et al. 1992). This mechanism produces multiple UGT1A isoforms having unique N-terminals and a common carboxy terminal. In recent years, splice variants formed by a mechanism different from this one have been identified and characterized (Lévesque et al. 2007; see also review by Guillemette et al. 2010). In this case, new exons 4 and 5 are used to produce unique UGT isoforms having new sequences: that is, their sequences coded in exons 4 and 5 differ from the isoforms hitherto known, whereas the sequences encoded by exons 1, 2, 3, and part of exon 4 are identical. The new splice variant i2 derived from the foregoing mechanism still retains the binding site of UDPGA, but it is functionally inactive. However, it can form hetero-oligomers with wild-type enzyme i1 to abolish or reduce the activity, which is further evidence supporting the concept that UGT function is markedly affected by oligomerization. The physiological significance of i2 remains unknown, but it may play a role in the organ-specific expression of UGT function through silencing UGT activity. The inhibitory factor of UGT often causes a drug–drug interaction (Kiang et al. 2005; references therein). For instance, azole antibiotics (Yong et al. 2005; Takeda et al. 2006; Uchaipichat et al. 2004) and benzodiazepines (Thomassin and Tephly 1990), which

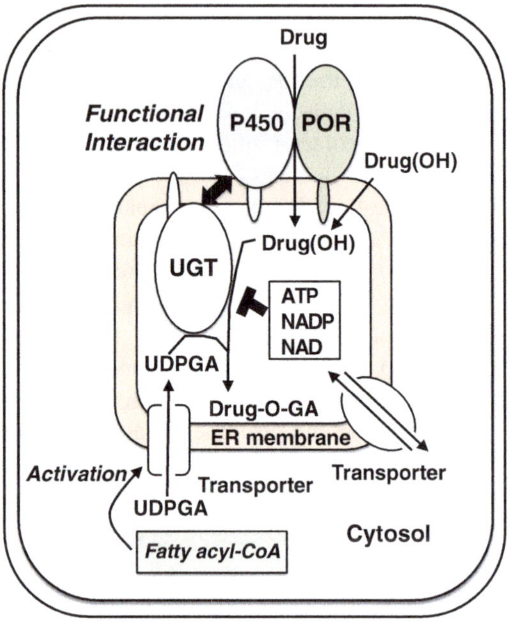

Fig. 18.1 Schematic representation of the functional interaction between P450 and UGT. *POR* NADPH cytochrome P450 oxidoreductase, *UDPGA* UDP-glucuronic acid. The picture is drawn on the basis of the following references: Muraoka et al. (2001), Takeda et al. (2005a, b), Muraoka et al. (2007), Nurrochmad et al. (2010), Ishii et al. (2012a)

are not UGT substrates, serve as UGT inhibitors. A number of foods and their ingredients are also known to exhibit inhibitory effects on UGTs (Wade 1986). However, whether endogenous substances modulate UGT function has not been fully characterized. The limited information reported so far involves UDP-*N*-acetylglucosamine (Pogell and Leloir 1961; Vanstapel and Blanckaert 1988) and UDP-xylose (Bossuyt and Blanckaert 1996), which are activators. Our previous studies have suggested that although long-chain fatty acyl-CoAs are also activators (Okamura et al. 2006; Nurrochmad et al. 2010; Ishii et al. 2010a), adenine nucleotides such as ATP have the ability to inhibit UGT (Nishimura et al. 2007; Ishii et al. 2012a) (Fig. 18.1).

Regarding the interindividual differences in UGT function, genetic polymorphism is undoubtedly one of the mechanisms. For example, interindividual variation in the tolerability of irinotecan, an anti-cancer drug, seems to be associated with a genetic polymorphism about UGT1A1 that detoxifies SN-38, an active metabolite derived from irinotecan (Iyer et al. 1998). However, large differences among individuals in terms of sensitivity to irinotecan seem not to be fully explained only by UGT1A1 polymorphism. Therefore, although we should keep the importance of polymorphism in mind, there must be other mechanisms involved in interindividual differences. The candidates would target transcriptional, post-transcriptional, and posttranslational regulation. For transcriptional regulation, certain transcription factors such as hepatic nuclear factor 1α (HNF1α) and CCAAT/enhancer-binding protein α (C/EBPα) are involved in the constitutive expression of UGTs (Hansen et al. 1998; Ishii et al. 2000; Gardner-Stephen and

Mackenzie 2008). Furthermore, inducible expression of UGTs is evoked by transcription factors such as aryl hydrocarbon receptors, the nuclear factor erythroid 2-related factor 2 (Nrf2), pregnan-X-receptor (PXR), the constitutive androstan receptor (CAR), and the peroxisome proliferator activated receptor-α (PPARα) (see reviews of Gardner-Stephen and Mackenzie 2008; Sugatani 2013). As for the posttranslational regulation of UGT, our laboratory has produced evidence that a number of dietary and endogenous compounds have the ability to modulate UGT catalysis (Ishii et al. 2010a, 2012a, b). In this review, we focus on the differences in UGT activity produced by protein–protein interaction with P450.

18.2 Protein–Protein Interaction Between P450 and UGT

Cumulative evidence suggests that UGT works as the homo- or hetero-oligomer (Gschaidmeier and Bock 1994; Koiwai et al. 1996; Ikushiro et al. 1997; Meech and Mackenzie 1997a; Ghosh et al. 2001; Ishii et al. 2001; Kurkela et al. 2004; Operaña and Tukey 2007; Nakajima et al. 2007; Fujiwara et al. 2010). Several studies have demonstrated that P450 isoforms also interact with each other to modulate their functions (Alston et al. 1991; Ikushiro et al. 1992; Yamazaki et al. 1997; Reed et al. 2010; Subramanian et al. 2009). We expanded these pieces of information to develop a new concept that UGT may interact with different drug-metabolizing enzymes such as P450 to alter its function. However, this hypothesis may sound very unlikely, as the membrane topology of these enzymes is very different, although the main body of P450 faces the cytosol, that of UGT is considered to be located in the luminal space of the ER (Fig. 18.1) (Edwards et al. 1991; Cosme and Johnson 2000; Williams et al. 2000; Meech and Mackenzie 1997b). Therefore, these enzymes have been thought to work separately, and there has been little interest in the interaction between P450 and UGT. However, a pilot study conducted in our laboratory predicted the presence of a P450–UGT interaction based on the following evidence: that is, UGTs as well as NADPH-P450 reductase (POR) in liver microsomes from phenobarbital-treated rats are associated with a CYP1A1-immobilized affinity column (Taura et al. 2000). Because calnexin and protein-disulfide isomerase, chaperones expressed in the ER, showed no or little affinity for the column, the foregoing chromatographic technique was thought to distinguish between the proteins that interact with CYP1A1 and other non-specific proteins. Microsomal epoxide hydrolase (mEH) was also trapped by the column. These lines of evidence suggest that UGT and mEH interact with CYP1A1 (Taura et al. 2000). It has been well established that CYP1A1 is barely expressed in the hepatic ER of phenobarbital-treated rats. Therefore, some UGT isoforms present in phenobarbital-treated rats may associate with CYP1A1 more favorably than the other P450s, including CYP2B1/2. However, to date, P450 isoform selectivity in the coupling with UGT remains largely unknown, although it has been examined to a small extent, as described later. Although our group showed that the P450–mEH interaction alters mEH function (Taura et al. 2002), we do not discuss this topic further in this chapter.

18.2.1 Protein–Protein Interactions Between Human CYP3A4 and UGTs

Our study demonstrated a CYP3A4-UGT2B7 interaction by an overlay assay using gluthathione-*S*-transferase (GST)-tagged CYP3A4 (Takeda et al. 2005a). The association of CYP3A4 and UGT2B7 has also been detected by co-immunoprecipitation using solubilized human liver microsomes and an anti-UGT antibody. Another group also showed that human UGT1A1, -1A6, and -2B7 were co-precipitated from solubilized human liver microsomes with CYP3A4 by anti-CYP3A4 antibody (Fremont et al. 2005). As human UGT1A7 was overlaid with GST-tagged CYP3A4 (Ishii et al. 2014), CYP3A4 must interact with multiple isoforms of UGT. However, as already mentioned, much more effort is needed to obtain a better understanding of the selectivity of UGT isoforms in the interaction with CYP3A4 and other P450s. The CYP3A4 region involved in the interaction with UGT2B7 has been investigated (Takeda et al. 2009). In this approach, we obtained an anti-CYP3A4 antibody incapable of reacting with the CYP3A4–UGT complex but capable of interfering with the complex formation. Therefore, the epitope of the antibody was suggested to overlap the CYP3A4 domain that is involved in the interaction with UGT2B7. This epitope was screened using a peptide library, and the result suggested that the epitope is located within the J-helix region (Takeda et al. 2009).

18.2.2 Protein–Protein Interactions of P450 and UGT in Rat Liver Microsomes

When rat liver microsomes are solubilized and then co-immunoprecipitated either with anti-CYP1A2, -2B2, -2C13, or -3A2 antibody, several UGT isoforms are precipitated (Ishii et al. 2007). Thus, the P450–UGT interaction appears to occur ubiquitously in an isoform-independent fashion. The affinity of their interaction varies among P450s, and UGTs seem to have a rather weaker affinity for CYP2C than that for CYP1A, -2B, and -3As. UGT1A1, -1A6, and -2B1, found in the CYP2C–UGT complex, can be easily dissociated by extensive washing of the immuno-precipitates. Thus, many isoforms of UGT would have an ability to interact with P450s even although the affinity varies. The UGTs in the CYP–UGT complex are catalytically active (Ishii et al. 2007), and then P450 seems not to be able to quench UGT function but is a modulator of UGT activity.

18.3 Modulation of UGT2B7 Function by CYP3A4, -2C9, and -1A2

It is of great interest whether P450 affects the UGT function through a protein–protein interaction. To address this issue, we coexpressed UGT2B7 and CYP3A4 in COS-1 cells (Takeda et al. 2005a). To minimize the variation in the results, we first established COS-1 cells stably expressing UGT2B7. Then, CYP3A4 was transiently coexpressed in the UGT2B7-expressing cells. The morphine glucuronidation activity was compared in the presence or absence of co-transfection of CYP3A4. In the kinetics of morphine-3-glucuronidation, CYP3A4 coexpression causes a marked increase (nearly tenfold) in K_m, whereas the V_{max} is little affected. When the purified preparation of recombinant CYP3A4 was added to UGT2B7-expressing microsomes, the formation of morphine-3-glucuronide (M-3-G) is suppressed in a CYP3A4 amount-dependent fashion. However, the formation of morphine-6-glucuronide (M-6-G), an active metabolite of morphine (Shimomura et al. 1971), was increased by CYP3A4. Then, the regioselectivity of UGT2B7-catalyzed morphine glucuronidation was altered by co-transfection with CYP3A4 (Takeda et al. 2005a). Furthermore, simultaneous expression of other major human P450s, CYP2C9 and -1A2, was carried out, and their effects on UGT2B7 catalysis were analyzed in a similar way to that just described. As a consequence, CYP2C9 and -1A2 hardly affect K_m, although they reduce the V_{max} to some extent (one half at most) (Takeda et al. 2005a). The suppression of M-3-G formation is reproduced when purified recombinant P450s are added to UGT2B7-expressing microsomes (Takeda et al. 2005b). Therefore, CYP3A4, -1A2, and -2C9 all have the ability to suppress M-3-G formation by UGT2B7. However, as the greatest change in UGT2B7 function is produced by CYP3A4 and enhancement of M-6-G formation is produced only by this P450, CYP3A4 may be the P450 isoform making the greater contribution to the modification of UGT2B7 function.

18.4 Modulation of the Function of UGT1A1 and Its Allelic Variants by CYP3A4

UGT1A1 is the most important isoform responsible for the glucuronidation of bilirubin (Ritter et al. 1992, 1993). In humans, a UGT1A1 deficiency causes severe hyperbilirubinemia, so-called Crigler–Najjar syndrome type I (Crigler and Najjar 1952; Ritter et al. 1993; Erps et al. 1994). Other polymorphisms of the *UGT1A1* gene are also known to cause hyperbilirubinemia, the malignancy of which is weaker than that of Crigler–Najjar syndrome. Among about 110 allelic variants reported so far, several are thought to increase the serum level of bilirubin. The most important alleles, which are related to moderate hyperbilirubinemia referred to as Gilbert syndrome (Gilbert and Lereboullet 1901), seem to be *UGT1A1*6* and **28*.

UGT1A1*6 is also an important factor causing hyperbilirubinemia in newborns in Japanese subjects (Jada et al. 2007; Minami et al. 2007; Han et al. 2006). UGT1A1*6 (G71R) exhibits impaired glucuronidation activity, and its allelic frequency in the Japanese population is about 15 % (Aono et al. 1993; Kurose et al. 2012). The allele of *UGT1A1*28* encodes for a wild-type protein, but this cannot be highly expressed because of the polymorphism in the number of TA repeats in the promoter region (Bosma et al. 1995). The *UGT1A1*28* allele is found to occur with a frequency of 12 % in the Japanese population (Kurose et al. 2012). It should be noted that the frequencies of *UGT1A1*6* and **28* are different from that of Gilbert syndrome (5 %) (Iyanagi 2007). The polymorphism of *UGT1A1* also affects drug efficacy and toxicity. For example, this UGT has the ability to detoxify SN-38, which is the active metabolite of irinotecan, an anti-cancer drug (Iyer et al. 1998). The genotyping of *UGT1A1* alleles (wild, *6, and *28), especially detecting *28, is expected to allow us to identify a safety regimen for irinotecan therapy (Minami et al. 2007; Kurose et al. 2012). However, some patients have a capacity to glucuronidate SN-38 comparable with normal subjects, although they carry the *UGT1A1*28* allele (Sai et al. 2004).

As expected, when glucuronidation activity is compared between wild-type UGT1A1*1 and its allelic variant UGT1A1*6 expressed in Sf9 cells, the V_{max} of UGT1A1*6 is markedly lower than that of the wild type. Coexpression of CYP3A4 enhances the UGT1A1-catalyzed glucuronidation of 4-methylumbelliferone (4-MU), SN-38, and the 3-hydroxy group of β-estradiol. An increase in V_{max} mainly contributes to this enhancement, and CYP3A4 coexpression has little effect on the K_m. As already described, the catalytic function of UGT1A1*6 is lesser than that of the wild-type enzyme. However, simultaneous expression of CYP3A4 renders UGT1A1*6 function comparable with that of UGT1A1*1 (Ishii et al. 2014). As CYP3A4 enhances the functions of both UGT1A1*1 and *6, the 71st amino acid residue of UGT1A1 seems not to be involved in the interaction with CYP3A4. From the observation that the enhancement of UGT1A1 function was achieved without any change in K_m, the UGT1A1 cavity for the substrate would retain its normal configuration even after binding to CYP3A4. It is known that there are marked interindividual differences, up to 40 fold, in the expression level of CYP3A4 (Shimada et al. 1994). It would be reasonable to suppose that a functional interaction between CYP3A4 and UGT1A1 is one of the reasons why some patients exhibit normal UGT1A1 function even although they carry the defective genotype.

18.5 Modulation of UGT1A6 Activity by CYP3A4

UGT1A6 is one of the most important isoforms catalyzing phenol glucuronidation. Serotonin, that is, 5-hydroxytryptamin, is a representative substrate of this UGT (Basu et al. 2004a; Hanioka et al. 2006). The kinetics of serotonin glucuronidation mediated by UGT1A6-expressing microsomes fit not only a Michaelis–Menten equation but also a substrate inhibition mode (Ishii et al. 2014). However, when CYP3A4 is coexpressed, both K_m and V_{max} are significantly increased. Furthermore, the kinetics fit only a Michalelis–Menten equation. Therefore, coexpression of CYP3A4 eliminates the substrate inhibition that occurs in the serotonin glucuronidation catalyzed by UGT1A6 at higher substrate concentrations (Ishii et al. 2014)

18.6 Modulation of the Function of UGT1A7 and Its Allelic Variants by CYP3A4

Human UGT1A7 is expressed in regions of the gastrointestinal tract, such as stomach and small intestine, and this isoform is involved in the glucuronidation of drugs (e.g., mycophenolic acid and SN-38), hormones (e.g., thyroxin), and carcinogens (e.g., benzo[a]pyrene metabolites) (Joy et al. 2010; Rosner et al. 2008; Emi et al. 2007; Basu et al. 2004a). Genetic polymorphisms are also known for UGT1A7. Of the allelic variants reported thus far, the frequency of UGT1A7*3 is estimated to be 35 % and 26 % in Caucasian and Japanese populations, respectively (Guillemette 2003; Huang et al. 2005). As UGT1A7*3 has been reported not to be involved in the glucuronidation of phenolic compounds (Strassburg et al. 2002), it has been suspected that this genotype may be a risk factor for cancer (Vogel et al. 2001; Strassburg et al. 2002; Ockenga et al. 2003). However, different research groups have claimed that UGT1A7*3 is capable of glucuronidating a phenolic substrate, SN-38, although its activity is lower than that of the wild type (Gagné et al. 2002; Villeneuve et al. 2003). In accordance with this, there are reports claiming that there is no significant relationship between UGT1A7*3 and cancer risk (Verlaan et al. 2005; te Morsche et al. 2008). Thus, the relationship between a change in the catalytic property of UGT1A7 and its relevance to a drug/carcinogen effect remains controversial.

Although the functions of UGT1A7*1, -*2, and -*3 are virtually identical at high concentrations of substrate and co-substrate, their catalytic properties differ at lower substrate concentrations (Ishii et al. 2014). The coexpression of CYP3A4 has a different effect on the kinetic properties of UGT1A7 allozymes. Namely, although the V_{max} and intrinsic clearance for 4-MU glucuronidation in UGT1A7*1 and *2 are enhanced by CYP3A4, those of *3 are reduced (Ishii et al. 2014). CYP3A4 has a similar effect on the glucuronidation of SN-38 and 4-hydroxybiphenyl. In a kinetic study involving varying the cofactor concentration (substrate, 4-MU), UGT1A7*2 and *3 exhibit greater K_m values than *1. Thus, it would be likely that N129K and R131K, which are common mutations in UGT1A7*2 and *3, affect UDPGA recognition. The intrinsic clearance of UGT1A7*1 remains apparently unchanged even after coexpression with CYP3A4

(Ishii et al. 2014), which appears to be the outcome of offsetting an increase in V_{max} by an increase in K_m. In contrast, double expression of CYP3A4 increases the intrinsic clearance of *2 and attenuates that of *3. Therefore, it would be reasonable to consider that the three nonsynonymous mutations present in UGT1A7*3, i.e., N129K, R131K, and W208R, affect not only UDPGA recognition but also a CYP3A4-dependent change in UGT function.

To the best of our knowledge, the exact concentration of UDPGA in the luminal space of the ER has not yet been determined. Conceivably, the UDPGA concentration within the ER may be maintained under in vivo conditions so that it meets the minimal requirement for UDPGA-utilizing proteins, such as UGT, present in the ER lumen. If this is true, lowering the affinity for UDPGA, such as in the case of UGT1A7*3, would result in reduced glucuronidation in an in vivo environment. On the other hand, the UDPGA concentration on the luminal side seems to be dynamically changed along with the different transportation of UDPGA from the cytosol to the ER. Long-chain fatty acyl CoAs enhance glucuronidation activity by increasing UDPGA uptake (Nurrochmad et al. 2010). From this evidence, the UDPGA concentration around UGTs does not always seem to be enough to allow them to exhibit their full function. In addition, the foregoing observation suggests that endogenous and exogenous chemicals capable of changing the UDPGA concentration in the ER lumen affect glucuronidation efficiency. It might be that many epidemiological studies on the relationship between UGT1A7 polymorphism and cancer risk were carried out under the influence of P450-UGT interactions and the factors modulating UDPGA concentration, which may be one of the reasons why there is the confusion or controversy described earlier.

18.7 Modulation of CYP3A4 Activity by UGT

To date, there has been little published information about whether UGT modulates P450 activity. Our preliminary study has suggested that UGT2B7 and -1A9 modify CYP3A4 function through a protein–protein interaction (Miyauchi et al. 2012). CYP3A4 is thought to contribute to the metabolism of about 50 % of drugs (Evans and Relling 1999). Although there are large interindividual differences (~40 fold) in the expression level of hepatic CYP3A4 (Shimada et al. 1994), differences in the in vivo clearance of drugs metabolized by this P450 are much less (8- to 10 fold) than the above difference (Lamba et al. 2002). Because of this discrepancy, it would appear that variations in CYP3A4 function involve a mechanism(s) distinct from its altered expression. This issue may also be explored by protein–protein interactions between CYP3A4 and UGTs. That is, an insect cell (Sf9) system coexpressing CYP3A4, NADPH-P450 reductase, and UGT isoform 2B7 or 1A9 exhibits a lowered CYP3A4 activity compared with Sf9 cells expressing only CYP3A4 and reductase (Miyauchi et al. 2012). This reduction of activity disappears on replacing the UGT of the ternary expression system by calnexin, a non-drug-metabolizing protein having the same topology as UGT. Thus, the CYP3A4–UGT interaction modifies the function of CYP3A4 as well as UGT. UGT2B7 and -1A9 have different effects on CYP3A4 kinetics. For example, although UGT2B7 reduces the V_{max}, UGT1A9 increases S_{50} without affecting the V_{max}. Hence, UGT seems to suppress CYP3A4 catalysis in an isoform-specific manner. Regarding the UGT domain that interacts with CYP3A4, truncation of the cytosolic tail of UGT2B7 reduces the suppressive effect on CYP3A4. Therefore, the cytosolic tail of UGT2B7 is involved in the interaction with CYP3A4. A protein–protein interaction between UGT and CYP3A4 may be one of the reasons there is a discrepancy between CYP3A4 levels and CYP3A4-mediated drug clearance. Further studies are necessary to identify the mechanism by which UGT2B7 and -1A9 modulate CYP3A4 activity.

18.8 Future Perspectives

Accumulating evidence from in vitro experiments (Table 18.1) has suggested a protein–protein interaction between P450 and UGT. Our current data suggest that the interaction also takes place in living cells, although its details are omitted in this review. However, it remains largely unknown how important this interaction is in in vivo situations. The significance of P450–UGT interactions in drug metabolism cannot be precisely evaluated until its role is clarified under in vivo conditions.

Table 18.1 Summary of the interactions between P450 and UGT described in the literature

UGT isoform(s)	P450 isoform(s)	Mode of interaction	Method used	References
UGT1A1/1A6/2B7	CYP3A4	UGT1A1. 1A6 and 2B7 in solubilized HLM were co-immunoprecipitated with CYP3A4 using anti-CYP3A4 antibody	Co-immunoprecipitation	Fremont et al. (2005)
UGT2B7	CYP3A4	UGT2B7 was overlayed with GST-tagged CYP3A4	Overlay assay	Takeda et al. (2005a)
UGT2B7	CYP3A4	UGT2B7 in solubilized HLM was co-immunoprecipitated with CYP3A4 using anti-UGT antibody	Co-immunoprecipitation	Takeda et al. (2005a)
UGT2B7	hCYP1A2/CYP2C9	Decrease in the V_{max} of UGT2B7-catalyzed M-3-G formation	Co-expression of UGT2B7 and CYP1A2/CYP2C9 in COS cells: kinetic analysis	Takeda et al. (2005a)
UGT2B7	hCYP1A2/CYP2C9	Decrease in the UGT2B7-catalyzed M-3-G formation	Addition of recombinant CYP1A2/CYP2C9 on UGT2B7-expressing microsomes	Takeda et al. (2005b)
UGT2B7	CYP3A4	Increase in the K_m of UGT2B7-catalyzed M-3-G formation	Co-expression of UGT2B7 and CYP3A4 in COS cells: kinetic analysis	Takeda et al. (2005a)
UGT2B7	CYP3A4	Decrease in the UGT2B7-catalyzed M-3-G formation but increase in the M-6-G formation	Addition of recombinant CYP3A4 on UGT2B7-expressing microsomes	Takeda et al. (2005a; b)

rUGTs	CYP1A2/2B2/3A2/2C13	CYP-UGT complex is catalytically active	Co-immunoprecipitation and measured the glucuronidation activity of the precipitate	Ishii et al. (2007)
rUGT1A1/1A6	CYP2C11/2C13	Interactions of the UGTs with CYP2C11/13 are rather weaker than that of CYP1A2/2B2/3A2	Co-immunoprecipitation and extensive washing	Ishii et al. (2007)
hUGT1A1*1	CYP3A4	Enhanced glucuronidation activity toward 4-MU, SN-38, and the 3-hydroxy group of β-estradiol	Co-transfection in Sf9 cells using recombinant baculovirus	Ishii et al. (2014)
hUGT1A1*6	CYP3A4	Restoration of glucuronidation activity to a level comparable with that of the wild-type UGT1A1*1	As above	Ishii et al. (2014)
hUGT1A6	CYP3A4	Kinetics shifted from substrate inhibition to Michaelis–Menten	As above	Ishii et al. (2014)
hUGT1A7*1/1A7*2	CYP3A4	Enhanced glucuronidation activity toward 4-MU, SN-38, and 4-hydroxybiphenyl	As above	Ishii et al. (2014)
hUGT1A7*3	CYP3A4	Reduced glucuronidation of 4-MU, SN-38, and 4-hydroxybiphenyl	As above	Ishii et al. (2014)
hUGT1A7*3	CYP3A4	Reduced affinity for UDPGA	As above	Ishii et al. (2014)

HLM human liver microsomes, *GST* glutathione-S-transferase, *4-MU* 4-methylumbelliferone, *h* human, *r* rat

Acknowledgments The studies carried out in our laboratory were supported in part by Grants in Aids from the Ministry of Technology, Education, Science, and Sports or the Japanese Society for the Promotion of Science. The authors thank collaborators Dr. Mackenzie in Flinders University, Dr. Nagata in Tohoku Pharmaceutical University, Dr. Yamazoe in the Graduate School of Pharmaceutical Sciences, Tohoku University (present address: Cabinet Office, Government of Japan), Dr. Ikushiro in Toyama Prefectural University, Dr. Taketomi in Hokkaido University School of Medicine, and Dr. Maehara in the Graduate School of Medical Science, Kyushu University. The authors are also very grateful to their former mentor and professor emeritus, the late Dr. Oguri. They also thank all the staff and graduate and undergraduate students in the Laboratory of Molecular Life Sciences, Kyushu University who were involved in the study.

References

Alston K, Robinson RC, Park SS, Gelboin HV, Friedman FK (1991) Interactions among cytochromes P-450 in the endoplasmic reticulum. Detection of chemically cross-linked complexes with monoclonal antibodies. J Biol Chem 266:735–739

Aono S, Yamada Y, Keino H, Hanada N, Nakagawa T, Sasaoka Y, Yazawa T, Sato H, Koiwai O (1993) Identification of defect in the genes for bilirubin UDP-glucuronosyl-transferase in a patient with Crigler-Najjar syndrome type II. Biochem Biophys Res Commun 197:1239–1244

Basu NK, Ciotti M, Hwang MS, Kole L, Mitra PS, Cho JW, Owens IS (2004a) Differential and special properties of the major human UGT1-encoded gastrointestinal UDP-glucuronosyltransferases enhance potential to control chemical uptake. J Biol Chem 279:1429–1441

Basu NK, Kubota S, Meselhy MR, Ciotti M, Chowdhury B, Hartori M, Owens IS (2004b) Gastrointestinally distributed UDP-glucuronosyltransferase 1A10, which metabolizes estrogens and nonsteroidal anti-inflammatory drugs, depends upon phosphorylation. J Biol Chem 279:28320–28329

Bosma PJ, Roy Chowdhury J, Bakker C, Gantla S, de Boer A, Oostra BA, Lindhout D, Tytgat GN, Jansen PL, Oude Elferink RP, Roy Chowdhury N (1995) The genetic basis of the reduced expression of bilirubin UDP-glucuronosyltransferase 1 in Gilbert's syndrome. N Engl J Med 333:1171–1175

Bossuyt X, Blanckaert N (1996) Uridine diphosphoxylose enhances hepatic microsomal UDP-glucuronosyltransferase activity by stimulating transport of UDP-glucuronic acid across the endoplasmic reticulum membrane. Biochem J 315:189–193

Cosme J, Johnson EF (2000) Engineering microsomal cytochrome P450 2C5 to be a soluble, monomeric enzyme. Mutations that alter aggregation, phospholipid dependence of catalysis, and membrane binding. J Biol Chem 275:2545–2553

Crigler JF, Najjar VA (1952) Congenital familial non-haemolytic jaundice with kernicterus. Paediatrics 10:169–180

Dutton GJ (1980) Glucuronidation of drugs and other compounds. CRC, Boca Raton

Edwards RJ, Murray BP, Singleton AM, Boobis AR (1991) Orientation of cytochrome P450 in the endoplasmic reticulum. Biochemistry 30:71–76

Emi Y, Ikushiro S, Kato Y (2007) Thyroxine-metabolizing rat uridine diphosphate-glucuronosyltransferase 1A7 is regulated by thyroid hormone receptor. Endocrinology 148:6124–6133

Erps LT, Ritter JK, Hersh JH, Blossom D, Martin NC, Owens IS (1994) Identification of two single base substitutions in the UGT1 gene locus which abolish bilirubin uridine diphosphate glucuronosyltransferase activity in vitro. J Clin Invest 93:564–570

Evans WE, Relling MV (1999) Pharmacogenomics: translating functional genomics into rational therapeutics. Science 286:487–491

Finel M, Kurkela M (2008) The UDP-glucuronosyltransferases as oligomeric enzymes. Curr Drug Metab 9:70–76

Fremont JJ, Wang RW, King CD (2005) Co-immunoprecipitation of UDP-glucuronosyltransferase (UGT) isoforms and cytochrome P450 3A4. Mol Pharmacol 67:260–262

Fretland AJ, Omiecinski CJ (2000) Epoxide hydrolases: biochemistry and molecular biology. Chem Biol Interact 129:41–59

Fujiwara R, Nakajima M, Oda S, Yamanaka H, Ikushiro SI, Sakaki T, Yokoi T (2010) Interactions between human UDP-glucuronosyltransferase (UGT) 2B7 and UGT1A enzymes. J Pharm Sci 99:442–454

Gagné JF, Montminy V, Belanger P, Journault K, Gaucher G, Guillemette C (2002) Common human UGT1A polymorphisms and the altered metabolism of irinotecan active metabolite 7-ethyl-10-hydroxycamptothecin (SN-38). Mol Pharmacol 62:608–617

Gardner-Stephen DA, Mackenzie PI (2008) Liver-enriched transcription factors and their role in regulating UDP glucuronosyltransferase gene expression. Curr Drug Metab 9:439–452

Ghosh SS, Sappal BS, Kalpana GV, Lee SW, Roy Chowdhury J, Roy Chowdhury N (2001) Homodimerization of human bilirubin-uridine-diphosphoglucuronate glucuronosyltransferase-1 (UGT1A1) and its functional implications. J Biol Chem 276:42108–42115

Gilbert A, Lereboullet P (1901) La cholémie simple familiale. Sem Med 21:241–245

Gonzalez FJ (2008) Regulation of hepatocyte nuclear factor 4 alpha-mediated transcription. Drug Metab Pharmacokinet 23:2–7

Gschaidmeier H, Bock KW (1994) Radiation inactivation analysis of microsomal UDP-glucuronosyltransferases catalysing mono- and diglucuronide formation of 3,6-dihydroxybenzo(a)pyrene and 3,6-dihydroxychrysene. Biochem Pharmacol 48:1545–1549

Guengerich FP (1999) Cytochrome P-450 3A4: regulation and role in drug metabolism. Annu Rev Pharmacol Toxicol 39:1–17

Guillemette C (2003) Pharmacogenomics of human UDP-glucuronosyltransferase enzymes. Pharmacogenomics J 3:136–158

Guillemette C, Lévesque E, Harvey M, Bellemare J, Ménard V (2010) UDP-glucuronosyltransferase (UGT) enzyme diversity: beyond gene duplication. Drug Metab Rev 42:24–44

Han JY, Lim HS, Shin ES, Yoo YK, Park YH, Lee JE, Jang IJ, Lee DH, Lee JS (2006) Comprehensive analysis of UGT1A polymorphisms predictive for pharmacokinetics and treatment outcome in patients with non-small-cell lung cancer treated with irinotecan and cisplatin. J Clin Oncol 24:2237–2244

Hanioka N, Takeda Y, Jinno H, Tanaka-Kagawa T, Naito S, Koeda A, Shimizu T, Nomura M, Narimatsu S (2006) Functional characterization of human and cynomolgus monkey UDP-glucuronosyltransferase 1A6 enzymes. Chem Biol Interact 164:136–145

Hansen AJ, Lee YH, Sterneck E, Gonzalez FJ, Mackenzie PI (1998) C/EBPalpha is a regulator of the UDP glucuronosyltransferase UGT2B1 gene. Mol Pharmacol 53:1027–1033

Hauser SC, Ziurys JC, Gollan JL (1988) A membrane transporter mediates access of uridine 5′-diphosphoglucuronic acid from the cytosol into the endoplasmic reticulum of rat hepatocytes: implications for glucuronidation reactions. Biochim Biophys Acta 967:149–157

Horecker BL (1950) Triphosphopyridine nucleotide-cytochrome *c* reductase in liver. J Biol Chem 183:593–605

Huang MJ, Yang SS, Lin MS, Huang CS (2005) Polymorphisms of uridine-diphospho-glucuronosyltransferase 1A7 gene in Taiwan Chinese. World J Gastroenterol 11:797–802

Ikushiro S, Kominami S, Takemori S (1992) Adrenal P-450scc modulates activity of P-45011 beta in liposomal and mitochondrial membranes. Implication of P-450scc in zone specificity of aldosterone biosynthesis in bovine adrenal. J Biol Chem 267:1464–1469

Ikushiro S, Emi Y, Iyanagi T (1997) Protein–protein interactions between UDP-glucuronosyltransferase isozymes in rat hepatic microsomes. Biochemistry 36:7154–7161
Ikushiro S, Emi Y, Iyanagi T (2002) Activation of glucuronidation through reduction of a disulfide bond in rat UDP-glucuronosyltransferase 1A6. Biochemistry 41:12813–12820
Ishii Y, Hansen A, Mackenzie PI (2000) Octamer transcription factor-1 enhances hepatic nuclear factor-1-alpha-mediated activation of the human UDP glucuronosyltransferase 2B7 promoter. Mol Pharmacol 57:940–947
Ishii Y, Miyoshi A, Watanabe R, Tsuruda K, Tsuda M, Yamaguchi-Nagamatsu Y, Yoshisue K, Tanaka M, Maji D, Ohgiya S, Oguri K (2001) Simultaneous expression of guinea pig UDP-glucuronosyltransferase 2B21 and 2B22 in COS-7 cells enhances UDP-glucuronosyltransferase 2B21-catalyzed morphine-6-glucuronide formation. Mol Pharmacol 60:1040–1048
Ishii Y, Iwanaga M, Nishimura Y, Takeda S, Ikushiro S, Nagata K, Yamazoe Y, Mackenzie PI, Yamada H (2007) Protein–protein interactions between rat hepatic cytochromes P450 (P450s) and UDP-glucuronosyltransferases (UGTs): evidence for the functionally active UGT in P450-UGT complex. Drug Metab Pharmacokinet 22:367–376
Ishii Y, Nurrochmad A, Yamada H (2010a) Modulation of UDP-glucuronosyltransferase activity by endogenous compounds. Drug Metab Pharmacokinet 25:134–148
Ishii Y, Takeda S, Yamada H (2010b) Modulation of UDP-glucuronosyltransferase activity by protein-protein association. Drug Metab Rev 42:145–158
Ishii Y, An K, Nishimura Y, Yamada H (2012a) ATP serves as an endogenous inhibitor of UDP-glucuronosyltransferase (UGT): a new insight into the latency of UGT. Drug Metab Dispos 40:2081–2089
Ishii Y, Iida N, Miyauchi Y, Mackenzie PI, Yamada H (2012b) Inhibition of morphine glucuronidation in the liver microsomes of rats and humans by monoterpenoid alcohols. Biol Pharm Bull 35:1811–1817
Ishii Y, Koba H, Kinoshita K, Oizaki T, Iwamoto Y, Takeda S, Miyauchi Y, Nishimura Y, Egoshi N, Taura F, Morimoto S, Ikushiro S, Nagata K, Yamazoe Y, Mackenzie PI, Yamada H (2014) Alteration of the function of the UDP-Glucuronosyltransferase 1A subfamily by cytochrome P450 3A4: different susceptibility for UGT isoforms and UGT1A1/7 variants. Drug Metab Dispos 42:229–238
Iyanagi T (1991) Molecular basis of multiple UDP-glucuronosyltransferase isoenzyme deficiencies in the hyperbilirubinemic rat (Gunn rat). J Biol Chem 266:24048–24052
Iyanagi T (2007) Molecular mechanism of phase I and phase II drug-metabolizing enzymes: implications for detoxification. Int Rev Cytol 260:35–112
Iyer L, King CD, Whitington PF, Green MD, Roy SK, Tephly TR, Coffman BL, Ratain MJ (1998) Genetic predisposition to the metabolism of irinotecan (CPT-11). Role of uridine diphosphate glucuronosyltransferase isoform 1A1 in the glucuronidation of its active metabolite (SN-38) in human liver microsomes. J Clin Invest 101:847–854
Jada SR, Lim R, Wong CI, Shu X, Lee SC, Zhou Q, Goh BC, Chowbay B (2007) Role of UGT1A1*6, UGT1A1*28 and ABCG2 c.421C > A polymorphisms in irinotecan-induced neutropenia in Asian cancer patients. Cancer Sci 98:1461–1467
Joy MS, Boyette T, Hu Y, Wang J, La M, Hogan SL, Stewart PW, Falk RJ, Dooley MA, Smith PC (2010) Effects of uridine diphosphate glucuronosyltransferase 2B7 and 1A7 pharmacogenomics and patient clinical parameters on steady-state mycophenolic acid pharmacokinetics in glomerulonephritis. Eur J Clin Pharmacol 66:1119–1130
Kiang TK, Ensom MH, Chang TK (2005) UDP-glucuronosyltransferases and clinical drug–drug interactions. Pharmacol Ther 106:97–132
Koiwai O, Aono S, Adachi Y, Kamisako T, Yasui Y, Nishizawa M, Sato H (1996) Crigler-Najjar syndrome type II is inherited both as a dominant and as a recessive trait. Hum Mol Genet 5:645–647

Kurkela M, Hirvonen J, Kostiainen R, Finel M (2004) The interactions between the N-terminal and C-terminal domains of the human UDP-glucuronosyltransferases are partly isoform-specific, and may involve both monomers. Biochem Pharmacol 68:2443–2450

Kurose K, Sugiyama E, Saito Y (2012) Population differences in major functional polymorphisms of pharmacokinetics/pharmacodynamics-related genes in Eastern Asians and Europeans: implications in the clinical trials for novel drug development. Drug Metab Pharmacokinet 27:9–54

Lamba JK, Lin YS, Schuetz EG, Thummel KE (2002) Genetic contribution to variable human CYP3A-mediated metabolism. Adv Drug Deliv Rev 54:1271–1294

Lévesque E, Girard H, Journaultm K, Lépinem J, Guillemettem C (2007) Regulation of the UGT1A1 bilirubin-conjugating pathway: role of a new splicing event at the UGT1A locus. Hepatology 45:128–138

Li D, Fournel-Gigleux S, Barré L, Mulliert G, Netter P, Magdalou J, Ouzzine M (2007) Identification of aspartic acid and histidine residues mediating the reaction mechanism and the substrate specificity of the human UDP-glucuronosyltransferases 1A. J Biol Chem 282:36514–36524

Mackenzie PI, Owens IS, Burchell B, Bock KW, Bairoch A, Bélanger A, Fournel-Gigleux S, Green M, Hum DW, Iyanagi T, Lancet D, Louisot P, Magdalou J, Roy Chowdhury J, Ritter JK, Schachter H, Tephly TR, Tipton KF, Nebert DW (1997) The UDP glycosyltransferase gene superfamily: recommended nomenclature update based on evolutionary divergence. Pharmacogenetics 7:255–269

Mackenzie PI, Bock KW, Burchell B, Guillemette C, Ikushiro S, Iyanagi T, Miners JO, Owens IS, Nebert DW (2005) Nomenclature update for the mammalian UDP glycosyltransferase (UGT) gene superfamily. Pharmacogenet Genomics 15:677–685

Mackenzie PI, Rogers A, Treloar J, Jorgensen BR, Miners JO, Meech R (2008) Identification of UDP glycosyltransferase 3A1 as a UDP-*N*-acetylglucosaminyltransferase. J Biol Chem 283:36205–36210

Meech R, Mackenzie PI (1997a) UDP-glucuronosyltransferase, the role of the amino terminus in dimerization. J Biol Chem 272:26913–26917

Meech R, Mackenzie PI (1997b) Structure and function of uridine diphosphate glucuronosyltransferases. Clin Exp Pharmacol Physiol 24:907–915

Meech R, Mackenzie PI (1998) Determinants of UDP glucuronosyltransferase membrane association and residency in the endoplasmic reticulum. Arch Biochem Biophys 356:77–85

Miley MJ, Zielinska AK, Keenan JE, Bratton SM, Radominska-Pandya A, Redinbo MR (2007) Crystal structure of the cofactor-binding domain of the human phase II drug-metabolism enzyme UDP-glucuronosyltransferase 2B7. J Mol Biol 369:498–511

Minami H, Sai K, Saeki M, Saito Y, Ozawa S, Suzuki K, Kaniwa N, Sawada J, Hamaguchi T, Yamamoto N, Shirao K, Yamada Y, Ohmatsu H, Kubota K, Yoshida T, Ohtsu A, Saijo N (2007) Irinotecan pharmacokinetics/pharmacodynamics and UGT1A genetic polymorphisms in Japanese: roles of UGT1A1*6 and *28. Pharmacogenet Genomics 17:497–504

Miyauchi Y, Ishii Y, Nagata K, Yamazoe Y, Mackenzie PI, and Yamada H (2012) UDP-Glucuronosyltransferase (UGT) 2B7 and 1A9 suppress cytochrome P450 3A4 function: evidence for the involvement of the cytosolic tail of UGT in the suppression. In: Abstracts of papers, 19th International Symposium on Microsomes and Drug Oxidations and 12th European Regional ISSX Meeting, Noordwijk aan Zee, The Netherlands, June 2012

Muraoka M, Kawakita M, Ishida N (2001) Molecular characterization of human UDP-glucuronic acid/UDP-*N*-acetylgalactosamine transporter, a novel nucleotide sugar transporter with dual substrate specificity. FEBS Lett 495:87–93

Muraoka M, Miki T, Ishida N, Hara T, Kawakita M (2007) Variety of nucleotide sugar transporters with respect to the interaction with nucleoside mono- and diphosphates. J Biol Chem 282:24615–24622

Nakajima M, Yamanaka H, Fujiwara R, Katoh M, Yokoi T (2007) Stereoselective glucuronidation of 5-(4′-hydroxyphenyl)-5-phenylhydantoin by human UDP-glucuronosyltransferase (UGT)

1A1, UGT1A9, and UGT2B15: effects of UGT–UGT interactions. Drug Metab Dispos 35:1679–1686

Nishimura Y, Maeda S, Ikushiro S, Mackenzie PI, Ishii Y, Yamada H (2007) Inhibitory effects of adenine nucleotides and related substances on UDP-glucuronosyltransferase: structure–effect relationships and evidence for an allosteric mechanism. Biochim Biophys Acta 1770:1557–1566

Nurrochmad A, Ishii Y, Nakanoh H, Inoue T, Horie T, Sugihara K, Ohta S, Taketomi A, Maehara Y, Yamada H (2010) Activation of morphine glucuronidation by fatty acyl-CoAs and its plasticity: a comparative study in humans and rodents including chimeric mice carrying human liver. Drug Metab Pharmacokinet 25:262–273

Ockenga J, Vogel A, Teich N, Keim V, Manns MP, Strassburg CP (2003) UDP glucuronosyltransferase (UGT1A7) gene polymorphisms increase the risk of chronic pancreatitis and pancreatic cancer. Gastroenterology 124:1802–1808

Oesch-Bartlomowicz B, Oesch F (2008) Phosphorylation of xenobiotic-metabolizing cytochromes P450. Anal Bioanal Chem 392:1085–1092

Okamura K, Ishii Y, Ikushiro S, Mackenzie PI, Yamada H (2006) Fatty acyl-CoA as an endogenous activator of UDP-glucuronosyltransferases. Biochem Biophys Res Commun 345:1649–1656

Operaña TN, Tukey RH (2007) Oligomerization of the UDP-glucuronosyltransferase 1A proteins: homo- and heterodimerization analysis by fluorescence resonance energy transfer and co-immunoprecipitation. J Biol Chem 282:4821–4829

Parsons JB, Lawrence AD, McLean KJ, Munro AW, Rigby SE, Warren MJ (2010) Characterisation of PduS, the pdu metabolosome corrin reductase, and evidence of substructural organisation within the bacterial microcompartment. PLoS One 5:e14009

Pogell BM, Leloir LF (1961) Nucleotide activation of liver microsomal glucuronidation. J Biol Chem 236:293–298

Radominska-Pandya A, Czernik PJ, Little JM, Battaglia E, Mackenzie PI (1999) Structural and functional studies of UDP-glucuronosyltransferases. Drug Metab Rev 31:817–899

Reed JR, Eyer M, Backes WL (2010) Functional interactions between cytochromes P450 1A2 and 2B4 require both enzymes to reside in the same phospholipid vesicle: evidence for physical complex formation. J Biol Chem 285:8942–8952

Ritter JK, Chen F, Sheen YY, Tran HM, Kimura S, Yeatman MT, Owens IS (1992) A novel complex locus UGT1 encodes human bilirubin, phenol, and other UDP-glucuronosyltransferase isozymes with identical carboxyl termini. J Biol Chem 267:3257–3261

Ritter JK, Yeatman MT, Kaiser C, Gridelli B, Owens IS (1993) A phenylalanine codon deletion at the UGT1 gene complex locus of a Crigler-Najjar type I patient generates a pH-sensitive bilirubin UDP-glucuronosyltransferase. J Biol Chem 268:23573–23579

Roizin L (1963) Mitochondria (pleomorpho metabolosomes) as histometabolic gradients. (Effects of prochlorperazine on the rat brain as revealed by electron microscope.) Dis Nerv Syst 24(4 pt 2):61–66

Rosner GL, Panetta JC, Innocenti F, Ratain MJ (2008) Pharmacogenetic pathway analysis of irinotecan. Clin Pharmacol Ther 84:393–402

Sai K, Saeki M, Saito Y, Ozawa S, Katori N, Jinno H, Hasegawa R, Kaniwa N, Sawada J, Komamura K, Ueno K, Kamakura S, Kitakaze M, Kitamura Y, Kamatani N, Minami H, Ohtsu A, Shirao K, Yoshida T, Saijo N (2004) UGT1A1 haplotypes associated with reduced glucuronidation and increased serum bilirubin in irinotecan-administered Japanese patients with cancer. Clin Pharmacol Ther 75:501–515

Shepherd SR, Baird SJ, Hallinan T, Burchell B (1989) An investigation of the transverse topology of bilirubin UDP-glucuronosyltransferase in rat hepatic endoplasmic reticulum. Biochem J 259:617–620

Shimada T, Yamazaki H, Mimura M, Inui Y, Guengerich FP (1994) Interindividual variations in human liver cytochrome P-450 enzymes involved in the oxidation of drugs, carcinogens and toxic chemicals: studies with liver microsomes of 30 Japanese and 30 Caucasians. J Pharmacol Exp Ther 270:414–423

Shimomura K, Kamata O, Ueki S, Ida S, Oguri K, Yoshimura H, Tsukamoto H (1971) Analgesic effect of morphine glucuronides. Tohoku J Exp Med 105:45–52

Strassburg CP, Vogel A, Kneip S, Tukey RH, Manns MP (2002) Polymorphisms of the human UDP-glucuronosyltransferase (UGT) 1A7 gene in colorectal cancer. Gut 50:851–856

Strittmatter CF, Ball EG (1951) A hemochromogen component of liver microsomes. Proc Natl Acad Sci USA 38:19–25

Subramanian M, Low M, Locuson CW, Tracy TS (2009) CYP2D6-CYP2C9 protein–protein interactions and isoform-selective effects on substrate binding and catalysis. Drug Metab Dispos 37:1682–1689

Sugatani J (2013) Function, genetic polymorphism, and transcriptional regulation of human UDP-glucuronosyltransferase (UGT) 1A1. Drug Metab Pharmacokinet 28:83–92

Takeda S, Ishii Y, Iwanaga M, Mackenzie PI, Nagata K, Yamazoe Y, Oguri K, Yamada H (2005a) Modulation of UDP-glucuronosyltransferase function by cytochrome P450: evidence for the alteration of UGT2B7-catalyzed glucuronidation of morphine by CYP3A4. Mol Pharmacol 67:665–672

Takeda S, Ishii Y, Mackenzie PI, Nagata K, Yamazoe Y, Oguri K, Yamada H (2005b) Modulation of UDP-glucuronosyltransferase 2B7 function by cytochrome P450s in vitro: differential effects of CYP1A2, CYP2C9 and CYP3A4. Biol Pharm Bull 28:2026–2027

Takeda S, Kitajima Y, Ishii Y, Nishimura Y, Mackenzie PI, Oguri K, Yamada H (2006) Inhibition of UDP-glucuronosyltransferase 2B7-catalyzed morphine glucuronidation by ketoconazole: dual mechanisms involving a novel noncompetitive mode. Drug Metab Dispos 34:1277–1282

Takeda S, Ishii Y, Iwanaga M, Nurrochmad A, Ito Y, Mackenzie PI, Nagata K, Yamazoe Y, Oguri K, Yamada H (2009) Interaction of cytochrome P450 3A4 and UDP-glucuronosyltransferase 2B7: evidence for protein–protein association and possible involvement of CYP3A4 J-helix in the interaction. Mol Pharmacol 75:956–964

Taura K, Yamada H, Hagino Y, Ishii Y, Mori M, Oguri K (2000) Interaction between cytochrome P450 and other drug-metabolizing enzymes: evidence for an association of CYP1A1 with microsomal epoxide hydrolase and UDP-glucuronosyltransferase. Biochem Biophys Res Commun 273:1048–1052

Taura K, Yamada H, Naito E, Ariyoshi N, Mori M, Oguri K (2002) Activation of microsomal epoxide hydrolase by interaction with cytochromes P450: kinetic analysis of the association and substrate-specific activation of epoxide hydrolase function. Arch Biochem Biophys 402:275–280

te Morsche RH, Drenth JP, Truninger K, Schulz HU, Kage A, Landt O, Verlaan M, Rosendahl J, Macek M Jr, Jansen JB, Witt H (2008) UGT1A7 polymorphisms in chronic pancreatitis: an example of genotyping pitfalls. Pharmacogenomics J 8:34–41

Thomassin J, Tephly TR (1990) Photoaffinity labeling of rat liver microsomal morphine UDP-glucuronosyltransferase by [^{3}H]flunitrazepam. Mol Pharmacol 38:294–298

Timsit YE, Negishi M (2007) CAR and PXR: the xenobiotic-sensing receptors. Steroids 72:231–246

Uchaipichat V, Mackenzie PI, Guo XH, Gardner-Stephen D, Galetin A, Houston JB, Miners JO (2004) Human UDP-glucuronosyltransferases: isoform selectivity and kinetics of 4-methylumbelliferone and 1-naphthol glucuronidation, effects of organic solvents, and inhibition by diclofenac and probenecid. Drug Metab Dispos 32:413–423

Vanstapel F, Blanckaert N (1988) Topology and regulation of bilirubin UDP-glucuronyltransferase in sealed native microsomes from rat liver. Arch Biochem Biophys 263:216–225

Verlaan M, Drenth JP, Truninger K, Koudova M, Schulz HU, Bargetzi M, Künzli B, Friess H, Cerny M, Kage A, Landt O, te Morsche RH, Rosendahl J, Luck W, Nickel R, Halangk J, Becker M, Macek M Jr, Jansen JB, Witt H (2005) Polymorphisms of UDP-glucuronosyltransferase 1A7 are not involved in pancreatic diseases. J Med Genet 42:e62

Villeneuve L, Girard H, Fortier LC, Gagné JF, Guillemette C (2003) Novel functional polymorphisms in the UGT1A7 and UGT1A9 glucuronidating enzymes in Caucasian and African-American subjects and their impact on the metabolism of 7-ethyl-10-hydroxycamptothecin and flavopiridol anticancer drugs. J Pharmacol Exp Ther 307:117–128

Vogel A, Kneip S, Barut A, Ehmer U, Tukey RH, Manns MP, Strassburg CP (2001) Genetic link of hepatocellular carcinoma with polymorphisms of the UDP-glucuronosyltransferase UGT1A7 gene. Gastroenterology 121:1136–1144

Wade AE (1986) Effects of dietary fat on drug metabolism. J Environ Pathol Toxicol Oncol 6:161–189

Williams PA, Cosme J, Sridhar V, Johnson EF, McRee DE (2000) Mammalian microsomal cytochrome P450 monooxygenase: structural adaptations for membrane binding and functional diversity. Mol Cell 5:121–131

Yamazaki H, Gillam EMJ, Dong MS, Johnson WW, Guengerich FP, Shimada T (1997) Reconstitution of recombinant cytochrome P450 2C10(2C9) and comparison with cytochrome P450 3A4 and other forms: effects of cytochrome P450-P450 and cytochrome P450-b5 interactions. Arch Biochem Biophys 342:329–337

Yokoi T, Nakajima M (2013) MicroRNAs as mediators of drug toxicity. Annu Rev Pharmacol Toxicol 53:377–400

Yong WP, Ramirez J, Innocenti F, Ratain MJ (2005) Effects of ketoconazole on glucuronidation by UDP-glucuronosyltransferase enzymes. Clin Cancer Res 11:6699–6704

Chapter 19
Control of Xeno/Endobiotics-Metabolizing Cytochrome P450s by MicroRNAs

Miki Nakajima

Abstract Human cytochrome P450s (P450s, CYPs) catalyze the metabolism of xenobiotics such as drugs, environmental chemicals, and pro-carcinogens as well as endobiotics such as steroids and bile acids. To understand the reasons for inter- and intraindividual variability of drug response and adverse reaction as well as xenobiotics-related toxicity, the mechanisms of transcriptional regulation and the genetic polymorphisms of P450s have been well studied. Recently, it was discerned that microRNAs (miRNAs), endogenous short noncoding RNAs 18–24 nucleotides in length, are involved in the posttranscriptional regulation of P450s through translational repression or mRNA degradation. It became clear that miRNAs regulate P450s expression not only by a direct mechanism but also by indirect mechanisms through the regulation of transcriptional factors or modulators of enzyme activities. The miRNA-dependent regulation of P450s confers variability in detoxification of drugs and metabolic activation of environmental chemicals. In addition, it plays roles in the homeostasis of endobiotics and cancer development or progression. Polymorphisms are present not only in the mRNA but also in miRNA sequences. The miRNA-related polymorphisms may cause gain- or loss of function, resulting in change of targets of mRNA expression. The miRNA-related polymorphisms join the pharmacogenetics. Further elucidation of the role of miRNAs in drug metabolism and pharmacokinetics and drug toxicology may offer new clues for pharmacotherapy.

Keywords Cancer • Drug metabolism • Interindividual variability • Posttranscriptional regulation • Steroid homeostasis • Transcription factors

M. Nakajima (✉)
Drug Metabolism and Toxicology, Faculty of Pharmaceutical Sciences,
Kanazawa University, Kakuma-machi, Kanazawa 920-1192, Japan
e-mail: nmiki@p.kanazawa-u.ac.jp

H. Yamazaki (ed.), *Fifty Years of Cytochrome P450 Research*,
DOI 10.1007/978-4-431-54992-5_19, © Springer Japan 2014

19.1 Introduction

A major part of the human genome DNA (~98 %) encodes noncoding RNAs, which include ribosomal RNAs, transfer RNAs, small nuclear RNA (snRNA), small nucleolar RNAs, long noncoding RNA (lncRNA), Piwi-interacting RNA (piRNA), and microRNAs (miRNAs) (Esteller 2011). Among them, miRNAs (18–25 nucleotides in length) are the most intensively studied noncoding RNAs. miRNAs are a major posttranscriptional regulator that downregulates gene expression through translational repression or mRNA degradation. miRNAs regulate every aspect of cellular processes such as differentiation, proliferation, and apoptosis as well as a large range of physiological processes including development, immune response, metabolism, tumor formation, and disease development (Kloosterman and Plasterk 2006). Generally, miRNAs play a role in fine-tuning of gene expression, forming intricate networks. Growing evidence shows that a variety of diseases are associated with the dysregulation of miRNAs. miRNAs are now considered as a new tool for diagnosis and therapy.

There is increasing evidence revealing the role of miRNAs in the regulation of cytochrome P450s (P450s, CYPs), a major family of drug-metabolizing enzymes as well as several factors controlling P450 expressions or activities. This chapter summarizes the current knowledge on the miRNA-dependent regulation of P450s in humans and rodents and their clinical and physiological significances.

19.2 Biogenesis, Function, and Nomenclature of miRNAs

To date, about 2,600 human miRNAs have been identified (miRBase, release 20, http://www.mirbase.org/). Genes encoding miRNA are scattered in all chromosomes in humans except for the Y chromosome (Kim and Nam 2006). miRNAs are first transcribed as long transcripts (250- to 4,000-base) called primary miRNAs (pri-miRNAs) having a stem-loop structure(s) (Fig. 19.1). After the pri-miRNA is transcribed, precursor miRNA (pre-miRNA) hairpin structure (70- to 100-base) is cleaved from the transcript by Drosha and DiGeorge syndrome critical region 8 protein (DGCR8) in the nucleus. In addition to the canonical pathway, the mirtron pathway also produces the pre-miRNA by using splicing machinery (spliceosome), bypassing the cleavage by Drosha (Curtis et al. 2012). The pre-miRNA is exported into the cytoplasm by Exportin 5, processed to miRNA duplex of about 22 nucleotides by Dicer or Argonaute protein Ago2, and then unwound into the single-strand form of mature miRNAs. The functional guide strand is loaded onto the RNA-induced silencing complex (RISC), composed of Dicer, TAR RNA-binding protein (TRBP), and Ago2 to form miRISC, and guides the complex to its target mRNAs (Fig. 19.1). In animals and humans, miRNAs bind to the miRNA recognition element (MRE) on the target mRNAs, which is usually located in the 3′-untranslated region (UTR), with partial complementarity (Fig. 19.2), although

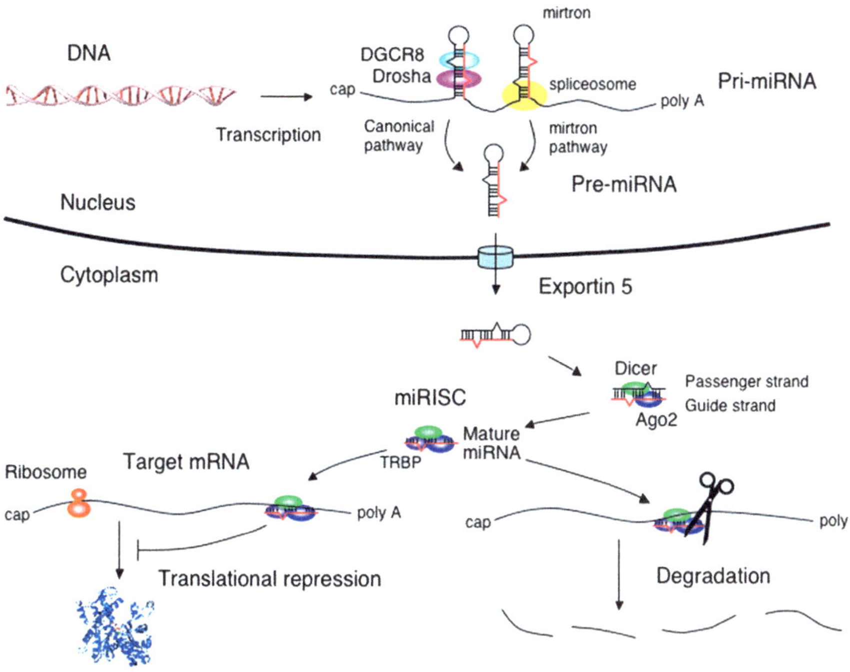

Fig. 19.1 Biogenesis and function of microRNAs (miRNAs). miRNA genes are transcribed to generate primary miRNAs (pri-miRNAs). The pri-miRNAs are processed by Drosha and DiGeorge syndrome critical region 8 protein (DGCR8) in the nucleus to produce precursor miRNAs (pre-miRNAs). In addition to the canonical pathway, pre-miRNAs are produced by a mirtron pathway bypassing the cleavage by Drosha. Exportin 5 exports the pre-miRNAs into cytoplasm. Dicer and Ago2 participate to produce miRNA duplexes and unwind into single-strand mature miRNA, forming miRISC. The miRISC binds to target mRNA to represses translation or facilitate mRNA degradation

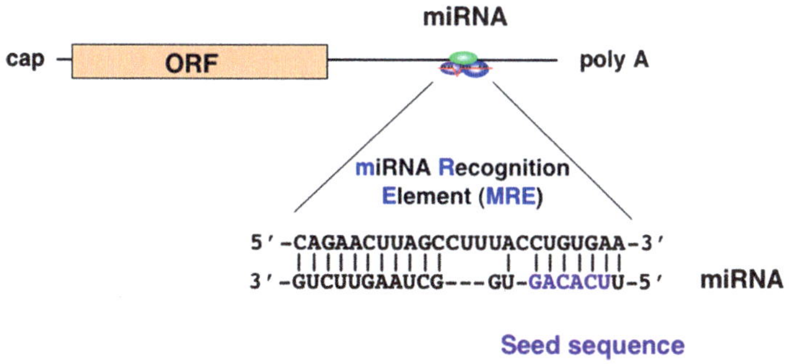

Fig. 19.2 miRNA binding to target mRNA. miRNAs bind to target mRNAs with partial complementarity. The miRNA recognition element (MRE) is usually located in the 3′-untranslated region (UTR), and sometime in the coding region. Perfect match of the seed sequence, the nucleotides 2–8 at the 5′-end of the miRNA, with the target mRNA is critical for the miRNA function. *ORF* open reading frame

there are examples in the coding region as well (Duursma et al. 2008). The seed sequence, the nucleotides 2–8 at the 5′-end of the miRNA (Fig. 19.2), is critical and sometimes sufficient to exert function of the miRNA (Lewis et al. 2005). Gene silencing by miRNAs includes multiple mechanisms (Chekulaeva and Filipowicz 2009; Fabian et al. 2010). The passenger strand, named miRNA*, is usually degraded, although it is sometimes functional.

miRNAs are assigned sequential numerical identifiers (Griffiths-Jones et al. 2006). Abbreviated three- or four-letter prefixes are used to designate the species, such that identifiers take the form hsa-miR-101 (for miR-101 in *Homo sapiens*). The hsa-miR-101 in human and mmu-miR-101 in mouse are orthologous. Paralogous miRNAs that differ at only one or two positions are given lettered suffixes (e.g., mmu-miR-10a and mmu-miR-10b). Distinct loci that give to identical mature miRNAs have numbered suffixes (hsa-miR-125b-1 and hsa-miR-125b-2). The preferred method of naming guide/passenger (miRNA/miRNA*) strands designates miRNAs from 5′- or 3′-arms (e.g., miR-378-5p and miR-378-3p) (Griffiths-Jones et al. 2006). In subsequent sections, the prefixes are omitted for simplification.

19.3 Prediction and Identification of miRNAs That Target a Given mRNA

Several computational programs (Table 19.1) are available to predict miRNAs that target a given mRNA. The predicted miRNAs vary significantly depending on the different algorithms used in the programs, because these algorithms place variable weight on complementarity to the miRNA seed sequence, evolutionary conservation of the MRE of the target gene, free energy of the miRNA–mRNA duplex, and accessibility of the target site. The false-positive rate of the predicted candidate targets of a given miRNA is thought to be 30–50 % (Alexiou et al. 2009). To validate specific miRNA–mRNA interaction, several lines of experiments are used. The most commonly employed technology is the use of luciferase assays using reporter constructs containing the MRE of the target mRNA downstream of the

Table 19.1 Computational programs to predict microRNA (miRNA) targets

Name	Website
miRWalk	http://www.umm.uni-heidelberg.de/apps/zmf/mirwalk/
MicroCosm	http://www.ebi.ac.uk/enright-srv/microcosm/htdocs/targets/v5/
miRanda	http://www.microrna.org/microrna/home.do
PicTar	http://pictar.mdc-berlin.de/
PITA	http://genie.weizmann.ac.il/pubs/mir07/index.html
RNA22	http://cbcsrv.watson.ibm.com/rna22.html
RNAhybrid	http://bibiserv.techfak.uni-bielefeld.de/rnahybrid/
TargetScan	http://www.targetscan.org/

reporter gene. The constructs are cotransfected into the cells with precursor miRNA or antisense oligonucleotide (AsO) for miRNA. Evidence can be established whether the reporter activity is significantly decreased or increased. Another effective method is the determination of target protein or mRNA expression after the overexpression or inhibition of the miRNA. In addition, an inverse relationship between the protein and miRNA levels is expected, for example, cancer versus normal tissues, or among individuals or cell lines.

19.4 Regulation of Cytochrome P450s by miRNAs

19.4.1 CYP1A1

CYP1A1 is mainly expressed in extrahepatic tissues including lung. CYP1A1 mediates bioactivation of pro-carcinogens and thus is associated with cancer risk. CYP1A1 expression is induced by polycyclic aromatic hydrocarbons (PAHs) through the activation of the aryl hydrocarbon receptor (AhR) and subsequent heterodimerization with aryl hydrocarbon receptor nuclear translocator (ARNT) and binding of the heterodimer to a response element in the 5′-flanking region. Choi et al. (2012) found that miR-892a negatively regulates CYP1A1 expression in a human breast cancer cell line. Overexpression or inhibition of miR-892a in MCF-7 cells resulted in the decrease or increase in CYP1A1 expression, respectively. Luciferase assays revealed that an MRE in the 3′-UTR that shows a perfect match with the seed sequence of miR-892a was functional. Interestingly, treatment of the MCF-7 cells with benzo[*a*]pyrene resulted in the decrease of miR-892a level, prominent induction of CYP1A1, and the decrease of cell viability. Overexpression of miR-892a restored the benzo[*a*]pyrene-induced loss of cell viability. These results suggest that miR-892a negatively regulates CYP1A1 function to protect the cells from chemical toxicity.

Oda et al. (2012) found that ARNT is regulated by miR-24 in human liver. Overexpression of miR-24 in human liver-derived cell lines HuH-7 and HepG2 cells significantly decreased the ARNT protein level but not the ARNT mRNA level, indicating its translational repression. The miR-24-dependent downregulation of ARNT decreased expression of CYP1A1. Thus, CYP1A1 is directly and indirectly regulated by miRNAs.

19.4.2 CYP1B1

CYP1B1 is expressed in ovarian, uterine, and breast tissues. CYP1B1 catalyzes the metabolic activation of a variety of pro-carcinogens including PAHs and aryl amines, and the metabolism of estradiol (E_2) to form 4-hydroxyestradiol, which causes DNA damage and contributes estrogen-dependent cancers.

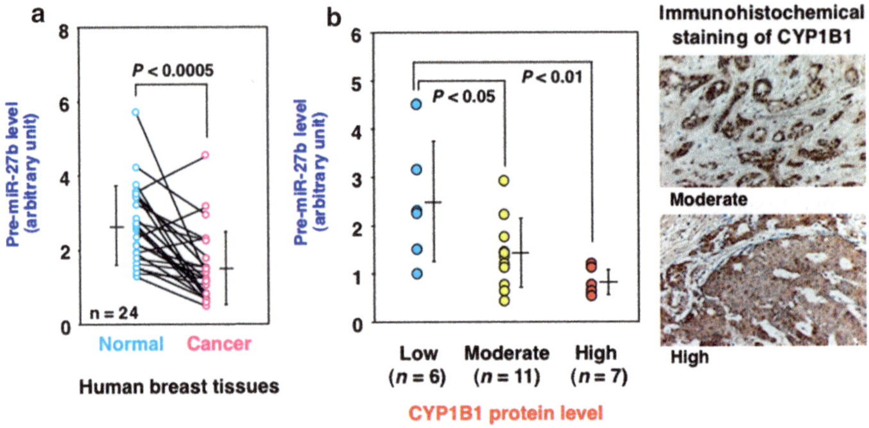

Fig. 19.3 Association of CYP1B1 regulated by miR-27b with cancer. miR-27b downregulates human CYP1B1. The miR-27b level was lower in breast cancer tissues than in normal tissues. (**a**) A significant inverse association was observed between miR-27b and CYP1B1 protein levels in breast cancer tissues. (**b**) The decreased miR-27b level would be one of the causes of high expression of CYP1B1 protein, catalyzing the metabolic activation of pro-carcinogens and estradiol

Tsuchiya et al. (2006) found that human CYP1B1 is regulated by miR-27b via translational repression: this was the first example that a P450 enzyme is regulated by miRNA. Luciferase assays revealed that an MRE in the 3′-UTR was functional. Interestingly, the CYP1B1 protein level was higher in breast cancer tissues than in adjacent normal tissues, whereas the miR-27b level was lower in breast cancer tissues than in adjacent normal tissues. A significant inverse association was observed in the expression levels of miR-27b and the CYP1B1 protein (Fig. 19.3). Thus, it was demonstrated that decreased miR-27b level would be one cause for the high expression of the CYP1B1 protein in cancer tissues. The miR-27b modulating CYP1B1 may trigger the initiation or progression of cancer.

Human CYP1B1 is transcriptionally regulated by estrogen receptor-α (ERα) (Tsuchiya et al. 2004). Human ERα is regulated by miR-206, and the activation of ERα results in the decreased expression of miR-206, showing mutually inhibitory regulation (Adams et al. 2007). miR-221, miR-222 (Zhao et al. 2008), and miR-22 (Xiong et al. 2010) were also reported to regulate human ERα expression. The miRNAs-dependent regulation of ERα may indirectly affect CYP1B1 expression, although experimental proof is required.

19.4.3 CYP2A3

Rat CYP2A3, which is expressed in the lung, catalyzes the metabolic activation of 4-(methylnitrosamino)-1-(3-pyridyl)-1-butanone (NNK). Kalscheuer et al. (2008) reported that chronic administration of NNK to F344 rats resulted in the increase of

CYP2A3 mRNA and protein levels and the decrease of miR-126* level in the lung. Luciferase assays revealed that an MRE in the 3′-UTR of CYP2A3 was functionally recognized by miR-126*. The miR-126*-dependent regulation of CYP2A3 seems to be associated with pulmonary tumorigenesis. Because the MRE sequence is conserved, human CYP2A13 expressed in the lung may also be regulated by miR-126*.

19.4.4 Cyp2b9

Mouse Cyp2b9 is a testosterone 16α-hydroxylase enzyme showing female-specific expression. It is appreciated that sex-dependent secretion of growth hormone, glucocorticoid hormone, and sex hormones are involved in the sexually dimorphic expression. Recently, evidence is growing to reveal that sex-based differences in some miRNAs underlie the sex-based differences in gene expression. Xie et al. (2013) found that miRNAs contributed to the sexually biased expression of Cyp2b9. They found that the expression of hepatic 18 miRNAs was higher in males than in females. Eight of 18 miRNAs (miR-139-3p, miR-1b-5p, miR-21, miR-291a-5p, miR-297a, miR-297b-3p, miR-467g, miR-667) decreased Cyp2b9 protein level in a Huh7 stable transformant. Thus, the lower expression of these miRNAs potentially contributes to the female-specific expression of Cyp2b9 in the liver.

19.4.5 CYP2C8, CYP2C9, and CYP2C19

CYP2C subfamily consists of four members in humans (CYP2C8, CYP2C9, CYP2C18, CYP2C19). CYP2C are exclusively expressed in the liver and are detected at lower levels in extrahepatic tissues. They are responsible for the metabolism of more than 20 % of all clinically prescribed drugs. Zhang et al. (2012) reported that overexpression of miR-103 and miR-107, which contain the identical seed sequence, into primary human hepatocytes resulted in the decrease of CYP2C8 protein but not CYP2C8 mRNA, indicating translational repression. Transfection of AsO for these miRNAs resulted in the increase of CYP2C8 protein level. Potential MREs were identified in the 3′-UTR of the CYP2C8, and a putative MRE was also found in the 3′-UTRs of CYP2C9 and CYP2C19, with a two- and one-nucleotide mismatch, respectively. Overexpression of miR-103 or miR-107 resulted in the decrease of CYP2C9 and CYP2C19 protein levels in human hepatocytes, although to a lesser degree than CYP2C8. The miR-103 and miR-107 are transcribed from the introns of the pantothenate kinase (PANK) genes, which are involved in the regulation of acetyl-CoA levels, cell stress, insulin sensitivity, lipid metabolism, and angiogenesis. It would be of interest to examine whether the conditions that affect the expression of the PANK genes and the miR-103/miR-107 family might alter the expression of CYP2C8 and CYP2C19 in human tissues.

19.4.6 CYP2E1

Human CYP2E1, mainly expressed in the liver, catalyzes the metabolism of low molecular weight xenobiotics, including drugs, organic solvents, and pro-carcinogens. CYP2E1 is induced by its own substrates, including isoniazid, ethanol, and acetone, through posttranscriptional or posttranslational mechanisms. Mohri et al. (2010) found that human CYP2E1 is regulated by miR-378 through a predicted MRE in the 3′-UTR. In a panel of 25 human liver tissues, no positive correlation was observed between the CYP2E1 protein level and the CYP2E1 mRNA level, supporting the posttranscriptional repression of CYP2E1. The miR-378 levels were inversely correlated with the CYP2E1 protein levels and the translational efficiency. Because the 3′-UTR of CYP2E1 is poorly conserved among humans, rats, and mice, the regulation by miR-378 should be specific to humans.

The miR-378 gene is within intron 1 of the gene encoding peroxisome proliferator-activated receptor-γ co-activator 1β (PGC1β), which regulates metabolism and mitochondrial biogenesis through the stimulation of nuclear hormone receptors and other transcription factors. The miR-378 expression is expected to be parallel with the PGC1β expression. It has been reported that the PGC1β is downregulated in diabetes and obesity but is upregulated by insulin treatment, whereas CYP2E1 is upregulated in diabetes and obesity but is downregulated by insulin treatment. Thus, the miR-378 may underlie the opposing results for PGC1β and CYP2E1. Recently, Carrer et al. (2012) reported that mice genetically lacking miR-378 and miR-378* are resistant to high-fat diet-induced obesity and exhibit enhanced mitochondrial fatty acid metabolism and elevated oxidative capacity in insulin target tissues. Thus, the miR-378 and miR-378* seem to be integral components of a regulatory circuit that functions under conditions of metabolic stress to control systemic energy homeostasis and the overall oxidative capacity of insulin target tissues.

Recently, Shukla et al. (2013) found the role of miRNAs in insulin-mediated downregulation of rat CYP2E1. Insulin treatment of primary rat hepatocytes increased the levels of miR-132 and miR-122, which are in a family clustered through phosphatidylinositol 3-kinase (PI3-K), protein kinase B (Akt)/mammalian target of rapamycin (mTOR) signaling, and these miRNAs bound to MRE in the 3′-UTR of rat CYP2E1. Thus, changes of miRNA expression might be the cause of the decreased CYP2E1 by insulin.

19.4.7 CYP2J2

Human CYP2J2 is primarily expressed in the heart and vessel endothelium and is also found in a variety of tissues including the liver, lung, kidney, and gastrointestinal tissues. CYP2J2 catalyzes the epoxidation of arachidonic acid into four regioisomers of *cis*-epoxyeicosatrienoic acid (EETs). EETs are related to cancer cell behavior and tumor pathogenesis. CYP2J2 is highly expressed in human tumors but not in adjacent normal tissues (Jiang et al. 2005). Chen et al. (2012) found that human CYP2J2 protein level in several cancer cell lines, including HeLa and MDA-MB-435, were significantly decreased by overexpression of let-7b, which is known as a tumor suppressor miRNA. Luciferase assays revealed that one of seven potential MREs was functional. Interestingly, an inverse correlation between the let-7b expression level and the CYP2J2 protein level in 18 sets of lung squamous tumor cancer and paired adjacent non-tumor tissues was demonstrated. The let-7b significantly inhibited the tumor phenotypes of tumor-xenografted mice by targeting CYP2J2. Thus, CYP2J2 was admitted into the targets of the tumor suppressor let-7b.

19.4.8 CYP3A4

Human CYP3A4, which is highly expressed in the liver and intestine, contributes to the metabolism of more than 50 % of clinically used drugs. Pan et al. (2009) reported that overexpression of miR-27b resulted in the decrease of CYP3A4 protein and mRNA levels in human colon cancer-derived LS180 and human pancreas cancer-derived PANC1 cells. It remains to be examined whether the miR-27b-dependent downregulation of CYP3A4 is functional in human tissues. Because CYP3A4 protein levels in multiple individual human liver samples were positively correlated with the CYP3A4 mRNA levels (Takagi et al. 2008), the posttranscriptional regulation by miRNA(s) would be minor in CYP3A4. Interestingly, the CYP3A4 mRNA and protein levels were positively correlated with protein levels of pregnane X receptor (PXR), which mainly contributes to the CYP3A4 transactivation. In contrast to CYP3A4, the PXR protein levels were not correlated with the PXR mRNA levels, suggesting the contribution of posttranscriptional regulation. Takagi et al. (2008) demonstrated that miR-148a repressed the translation of human PXR through binding to an MRE in the 3′-UTR. Although a potential MRE for the miR-148a was also found in the 3′-UTR of CYP3A4, it was not functional. In the panel of human livers, the miR-148a levels were inversely correlated with the translational efficiency of PXR. The miR-148a-dependent downregulation of PXR affected the constitutive and inducible expression of CYP3A4.

19.4.9 CYP19A1

CYP19A1, which is known as aromatase, is a key enzyme in estrogen biosynthesis. Panda et al. (2012) reported that miR-98, whose aberrant expression is associated with the endometrial transition from a normal to cancerous state, downregulates CYP19A1 expression in human endometrial adenocarcinoma-derived Ishikawa cells through binding to an MRE in the 3′-UTR. Shibahara et al. (2012) found by luciferase assays that let-7f, which is known as tumor suppressor miRNA, functionally binds to the 3′-UTR of human CYP19A1. Treatment of MCF-7 cells cocultured with primary breast cancer stromal cells with letrozole, an aromatase inhibitor, resulted in the decrease of CYP19A1 mRNA and the increase of let-7f. These results suggest that aromatase inhibitors may exert their tumor-suppressing effects upon breast cancer cells by suppressing CYP19A1 expression via restoration of let-7f. An inverse correlation between let-7 expression and the CYP19A1 mRNA and protein levels in 11 breast cancer tissues supported the clinical significance of this regulatory mechanism. Kumar et al. (2013) reported that miR-19b and miR-106a downregulate CYP19A1 expression in human placental trophoblasts. Overexpression of c-Myc proto-oncogene resulted in the increase of these miRNAs, the decrease of CYP19A1, and decreased trophoblast differentiation, suggesting that aberrant regulation of these miRNAs may contribute to the pathogenesis of preeclampsia.

19.4.10 CYP24A1

CYP24A1 catalyzes the inactivation of 1,25-dihydroxyvitamin D_3 (calcitriol), a biologically active metabolite of vitamin D_3. Calcitriol is a regulator of calcium homeostasis and has received much attention for its antitumor activity. Most of the biological effects of calcitriol are elicited by its binding to the vitamin D receptor (VDR). CYP24A1 is transcriptionally regulated by VDR. Human CYP24A1 (Komagata et al. 2009) and VDR (Mohri et al. 2009) are both regulated by miR-125b, which is known as a tumor suppressor gene. The decreased level of miR-125b in cancer tissues would link with the increased levels of CYP24A1 and VDR in cancer tissues. An increase in VDR in cancer tissues would augment the anti-tumor effects of calcitriol, whereas an increase of CYP24A1 would attenuate the anti-tumor effects. The antiproliferative effects of calcitriol in MCF-7 cells were significantly abolished by the overexpression of miR-125b, suggesting that miR-125b has a larger impact on the regulation of VDR rather than on the regulation of CYP24A1 in the cell system.

19.5 miRNA-Dependent Regulation of Other Factors Affecting P450 Expression and/or Activity

19.5.1 HNF4α

Hepatocyte nuclear factor 4α (HNF4α) is a master regulator of drug-metabolizing enzymes, drug transporters, and genes involved in the synthesis/metabolism of bile acids, fatty acids, cholesterol, glucose, and urea. HNF4α regulates gene expression not only via direct binding to the gene's regulatory sequences but also through the regulation of other transcriptional factors. Expression of wide ranges of P450s depends on HNF4α. Takagi et al. (2010) found that human HNF4α is regulated by miR-24 and miR-34a. Interestingly, the miR-34a recognizes an MRE in the 3′-UTR and causes translational repression, whereas the miR-24 recognizes an MRE in the coding region and causes mRNA degradation. The downregulation of HNF4α by these miRNAs resulted in the decrease of various target genes such as CYP7A1, CYP8B1 (bile acid-synthesizing enzymes), CYP27A1 (a cholesterol-metabolizing enzyme), and phosphoenolpyruvate carboxykinase (PEPCK, a gluconeogenetic enzyme). The miRNAs were added as a key member into the feedback loop for bile acid synthesis. After the report, the miR-34a-dependent downregulation of HNF4α was supported by experiments by other two research groups (Ramamoorthy et al. 2012; Wang and Burke 2013), although they suggested that miR-449a and miR-34c-5p, having an identical seed sequence, also downregulate HNF4α. Interestingly, Ramamoorthy et al. (2012) demonstrated that a single nucleotide polymorphism (SNP) in the miR-34a target site of HNF4α gene (found in only African-Americans, 4.6 % frequency) disrupted the binding of miR-34a. Validation of the clinical significance of the SNP in drug metabolism and pharmacokinetics in vivo is required.

19.5.2 PPARα and PPARγ

Peroxisome proliferator-activated receptors (PPARs) are a family of ligand-activated transcription factors implicated in adipocyte differentiation. The PPAR family includes three members: PPARα, PPARγ, and PPARβ/δ. PPARα, which is mainly expressed in the liver, regulates genes involved in fatty acid transport, catabolism, and energy homeostasis. PPARs are also involved in transactivation of multiple P450s, not only CYP4A but also CYP1A, CYP2A, CYP2B, CYP2C, CYP2E, CYP2J, and CYP3A. Kida et al. (2011) found that miR-21 and miR-27b downregulate PPARα expression in human liver-derived HuH7 cells. In a panel of human liver samples, PPARα protein levels were not positively correlated with the PPARα mRNA levels, whereas they were inversely correlated with the miR-21 levels, indicating clinical significance of the miRNA-dependent regulation in human liver. The miRNAs-dependent regulation of PPARα affected the expression of downstream genes such as acyl-CoA synthetase. miR-27b also regulates PPARγ, which is predominantly expressed in intestine and adipose tissue, and triggers adipocyte differentiation and promotes lipid storage (Karbiener et al. 2009; Lin et al. 2009). Thus, miR-27b seems to be one of the important miRNAs in adipogenesis and energy homeostasis.

Lee et al. (2010) reported that adipogenesis-regulating miR-130 downregulates PPARγ. Interestingly, PPARγ mRNA was high in obese women but low in lean women, whereas miR-130 was low in obese women but high in lean women. Thus, it seems that miR-130 reduces adipogenesis by repressing biosynthesis and perturbations in this regulation is linked to human obesity.

19.5.3 Nrf2 and Keap1

NF-E2-related factor 2 (Nrf2) is a redox-sensitive transcriptional factor regulating the expression of several detoxifying enzymes including CYP1 and CYP2A, through binding to antioxidant response element (ARE) in the promoter region of target genes, after the release from Kelch-like ECH-associated protein 1 (Keap1). Yang et al. (2011) reported that overexpression of miR-28 decreased Nrf2 protein and mRNA levels in MCF-7 cells, without affecting Keap1 protein level. Luciferase assays revealed that an MRE in the 3′-UTR was functional. Because the overexpression of miR-28 resulted in the increase of ability of colony formation of MCF-7 cells, miR-28 may control breast cancer progression by targeting the Nrf2 pathway. Eades et al. (2011) reported that miR-200a-dependent downregulation of Keap1 facilitated Nrf2 nuclear translation and transactivation of the Nrf2 target gene. In breast cancer, the miR-200 family is downregulated through DNA hypermethylation and histone deacetylation. Treatment of breast cancer cell lines with histone deacetylase inhibitor vorinostat resulted in increase of the miR-200a level, decrease of the Keap1 protein level, and increased localization of Nrf2 to promoters of target genes. In addition to these miRNAs, other miRNAs including miR-144, miR-34a, miR-132, and miR-93 have been reported to downregulate Nrf2 (Cheng et al. 2013; Singh et al. 2013). Thus, miRNAs regulate cellular redox homeostasis through the modulation of Nrf2-driven antioxidant gene expression as well as enzymes that generate reactive oxygen species, although the effects of the miRNA-dependent downregulation of Nrf2 system on P450 expression have not yet been examined.

19.5.4 Cytochrome b_5

Cytochrome b_5 (b_5), a hemoprotein localized to the endoplasmic reticulum membrane, plays an important role in the modulation of P450 catalysis through electron donation and allosteric modification. In spite of its nearly obligatory role for P450 catalysis, the regulation of human b_5 is not fully understood. There is no positive correlation between the b_5 protein and mRNA levels in human liver, suggesting posttranscriptional regulation (Takahashi et al. 2013). Takahashi et al. (2013) found that miR-223 downregulates human b_5 through binding to an MRE in the 3′-UTR. Interestingly, it was demonstrated that the miR-223-dependent downregulation of b_5 decreased CYP3A4-catalyzed testosterone 6β-hydroxylation activity and CYP2E1-catalyzed chlorzoxazone 6-hydroxylase activity with no changes of protein expression of these P450s. However, it did not affect CYP1A2-catalyzed 7-ethoxyresorufin *O*-deethylase activity, supporting the previous finding that CYP1A2 is a b_5-independent isoform. This regulatory mechanism might be an additional factor for the interindividual and intraindividual variability in drug metabolism potencies.

19.5.5 Cytochrome P450 Oxidoreductase

Cytochrome P450 oxidoreductase (POR) is another electron-transfer component for P450s. Dong et al. (2014) reported that miR-214 downregulates POR in human hepatoma-derived Bel7402 cells and rat normal hepatocyte BRL cells through binding to an MRE in the 3′-UTR. They reported that ethanol treatment of the cells in vitro and rats in vivo increased miR-214, decreased POR protein levels and activities, and induced oxidative stress. Because POR acts as an antioxidant defense enzyme through association with heme oxygenase-1 (HO-1), the authors postulated that alcohol-induced oxidative stress might in part result from the repression of POR and in turn the dysfunction of HO-1.

19.6 miRNA-Related Pharmacogenetics

Knowledge of pharmacogenetics relevant to drug metabolism has been increasing. miRNA-related polymorphisms joined in this area. Genome-wide analyses have revealed that many SNPs exist in the miRNA-binding sites in the 3′-UTR and the seed sequence of miRNAs, possibly interfering or strengthening the binding of the miRNA to target mRNAs. In addition, SNPs in pri-miRNAs or pre-miRNAs may affect the processing to alter the level of mature miRNAs. Such gain- or loss of function of miRNAs would result in changes in gene expression. Actually, some miRNA-related polymorphisms are associated with diseases (Sethupathy and Collins 2008). In silico analysis has revealed potential miRNAs that may bind to the sequence around the SNP in the 3′-UTR of CYP1A2, CYP2A6, CYP2C8, CYP2C9, CYP2D6, CYP2E1, CYP3A4, and CYP3A5, destroying or creating target sites

(Ramamoorthy and Skaar 2011; Wei et al. 2012). Experimental proof is needed for the significance of the miRNA-related SNPs in the interindividual variability of drug metabolism.

19.7 miRNAs and Disease or Drug Response

Growing evidence reveals that dysregulation of miRNAs is associated with a variety of diseases (http://202.38.126.151/hmdd/mirna/md/) (Lu et al. 2008). Therefore, miRNAs became new targets of diagnosis and therapy. miRNA expressions are also modified by exogenous and endogenous compounds such as drug, environmental chemicals, steroids, and bile acid, although the modification might be relatively less than that found in disease. The changes of miRNA expression would occur at various stages including epigenetic regulation, transcriptional regulation, and processing. A cross-linking study between miRNA modification/manipulation, miRNA-dependent regulation of drug-metabolizing enzymes, and polymorphisms will allow us realize more optimized pharmacotherapy.

19.8 Conclusions and Future Perspectives

Since the first report in 2006 for CYP1B1, it became clear during less than 10 years that miRNAs regulate the expressions and activities of a variety of P450s by direct and indirect mechanisms (Fig. 19.4), and such evidence will continue to grow. The miRNA-dependent regulation of P450s confers inter- or intraindividual differences in xenobiotic metabolism, the homeostasis of endobiotics, and cancer development or progression. To identify miRNA–target mRNA pairs, in vitro studies are usually performed, such as luciferase assays and miRNA overexpression or inhibition experiments in cells. To test whether individual mRNAs are true endogenous targets of miRNA under endogenous conditions, evaluations in vivo are warranted. Some pharmaceutical companies have initiated the development of therapeutic candidates using miRNA mimics or miRNA inhibitors for several diseases. There are several ongoing clinical trials. Because an individual miRNA could regulate several genes and pathways simultaneously, miRNA modulation could produce a powerful outcome, although attention must be paid to the possibility that miRNA manipulation may cause unanticipated effects. Metabolome analysis with plasma or urine samples under in vivo miRNA manipulation will provide some clues for understanding the role of the individual miRNA in metabolism and for facilitating our success in therapy.

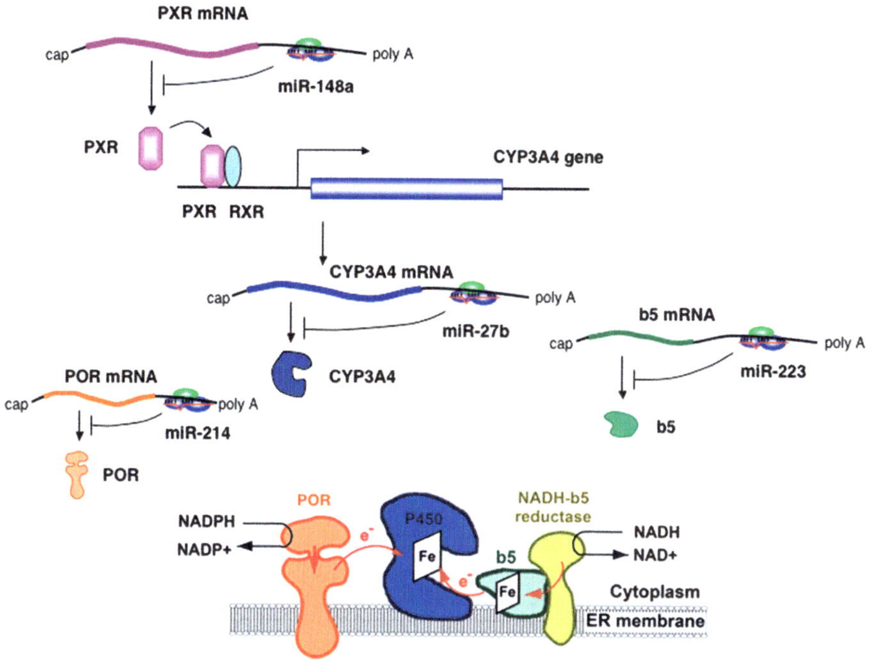

Fig. 19.4 An example of P450 regulated by multiple miRNAs at various phases. The miR-148a-dependent downregulation of pregnane X receptor (PXR) indirectly affects CYP3A4 expression. miR-27b directly regulated CYP3A4. miR-223-dependent downregulation of cytochrome b_5 (b_5) modulates CYP3A4 activity. miR-214 negatively regulates P450 oxidoreductase (POR) expression. miRNAs control P450s expression and activities conferring variability in the metabolism and xenobiotics and endobiotics. *ER* endoplasmic reticulum

References

Adams BD, Furneaux H, White BA (2007) The micro-ribonucleic acid (miRNA) miR-206 targets the human estrogen receptor-α (ERα) and represses ERα messenger RNA and protein expression in breast cancer cell lines. Mol Endocrinol 21:1132–1147

Alexiou P, Maragkakis M, Papadopoulos GL, Reczko M, Hatzigeorgiou AG (2009) Lost in translation: an assessment and perspective for computational microRNA target identification. Bioinformatics 25:3049–3055

Carrer M, Liu N, Grueter CE, Williams AH, Frisard MI, Hulver MW, Bassel-Duby R, Olson EN (2012) Control of mitochondrial metabolism and systemic energy homeostasis by microRNAs 378 and 378*. Proc Natl Acad Sci USA 109:15330–15335

Chekulaeva M, Filipowicz W (2009) Mechanisms of miRNA-mediated post-transcriptional regulation in animal cells. Curr Opin Cell Biol 21:452–460

Chen F, Chen C, Yang S, Gong W, Wang Y, Cianflone K, Tang J, Wang DW (2012) Let-7b inhibits human cancer phenotype by targeting cytochrome P450 epoxygenase 2J2. PLoS One 7:e39197

Cheng X, Ku CH, Siow RC (2013) Regulation of the Nrf2 antioxidant pathway by microRNAs: new players in micromanaging redox homeostasis. Free Radic Biol Med 64:4–11

Choi YM, An S, Lee EM, Kim K, Choi SJ, Kim JS, Jang HH, An IS, Bae S (2012) CYP1A1 is a target of miR-892a-mediated post-transcriptional repression. Int J Oncol 41:331–336

Curtis HJ, Sibley CR, Wood MJ (2012) Mirtrons, an emerging class of atypical miRNA. Wiley Interdiscipl Rev RNA 3:617–632

Dong X, Liu H, Chen F, Li D, Zhao Y (2014) MiR-214 promotes the alcohol-induced oxidative stress via down-regulation of glutathione reductase and cytochrome P450 oxidoreductase in liver cells. Alcohol Clin Exp Res 38(1):68–77

Duursma AM, Kedde M, Schrier M, le Sage C, Agami R (2008) miR-148 targets human DNMT3b protein coding region. RNA 14:872–877

Eades G, Yang M, Yao Y, Zhang Y, Zhou Q (2011) miR-200a regulates Nrf2 activation by targeting Keap1 mRNA in breast cancer cells. J Biol Chem 286:40725–40733

Esteller M (2011) Non-coding RNAs in human disease. Nat Rev Genet 12:861–874

Fabian MR, Sonenberg N, Filipowicz W (2010) Regulation of mRNA translation and stability by microRNAs. Annu Rev Biochem 79:351–379

Griffiths-Jones S, Grocock RJ, van Dongen S, Bateman A, Enright AJ (2006) miRBase: microRNA sequences, targets and gene nomenclature. Nucleic Acids Res 34(Database Issue):D140–D144

Jiang JG, Chen CL, Card JW, Yang S, Chen JX, Fu XN, Ning YG, Xiao X, Zeldin DC, Wang DW (2005) Cytochrome P450 2J2 promotes the neoplastic phenotype of carcinoma cells and is up-regulated in human tumors. Cancer Res 65:4707–4715

Kalscheuer S, Zhang X, Zeng Y, Upadhyaya P (2008) Differential expression of microRNAs in early-stage neoplastic transformation in the lungs of F344 rats chronically treated with the tobacco carcinogen 4-(methylnitrosamino)-1-(3-pyridyl)-1-butanone. Carcinogenesis (Oxf) 29:2394–2399

Karbiener M, Fischer C, Nowitsch S, Opriessnig P, Papak C, Ailhaud G, Dani C, Amri EZ, Scheideler M (2009) microRNA miR-27b impairs human adipocyte differentiation and targets PPARγ. Biochem Biophys Res Commun 390:247–251

Kida K, Nakajima M, Mohri T, Oda Y, Takagi S, Fukami T, Yokoi T (2011) PPARα is regulated by miR-21 and miR-27b in human liver. Pharm Res 28:2467–2476

Kim VN, Nam JW (2006) Genomics of microRNA. Trends Genet 22:165–173

Kloosterman WP, Plasterk RH (2006) The diverse functions of microRNAs in animal development and disease. Dev Cell 11:441–450

Komagata S, Nakajima M, Takagi S, Mohri T, Taniya T, Yokoi T (2009) Human CYP24 catalyzing the inactivation of calcitriol is post-transcriptionally regulated by miR-125b. Mol Pharmacol 76:702–709

Kumar P, Luo Y, Tudela C, Alexander JM, Mendelson CR (2013) The c-Myc-regulated microRNA-17-92 (miR-17-92) and miR-106a-363 clusters target hCYP19A1 and hGCM1 to inhibit human trophoblast differentiation. Mol Cell Biol 33:1782–1796

Lee EK, Lee MJ, Abdelmohsen K, Kim W, Kim MM, Srikantan S, Martindale JL, Hutchison ER, Kim HH, Marasa BS, Selimyan R, Egan JM, Smith SR, Fried SK, Gorospe M (2010) miR-130 suppresses adipogenesis by inhibiting peroxisome proliferator-activated receptor γ expression. Mol Cell Biol 31:626–638

Lewis BP, Burge CB, Bartel DP (2005) Conserved seed pairing, often flanked by adenosines, indicates that thousands of human genes are microRNA targets. Cell 120:15–20

Lin Q, Gao Z, Alarcon RM, Ye J, Yun Z (2009) A role of miR-27 in the regulation of adipogenesis. FEBS J 276:2348–2358

Lu M, Zhang Q, Deng M, Miao J, Guo Y, Gao W, Cui Q (2008) An analysis of human microRNA and disease associations. PLoS One 3:e3420

Mohri T, Nakajima M, Takagi S, Komagata S, Yokoi T (2009) MicroRNA regulates human vitamin D receptor. Int J Cancer 125:1328–1333

Mohri T, Nakajima M, Fukami T, Takamiya M, Aoki Y, Yokoi T (2010) Human CYP2E1 is regulated by miR-378. Biochem Pharmacol 79:1045–1052

Oda Y, Nakajima M, Mohri T, Takamiya M, Aoki Y, Fukami T, Yokoi T (2012) Aryl hydrocarbon receptor nuclear translocator in human liver is regulated by miR-24. Toxicol Appl Pharmacol 260:222–231

Pan Y-Z, Gao W, Yu A-M (2009) MicroRNAs regulate CYP3A4 expression via direct and indirect targeting. Drug Metab Dispos 37:2112–2117

Panda H, Chuang TD, Luo X, Chegini N (2012) Endometrial miR-181a and miR-98 expression is altered during transition from normal into cancerous state and target PGR, PGRMC1, CYP19A1, DDX3X, and TIMP3. J Clin Endocrinol Metab 97:E1316–E1326

Ramamoorthy A, Skaar TC (2011) In silico identification of microRNAs predicted to regulate the drug metabolizing cytochrome P450 genes. Drug Metab Lett 5:126–131

Ramamoorthy A, Li L, Gaedigk A, Bradford LD, Benson EA, Flockhart DA, Skaar TC (2012) In silico and in vitro identification of microRNAs that regulate hepatic nuclear factor 4α expression. Drug Metab Dispos 40:726–733

Sethupathy P, Collins FS (2008) MicroRNA target site polymorphisms and human disease. Trends Genet 24:489–497

Shibahara Y, Miki Y, Onodera Y, Hata S, Chan MS, Yiu CC, Loo TY, Nakamura Y, Akahira J, Ishida T, Abe K, Hirakawa H, Chow LW, Suzuki T, Ouchi N, Sasano H (2012) Aromatase inhibitor treatment of breast cancer cells increases the expression of let-7f, a microRNA targeting CYP19A1. J Pathol 227:357–366

Shukla U, Tumma N, Gratsch T, Dombkowski A, Novak RF (2013) Insights into insulin-mediated regulation of CYP2E1: miR-132/-212 targeting of CYP2E1 and role of phosphatidylinositol 3-kinase, Akt (protein kinase B), mammalian target of rapamycin signaling in regulating miR-132/-212 and miR-122/-181a expression in primary cultured rat hepatocytes. Drug Metab Dispos 41:1769–1777

Singh B, Ronghe AM, Chatterjee A, Bhat NK, Bhat HK (2013) MicroRNA-93 regulates NRF2 expression and is associated with breast carcinogenesis. Carcinogenesis (Oxf) 34:1165–1172

Takagi S, Nakajima M, Mohri T, Yokoi T (2008) Post-transcriptional regulation of human pregnane X receptor by micro-RNA affects the expression of cytochrome P450 3A4. J Biol Chem 283:9674–9680

Takagi S, Nakajima M, Kida K, Yamaura Y, Fukami T, Yokoi T (2010) MicroRNAs regulate human hepatocyte nuclear factor 4α, modulating the expression of metabolic enzymes and cell cycle. J Biol Chem 285:4415–4422

Takahashi K, Oda Y, Toyoda Y, Fukami T, Yokoi T, Nakajima M (2014) Regulation of cytochrome b_5 expression by miR-223 in human liver: effects on cytochrome P450 activities. Pharm Res 31:780–794

Tsuchiya Y, Nakajima M, Kyo S, Kanaya T, Inoue M, Yokoi T (2004) Human CYP1B1 is regulated by estradiol via estrogen receptor. Cancer Res 64:3119–3125

Tsuchiya Y, Nakajima M, Takagi S, Taniya T, Yokoi T (2006) MicroRNA regulates the expression of human cytochrome P450 1B1. Cancer Res 66:9090–9098

Wang Z, Burke PA (2013) The role of microRNAs in hepatocyte nuclear factor-4alpha expression and transactivation. Biochim Biophys Acta 1829:436–442

Wei R, Yang F, Urban TJ, Li L, Chalasani N, Flockhart DA, Liu W (2012) Impact of the interaction between 3′-UTR SNPs and microRNA on the expression of human xenobiotic metabolism enzyme and transporter genes. Front Genet 3:248

Xie X, Miao L, Yao J, Feng C, Li C, Gao M, Liu M, Gong L, Wang Y, Qi X, Ren J (2013) Role of multiple microRNAs in the sexually dimorphic expression of Cyp2b9 in mouse liver. Drug Metab Dispos 41:1732–1737

Xiong J, Yu D, Wei N, Fu H, Cai T, Huang Y, Wu C, Zheng X, Du Q, Lin D, Liang Z (2010) An estrogen receptor α suppressor, microRNA-22, is downregulated in estrogen receptor α-positive human breast cancer cell lines and clinical samples. FEBS J 277:1684–1694

Yang M, Yao Y, Eades G, Zhang Y, Zhou Q (2011) MiR-28 regulates Nrf2 expression through a Keap1-independent mechanism. Breast Cancer Res Treat 129:983–991

Zhang SY, Surapureddi S, Coulter S, Ferguson SS, Goldstein JA (2012) Human CYP2C8 is post-transcriptionally regulated by microRNAs 103 and 107 in human liver. Mol Pharmacol 82:529–540

Zhao JJ, Lin J, Yang H, Kong W, He L, Ma X, Coppola D, Cheng JQ (2008) MicroRNA-221/222 negatively regulates estrogen receptor α and is associated with tamoxifen resistance in breast cancer. J Biol Chem 283:31079–31086

Chapter 20
The Pharmacogenomics of Cytochrome P450s: From Molecular to Clinical Application

Su-Jun Lee and Jae-Gook Shin

Abstract Polymorphisms in genes encoding cytochrome P450s (CYPs) cause interindividual variability in the metabolism of a wide range of structurally diverse substrates including drugs, hormones, lipids, and environmental pollutants. Because P450s account for approximately 75 % of the total drug metabolisms, genetic polymorphisms in P450-encoding genes have garnered a good deal of attention in the context of pharmacogenomics research, resulting in the discovery of a number of clinically relevant P450 genetic polymorphisms. Advances in our understanding of P450 pharmacogenomics enhance the therapeutic efficacy of drugs and reduce their toxicities. In this chapter, we review the pharmacogenomics of major P450s from molecular discovery to clinical application, and discuss the clinical implications of genotype-guided personalized pharmacotherapy.

Keywords Allele frequency • Clinical implication • Cytochrome P450 • Drug metabolism • Drug response • Genetic polymorphism • Molecular functionality • Pharmacodynamics • Pharmacogenomics • Pharmacokinetics

20.1 Prologue

Cytochrome P450 enzymes (P450s, CYPs) are a superfamily of heme-thiolate enzymes involved in the metabolism of both xenobiotics and endogenous compounds. Since the genetic polymorphism of P450s has been first reported for the phenotype of debrisoquine (Mahgoub et al. 1977) and sparteine, both well-known substrates of CYP2D6 (in the thesis of Dr. Eichelbaum et al. 1975), numerous scientific findings have cemented the important position of P450 in biochemical and medical science. The pharmacogenetics and pharmacogenomics of drug-metabolizing

---

S.-J. Lee • J.-G. Shin (✉)
Department of Pharmacology, Inje University College of Medicine, Busan, South Korea
e-mail: phshinjg@gmail.com

H. Yamazaki (ed.), *Fifty Years of Cytochrome P450 Research*,
DOI 10.1007/978-4-431-54992-5_20, © Springer Japan 2014

P450 enzymes have historically taken center stage within the new science of personalized medicine, and the P450 enzyme has become the most extensively studied target gene in the field.

The genetic variants in P450s cause altered enzyme activity of oxidative metabolism of xenobiotics. Accordingly, genetic polymorphisms of P450s are associated with the interindividual variation of drug or metabolite concentrations and clinical responses such as efficacy and toxicity. Additionally, P450s also catalyze many endogenous compounds as their substrates, leading to reported genetic polymorphism involvement in the pathogenesis of diseases such as increased susceptibility to both bladder cancer and Parkinson's disease, although these findings require further validation. In this chapter, we focus primarily on the clinical relevance of genetic polymorphisms of drug response, rather than their association with disease development.

Owing to advances in molecular genomics and sequencing technology, reports of novel functional genetic variants of P450s in different ethnic populations have been accumulated. Unsurprisingly, since 1999 more than 660 alleles of functionally validated genetic variants from 29 P450 genes in the cytochrome P450 have been published on the Allele Nomenclature website to date (http://www.cypalleles.ki.se). Although somewhat limited to specific therapeutic drugs, pharmacogenetic biomarkers have already been applied to the genotype-guided personalized pharmacotherapy in clinical practice. The clinical relevance and clinical utility of genomic biomarkers should therefore be proven to explore avenues for their application in personalized pharmacotherapy. A table describing the valid genomic biomarkers that are currently part of the U.S. Food and Drug Administration (FDA)-approved drug labels can be found on the FDA website at the following address: http://www.fda.gov/Drugs/ScienceResearch/ResearchAreas/Pharmaco genetics/ucm083378.htm. Among the 120 drugs with pharmacogenomics biomarkers presently listed on this website, about half have been identified to be significantly influenced by P450 genetic polymorphisms, enabling us to see the extent to which P450 genetic biomarkers are the most extensively used predictive genetic biomarker for personalized medicine to date. In this chapter, we primarily focus on the major functional allelic variants with molecular aspects and their clinical relevance as genetic biomarkers.

20.2 CYP2C9 Pharmacogenomics

20.2.1 Molecular Genomics of CYP2C9

Of the four human *CYP2C* genes, CYP2C9 is the most abundantly expressed in the liver (Goldstein and de Morais 1994), metabolizing about 15 % of clinically used drugs (Rendic and Di Carlo 1997), including the anti-diabetic glipizide; the anti-convulsant phenytoin; the anti-coagulant warfarin; the anti-hypertensive drug

Table 20.1 Major genetic variants of *CYP2C9* and their molecular consequences

			Allele frequency (%)		
Allele	Effect	Consequence	Asian[a]	Caucasian[b]	African-American[c]
*CYP2C9*1*	Wild-type	Normal	95.3	80.4	95.4
*CYP2C9*2*	R144C	Decreased activity	1.23	11.3	3.0
*CYP2C9*3*	I359L	Decreased activity	3.44	8.3	1.6

When the level of any variant was not determined, the residual frequency was defaulted to *1
[a]Calculated from data on 3,804 alleles gathered from seven ethnically distinct populations: Korean, Chinese, Japanese, Taiwanese, Indian, Malay, and Vietnamese
[b]Calculated from data on 3,028 alleles gathered from five ethnically distinct populations: American, British, Spanish, Turkish, and Italian
[c]Calculated from data on 500 alleles gathered from two ethnically distinct populations: African-American and Ethiopian

losartan; the diuretic torasemide; and nonsteroidal antiinflammatory drugs (NSAIDs) such as ibuprofen, diclofenac, piroxicam, tenoxicam, and mefenamic acid. Several polymorphisms in *CYP2C9* have been shown to alter CYP2C9 activity, causing interindividual variation in the metabolism and disposition of the drugs just mentioned. No-activity alleles (null variants) include *CYP2C9*6*, **15*, and **25* (Kidd et al. 2001; Maekawa et al. 2006; Zhao et al. 2004). Human *CYP2C9* variants identified in vivo and characterized in vitro at the molecular level include *CYP2C9*2*, **3*, **4*, **5*, **6*, **11*, **13*, and **14* (Allabi et al. 2004; King et al. 2004; Lee et al. 2002; Si et al. 2004). The two most common (and clinically important) alleles are *CYP2C9*2* and *CYP2C9*3* (Dickmann et al. 2001; Goldstein 2001; Scordo et al. 2001; Taube et al. 2000). Ethnic differences in the frequency of occurrence of major functional alleles are summarized in Table 20.1. The frequency of *CYP2C9*2* is 11 % in whites (Aynacioglu et al. 1999; Dickmann et al. 2001; Garcia-Martin et al. 2001; Scordo et al. 2001; Taube et al. 2000) and about 3 % in blacks (Scordo et al. 2001; Sullivan-Klose et al. 1996). However, *CYP2C9*2* is very rare or absent in Asians (Jose et al. 2005; Lee et al. 2010; Lee et al. 2005; Nasu et al. 1997; Ngow et al. 2009; Sullivan-Klose et al. 1996; Wang et al. 1995; Yoon et al. 2001). *CYP2C9*3* is found in 8 % of whites (Aynacioglu et al. 1999; Dickmann et al. 2001; Garcia-Martin et al. 2001; Scordo et al. 2001; Taube et al. 2000), less than 2 % of blacks (Scordo et al. 2001; Sullivan-Klose et al. 1996), and 3 % of Asians (Jose et al. 2005; Lee et al. 2010; Lee et al. 2005; Nasu et al. 1997; Ngow et al. 2009; Sullivan-Klose et al. 1996; Wang et al. 1995; Yoon et al. 2001). *CYP2C9*2* carries an Arg144Cys substitution located on the exterior of the enzyme, affecting the interaction with NADPH-cytochrome P450 reductase. Depending on the substrate, CYP2C9.2 exhibits 70–90 % of the activity of the wild-type enzyme (CLint) in various in vitro systems (Lee et al. 2002). *CYP2C9*3* features an Ile359Leu change located in the substrate recognition site and exhibits 10–30 % of the activity of the wild-type enzyme when metabolizing various substrate drugs. In most studies, the *CYP2C9*3* variant consistently metabolized CYP2C9-substrate drugs to a lesser extent than the *CYP2C9*2* variant (Lee et al. 2002), suggesting that the *CYP2C9*3* allele is more defective than the

*CYP2C9*2* in terms of drug pharmacokinetics and pharmacodynamics. The underlying cause of the reduced activities of both the *CYP2C9*2* and **3* variants has been suggested to be attributable in part to alterations in coupling of the P450 reaction cycle, which includes the P450 enzyme, the electron-transfer system, molecular oxygen, the substrate, and water (Wei et al. 2007). The metabolic activities of *CYP2C9* variant alleles are substrate specific (Wei et al. 2007), suggesting that a particular *CYP2C9* allele can be associated with a unique pattern of drug metabolism. Although characterization of *CYP2C9* variants in vitro is useful to predict drug responses in vivo, it remains difficult to definitively describe how variants metabolize drugs in vivo, particularly if the variant occurs at a low frequency. Further, no single in vitro system has been adopted as a standard protocol for evaluation of functional changes in drug metabolism. Therefore, comparisons of the activities of P450 variants with those of the wild-type enzyme should be interpreted with great caution. This caveat is true not only of CYP2C9 but also of all other P450s.

20.2.2 Clinical Relevance of CYP2C9 Pharmacogenomics

A total of 57 alleles in CYP2C9 have been described; see http://www.cypalleles.ki.se/cyp2c9.htm. The clinical implications of such polymorphisms have been highlighted by study of the two principal variants, *CYP2C9*2* and **3*. The most significant clinical problems are the dose-related toxicities of warfarin and phenytoin in humans with CYP2C9 "poor metabolizer" (PM) genotypes (Kirchheiner and Brockmoller 2005; Ninomiya et al. 2000). A PM genotype is defined as possession of two defective CYP2C9 alleles, for example, *CYP2C9*2/*2*, *CYP2C9*2/*3*, or *CYP2C9*3/*3*.

Warfarin is commonly prescribed to prevent thromboembolism. However, the therapeutic index is narrow, and the dose must fall within the safe range of the international normalized ratio (INR). Although the thrombosis risk increases if subtherapeutic doses of warfarin are given, the bleeding risk rises significantly when the INR exceeds 4. Warfarin is a racemic mixture of the *R*- and *S*-compounds. *S*-warfarin is the more potent and is metabolized exclusively by CYP2C9 to yield (inactive) 7-hydroxy warfarin, whereas *R*-warfarin is metabolized by all of CYP1A1, CYP2C19, and CYP3A4 (Steward et al. 1997). The *CYP2C9* genotype has been associated with bleeding risk caused by excessive over-anticoagulant action. Those patients with *CYP2C9*2* or **3* alleles are at significantly higher risk of bleeding and require prolonged hospitalization because enzyme function is impaired. Therefore, patients with the *CYP2C9*2* and **3* alleles require lower warfarin doses than do others. It is important to determine if patients express these alleles, particularly to guide the choice of the starting dose of warfarin (Kaminsky and Zhang 1997; Mannucci 1999; Taube et al. 2000). The stabilized daily dose of warfarin varies with genotype, averaging 7.9 mg/day for *CYP2C9*1/*1* patients and 2.2 mg/day for those who are *CYP2C9*1/*3* (Loebstein et al. 2001; Ogg et al. 1999).

The required warfarin dose varies with ethnicity. The International Warfarin Pharmacogenomics Consortium examined pooled genotypic and phenotypic data from 5,700 warfarin-treated patients, and found that the average daily dose was 5 mg/day for *CYP2C9*1/*1* Caucasians, 3.2 mg/day for *CYP2C9*1/*1* Asians, 1.1 mg/day for *CYP2C9*3/*3* Caucasians, and 2.0 mg/day for *CYP2C9*3/*3* Asians. The average warfarin doses in Caucasian, African-American, and Asian heterozygotes (*CYP2C9*1/*3*) were 3.4, 5.0, and 2.3 mg/day, respectively (Limdi et al. 2010). Also, African-Americans with the *CYP2C9*5* (D360E), **6* (frameshift; null allele), **8* (R150H), or **11* (R335W) mutations had lower warfarin requirements than did those with the **1/*1* genotype (median dose, 5.0 mg/day vs. 6.1 mg/day; $P = 0.004$) (Cavallari et al. 2010). One meta-analysis found that the *CYP2C9*2* and **3* variations explained about 12 % of total variability in warfarin dose requirement and *VKORC1* polymorphisms about 25 % (Au and Rettie 2008). The human hepatic expression level of CYP2C9 varied with age (Koukouritaki et al. 2004), suggesting that the effect of *CYP2C9* genotype on warfarin requirement changes over the lifespan. A recent study found that CYP2C9 variation was of minimal importance in terms of the warfarin dose required by pediatric patients (Nowak-Gottl et al. 2010), again showing that age is important when constructing an algorithm to be used to calculate a warfarin dose. The FDA has approved a pharmacogenetic test to improve the safety profile of warfarin and to aid in choosing the initial dose. A comprehensive dosing model incorporating clinical factors and profiles of genes associated with warfarin metabolism is required in addition to *CYP2C9* genetics. In fact, several algorithms allowing clinicians to determine initial warfarin doses are currently available. However, most algorithms feature only common *CYP2C9* variants, excluding low-frequency alleles.

Phenytoin is metabolized by both CYP2C9 and CYP2C19, but CYP2C9 eliminates 80–90 % of the drug (Giancarlo et al. 2001; Miners and Birkett 1998). Phenytoin also has a narrow therapeutic index, and changes in CYP2C9 activity caused by mutation can trigger intoxication of the nervous system (Kidd et al. 2001; Ninomiya et al. 2000). For example, a patient homozygous for *CYP2C9*6* (frameshift; a null allele) suffered a medical emergency caused by severe phenytoin toxicity because of impaired drug metabolism; the phenytoin half-life in this patient was 13 days (thus very long) (Kidd et al. 2001). Patients with a defective *CYP2C9* allele require 30 % less phenytoin, in the steady state, than do those with the *CYP2C9*1/*1* genotype (van der Weide et al. 2001). Similarly, the drug has a significantly longer half-life (threefold) in patients with the *CYP2C9*3/*3* genotype compared to those who are *CYP2C9*1/*1* (Kidd et al. 1999). In general, the *CYP2C9* genotype is relevant in those treated with phenytoin. A number of *CYP2C9* variants differing in terms of activity are known, but the *CYP2C9*2* and **3* alleles are the best studied. The effects of other variants on the clinical outcomes of phenytoin-treated patients remain to be determined.

CYP2C9 is a major contributor to the metabolism and disposition of sulfonylureas, including the anti-diabetic drugs glibenclamide and glimepiride. Patients with the *CYP2C9*3* genotype have higher drug AUCs (areas under the curve) and reduced drug clearance rates (Kirchheiner et al. 2002). However, the clinical consequences of *CYP2C9* polymorphisms (for example, variation in blood glucose levels) remain controversial (Kirchheiner et al. 2002; Niemi et al. 2002), because sulfonylurea pharmacodynamics are affected by many different genes and physiological factors (Xu et al. 2009). Further studies are required to understand the interplay of pharmacokinetics and pharmacodynamics in this context, in turn allowing the doses of sulfonylurea drugs to be modified in diabetic patients differing in *CYP2C9* genotype.

20.3 CYP2C19 Pharmacogenomics

20.3.1 Molecular Genomics of CYP2C19

CYP2C19 metabolizes about 10 % of all drugs given to humans, including proton pump inhibitors (PPIs); the anti-convulsant mephenytoin; the anti-malarial proguanil; the anxiolytic drug diazepam; antidepressants including citalopram, imipramine, amitriptyline, and clomipramine; and the anti-platelet agent clopidogrel. A number of *CYP2C19* polymorphisms alter CYP2C19 activity, resulting in interindividual variability in the plasma concentrations of drugs that are substrates of CYP2C19. The most common and clinically well-characterized variants are *CYP2C19*2*, *CYP2C19*3*, and *CYP2C19*17* (Table 20.2) (Lee 2012). *CYP2C19*2* carries a single base-pair mutation (G to A) in exon 5 that produces a splicing defect, resulting in a truncated nonfunctional protein (De Morais et al. 1994a). *CYP2C19*3* has a single base-pair mutation (G to A) in exon 4,

Table 20.2 Major genetic variants of CYP2C19 and their molecular consequences

			Allele frequency (%)		
Allele	Effect	Consequence	Asian[a]	Caucasian[b]	African-American[c]
*CYP2C19*1*	Wild-type	Normal	61.5	69.6	80.2
*CYP2C19*2*	Splicing defect	Absent	28.0	14.9	22.2
*CYP2C19*3*	Stop codon	Absent	9.3	0.7	0
*CYP2C19*17*	Promoter variant	Increased transcription	1.8	20.2	17.9

When the level of any variant was not determined, the residual frequency was defaulted to *1. Some reports genotyped variants *2 and *3, but not *17, and frequencies were calculated from data on genotyped subjects only. For example, the frequency of *17 in Asia was derived from analysis of 1,172 alleles in Korean, Chinese, and Japanese populations; no data were available from India or Vietnam

[a]Calculated from data on 1,678 alleles gathered from five ethnically distinct populations: Korean, Chinese, Japanese, Indian, and Vietnamese

[b]Calculated from data on 2,856 alleles gathered from five ethnically distinct populations: American, British, Spanish, Turkish, and Italian

[c]Calculated from data on 682 alleles gathered from two ethnically distinct populations: African-American and Ethiopian

which creates a premature stop codon (De Morais et al. 1994b). Other null variant alleles include *CYP2C19*4*, which has a mutation (A to G) in the initiation codon, creating a GTG start (Ferguson et al. 1998). *CYP2C19*7* carries a T-to-A mutation in the invariant GT doublet at the 5′-donor splice site of intron 5 (Ibeanu et al. 1999). Several *CYP2C19* variants have been identified by sequencing the *CYP2C19* genes of individuals identified by clinicians as PMs or outliers in terms of their phenotypic reactions to CYP2C19-substrate drugs, such as mephenytoin or omeprazole (Balian et al. 1995; Wilkinson et al. 1989). Functional variants discovered in such studies include *CYP2C19*5*, **6*, **8*, **16*, and **26* (Ibeanu et al. 1999; Ibeanu et al. 1998a, b; Lee et al. 2009a, b; Morita et al. 2004; Xiao et al. 1997). Of the defective *CYP2C19* alleles, *CYP2C19*2* and **3* contribute to most PM genotypes (Desta et al. 2002; Lee 2012). The PM trait is inherited in an autosomal recessive manner, and PM genotypes differ with ethnicity. PMs constitute 3–6 % of whites and blacks, 13–23 % of Asians, and 38–79 % of Polynesians and Micronesians (46, 59, and 60 %) (Kaneko et al. 1999; Lee 2012; Xie et al. 2001). Interindividual variations in the omeprazole plasma concentration are also observed in subjects homozygous for *CYP2C19*1*, and part of this variation was explained by the discovery of *CYP2C19*17* (Sim et al. 2006), which carries a C-to-T change in the 5′-flanking region of *CYP2C19*. The change increases transcriptional activity. Thus, individuals who are *CYP2C19*17/*17* rapidly clear CYP2C19-substrate drugs (Ingelman-Sundberg et al. 2007; Rudberg et al. 2008). Individuals homozygous for *CYP2C19*17* are likely to suffer therapeutic failure, compared to EM (extensive metabolizer) patients (*CYP2C19*1/*1*), upon treatment with certain PPIs and antidepressants.

20.3.2 Clinical Relevance of CYP2C19 Pharmacogenomics

The most extensively studied and clinically relevant polymorphic variants of *CYP2C19* are *CYP2C19*2*, **3*, and **17*. These polymorphisms significantly affect the metabolism of PPIs, diazepam, and clopidogrel in patients with loss-of-function (LOF) alleles. Individual variations in the plasma concentrations of various PPIs can be explained by changes in the levels of CYP2C19 activity, which are greatly influenced by *CYP2C19* polymorphisms. In general, patients with *CYP2C19* LOF alleles have higher plasma concentrations of PPIs and enhanced suppression of gastric acid production. Although the PPIs omeprazole, esomeprazole, pantoprazole, lansoprazole, and rabeprazole are all metabolized by CYP2C19, differences in metabolism and pharmacokinetics can translate into differences in clinical outcomes. For example, the AUC ratios in PMs and EMs differ in the decreasing order of omeprazole, pantoprazole, lansoprazole, and rabeprazole (Funck-Brentano et al. 1997; Furuta et al. 2004). Although metabolism of all these PPIs is affected by the *CYP2C19* genotype, rabeprazole metabolism appears to be less influenced by this genotype than is the metabolism of other PPIs. Support for this contention was afforded by recent studies showing that the presence of

CYP2C19 LOF alleles was associated with enhanced eradication of *Helicobacter pylori* in patients taking PPI-based triple therapies including omeprazole or lansoprazole, but not when rabeprazole or esomeprazole was used (Serrano et al. 2012; Tang et al. 2013). The *CYP2C19* genotype affects the cure rates of *H. pylori* infection in peptic ulcer patients. For example, the cure rates of peptic or duodenal ulcers in Japanese patients given dual omeprazole and amoxicillin therapy were 100 % in *CYP2C19* PMs, 60 % in patients heterozygous for one LOF, and 29 % in individuals homozygous for *CYP2C19*1* (Furuta et al. 1998). Both earlier and recent studies show that possession of *CYP2C19* LOF alleles aids in treatment of gastric ulcers and gastroesophageal reflux disease because the plasma concentrations of prescribed PPIs are higher in such patients (Furuta et al. 2004).

CYP2C19 metabolizes diazepam and the antidepressants citalopram, escitalopram, and sertraline. The plasma half-life of diazepam is about fourfold longer in PM than EM patients (Qin et al. 1999; Wan et al. 1996). Many Hong Kong physicians routinely prescribe lower doses of diazepam for Chinese than Caucasian patients (Kumana et al. 1987) because the *CYP2C19*2* and **3* alleles occur more frequently in Asians. Diazepam may be toxic to PMs, and the dose must thus be carefully controlled, particularly in Asians.

Depending on the substrate drug, an increase in CYP2C19 activity can cause toxicity or improve efficacy. Clopidogrel is an antiplatelet agent and requires hepatic bioactivation into an active thiol metabolite, which irreversibly binds to platelet P2Y12 protein, inhibiting platelet activation and subsequent aggregation. Most clopidogrel (85 %) is inactivated by esterases and the remainder (15 %) is available for bioactivation into the active metabolite. The major clopidogrel activators are CYP2C19, CYP3A4, and CYP2B6 (Ancrenaz et al. 2012; Kazui et al. 2010). Marked interindividual variations in responses to clopidogrel are evident on ex vivo platelet aggregation testing (Combescure et al. 2010). Reduced inhibition of platelet aggregation by clopidogrel has been associated with increased risks of occurrence of major cardiac events, including infarction and thrombosis (Bonello et al. 2010). Variation in the clopidogrel response is largely attributable to differences in the pharmacokinetics of the active clopidogrel metabolite, which suggests that PM individuals may be inadequately protected from thrombotic risk by clopidogrel. Indeed, higher incidences of stent thrombosis and major adverse cardiovascular events were recorded in PM or IM (intermediate metabolizer) patients than in EM individuals (Mega et al. 2009). However, the protection afforded by clopidogrel against myocardial infarction was elevated in patients with the *CYP2C19*17* allele (Tiroch et al. 2010), but this was associated with an increased risk of bleeding because of the high rate of biotransformation into the active metabolite (Sibbing et al. 2010). The relevance of *CYP2C19* genotype to clopidogrel therapy is receiving a great deal of attention. In general, clopidogrel-treated patients with *CYP2C19* LOF alleles exhibit a higher rate of major adverse cardiac events and reduced inhibition of platelet activity compared to EM patients (Hulot et al. 2006). *CYP2C19* PMs may not benefit from clopidogrel. Alternative drugs, such as prasugrel, should be considered. No definitive guidelines on clopidogrel dose adjustment in patients differing in *CYP2C19* genotype have been formulated. Further large-scale tests of genotype-guided clopidogrel therapy, compared with other therapy options, are required.

20.4 CYP2D6 Pharmacogenomics

20.4.1 Molecular Genomics of CYP2D6

CYP2D6 metabolizes approximately 20–25 % of all currently prescribed drugs (Brockmoller et al. 2000; Cascorbi 2003). However, the hepatic level of the enzyme is relatively low, comprising only about 2 % of all liver P450 content (Shimada et al. 1994). CYP2D6 metabolizes drugs of many different classes including antipsychotics (haloperidol, clozapine, and risperidone), anti-arrhythmic agents (flecainide and perphenazine), β-adrenoreceptor antagonists (metoprolol, propranilol, bupralol, carvedilol), opioids (codeine, tramadol), tricyclic antidepressants (imipramine, clomipramine, nortriptyline, amitriptyline), and the estrogen-receptor antagonist tamoxifen. Triggers of CYP2D6 induction have not been described to date. Genetic polymorphisms create a great deal of interindividual variation in enzyme activity, ranging from the complete lack of functional protein to overexpression of protein encoded by several active alleles. Four CYP2D6 phenotypes are known. UMs (ultra-rapid metabolizers) carry more than two functional alleles; EMs carry at least one functional allele; IMs two alleles of reduced function or one such allele and a null allele; and PMs two nonfunctional alleles. Ethnic differences in phenotype are evident. UMs occur principally in North Africa and Oceania, IMs in Asia, and PMs in white populations (Xie et al. 2001). *CYP2D6* is one of the most polymorphic P450 genes. Currently, 105 *CYP2D6* variant alleles are recognized by the Human Cytochrome P450 Allele Nomenclature Committee (http://www.cypalleles.ki.se/cyp2d6.htm). Of these, nonfunctional alleles include **3*, **4*, **5*, **6*, **7*, **8*, **11*, **12*, **13*, **14*, **15*, **16*, **18*, **19*, **20*, **21*, **31*, **38*, **40*, **42*, **44*, **56*, **60*, **62*, **100*, and **101* (Broly et al. 1995; Chida et al. 1999; Daly et al. 1996; Daly et al. 1995; Evert et al. 1994; Gaedigk et al. 1991; Gaedigk et al. 2002; Gaedigk et al. 2003; Gough et al. 1990; Hanioka et al. 1990; Ji et al. 2002; Kagimoto et al. 1990; Klein et al. 2007; Leathart et al. 1998; Lee et al. 2009a, b; Li et al. 2006; Marez-Allorge et al. 1999; Marez et al. 1996; Marez et al. 1997; Marez et al. 1995; Panserat et al. 1995; Sachse et al. 1997; Sachse et al. 1999; Saxena et al. 1994; Shimada et al. 2001; Steen et al. 1995; Wang et al. 1999; Yamazaki et al. 2003; Yokoi et al. 1996; Yokota et al. 1993).

The clinical impact of nonfunctional alleles is greater than that of any functional allele, and most nonfunctional alleles impact the PM phenotype. Of nonfunctional variants, the most frequently studied alleles are *CYP2D6*3* (a frameshift),

Table 20.3 Major genetic variants of *CYP2D6* and their molecular consequences

Allele	Effect	Consequence	Allele frequency (%)		
			Asian[a]	Caucasian[b]	African-American[c]
*CYP2D6*1*	Wild-type	Normal	23–43	35–40	28–56
*CYP2D6*2*	R269C, S486T	Normal	12–20	22–34	10–78
*CYP2D6*3*	Frameshift	No activity	0	1.7	0.2
*CYP2D6*4*	Splicing defect	Absent	2.0	14.0	5.3
*CYP2D6*5*	Gene deletion	Absent	5.7	2.6	4.1
*CYP2D6*10*	P34S, S486T	Decreased activity	48.4	6.1	5.5
*CYP2D6*17*	T107I, R269C, S486T	Decreased activity	0.2	0.7	12.0
*CYP2D6*1xN*	Gene multiplication	Increased activity	0.2	0.8	0
*CYP2D6*2xN*	Gene multiplication	Increased activity	0.9	1.9	9.9

When the level of any variant was not determined, the residual frequency was defaulted to *1
[a]Calculated from data on 2,486 alleles gathered from five ethnically distinct populations: Korean, Chinese, Japanese, Taiwanese, and Malay
[b]Calculated from data on 2,154 alleles gathered from four ethnically distinct populations: American, Spanish, Turkish, and Italian
[c]Calculated from data on 566 alleles gathered from three ethnically distinct populations: African-American, Mozabite, and Ethiopian

**4* (a splice variant), and **5* (a deletion) (Table 20.3). The most common CYP2D6 variants exhibiting reduced function are *CYP2D6*10*, **17*, and **41*. Some variants differ in terms of ethnic frequencies. For example, *CYP2C6*3* is found in 2–5 % of whites, <1 % of blacks, and not yet in Asians. *CYP2D6*4* occurs in about 14 % of whites, <5 % of blacks, and <2 % of Asians. This varying frequency of *CYP2D6*4* is the major cause of ethnic differences in PM status. In general, PM phenotype is found in 5–10 % of white, 0–19 % of black, and <1 % of Asian patients (Gaedigk 2000; Meyer and Zanger 1997). However, the frequency of *CYP2D6*5* is similar in different ethnic groups, ranging from 3 % to 6 % (Eichelbaum et al. 2006; Ingelman-Sundberg 2005). The *CYP2D6*10* allele is most commonly found in Asians (up to 63 % of such populations), which means that a greater proportion of Asians (compared to other populations) are IMs. The prevalence of the UM phenotype is greatest in Northeast Africa and Oceania. The frequency of *CYP2D6*2xN* is about 28 % in the Mozabite population of Algeria and about 14 % in Ethiopians (Gaedigk et al. 2007; Sistonen et al. 2007). Genetic variants associated with reduced levels of CYP2D6 activity include *CYP2D6*10*, **14*, **17*, **18*, **36*, **41*, **47*, **49*, **50*, **51*, **54*, **55*, **57*, **59*, **62*, and **72*. Alleles associated with activity levels similar to the wild type include **27*, **39*, and **48* (Sakuyama et al. 2008). As most variant alleles have been characterized in different laboratories using various protocols, precise quantitation of variant activities (compared to that of the wild-type protein) remains difficult. As also mentioned in the context of CYP2C9, comparisons between the activities of CYP2D6 variants and the wild type should be interpreted with great caution.

20.4.2 Clinical Relevance of CYP2D6 Pharmacogenomics

Individuals carrying duplications or other multiples of active *CYP2D6* would be expected to exhibit significantly enhanced clearance rates of CYP2D6-substrate drugs, resulting in a lack of therapeutic effect even when standard doses are given. In contrast, other drugs, including codeine and tramadol, are effectively biotransformed into the active forms in UM individuals (with UM genotypes) and are associated with an increased risk of attainment of toxic plasma concentrations of morphine and *O*-desmethyltramadol, respectively. However, *CYP2D6* PMs have lower plasma levels of active drugs, associated with minimal analgesic effects at the usual doses. As already described, many drugs are metabolized by CYP2D6, and polymorphisms in this gene influence drug responses. *CYP2D6* variants are associated with variations in the plasma concentrations of antidepressants, antipsychotics, and selective serotonin reuptake inhibitors (Baumann et al. 2004; Kirchheiner et al. 2004; Thuerauf and Lunkenheimer 2006). For example, *CYP2D6* PM genotypes have been associated with higher plasma concentrations of tricyclic antidepressants (amitriptyline, protriptyline, nortriptyline, imipramine) (Bijl et al. 2008; Kwadijk-de Gijsel et al. 2009); antipsychotic drugs (perphenazine, risperidone) (Linnet and Wiborg 1996; Scordo et al. 1999); and selective serotonin reuptake inhibitors (paroxetine, fluoxetine) (Fjordside et al. 1999; Sindrup et al. 1992); associated with higher incidences of undesirable side effects. Similarly, *CYP2D6* UMs should be given higher doses of such drugs to allow therapeutic plasma concentrations to be attained (Kawanishi et al. 2004; Rau et al. 2004). It has been suggested that physicians should consider reducing the dose of tricyclic antidepressants by about 50 % for PM patients and increasing the dose by 140–180 % for UMs (Kirchheiner et al. 2004; Kitzmiller et al. 2011). However, routine testing of the *CYP2D6* genotype has not been approved by the FDA, because the clinical significance of this genotype in terms of responses to antidepressants and antipsychotics has not yet been validated. Interestingly, some CYP2D6-substrate drugs also inhibit the CYP2D6 enzyme and thus either mask or mimic genotypic effects. Therefore, depending on the severity of inhibition, co-treatment with CYP2D6 inhibitors can either reduce therapeutic effects or allow unmetabolized drugs to reach toxic proportions. The drugs known to strongly inhibit CYP2D6 in vivo include quinidine, paroxetine, and fluoxetine.

CYP2D6 is the principal enzyme metabolizing tamoxifen into the active form that is critical to exert antitumor activity in breast cancer patients (Dehal and Kupfer 1997; Jordan et al. 1977). Accumulated evidence supports the notion that a relationship between the *CYP2D6* genotype and the response to tamoxifen is clinically relevant (Dieudonne et al. 2010; Madlensky et al. 2011). Treatment of *CYP2D6* PMs with tamoxifen has been associated with poor clinical outcomes, including a significantly increased incidence of recurrence, shorter relapse-free survival, and reduced event-free survival (Goetz et al. 2005; Lim et al. 2007; Schroth et al. 2007). Similarly, patients with the *CYP2D6* UM genotype had significantly longer relapse-free intervals, less cancer recurrence, and longer event-free survival than did *CYP2D6* EMs (Flockhart 2008). As for CYP2D6, CYP2C19 contributes to formation of the active forms of tamoxifen (4-hydroxytamoxifen and endoxifen). Therefore, *CYP2D6* EM patients with the *CYP2C19*17* genotype experienced more favorable clinical outcomes than did carriers of the *CYP2C9*1*, **2*, and **3* alleles (Schroth et al. 2007).

Codeine is metabolized to morphine by CYP2D6 (Xu et al. 1995), and morphine toxicity can thus develop in *CYP2D6* UM patients (Dalen et al. 1997; Koren et al. 2006). A relationship between *CYP2D6* genotype and morphine toxicity was evidenced by a case report on a breast-fed infant who succumbed to morphine toxicity. Both the mother and the infant were *CYP2D6* UMs, resulting in high morphine concentrations in breast milk and blood caused by rapid and extensive conversion of codeine to morphine (Koren et al. 2006). Indeed, severe respiratory depression and abdominal pain have been reported in *CYP2D6* UM patients treated with codeine-containing analgesics (Dalen et al. 1997; Gasche et al. 2004), further supporting the clinical relevance of the *CYP2D6* genotype in terms of the response to codeine. *CYP2D6* UMs have an approximately 50 % higher concentration of blood morphine compared to EMs. However, individuals with defective *CYP2D6* alleles had very low or undetectable morphine levels and experienced minimal analgesic relief when given the usual dose of codeine (Caraco et al. 1996; Lotsch et al. 2004).

Tramadol is a prodrug that is converted to the pharmacologically active *O*-desmethyltramadol by CYP2D6, various alleles of which are known to be associated with the *O*-desmethyltramadol concentration, causing treatment efficacy to vary (Pedersen et al. 2006; Stamer et al. 2007). As was also true of codeine, *CYP2D6* PM patients experienced fewer analgesic effects than did EM patients (Fliegert et al. 2005; Poulsen et al. 1996; Stamer et al. 2003). *CYP2D6* UM patients had higher concentrations of *O*-desmethyltramadol and experienced better pain control than do EM patients, but increased incidences of side effects including nausea and respiratory depression were reported in such patients (Kirchheiner et al. 2008; Stamer et al. 2008).

Many beta-blockers are metabolized by CYP2D6, and the pharmacokinetic parameters of drugs such as carvedilol, metoprolol, propranolol, and timolol are influenced by *CYP2D6* polymorphisms (Mehvar and Brocks 2001; Zhou 2009). However, the clinical significance of such effects remains unclear. Also, beta-blockers have wide therapeutic indices. Therefore, beta-blocker-metabolizing genes may have minimal effects on drug responses. However, polymorphisms in the gene encoding the beta-1-adrenergic receptor gene affected the clinical responses of hypertensive patients to metoprolol (Johnson et al. 2003). Combination analysis of polymorphisms in genes encoding the drug-metabolizing enzyme (CYP2D6), and drug response genes, would enable clinicians to incorporate genotypic data into clinical practice with confidence.

20.5 Molecular Genomics and Clinical Relevance of Other P450s

20.5.1 CYP1A2

CYP1A2 is involved in the metabolism of commonly prescribed drugs including clozapine, theophylline, olanzapine, imipramine, and paracetamol. The enzyme is expressed predominantly in the liver, comprising about 10 % of total hepatic P450 (Shimada et al. 1994; Zanger et al. 2008). The *CYP1A2*1F* (−163C to A in intron 1) allele is subject to enhanced induction by caffeine, smoking, and omeprazole treatment, compared to *CYP1A2*1A* (Bondolfi et al. 2005; Han et al. 2002; Sachse et al. 1999); this is associated with increased metabolism of CYP1A2-substrate drugs, altering therapeutic outcomes (Eap et al. 2004) and increasing the risk of development of certain cancers (Moonen et al. 2005; Saebo et al. 2008). For example, the plasma concentrations of clozapine are reduced in smokers, and smoking cessation caused plasma levels to rise and side effects to develop (Bondolfi et al. 2005).

20.5.2 CYP2B6

CYP2B6 is involved in the metabolism of artemisinin, ketamine, methadone, bupropion, efavirenz, nevirapine, cyclophosphamide, and ifosfamide (Turpeinen et al. 2006; Turpeinen and Zanger 2012). *CYP2B6* is one of the most polymorphic enzymes, and ethnic differences in the frequencies of variants are evident. The most common functionally defective *CYP2B6* variant is *CYP2B6*6* (15–60 % of all mutations in different ethnic populations) (Zanger and Klein 2013). CYP2B6*6 contains two amino acid changes (Q172H and K262R), causing an approximately 75 % decrease in hepatic expression level (Desta et al. 2007; Lang et al. 2001) and substrate-specific differences compared to the wild type (Hesse et al. 2004; Xie et al. 2003). The other important allele, *CYP2B6*18* (I328T; 4–12 % of all mutations) occurs predominantly in Africans. Central nervous system side effects have been reported in a number of human immunodeficiency virus (HIV) patients taking efavirenz, suggested to be attributable to variations in the plasma concentration of the drug (Telenti and Zanger 2008). Patients (particularly in Africa, where both *CYP2B6*6* and **18* are common) would benefit from genotyping to minimize the adverse effects of efavirenz.

20.5.3 CYP3As

CYP3A4 and CYP3A5 are involved in the metabolism of more than half of all currently used drugs (Guengerich 1999; Lamba et al. 2002). Although large interindividual variations in the blood levels and clearance rates of

CYP3A4-substrate drugs are apparent, the genetic polymorphisms responsible for such variations have not yet been identified. A rare null allele of *CYP3A4* is *CYP3A4*20* (Westlind-Johnsson et al. 2006). *CYP3A4* alleles associated with significant decreases in drug metabolic activity include **8*, **11*, **13*, **16*, and **17*; all occur at very low frequencies (Lee and Goldstein 2005). *CYP3A5* is more polymorphic than is *CYP3A4*. Of the null alleles of *CYP3A5* (**3*, **5*, **6*, **7*), *CYP3A5*3* is the most common (90 % in whites, 75 % in Asians, and 20–50 % in blacks) (Kuehl et al. 2001). *CYP3A5*6* and **7* are most common in blacks (17 % and 6 % of all mutations, respectively) but are not found in whites or Asians (Hustert et al. 2001; Lee et al. 2003). The *CYP3A5*3* allele is caused by a single nucleotide polymorphism (SNP) in intron 3, creating a cryptic acceptor splice site that results in production of a truncated nonfunctional protein. Individuals carrying *CYP3A5*3* exhibit greatly reduced expression of the CYP3A5 protein (Kuehl et al. 2001). The pharmacokinetics of tacrolimus and cyclosporine vary greatly in renal transplant recipients (Schroeder et al. 1996; Staatz and Tett 2004). Several studies have found that clearance of orally administered tacrolimus in renal transplant recipients who are *CYP3A5*3/*3* is lower than in *CYP3A5*1* heterozygotes or homozygotes, suggesting that an initial low dose of tacrolimus should be given to *CYP3A5*3/*3* patients (Macphee et al. 2005; MacPhee et al. 2004; Thervet et al. 2010). However, the contributions of *CYP3A4* genetic polymorphisms to drug pharmacokinetic parameters remain unclear. Further work on the CYP3As would enhance drug efficacy and eliminate side effects.

20.6 Epilogue

A number of P450 polymorphisms have been evaluated for their clinical relevance. In particular, PM genotypes of *CYP2C9*, *19*, and *2D6* are associated with significant differences in the pharmacokinetics and pharmacodynamics of many drugs compared to the wild type (Table 20.4). However, the contribution of each genotype to drug metabolism and the clinical significance vary for different drugs, because most drugs metabolized by P450 are also acted upon, to some extent, by other enzymes. Comprehensive pharmacogenomic information on P450 enzymes, drug target genes, drug transporters, and clinical factors is required to understand interindividual variations in drug responses. Another emerging challenge is how to integrate and translate increasing amounts of pharmacogenomic data from P450s and others. The ever-increasing number of genes involved in drug metabolism, and the massive amounts of genetic information that are being collected, will soon overload the ability of a physician to translate such data into wise clinical decision making. Also, the interpretation of pharmacogenomic data may vary among both individual physicians and hospitals, creating inconsistencies in the clinical application of such data. For pharmacogenomic data from P450s and others to be implemented in clinical settings, a systematically constructed decision-support tool is needed to ensure consistency in data interpretation.

Table 20.4 Clinical implications of major genetic polymorphisms in cytochrome P450s (CYPs)

Drug class	CYP	Clinical implication
Cardiovascular drugs		
Warfarin	CYP2C9	Increased risk of bleeding in PMs attributable to increased plasma levels of unmetabolized *S*-warfarin
Clopidogrel	CYP2C19	Decreased platelet aggregation in PMs attributable to low levels of the active metabolite
Carvedilol[a]	CYP2D6	Increased plasma levels in PMs attributable to low-level clearance
Metoprolol[a]	CYP2D6	Increased plasma levels in PMs attributable to low-level clearance
Propranolol[a]	CYP2D6	Increased plasma levels in PMs attributable to low-level clearance
Timolol[a]	CYP2D6	Increased plasma levels in PMs attributable to low-level clearance
Analgesic drugs		
Codeine	CYP2D6	Increased risk of abdominal pain and life-threatening respiratory depression in UMs attributable to excessively high concentrations of the active metabolite (morphine)
		Reduced analgesic effects in PMs
Tramadol	CYP2D6	Increased frequency of nausea attributable to excessively high concentrations of the active metabolite
		Reduced analgesic effects in PMs
NSAIDs in general[a]	CYP2C9	Increased risk of gastrointestinal bleeding inpatients carrying alleles *2 or *3
Antidepressants and antipsychotics		
Imipramine[a], Trimipramine[a]	CYP2D6	Increased plasma levels in PMs attributable to low-level clearance
Doxepin[a], Nortriptyline[a]	CYP2D6	Increased plasma levels in PMs attributable to low-level clearance
Fluoxetine[a], Paroxetine[a]	CYP2D6	Increased plasma levels in PMs attributable to low-level clearance
Perphenazine[a], Thioridazine[a]	CYP2D6	Increased plasma levels in PMs attributable to low-level clearance
Risperidone[a], Olanzapine[a]	CYP2D6	Increased plasma levels in PMs attributable to low-level clearance
Clozapine[a], Haloperidol[a]	CYP2D6	Increased plasma levels in PMs attributable to low-level clearance
Diazepam	CYP2C19	Increased risk of sedation and unconsciousness in PMs attributable to an increase in the half-life of the parent drug

(continued)

Table 20.4 (continued)

Drug class	CYP	Clinical implication
Proton pump inhibitors		
Omeprazole	CYP2C19	Improved cure rates in PMs attributable to increased half-life of the parent drug
Lansoprazole	CYP2C19	Improved cure rates in PMs attributable to increased half-life of the parent drug
Other drugs		
Tamoxifen	CYP2D6	Poor drug efficacy in PMs attributable to low levels of active metabolite
Phenytoin	CYP2C9	Increased risk of ataxia, unconsciousness, and mental confusion in PMs attributable to increased plasma drug concentration
Glimepiride[a], glyburide[a]	CYP2C9	Increased plasma concentration in PMs attributable to low-level clearance
Tolterodine[a]	CYP2D6	Increased plasma concentration in PMs attributable to low-level clearance
Tacrolimus[a]	CYP3A5	Increased risk of nephrotoxicity in renal transplant recipients who are *CYP3A5*3/*3*
Saquinavir[a]	CYP3A5	Increased plasma concentration in individuals who are CYP3A5*3/*3
Efavirenz[a]	CYP2B6	Increased plasma concentration in individuals who are CYP2B6*6/*6

PM poor metabolizer genotypes, *UM* ultrarapid metabolizer genotypes

[a]The clinical significance of these observations is unclear, although the pharmacokinetic data correlate with genotype

References

Allabi AC, Gala JL, Horsmans Y, Babaoglu MO, Bozkurt A, Heusterspreute M, Yasar U (2004) Functional impact of CYP2C95, CYP2C96, CYP2C98, and CYP2C911 in vivo among black Africans. Clin Pharmacol Ther 76:113–118

Ancrenaz V, Desmeules J, James R, Fontana P, Reny JL, Dayer P, Daali Y (2012) The paraoxonase-1 pathway is not a major bioactivation pathway of clopidogrel in vitro. Br J Pharmacol 166:2362–2370

Au N, Rettie AE (2008) Pharmacogenomics of 4-hydroxycoumarin anticoagulants. Drug Metab Rev 40:355–375

Aynacioglu AS, Brockmoller J, Bauer S, Sachse C, Guzelbey P, Ongen Z, Nacak M, Roots I (1999) Frequency of cytochrome P450 CYP2C9 variants in a Turkish population and functional relevance for phenytoin. Br J Clin Pharmacol 48:409–415

Balian JD, Sukhova N, Harris JW, Hewett J, Pickle L, Goldstein JA, Woosley RL, Flockhart DA (1995) The hydroxylation of omeprazole correlates with *S*-mephenytoin metabolism: a population study. Clin Pharmacol Ther 57:662–669

Baumann P, Hiemke C, Ulrich S, Eckermann G, Gaertner I, Gerlach M, Kuss HJ, Laux G, Muller-Oerlinghausen B, Rao ML, Riederer P, Zernig G (2004) The AGNP-TDM expert group consensus guidelines: therapeutic drug monitoring in psychiatry. Pharmacopsychiatry 37:243–265

Bijl MJ, Visser LE, Hofman A, Vulto AG, van Gelder T, Stricker BH, van Schaik RH (2008) Influence of the CYP2D6*4 polymorphism on dose, switching and discontinuation of antidepressants. Br J Clin Pharmacol 65:558–564

Bondolfi G, Morel F, Crettol S, Rachid F, Baumann P, Eap CB (2005) Increased clozapine plasma concentrations and side effects induced by smoking cessation in 2 CYP1A2 genotyped patients. Ther Drug Monit 27:539–543

Bonello L, Tantry US, Marcucci R, Blindt R, Angiolillo DJ, Becker R, Bhatt DL, Cattaneo M, Collet JP, Cuisset T, Gachet C, Montalescot G, Jennings LK, Kereiakes D, Sibbing D, Trenk D, Van Werkum JW, Paganelli F, Price MJ, Waksman R, Gurbel PA (2010) Consensus and future directions on the definition of high on-treatment platelet reactivity to adenosine diphosphate. J Am Coll Cardiol 56:919–933

Brockmoller J, Kirchheiner J, Meisel C, Roots I (2000) Pharmacogenetic diagnostics of cytochrome P450 polymorphisms in clinical drug development and in drug treatment. Pharmacogenomics 1:125–151

Broly F, Marez D, Lo Guidice JM, Sabbagh N, Legrand M, Boone P, Meyer UA (1995) A nonsense mutation in the cytochrome P450 CYP2D6 gene identified in a Caucasian with an enzyme deficiency. Hum Genet 96:601–603

Caraco Y, Sheller J, Wood AJ (1996) Pharmacogenetic determination of the effects of codeine and prediction of drug interactions. J Pharmacol Exp Ther 278:1165–1174

Cascorbi I (2003) Pharmacogenetics of cytochrome p4502D6: genetic background and clinical implication. Eur J Clin Invest 33(suppl 2):17–22

Cavallari LH, Langaee TY, Momary KM, Shapiro NL, Nutescu EA, Coty WA, Viana MA, Patel SR, Johnson JA (2010) Genetic and clinical predictors of warfarin dose requirements in African Americans. Clin Pharmacol Ther 87:459–464

Chida M, Yokoi T, Nemoto N, Inaba M, Kinoshita M, Kamataki T (1999) A new variant CYP2D6 allele (CYP2D6*21) with a single base insertion in exon 5 in a Japanese population associated with a poor metabolizer phenotype. Pharmacogenetics 9:287–293

Combescure C, Fontana P, Mallouk N, Berdague P, Labruyere C, Barazer I, Gris JC, Laporte S, Fabbro-Peray P, Reny JL (2010) Clinical implications of clopidogrel non-response in cardiovascular patients: a systematic review and meta-analysis. J Thromb Haemost 8:923–933

Dalen P, Frengell C, Dahl ML, Sjoqvist F (1997) Quick onset of severe abdominal pain after codeine in an ultrarapid metabolizer of debrisoquine. Ther Drug Monit 19:543–544

Daly AK, Leathart JB, London SJ, Idle JR (1995) An inactive cytochrome P450 CYP2D6 allele containing a deletion and a base substitution. Hum Genet 95:337–341

Daly AK, Fairbrother KS, Andreassen OA, London SJ, Idle JR, Steen VM (1996) Characterization and PCR-based detection of two different hybrid CYP2D7P/CYP2D6 alleles associated with the poor metabolizer phenotype. Pharmacogenetics 6:319–328

De Morais SM, Wilkinson GR, Blaisdell J, Meyer UA, Nakamura K, Goldstein JA (1994a) Identification of a new genetic defect responsible for the polymorphism of (S)-mephenytoin metabolism in Japanese. Mol Pharmacol 46:594–598

de Morais SM, Wilkinson GR, Blaisdell J, Nakamura K, Meyer UA, Goldstein JA (1994b) The major genetic defect responsible for the polymorphism of *S*-mephenytoin metabolism in humans. J Biol Chem 269:15419–15422

Dehal SS, Kupfer D (1997) CYP2D6 catalyzes tamoxifen 4-hydroxylation in human liver. Cancer Res 57:3402–3406

Desta Z, Zhao X, Shin JG, Flockhart DA (2002) Clinical significance of the cytochrome P450 2C19 genetic polymorphism. Clin Pharmacokinet 41:913–958

Desta Z, Saussele T, Ward B, Blievernicht J, Li L, Klein K, Flockhart DA, Zanger UM (2007) Impact of CYP2B6 polymorphism on hepatic efavirenz metabolism in vitro. Pharmacogenomics 8:547–558

Dickmann LJ, Rettie AE, Kneller MB, Kim RB, Wood AJ, Stein CM, Wilkinson GR, Schwarz UI (2001) Identification and functional characterization of a new CYP2C9 variant (CYP2C9*5) expressed among African Americans. Mol Pharmacol 60:382–387

Dieudonne AS, Van Belle V, Neven P (2010) Association between CYP2D6 polymorphisms and breast cancer outcomes. JAMA 303:516–517, author reply 517

Eap CB, Bender S, Jaquenoud Sirot E, Cucchia G, Jonzier-Perey M, Baumann P, Allorge D, Broly F (2004) Nonresponse to clozapine and ultrarapid CYP1A2 activity: clinical data and analysis of CYP1A2 gene. J Clin Psychopharmacol 24:214–219

Eichelbaum M, Spannbrucker N, Dengler HJ (1975) Proceedings: N-oxidation of sparteine in man and its interindividual differences. Naunyn Schmiedebergs Arch Pharmacol 287:R94

Eichelbaum M, Ingelman-Sundberg M, Evans WE (2006) Pharmacogenomics and individualized drug therapy. Annu Rev Med 57:119–137

Evert B, Griese EU, Eichelbaum M (1994) Cloning and sequencing of a new non-functional CYP2D6 allele: deletion of T1795 in exon 3 generates a premature stop codon. Pharmacogenetics 4:271–274

Ferguson RJ, De Morais SM, Benhamou S, Bouchardy C, Blaisdell J, Ibeanu G, Wilkinson GR, Sarich TC, Wright JM, Dayer P, Goldstein JA (1998) A new genetic defect in human CYP2C19: mutation of the initiation codon is responsible for poor metabolism of S-mephenytoin. J Pharmacol Exp Ther 284:356–361

Fjordside L, Jeppesen U, Eap CB, Powell K, Baumann P, Brosen K (1999) The stereoselective metabolism of fluoxetine in poor and extensive metabolizers of sparteine. Pharmacogenetics 9:55–60

Fliegert F, Kurth B, Gohler K (2005) The effects of tramadol on static and dynamic pupillometry in healthy subjects: the relationship between pharmacodynamics, pharmacokinetics and CYP2D6 metaboliser status. Eur J Clin Pharmacol 61:257–266

Flockhart D (2008) CYP2D6 genotyping and the pharmacogenetics of tamoxifen. Clin Adv Hematol Oncol 6:493–494

Funck-Brentano C, Becquemont L, Lenevu A, Roux A, Jaillon P, Beaune P (1997) Inhibition by omeprazole of proguanil metabolism: mechanism of the interaction in vitro and prediction of in vivo results from the in vitro experiments. J Pharmacol Exp Ther 280:730–738

Furuta T, Ohashi K, Kamata T, Takashima M, Kosuge K, Kawasaki T, Hanai H, Kubota T, Ishizaki T, Kaneko E (1998) Effect of genetic differences in omeprazole metabolism on cure rates for *Helicobacter pylori* infection and peptic ulcer. Ann Intern Med 129:1027–1030

Furuta T, Shirai N, Sugimoto M, Ohashi K, Ishizaki T (2004) Pharmacogenomics of proton pump inhibitors. Pharmacogenomics 5:181–202
Gaedigk A (2000) Interethnic differences of drug-metabolizing enzymes. Int J Clin Pharmacol Ther 38:61–68
Gaedigk A, Blum M, Gaedigk R, Eichelbaum M, Meyer UA (1991) Deletion of the entire cytochrome P450 CYP2D6 gene as a cause of impaired drug metabolism in poor metabolizers of the debrisoquine/sparteine polymorphism. Am J Hum Genet 48:943–950
Gaedigk A, Bradford LD, Marcucci KA, Leeder JS (2002) Unique CYP2D6 activity distribution and genotype-phenotype discordance in black Americans. Clin Pharmacol Ther 72:76–89
Gaedigk A, Ndjountche L, Gaedigk R, Leeder JS, Bradford LD (2003) Discovery of a novel nonfunctional cytochrome P450 2D6 allele, CYP2D642, in African American subjects. Clin Pharmacol Ther 73:575–576
Gaedigk A, Ndjountche L, Divakaran K, Dianne Bradford L, Zineh I, Oberlander TF, Brousseau DC, McCarver DG, Johnson JA, Alander SW, Wayne Riggs K, Steven Leeder J (2007) Cytochrome P4502D6 (CYP2D6) gene locus heterogeneity: characterization of gene duplication events. Clin Pharmacol Ther 81:242–251
Garcia-Martin E, Martinez C, Ladero JM, Gamito FJ, Agundez JA (2001) High frequency of mutations related to impaired CYP2C9 metabolism in a Caucasian population. Eur J Clin Pharmacol 57:47–49
Gasche Y, Daali Y, Fathi M, Chiappe A, Cottini S, Dayer P, Desmeules J (2004) Codeine intoxication associated with ultrarapid CYP2D6 metabolism. N Engl J Med 351:2827–2831
Giancarlo GM, Venkatakrishnan K, Granda BW, von Moltke LL, Greenblatt DJ (2001) Relative contributions of CYP2C9 and 2C19 to phenytoin 4-hydroxylation in vitro: inhibition by sulfaphenazole, omeprazole, and ticlopidine. Eur J Clin Pharmacol 57:31–36
Goetz MP, Rae JM, Suman VJ, Safgren SL, Ames MM, Visscher DW, Reynolds C, Couch FJ, Lingle WL, Flockhart DA, Desta Z, Perez EA, Ingle JN (2005) Pharmacogenetics of tamoxifen biotransformation is associated with clinical outcomes of efficacy and hot flashes. J Clin Oncol 23:9312–9318
Goldstein JA (2001) Clinical relevance of genetic polymorphisms in the human CYP2C subfamily. Br J Clin Pharmacol 52:349–355
Goldstein JA, de Morais SM (1994) Biochemistry and molecular biology of the human CYP2C subfamily. Pharmacogenetics 4:285–299
Gough AC, Miles JS, Spurr NK, Moss JE, Gaedigk A, Eichelbaum M, Wolf CR (1990) Identification of the primary gene defect at the cytochrome P450 CYP2D locus. Nature (Lond) 347:773–776
Guengerich FP (1999) Cytochrome P-450 3A4: regulation and role in drug metabolism. Annu Rev Pharmacol Toxicol 39:1–17
Han XM, Ouyang DS, Chen XP, Shu Y, Jiang CH, Tan ZR, Zhou HH (2002) Inducibility of CYP1A2 by omeprazole in vivo related to the genetic polymorphism of CYP1A2. Br J Clin Pharmacol 54:540–543
Hanioka N, Kimura S, Meyer UA, Gonzalez FJ (1990) The human CYP2D locus associated with a common genetic defect in drug oxidation: a G1934-A base change in intron 3 of a mutant CYP2D6 allele results in an aberrant 3′ splice recognition site. Am J Hum Genet 47:994–1001
Hesse LM, He P, Krishnaswamy S, Hao Q, Hogan K, von Moltke LL, Greenblatt DJ, Court MH (2004) Pharmacogenetic determinants of interindividual variability in bupropion hydroxylation by cytochrome P450 2B6 in human liver microsomes. Pharmacogenetics 14:225–238
Hulot JS, Bura A, Villard E, Azizi M, Remones V, Goyenvalle C, Aiach M, Lechat P, Gaussem P (2006) Cytochrome P450 2C19 loss-of-function polymorphism is a major determinant of clopidogrel responsiveness in healthy subjects. Blood 108:2244–2247
Hustert E, Haberl M, Burk O, Wolbold R, He YQ, Klein K, Nuessler AC, Neuhaus P, Klattig J, Eiselt R, Koch I, Zibat A, Brockmoller J, Halpert JR, Zanger UM, Wojnowski L (2001) The genetic determinants of the CYP3A5 polymorphism. Pharmacogenetics 11:773–779

Ibeanu GC, Blaisdell J, Ghanayem BI, Beyeler C, Benhamou S, Bouchardy C, Wilkinson GR, Dayer P, Daly AK, Goldstein JA (1998a) An additional defective allele, CYP2C19*5, contributes to the *S*-mephenytoin poor metabolizer phenotype in Caucasians. Pharmacogenetics 8:129–135

Ibeanu GC, Goldstein JA, Meyer U, Benhamou S, Bouchardy C, Dayer P, Ghanayem BI, Blaisdell J (1998b) Identification of new human CYP2C19 alleles (CYP2C19*6 and CYP2C19*2B) in a Caucasian poor metabolizer of mephenytoin. J Pharmacol Exp Ther 286:1490–1495

Ibeanu GC, Blaisdell J, Ferguson RJ, Ghanayem BI, Brosen K, Benhamou S, Bouchardy C, Wilkinson GR, Dayer P, Goldstein JA (1999) A novel transversion in the intron 5 donor splice junction of CYP2C19 and a sequence polymorphism in exon 3 contribute to the poor metabolizer phenotype for the anticonvulsant drug *S*-mephenytoin. J Pharmacol Exp Ther 290:635–640

Ingelman-Sundberg M (2005) Genetic polymorphisms of cytochrome P450 2D6 (CYP2D6): clinical consequences, evolutionary aspects and functional diversity. Pharmacogenomics J 5:6–13

Ingelman-Sundberg M, Sim SC, Gomez A, Rodriguez-Antona C (2007) Influence of cytochrome P450 polymorphisms on drug therapies: pharmacogenetic, pharmacoepigenetic and clinical aspects. Pharmacol Ther 116:496–526

Ji L, Pan S, Marti-Jaun J, Hanseler E, Rentsch K, Hersberger M (2002) Single-step assays to analyze CYP2D6 gene polymorphisms in Asians: allele frequencies and a novel *14B allele in mainland Chinese. Clin Chem 48:983–988

Johnson JA, Zineh I, Puckett BJ, McGorray SP, Yarandi HN, Pauly DF (2003) Beta 1-adrenergic receptor polymorphisms and antihypertensive response to metoprolol. Clin Pharmacol Ther 74:44–52

Jordan VC, Collins MM, Rowsby L, Prestwich G (1977) A monohydroxylated metabolite of tamoxifen with potent antioestrogenic activity. J Endocrinol 75:305–316

Jose R, Chandrasekaran A, Sam SS, Gerard N, Chanolean S, Abraham BK, Satyanarayanamoorthy K, Peter A, Rajagopal K (2005) CYP2C9 and CYP2C19 genetic polymorphisms: frequencies in the south Indian population. Fundam Clin Pharmacol 19:101–105

Kagimoto M, Heim M, Kagimoto K, Zeugin T, Meyer UA (1990) Multiple mutations of the human cytochrome P450IID6 gene (CYP2D6) in poor metabolizers of debrisoquine. Study of the functional significance of individual mutations by expression of chimeric genes. J Biol Chem 265:17209–17214

Kaminsky LS, Zhang ZY (1997) Human P450 metabolism of warfarin. Pharmacol Ther 73:67–74

Kaneko A, Lum JK, Yaviong L, Takahashi N, Ishizaki T, Bertilsson L, Kobayakawa T, Bjorkman A (1999) High and variable frequencies of CYP2C19 mutations: medical consequences of poor drug metabolism in Vanuatu and other Pacific islands. Pharmacogenetics 9:581–590

Kawanishi C, Lundgren S, Agren H, Bertilsson L (2004) Increased incidence of CYP2D6 gene duplication in patients with persistent mood disorders: ultrarapid metabolism of antidepressants as a cause of nonresponse. A pilot study. Eur J Clin Pharmacol 59:803–807

Kazui M, Nishiya Y, Ishizuka T, Hagihara K, Farid NA, Okazaki O, Ikeda T, Kurihara A (2010) Identification of the human cytochrome P450 enzymes involved in the two oxidative steps in the bioactivation of clopidogrel to its pharmacologically active metabolite. Drug Metab Dispos 38:92–99

Kidd RS, Straughn AB, Meyer MC, Blaisdell J, Goldstein JA, Dalton JT (1999) Pharmacokinetics of chlorpheniramine, phenytoin, glipizide and nifedipine in an individual homozygous for the CYP2C9*3 allele. Pharmacogenetics 9:71–80

Kidd RS, Curry TB, Gallagher S, Edeki T, Blaisdell J, Goldstein JA (2001) Identification of a null allele of CYP2C9 in an African-American exhibiting toxicity to phenytoin. Pharmacogenetics 11:803–808

King BP, Khan TI, Aithal GP, Kamali F, Daly AK (2004) Upstream and coding region CYP2C9 polymorphisms: correlation with warfarin dose and metabolism. Pharmacogenetics 14:813–822

Kirchheiner J, Brockmoller J (2005) Clinical consequences of cytochrome P450 2C9 polymorphisms. Clin Pharmacol Ther 77:1–16

Kirchheiner J, Brockmoller J, Meineke I, Bauer S, Rohde W, Meisel C, Roots I (2002) Impact of CYP2C9 amino acid polymorphisms on glyburide kinetics and on the insulin and glucose response in healthy volunteers. Clin Pharmacol Ther 71:286–296

Kirchheiner J, Nickchen K, Bauer M, Wong ML, Licinio J, Roots I, Brockmoller J (2004) Pharmacogenetics of antidepressants and antipsychotics: the contribution of allelic variations to the phenotype of drug response. Mol Psychiatry 9:442–473

Kirchheiner J, Keulen JT, Bauer S, Roots I, Brockmoller J (2008) Effects of the CYP2D6 gene duplication on the pharmacokinetics and pharmacodynamics of tramadol. J Clin Psychopharmacol 28:78–83

Kitzmiller JP, Groen DK, Phelps MA, Sadee W (2011) Pharmacogenomic testing: relevance in medical practice: why drugs work in some patients but not in others. Cleve Clin J Med 78:243–257

Klein K, Tatzel S, Raimundo S, Saussele T, Hustert E, Pleiss J, Eichelbaum M, Zanger UM (2007) A natural variant of the heme-binding signature (R441C) resulting in complete loss of function of CYP2D6. Drug Metab Dispos 35:1247–1250

Koren G, Cairns J, Chitayat D, Gaedigk A, Leeder SJ (2006) Pharmacogenetics of morphine poisoning in a breastfed neonate of a codeine-prescribed mother. Lancet 368:704

Koukouritaki SB, Manro JR, Marsh SA, Stevens JC, Rettie AE, McCarver DG, Hines RN (2004) Developmental expression of human hepatic CYP2C9 and CYP2C19. J Pharmacol Exp Ther 308:965–974

Kuehl P, Zhang J, Lin Y, Lamba J, Assem M, Schuetz J, Watkins PB, Daly A, Wrighton SA, Hall SD, Maurel P, Relling M, Brimer C, Yasuda K, Venkataramanan R, Strom S, Thummel K, Boguski MS, Schuetz E (2001) Sequence diversity in CYP3A promoters and characterization of the genetic basis of polymorphic CYP3A5 expression. Nat Genet 27:383–391

Kumana CR, Lauder IJ, Chan M, Ko W, Lin HJ (1987) Differences in diazepam pharmacokinetics in Chinese and white Caucasians: relation to body lipid stores. Eur J Clin Pharmacol 32:211–215

Kwadijk-de Gijsel S, Bijl MJ, Visser LE, van Schaik RH, Hofman A, Vulto AG, van Gelder T, Ch Stricker BH (2009) Variation in the CYP2D6 gene is associated with a lower serum sodium concentration in patients on antidepressants. Br J Clin Pharmacol 68:221–225

Lamba JK, Lin YS, Schuetz EG, Thummel KE (2002) Genetic contribution to variable human CYP3A-mediated metabolism. Adv Drug Deliv Rev 54:1271–1294

Lang T, Klein K, Fischer J, Nussler AK, Neuhaus P, Hofmann U, Eichelbaum M, Schwab M, Zanger UM (2001) Extensive genetic polymorphism in the human CYP2B6 gene with impact on expression and function in human liver. Pharmacogenetics 11:399–415

Leathart JB, London SJ, Steward A, Adams JD, Idle JR, Daly AK (1998) CYP2D6 phenotype-genotype relationships in African-Americans and Caucasians in Los Angeles. Pharmacogenetics 8:529–541

Lee SJ (2012) Clinical application of CYP2C19 pharmacogenetics toward more personalized medicine. Front Genet 3:318

Lee SJ, Goldstein JA (2005) Functionally defective or altered CYP3A4 and CYP3A5 single nucleotide polymorphisms and their detection with genotyping tests. Pharmacogenomics 6:357–371

Lee CR, Goldstein JA, Pieper JA (2002) Cytochrome P450 2C9 polymorphisms: a comprehensive review of the in-vitro and human data. Pharmacogenetics 12:251–263

Lee SJ, Usmani KA, Chanas B, Ghanayem B, Xi T, Hodgson E, Mohrenweiser HW, Goldstein JA (2003) Genetic findings and functional studies of human CYP3A5 single nucleotide polymorphisms in different ethnic groups. Pharmacogenetics 13:461–472

Lee SS, Kim KM, Thi-Le H, Yea SS, Cha IJ, Shin JG (2005) Genetic polymorphism of CYP2C9 in a Vietnamese Kinh population. Ther Drug Monit 27:208–210

Lee SJ, Kim WY, Kim H, Shon JH, Lee SS, Shin JG (2009a) Identification of new CYP2C19 variants exhibiting decreased enzyme activity in the metabolism of S-mephenytoin and omeprazole. Drug Metab Dispos 37:2262–2269

Lee SJ, Lee SS, Jung HJ, Kim HS, Park SJ, Yeo CW, Shin JG (2009b) Discovery of novel functional variants and extensive evaluation of CYP2D6 genetic polymorphisms in Koreans. Drug Metab Dispos 37:1464–1470

Lee SJ, Jang YJ, Cha EY, Kim HS, Lee SS, Shin JG (2010) A haplotype of CYP2C9 associated with warfarin sensitivity in mechanical heart valve replacement patients. Br J Clin Pharmacol 70:213–221

Li L, Pan RM, Porter TD, Jensen NS, Silber P, Russo G, Tine JA, Heim J, Ring B, Wedlund PJ (2006) New cytochrome P450 2D6*56 allele identified by genotype/phenotype analysis of cryopreserved human hepatocytes. Drug Metab Dispos 34:1411–1416

Lim HS, Ju Lee H, Seok Lee K, Sook Lee E, Jang IJ, Ro J (2007) Clinical implications of CYP2D6 genotypes predictive of tamoxifen pharmacokinetics in metastatic breast cancer. J Clin Oncol 25:3837–3845

Limdi NA, Wadelius M, Cavallari L, Eriksson N, Crawford DC, Lee MT, Chen CH, Motsinger-Reif A, Sagreiya H, Liu N, Wu AH, Gage BF, Jorgensen A, Pirmohamed M, Shin JG, Suarez-Kurtz G, Kimmel SE, Johnson JA, Klein TE, Wagner MJ (2010) Warfarin pharmacogenetics: a single VKORC1 polymorphism is predictive of dose across 3 racial groups. Blood 115:3827–3834

Linnet K, Wiborg O (1996) Steady-state serum concentrations of the neuroleptic perphenazine in relation to CYP2D6 genetic polymorphism. Clin Pharmacol Ther 60:41–47

Loebstein R, Yonath H, Peleg D, Almog S, Rotenberg M, Lubetsky A, Roitelman J, Harats D, Halkin H, Ezra D (2001) Interindividual variability in sensitivity to warfarin: nature or nurture? Clin Pharmacol Ther 70:159–164

Lotsch J, Skarke C, Liefhold J, Geisslinger G (2004) Genetic predictors of the clinical response to opioid analgesics: clinical utility and future perspectives. Clin Pharmacokinet 43:983–1013

MacPhee IA, Fredericks S, Tai T, Syrris P, Carter ND, Johnston A, Goldberg L, Holt DW (2004) The influence of pharmacogenetics on the time to achieve target tacrolimus concentrations after kidney transplantation. Am J Transplant 4:914–919

Macphee IA, Fredericks S, Mohamed M, Moreton M, Carter ND, Johnston A, Goldberg L, Holt DW (2005) Tacrolimus pharmacogenetics: the CYP3A5*1 allele predicts low dose-normalized tacrolimus blood concentrations in whites and South Asians. Transplantation 79:499–502

Madlensky L, Natarajan L, Tchu S, Pu M, Mortimer J, Flatt SW, Nikoloff DM, Hillman G, Fontecha MR, Lawrence HJ, Parker BA, Wu AH, Pierce JP (2011) Tamoxifen metabolite concentrations, CYP2D6 genotype, and breast cancer outcomes. Clin Pharmacol Ther 89:718–725

Maekawa K, Fukushima-Uesaka H, Tohkin M, Hasegawa R, Kajio H, Kuzuya N, Yasuda K, Kawamoto M, Kamatani N, Suzuki K, Yanagawa T, Saito Y, Sawada J (2006) Four novel defective alleles and comprehensive haplotype analysis of CYP2C9 in Japanese. Pharmacogenet Genomics 16:497–514

Mahgoub A, Idle JR, Dring LG, Lancaster R, Smith RL (1977) Polymorphic hydroxylation of debrisoquine in man. Lancet 2:584–586

Mannucci PM (1999) Genetic control of anticoagulation. Lancet 353:688–689

Marez D, Sabbagh N, Legrand M, Lo-Guidice JM, Boone P, Broly F (1995) A novel CYP2D6 allele with an abolished splice recognition site associated with the poor metabolizer phenotype. Pharmacogenetics 5:305–311

Marez D, Legrand M, Sabbagh N, Lo-Guidice JM, Boone P, Broly F (1996) An additional allelic variant of the CYP2D6 gene causing impaired metabolism of sparteine. Hum Genet 97:668–670

Marez D, Legrand M, Sabbagh N, Lo Guidice JM, Spire C, Lafitte JJ, Meyer UA, Broly F (1997) Polymorphism of the cytochrome P450 CYP2D6 gene in a European population:

characterization of 48 mutations and 53 alleles, their frequencies and evolution. Pharmacogenetics 7:193–202
Marez-Allorge D, Ellis SW, Lo Guidice JM, Tucker GT, Broly F (1999) A rare G2061 insertion affecting the open reading frame of CYP2D6 and responsible for the poor metabolizer phenotype. Pharmacogenetics 9:393–396
Mega JL, Close SL, Wiviott SD, Shen L, Hockett RD, Brandt JT, Walker JR, Antman EM, Macias W, Braunwald E, Sabatine MS (2009) Cytochrome p-450 polymorphisms and response to clopidogrel. N Engl J Med 360:354–362
Mehvar R, Brocks DR (2001) Stereospecific pharmacokinetics and pharmacodynamics of beta-adrenergic blockers in humans. J Pharm Pharm Sci 4:185–200
Meyer UA, Zanger UM (1997) Molecular mechanisms of genetic polymorphisms of drug metabolism. Annu Rev Pharmacol Toxicol 37:269–296
Miners JO, Birkett DJ (1998) Cytochrome P4502C9: an enzyme of major importance in human drug metabolism. Br J Clin Pharmacol 45:525–538
Moonen H, Engels L, Kleinjans J, Kok T (2005) The CYP1A2-164A→C polymorphism (CYP1A2*1F) is associated with the risk for colorectal adenomas in humans. Cancer Lett 229:25–31
Morita J, Kobayashi K, Wanibuchi A, Kimura M, Irie S, Ishizaki T, Chiba K (2004) A novel single nucleotide polymorphism (SNP) of the CYP2C19 gene in a Japanese subject with lowered capacity of mephobarbital 4′-hydroxylation. Drug Metab Pharmacokinet 19:236–238
Nasu K, Kubota T, Ishizaki T (1997) Genetic analysis of CYP2C9 polymorphism in a Japanese population. Pharmacogenetics 7:405–409
Ngow HA, Wan Khairina WM, Teh LK, Lee WL, Harun R, Ismail R, Salleh MZ (2009) CYP2C9 polymorphism: prevalence in healthy and warfarin-treated Malay and Chinese in Malaysia. Singapore Med J 50:490–493
Niemi M, Cascorbi I, Timm R, Kroemer HK, Neuvonen PJ, Kivisto KT (2002) Glyburide and glimepiride pharmacokinetics in subjects with different CYP2C9 genotypes. Clin Pharmacol Ther 72:326–332
Ninomiya H, Mamiya K, Matsuo S, Ieiri I, Higuchi S, Tashiro N (2000) Genetic polymorphism of the CYP2C subfamily and excessive serum phenytoin concentration with central nervous system intoxication. Ther Drug Monit 22:230–232
Nowak-Gottl U, Dietrich K, Schaffranek D, Eldin NS, Yasui Y, Geisen C, Mitchell LG (2010) In pediatric patients, age has more impact on dosing of vitamin K antagonists than VKORC1 or CYP2C9 genotypes. Blood 116:6101–6105
Ogg MS, Brennan P, Meade T, Humphries SE (1999) CYP2C9*3 allelic variant and bleeding complications. Lancet 354:1124
Panserat S, Mura C, Gerard N, Vincent-Viry M, Galteau MM, Jacoz-Aigrain E, Krishnamoorthy R (1995) An unequal cross-over event within the CYP2D gene cluster generates a chimeric CYP2D7/CYP2D6 gene which is associated with the poor metabolizer phenotype. Br J Clin Pharmacol 40:361–367
Pedersen RS, Damkier P, Brosen K (2006) Enantioselective pharmacokinetics of tramadol in CYP2D6 extensive and poor metabolizers. Eur J Clin Pharmacol 62:513–521
Poulsen L, Arendt-Nielsen L, Brosen K, Sindrup SH (1996) The hypoalgesic effect of tramadol in relation to CYP2D6. Clin Pharmacol Ther 60:636–644
Qin XP, Xie HG, Wang W, He N, Huang SL, Xu ZH, Ou-Yang DS, Wang YJ, Zhou HH (1999) Effect of the gene dosage of CgammaP2C19 on diazepam metabolism in Chinese subjects. Clin Pharmacol Ther 66:642–646
Rau T, Wohlleben G, Wuttke H, Thuerauf N, Lunkenheimer J, Lanczik M, Eschenhagen T (2004) CYP2D6 genotype: impact on adverse effects and nonresponse during treatment with antidepressants: a pilot study. Clin Pharmacol Ther 75:386–393
Rendic S, Di Carlo FJ (1997) Human cytochrome P450 enzymes: a status report summarizing their reactions, substrates, inducers, and inhibitors. Drug Metab Rev 29:413–580

Rudberg I, Mohebi B, Hermann M, Refsum H, Molden E (2008) Impact of the ultrarapid CYP2C19*17 allele on serum concentration of escitalopram in psychiatric patients. Clin Pharmacol Ther 83:322–327

Sachse C, Brockmoller J, Bauer S, Roots I (1997) Cytochrome P450 2D6 variants in a Caucasian population: allele frequencies and phenotypic consequences. Am J Hum Genet 60:284–295

Sachse C, Brockmoller J, Bauer S, Roots I (1999) Functional significance of a C→A polymorphism in intron 1 of the cytochrome P450 CYP1A2 gene tested with caffeine. Br J Clin Pharmacol 47:445–449

Saebo M, Skjelbred CF, Brekke Li K, Bowitz Lothe IM, Hagen PC, Johnsen E, Tveit KM, Kure EH (2008) CYP1A2 164A→C polymorphism, cigarette smoking, consumption of well-done red meat and risk of developing colorectal adenomas and carcinomas. Anticancer Res 28:2289–2295

Sakuyama K, Sasaki T, Ujiie S, Obata K, Mizugaki M, Ishikawa M, Hiratsuka M (2008) Functional characterization of 17 CYP2D6 allelic variants (CYP2D6.2, 10, 14A-B, 18, 27, 36, 39, 47–51, 53–55, and 57). Drug Metab Dispos 36:2460–2467

Saxena R, Shaw GL, Relling MV, Frame JN, Moir DT, Evans WE, Caporaso N, Weiffenbach B (1994) Identification of a new variant CYP2D6 allele with a single base deletion in exon 3 and its association with the poor metabolizer phenotype. Hum Mol Genet 3:923–926

Schroeder TJ, Shah M, Hariharan S, First MR (1996) Increased resources are required in patients with low cyclosporine bioavailability. Transplant Proc 28:2151–2155

Schroth W, Antoniadou L, Fritz P, Schwab M, Muerdter T, Zanger UM, Simon W, Eichelbaum M, Brauch H (2007) Breast cancer treatment outcome with adjuvant tamoxifen relative to patient CYP2D6 and CYP2C19 genotypes. J Clin Oncol 25:5187–5193

Scordo MG, Spina E, Facciola G, Avenoso A, Johansson I, Dahl ML (1999) Cytochrome P450 2D6 genotype and steady state plasma levels of risperidone and 9-hydroxyrisperidone. Psychopharmacology (Berl) 147:300–305

Scordo MG, Aklillu E, Yasar U, Dahl ML, Spina E, Ingelman-Sundberg M (2001) Genetic polymorphism of cytochrome P450 2C9 in a Caucasian and a black African population. Br J Clin Pharmacol 52:447–450

Serrano D, Torrado S, Torrado-Santiago S, Gisbert JP (2012) The influence of CYP2C19 genetic polymorphism on the pharmacokinetics/pharmacodynamics of proton pump inhibitor-containing *Helicobacter pylori* treatments. Curr Drug Metab 13:1303–1312

Shimada T, Yamazaki H, Mimura M, Inui Y, Guengerich FP (1994) Interindividual variations in human liver cytochrome P-450 enzymes involved in the oxidation of drugs, carcinogens and toxic chemicals: studies with liver microsomes of 30 Japanese and 30 Caucasians. J Pharmacol Exp Ther 270:414–423

Shimada T, Tsumura F, Yamazaki H, Guengerich FP, Inoue K (2001) Characterization of (+/−)-bufuralol hydroxylation activities in liver microsomes of Japanese and Caucasian subjects genotyped for CYP2D6. Pharmacogenetics 11:143–156

Si D, Guo Y, Zhang Y, Yang L, Zhou H, Zhong D (2004) Identification of a novel variant CYP2C9 allele in Chinese. Pharmacogenetics 14:465–469

Sibbing D, Koch W, Gebhard D, Schuster T, Braun S, Stegherr J, Morath T, Schomig A, von Beckerath N, Kastrati A (2010) Cytochrome 2C19*17 allelic variant, platelet aggregation, bleeding events, and stent thrombosis in clopidogrel-treated patients with coronary stent placement. Circulation 121:512–518

Sim SC, Risinger C, Dahl ML, Aklillu E, Christensen M, Bertilsson L, Ingelman-Sundberg M (2006) A common novel CYP2C19 gene variant causes ultrarapid drug metabolism relevant for the drug response to proton pump inhibitors and antidepressants. Clin Pharmacol Ther 79:103–113

Sindrup SH, Brosen K, Gram LF, Hallas J, Skjelbo E, Allen A, Allen GD, Cooper SM, Mellows G, Tasker TC et al (1992) The relationship between paroxetine and the sparteine oxidation polymorphism. Clin Pharmacol Ther 51:278–287

Sistonen J, Sajantila A, Lao O, Corander J, Barbujani G, Fuselli S (2007) CYP2D6 worldwide genetic variation shows high frequency of altered activity variants and no continental structure. Pharmacogenet Genomics 17:93–101

Staatz CE, Tett SE (2004) Clinical pharmacokinetics and pharmacodynamics of tacrolimus in solid organ transplantation. Clin Pharmacokinet 43:623–653

Stamer UM, Lehnen K, Hothker F, Bayerer B, Wolf S, Hoeft A, Stuber F (2003) Impact of CYP2D6 genotype on postoperative tramadol analgesia. Pain 105:231–238

Stamer UM, Musshoff F, Kobilay M, Madea B, Hoeft A, Stuber F (2007) Concentrations of tramadol and *O*-desmethyltramadol enantiomers in different CYP2D6 genotypes. Clin Pharmacol Ther 82:41–47

Stamer UM, Stuber F, Muders T, Musshoff F (2008) Respiratory depression with tramadol in a patient with renal impairment and CYP2D6 gene duplication. Anesth Analg 107:926–929

Steen VM, Andreassen OA, Daly AK, Tefre T, Borresen AL, Idle JR, Gulbrandsen AK (1995) Detection of the poor metabolizer-associated CYP2D6(D) gene deletion allele by long-PCR technology. Pharmacogenetics 5:215–223

Steward DJ, Haining RL, Henne KR, Davis G, Rushmore TH, Trager WF, Rettie AE (1997) Genetic association between sensitivity to warfarin and expression of CYP2C9*3. Pharmacogenetics 7:361–367

Sullivan-Klose TH, Ghanayem BI, Bell DA, Zhang ZY, Kaminsky LS, Shenfield GM, Miners JO, Birkett DJ, Goldstein JA (1996) The role of the CYP2C9-Leu359 allelic variant in the tolbutamide polymorphism. Pharmacogenetics 6:341–349

Tang XF, Wang J, Zhang JH, Meng XM, Xu B, Qiao SB, Wu YJ, Chen J, Wu Y, Chen JL, Gao RL, Yuan JQ, Yang YJ (2013) Effect of the CYP2C19 2 and 3 genotypes, ABCB1 C3435T and PON1 Q192R alleles on the pharmacodynamics and adverse clinical events of clopidogrel in Chinese people after percutaneous coronary intervention. Eur J Clin Pharmacol 69:1103–1112

Taube J, Halsall D, Baglin T (2000) Influence of cytochrome P-450 CYP2C9 polymorphisms on warfarin sensitivity and risk of over-anticoagulation in patients on long-term treatment. Blood 96:1816–1819

Telenti A, Zanger UM (2008) Pharmacogenetics of anti-HIV drugs. Annu Rev Pharmacol Toxicol 48:227–256

Thervet E, Loriot MA, Barbier S, Buchler M, Ficheux M, Choukroun G, Toupance O, Touchard G, Alberti C, Le Pogamp P, Moulin B, Le Meur Y, Heng AE, Subra JF, Beaune P, Legendre C (2010) Optimization of initial tacrolimus dose using pharmacogenetic testing. Clin Pharmacol Ther 87:721–726

Thuerauf N, Lunkenheimer J (2006) The impact of the CYP2D6-polymorphism on dose recommendations for current antidepressants. Eur Arch Psychiatry Clin Neurosci 256:287–293

Tiroch KA, Sibbing D, Koch W, Roosen-Runge T, Mehilli J, Schomig A, Kastrati A (2010) Protective effect of the CYP2C19 *17 polymorphism with increased activation of clopidogrel on cardiovascular events. Am Heart J 160:506–512

Turpeinen M, Zanger UM (2012) Cytochrome P450 2B6: function, genetics, and clinical relevance. Drug Metabol Drug Interact 27:185–197

Turpeinen M, Raunio H, Pelkonen O (2006) The functional role of CYP2B6 in human drug metabolism: substrates and inhibitors in vitro, in vivo and in silico. Curr Drug Metab 7:705–714

van der Weide J, Steijns LS, van Weelden MJ, de Haan K (2001) The effect of genetic polymorphism of cytochrome P450 CYP2C9 on phenytoin dose requirement. Pharmacogenetics 11:287–291

Wan J, Xia H, He N, Lu YQ, Zhou HH (1996) The elimination of diazepam in Chinese subjects is dependent on the mephenytoin oxidation phenotype. Br J Clin Pharmacol 42:471–474

Wang SL, Huang J, Lai MD, Tsai JJ (1995) Detection of CYP2C9 polymorphism based on the polymerase chain reaction in Chinese. Pharmacogenetics 5:37–42

Wang SL, Lai MD, Huang JD (1999) G169R mutation diminishes the metabolic activity of CYP2D6 in Chinese. Drug Metab Dispos 27:385–388

Wei L, Locuson CW, Tracy TS (2007) Polymorphic variants of CYP2C9: mechanisms involved in reduced catalytic activity. Mol Pharmacol 72:1280–1288

Westlind-Johnsson A, Hermann R, Huennemeyer A, Hauns B, Lahu G, Nassr N, Zech K, Ingelman-Sundberg M, von Richter O (2006) Identification and characterization of CYP3A4*20, a novel rare CYP3A4 allele without functional activity. Clin Pharmacol Ther 79:339–349

Wilkinson GR, Guengerich FP, Branch RA (1989) Genetic polymorphism of *S*-mephenytoin hydroxylation. Pharmacol Ther 43:53–76

Xiao ZS, Goldstein JA, Xie HG, Blaisdell J, Wang W, Jiang CH, Yan FX, He N, Huang SL, Xu ZH, Zhou HH (1997) Differences in the incidence of the CYP2C19 polymorphism affecting the *S*-mephenytoin phenotype in Chinese Han and Bai populations and identification of a new rare CYP2C19 mutant allele. J Pharmacol Exp Ther 281:604–609

Xie HG, Kim RB, Wood AJ, Stein CM (2001) Molecular basis of ethnic differences in drug disposition and response. Annu Rev Pharmacol Toxicol 41:815–850

Xie HJ, Yasar U, Lundgren S, Griskevicius L, Terelius Y, Hassan M, Rane A (2003) Role of polymorphic human CYP2B6 in cyclophosphamide bioactivation. Pharmacogenomics J 3:53–61

Xu BQ, Aasmundstad TA, Bjorneboe A, Christophersen AS, Morland J (1995) Ethylmorphine *O*-deethylation in isolated rat hepatocytes. Involvement of codeine *O*-demethylation enzyme systems. Biochem Pharmacol 49:453–460

Xu H, Murray M, McLachlan AJ (2009) Influence of genetic polymorphisms on the pharmacokinetics and pharmaco-dynamics of sulfonylurea drugs. Curr Drug Metab 10:643–658

Yamazaki H, Kiyotani K, Tsubuko S, Matsunaga M, Fujieda M, Saito T, Miura J, Kobayashi S, Kamataki T (2003) Two novel haplotypes of CYP2D6 gene in a Japanese population. Drug Metab Pharmacokinet 18:269–271

Yokoi T, Kosaka Y, Chida M, Chiba K, Nakamura H, Ishizaki T, Kinoshita M, Sato K, Gonzalez FJ, Kamataki T (1996) A new CYP2D6 allele with a nine base insertion in exon 9 in a Japanese population associated with poor metabolizer phenotype. Pharmacogenetics 6:395–401

Yokota H, Tamura S, Furuya H, Kimura S, Watanabe M, Kanazawa I, Kondo I, Gonzalez FJ (1993) Evidence for a new variant CYP2D6 allele CYP2D6J in a Japanese population associated with lower in vivo rates of sparteine metabolism. Pharmacogenetics 3:256–263

Yoon YR, Shon JH, Kim MK, Lim YC, Lee HR, Park JY, Cha IJ, Shin JG (2001) Frequency of cytochrome P450 2C9 mutant alleles in a Korean population. Br J Clin Pharmacol 51:277–280

Zanger UM, Klein K (2013) Pharmacogenetics of cytochrome P450 2B6 (CYP2B6): advances on polymorphisms, mechanisms, and clinical relevance. Front Genet 4:24

Zanger UM, Turpeinen M, Klein K, Schwab M (2008) Functional pharmacogenetics/genomics of human cytochromes P450 involved in drug biotransformation. Anal Bioanal Chem 392:1093–1108

Zhao F, Loke C, Rankin SC, Guo JY, Lee HS, Wu TS, Tan T, Liu TC, Lu WL, Lim YT, Zhang Q, Goh BC, Lee SC (2004) Novel CYP2C9 genetic variants in Asian subjects and their influence on maintenance warfarin dose. Clin Pharmacol Ther 76:210–219

Zhou SF (2009) Polymorphism of human cytochrome P450 2D6 and its clinical significance: Part I. Clin Pharmacokinet 48:689–723

Chapter 21
Cytochrome P450 Polymorphisms of Clinical Importance

Allan E. Rettie and Kenneth E. Thummel

Abstract Human cytochrome P450 enzymes catalyze diverse metabolic reactions that are critical to the function of important endogenous compounds and elimination of xenobiotics, including therapeutic drugs and environmental toxins. The body relies on these metabolic reactions for the production of steroids and bile salts, to help control plasma and tissue levels of key fat-soluble vitamins and hormones, and to eliminate therapeutic agents, pesticides, and a host of other environmental agents to which humans are exposed. Because natural variation in P450 genes has the potential to disrupt critical biochemical pathways, carriers of alleles that lead to altered enzymatic function can cause disease as well as negatively impact therapeutic drug response. These two sequelae are the focus of this chapter, which reviews significant rare and common variations in two groups of P450 gene subfamilies: first, CYP11A, CYP17A, CYP21A, CYP2R, CYP24A, CYP27A, CYP27B, CYP4V, and CYP1B, and second, CYP3A, CYP2C, CYP2B, CYP2A, CYP2E, and CYP1A. Emphasis is placed on select rare and common polymorphisms in each of these P450 genes and the biochemical or metabolic elimination pathways that are disrupted to elicit their pathological or therapeutic consequences.

Keywords Cytochrome P450 • Drug metabolism • Inborn errors of metabolism • Pharmacogenetics • Pharmacogenomics • Steroidogenesis

A.E. Rettie (✉) • K.E. Thummel
Departments of Medicinal Chemistry and Pharmaceutics, School of Pharmacy, University of Washington, Box 357610, Seattle, WA 98195-7610, USA
e-mail: rettie@uw.edu

H. Yamazaki (ed.), *Fifty Years of Cytochrome P450 Research*,
DOI 10.1007/978-4-431-54992-5_21,

21.1 Introduction

This book commemorates the groundbreaking studies of the late 1950s and early 1960s that provided the first description of the hemoprotein cytochrome P450 (Omura and Sato 1964). These seminal efforts laid the foundation for a half century of exciting progress on the biochemistry, biophysics, and clinical aspects of these unique catalysts of mostly phase I oxidative reactions that involve both endobiotic and xenobiotic substrates. This era was also notable for the first recognition of genetic control of a phase I drug metabolism process, that of the N-acetylation of isoniazid (Evans et al. 1960). These researchers introduced the phenotypic terms 'fast' and slow' acetylators, and it soon became apparent that slow acetylators were at risk of toxicities because of the impaired elimination of several drugs including isoniazid and especially the sulfonamides (Shear et al. 1986). The molecular mechanism underlying the slow acetylator phenotype was not solved until the early 1990s with the identification of polymorphisms in the *NAT2* gene (Vatsis et al. 1991; Blum et al. 1991). Large phenotypic differences in drug response that were secondary to impaired P450-dependent metabolic clearance were first described in the late 1970s for the anti-hypertensive drug debrisoquine, and the oxytocic sparteine, with the molecular basis for these 'CYP2D6 polymorphisms' elucidated in the late 1980s (Meyer 2004). In this chapter, we address several of the translational implications of these versatile enzymes by reviewing cytochrome P450 polymorphisms that have important clinical repercussions for drug response and human disease.

21.2 Disease: P450 Polymorphisms Affecting Essential Homeostatic Processes

The majority of human P450 families are constituted of enzymes that act primarily on important classes of endogenous compounds, notably steroids, bile acids, vitamins, and fatty acids (Table 21.1). Cholesterol is the key molecule that unites the sterol and bile acid synthesis pathways, which involve many different P450 enzymes. Genetic defects in several of these enzymes underlie a variety of metabolic disorders, as discussed next.

21.2.1 CYP21A2 and Congenital Adrenal Hyperplasia

Congenital adrenal hyperplasia (CAH) is a group of autosomal recessive disorders that are caused by enzymatic deficiencies in steroidogenic enzymes present in the adrenal gland and gonads. The most common form of CAH (>90 % of cases) is caused by a deficiency of steroid 21-hydroxylation, which is catalyzed by CYP21A2.

Table 21.1 Classification of the 57 human cytochrome P450 enzymes on the basis of preferred substrates

Xenobiotics	Steroids	Eicosanoids/fatty acids	'Orphans'
CYP1A1	CYP11A1	CYP2U1	CYP2A7
CYP1A2	**CYP11B1**	CYP4A11	CYP2S1
CYP1B1	**CYP11B2**	CYP4F3	CYP2W1
CYP2A6	**CYP17A1**	CYP4F8	CYP3A43
CYP2A13	CYP19A2	CYP4F12	CYP4A22
CYP2B6	**CYP21A2**	CYP5A1	CYP4F22
CYP2C8		CYP8A1	**CYP4V2**
CYP2C9			CYP4X1
CYP2C18	Bile acids	Vitamins	CYP4Z1
CYP2C19	**CYP7A1**	**CYP2R1**	CYP20A1
CYP2D6	**CYP7B1**	CYP4F2	CYP27C1
CYP2E1	CYP8B1	CYP4F11	
CYP2F1	CYP27A1	**CYP24A1**	
CYP2J2	CYP39A1	CYP26A1	
CYP3A4	CYP46A1	CYP26B1	
CYP3A5	CYP51A1	CYP26C1	
CYP3A7		**CYP27B1**	
CYP4B1			

'Orphan' designates a P450 for which substrate specificity remains to be established. Enzymes may belong to one or more categories. Enzymes shown in *bold* are those discussed in the text

CYP21A2 is a microsomal enzyme found exclusively in the adrenal gland. This particular enzyme has played a pivotal role in the history of P450 biochemistry, in that bovine adrenal microsomes were first used to demonstrate the NADPH and oxygen dependence of progesterone 21-hydroxylation and its inhibition by carbon monoxide (Ryan and Engel 1957), as well as the photochemical action spectrum of cytochrome P450 (Cooper et al. 1965). Genetic mutations that abrogate or otherwise impair CYP21A2 activity result in reduced production of glucocorticoids and mineralocorticoids, cortisol and aldosterone, respectively, with shunting to increased formation of androgens.

Three types of 21-hydroxylase deficiency are recognized: two 'classical' and one 'nonclassical' form. Classical CAH is further subdivided into severe salt-wasting (SW) CAH and the simple virilizing (SV) form of the disease. The nonclassical (NC) form has a much milder phenotype of hirsutism and decreased fertility, although some individuals exhibit no symptoms (http://ghr.nlm.nih.gov/condition/21-hydroxylase-deficiency). In untreated classical CAH, an excess of androgens prenatally causes genital ambiguity in females, and both sexes exhibit postnatal virilization. Aldosterone deficiency in CAH that is attributable to complete loss of 21-hydroxylase activity causes SW, which is typified by hyponatremia and hypokalemia that can be life threatening if not treated with glucocorticoids. Less functionally deleterious defects in CYP21A2 lead to the SV form, whereas mutations with only a modest impact on catalytic activity are associated with NC (Wilson et al. 2007).

The classical and nonclassical forms have a worldwide incidence of about 1:15,000, and 1:1,000, respectively (http://ghr.nlm.nih.gov/condition/21-hydroxy lase-deficiency). However, there are pronounced interethnic differences in the prevalence of acquired mutations, with the Yup'ik Eskimos of Alaska exhibiting the highest rate of classical CAH, 1 in 282 live births. This high frequency in the Yup'ik people was identified in the first newborn screening program to monitor 17 α-hydroxyprogestrone in whole blood by radioimmunoassay (Pang et al. 1982). Today, prenatal screening is possible using reverse transcription-polymerase chain reaction (RT-PCR) to identify CYP21A2 polymorphisms in the first trimester, but the molecular genetics of CAH is complex, and ever more comprehensive mutation analysis protocols continue to emerge (Xu et al. 2013; Coeli-Lacchini et al. 2013).

The *CYP21A2* gene resides within the HLA locus on the short arm of chromosome 6, in a complex region known as the RCCX module that is composed of the *RP-C4-CYP21-TNX* genes and their respective pseudogenes. The HLA locus is highly recombinogenic, leading to the large gene deletions and gene conversions that often underlie the SW form of the disease. Numerous point mutations (which are actually microconversions from the *CYP21A2P* pseudogene) that cause amino acid substitutions in CYP21A2 are more commonly found in the virilizing form of the disease, whereas some relatively benign mutations such as P30L and V281L tend to appear in patients with the NC form (Coeli-Lacchini et al. 2013; Wilson et al. 2007). Although the human enzyme has not yet been crystallized, the structure of bovine CYP21A2 is available and useful for analyzing the likely functional consequences of these mutations because about 80 % of the disease-causing variants are conserved across bovine and human enzymes (Zhao et al. 2012).

21.2.2 CYP11B1 and CYP11B2

Two CYP11B enzymes are found in the adrenal gland, but in contrast to CYP21A2, these are mitochondrial proteins that require adrenodoxin and adrenodoxin reductase to transfer electrons from NADPH to the catalytic heme center. These two enzymes, purified in the late 1970s (Katagiri et al. 1976) and cloned in the early 1990s (Okamoto and Nonaka 1992), are localized to histologically discrete cellular layers of the adrenal cortex. CYP11B1 and CYP11B2 catalyze 11β-hydroxylation reactions that involve the 11-deoxy precursors of corticosterone and cortisol, respectively, downstream of the CYP21A2 reaction step. CYP11B2 also uniquely possesses 18-hydroxylase and 18-oxidase activities and thus is the sole enzyme responsible for the synthesis of aldosterone from corticosterone.

Congenital steroid 11β-hydroxylase deficiency caused by mutations in *CYP11B1* is the second most common cause of CAH, accounting for 5–8 % of all cases (Speiser and White 2003); this, however, is a comparatively rare cause of the disease in that it affects only 1:200,000 in the Caucasian population

(White et al. 1994). Virilization and hypertension resulting from shunting to the androgen pathway and accumulation of deoxycorticosterone, respectively, are the prominent clinical features of CYP11B1 deficiency (Nimkarn and New 2010). More than 80 missense/nonsense/splicing mutations, indels, and complex rearrangements of the *CYP11B1* gene are listed in the Human Gene Mutation (HGM) database (www.hgmd.cf.ac.uk). The *CYP11B1* and *CYP11B2* genes are about 95 % identical and are located close to each other on the long arm of chromosome 8. Should unequal crossover occur, an in-frame chimeric 11B1-B2 protein can be produced that causes glucocorticoid-remediable hyperaldosteronism and hypertension (Pascoe et al. 1992; Lifton et al. 1992).

21.2.3 CYP17A1

CYP17A1 is a microsomal monooxygenase found primarily in the adrenals and gonads. The enzyme plays a critical role in the formation of several steroid hormones because its catalytic actions occur at the crossroads between synthesis of androgens/estrogens and glucocorticoids/mineralocorticoids. CYP17A1 has also attracted much attention because it is a dual-function enzyme that catalyzes both steroid 17α-hydroxylation and 17,20-lyase reactions (Gilep et al. 2011). Although the former is a conventional monooxygenase activity, the lyase reaction that generates androstenedione and dehydroepiandrostenedione is unusual, and the mechanistic details for C–C bond cleavage in the first committed step of androgen synthesis continue to be debated (Mak et al. 2014). CYP17A1 is also notable because this enzyme served as the template for the first successful high-yield expression of a catalytically active P450 in *Escherichia coli* following modifications to its N-terminal sequence (Barnes et al. 1991). This advance propelled bacterial expression to the forefront of P450 purification and paved the way for large-scale isolation and ultimate crystallization of eukaryotic P450s.

Although more than 100 types of mutations of CYP17A1 are listed in the HGM database, CYP17A1 deficiency is a very rare cause of CAH. Interestingly, point mutations in the presumptive cytochrome b_5-binding domain of CYP17A1 diminish the 17,20-lyase activity of the enzyme without affecting the 17α-hydroxylase activity; however, 17,20-lyase deficiency alone is even more rare than 17-α-hydroxylase deficiency (Geller et al. 1997). Also of clinical note, CYP17A1 was recently crystallized in the presence of inhibitors that are used clinically to treat prostate cancer which proliferates in response to androgens (DeVore and Scott 2012). The availability of ligand-bound structures of the enzyme may aid the future design of more selective inhibitors that target only the lyase activity and thereby minimize side effects in their treatment of prostate cancer and possibly breast cancer (Pikuleva and Waterman 2013).

21.2.4 CYP7A1 and CYP7B1 in Bile Salt Formation

Conversion to bile salts is the dominant pathway of cholesterol catabolism; this occurs in the liver and involves microsomal P450s from the CYP7 and CYP8 families. CYP7A1 catalyzes the initiating and rate-limiting step in the classic pathway of bile salt formation, which in nearly all mammals is hydroxylation on the B-ring to form 7α-hydroxycholesterol. In contrast, CYP7B1 catalyzes 7-α-hydroxylation of oxysterols formed by 25- and 27-hydroxylation of cholesterol from the alternative pathway of bile salt formation. Bile salts serve key physiological roles in promoting absorption of dietary fat and fat-soluble vitamins, stimulating bile acid flow and FXR- and M-BAR/TGR5 receptor signaling (Monte 2009). Indeed, ablation of the *CYP7a1* gene in mice causes substantial postnatal lethality that is offset in survivors by upregulation of CYP7B1 and utilization of this oxysterol pathway of bile salt formation to eventually restore about 30 % of the bile acid pool in adult null mice (Schwarz et al. 1996). *CYP7b1* null mice are viable, but appear to exhibit altered oxysterol metabolism, primarily in extrahepatic tissues (Rose et al. 2001).

Genes for both human enzymes are located on the long arm of chromosome 8. Inherited CYP7A1 deficiency is an extremely rare condition, with clinical characteristics reported to date only for three members of a single family (Pullinger et al. 2002). Complete loss of CYP7A1 enzyme caused by a frameshift mutation was associated with an average total cholesterol levels in excess of 300 mg/dl and LDL cholesterol above 180 mg/dl. This extreme hypercholesteremia responded to aggressive statin therapy plus niacin.

CYP7B1 deficiency is associated with different phenotypes in children and adults. Neonatal cholestasis caused by CYP7B1 deficiency is an extremely rare condition associated with severe liver damage and a poor prognosis (Setchell et al. 1998). Only a handful of cases have been reported, with an outcome of successful liver transplantation in one (Mizuochi et al. 2011) and symptom amelioration by treating with chenodeoxycholic acid in another (Ueki et al. 2008). In adults, there are numerous reports of mutations in the *CYP7B1* gene causing the neurodegenerative disease, spastic paraplegia type 5, which presents as progressive weakness and spasticity in the lower limbs (Stiles et al. 2009). Other P450 enzymes—CYP39A1, CYP8B1, CYP46A1, and CYP27A1—also catalyze early steps in bile acid synthesis, but genetic deficiencies of these enzymes are not clearly associated with human disease.

21.2.5 CYP2R1, CYP24A1 and CYP27B1 in Vitamin D Homeostasis

Vitamin D is an essential hormone that has multiple functions in the body, most notably the maintenance of mineral (calcium and phosphate) homeostasis (DeLuca 2008).

The most prevalent form in humans, vitamin D_3 (cholecalciferol), is produced in the skin from the interaction of UVB light with 7-dehydrocholesterol. It can also be absorbed from the intestinal lumen following ingestion of foods enriched in vitamin D_3 (from animal tissue). Another form of the hormone, vitamin D_2 (ergocalciferol), is obtained from the ingestion of plant material, but it contributes little to the total body burden of the hormone, except in individuals taking vitamin D_2 supplements.

Vitamin D must be metabolized by cytochrome P450 enzymes into the biologically active form of the hormone 1α,25-dihydroxy vitamin D (Jones et al. 2014). The first oxidation step occurs primarily in the liver, generating the major circulating form of the hormone, 25-hydroxy vitamin D (25(OH)D) from vitamin D_3 or D_2. Vitamin D 25-hydroxylation was originally attributed to a mitochondrial enzyme, CYP27A1, which also metabolizes cholesterol, but more recent investigations have demonstrated that CYP2R1, an enzyme in the endoplasmic reticulum, is the principal catalyst of the reaction.

The *CYP2R1* gene is found on chromosome 11p15.2. It was first described as one of several orphan P450 enzymes with unknown function, although highly conserved in vertebrate species (Nelson 2003). That designation changed with the discovery that its expression in cell models stimulated the activation of VDR and target gene transcription (Cheng et al. 2003), and that it exhibited a 25-hydroxylase intrinsic clearance that was much higher than that of CYP27A1 (Shinkyo et al. 2004). In a short time, a report of severe 25(OH)D_3 deficiency in an individual homozygous for a rare defective allele of the *CYP2R1* gene (L99P) brought further reconsideration and assignment of CYP2R1 as the principal catalyst for the 25-hydroxylation reaction in humans (Cheng et al. 2004). Subsequent genome-wide and candidate gene studies have consistently identified an association between common variants in the *CYP2R1* gene and plasma 25(OH)D_3 concentrations in adults and children (Ahn et al. 2010; Bu et al. 2010; Lasky-Su et al. 2012; Wang et al. 2010). Findings from a gene knockout study in mice confirmed that CYP2R1 is the predominant, but not exclusive, source of 25(OH) D_3 (Zhu et al. 2013). Not surprisingly, common variants in the *CYP27A1* gene appear to have no impact on 25(OH)D_3 plasma concentrations, consistent with its limited role in the bioactivation reaction. The rare coding variant (L99P) associated with vitamin D deficiency and rickets (Cheng et al. 2004) is found in the B helix of the enzyme and appears critical for proper protein folding and stability (Strushkevich et al. 2008). With regard to more common variants associated with low plasma 25(OH)D_3 concentrations, there is little known about how they might cause reduced hepatic CYP2R1 activity. Most are in the 5′-flanking region of the gene or are synonymous coding single nucleotide polymorphisms (SNPs) (Bu et al. 2010). Presumably, if causal for a change in enzyme function, they would likely affect gene transcription (change in enhancer/repressor activity) or the stability of the transcript. They may also simply be in linkage disequilibrium with another, unidentified causal SNP.

The second step in vitamin D bioactivation involves 1α-hydroxylation and is catalyzed exclusively by CYP27B1. The gene encoding CYP27B1 is found on chromosome 12q14.1. The kidney is the major site of 25(OH)D_3 1 α-hydroxylation, where the expression of CYP27B1 is tightly regulated by parathyroid hormone and its principal secretory trigger, plasma calcium concentration. More recent studies have demonstrated CYP27B1 expression and 1α-hydroxylase activity in extra-renal tissues, including the small intestine (Balesaria et al. 2009) and bone (Anderson et al. 2010), where it is thought to confer paracrine effects from 25(OH)D_3 (Haussler et al. 2013).

Given the essential role of the 1α-hydroxylation reaction in human health, it is not surprising that function-disrupting *CYP27B1* gene variation is relatively rare, but when it occurs is associated with vitamin D-dependent rickets type I (Durmaz et al. 2012; Wang et al. 1998). In most cases, the putatively causal mutation introduces amino acid changes that lead to a loss of function from aberrant mRNA splicing or frameshift changes resulting in a truncated protein. With regard to more common *CYP27B1* variation, some investigators have identified a significant association with plasma 25(OH)D_3 concentration, but there has been only limited replication in the literature and no mechanistic insight if the associations are real (Signorello et al. 2011; McGrath et al. 2010).

The kidney is also a principal site of vitamin D inactivation, primarily from the enzymatic activity of CYP24. This enzyme is encoded by *CYP24A1* in humans, a gene found on chromosome 20q13. It catalyzes the 24R- and 23S-hydroxylation of 25(OH)D_3 and 1α,25$(OH)_2D_3$, to metabolites that are processed further to terminal downstream products, calcitrioic acid and a lactone product (for 1α,25$(OH)_2D_3$), respectively (Jones et al. 2012). The expression of CYP24 is controlled by 1α,25$(OH)_2D_3$ through activation of the vitamin D receptor (VDR). In addition to the kidney, the CYP24 transcript and protein can be found in many of the tissues in which the hormone is thought to elicit important biological effects. Because of the critical role that CYP24 plays in terminating the biological effects of 1α,25$(OH)_2D_3$ and 25(OH)D_3, mutations in the *CYP24A1* gene have the potential for causing morbidity and mortality from an excess of vitamin D effects in body tissues. Indeed, recent investigations of patients with severe hypercalcemia revealed that they invariably are homozygous or compound heterozygous for deleterious missense or nonsense *CYP24A1* mutations (Dinour et al. 2013; Schlingmann et al. 2011). Moreover, it has been suggested that, "on the basis of dbSNP data, the frequency of predicted deleterious bi-allelic *CYP24A1* variants in the general population is estimated to be as high as 4–20 %" (Nesterova et al. 2013). If borne out, this would represent a remarkable advance for genomic medicine and our understanding of the etiology of chronic kidney disease. However, it is also possible that many of the uncharacterized *CYP24A1* variants reported in dbSNP could have a more benign phenotype, compared to those identified in patients with known mineral pathology. One such common variant could be behind the significant association reported between plasma 25(OH)D_3 concentration and a tag SNP in the vicinity of the *CYP24A1* gene identified recently in a genome-wide study of 30,000 adults of European descent (Wang et al. 2010).

21.2.6 *CYP1B1 and CYP4V2 in Genetic Eye Diseases*

The preceding sections covered exclusively inherited metabolic disorders of P450-dependent steroidogenesis or vitamin D activation. These enzymatic defects occur in well-established pathways that enable a good mechanistic appreciation of disease etiology because the underlying pharmacology is relatively well understood. This basis is in stark contrast to that of two eye diseases, glaucoma and Bietti's corneoretinal dystrophy, respectively, that are strongly associated with mutations in the *CYP1B1* and *CYP4V2* genes because the endogenous substrates involved and the metabolic pathways that are affected remain to be established.

CYP1B1 was not discovered until the early 1990s, when Jefcoate and coworkers reported on a new form of P450 in mouse embryo fibroblasts and rat adrenal gland that catalyzed formation of novel polycyclic aromatic hydrocarbon metabolites (Otto et al. 1991; Pottenger et al. 1991). The enzyme was subsequently cloned and shown to reside on the short arm of human chromosome 2 (Sutter et al. 1994). In healthy human tissue, despite oftentimes abundant transcript levels, CYP1B1 protein content is extremely low or undetectable (Murray et al. 2001). The enzyme was undetectable in normal human liver (Edwards et al. 1998), although a very low median level of 1 pmol/mg was reported in lung microsomes from non-smokers that increased two- to fourfold in individuals who had been exposed chronically to cigarette smoke (Kim et al. 2004). CYP1B1 is found in practically all tumors and has been referred to as a potential tumor biomarker (Murray et al. 1997). The discrepancy between transcript and protein levels likely reflects a posttranscriptional mechanism of regulation by microRNAs (Devlin et al. 2010).

Given the early focus on pro-carcinogen metabolism by CYP1B1 it was rather surprising when in 1997 an association was reported between *CYP1B1* mutations and primary congenital glaucoma (PCG) (Stoilov et al. 1997). This association was quickly reproduced and extended to other types of glaucoma. Indeed, a recent review identified more than 140 distinct mutations associated with PCG alone in more than 500 patients (Li et al. 2011). The analysis by these workers demonstrated pronounced ethnic differences in the major mutations and a predominance of just three coding mutations (G61, R368H, and R469W) that have been identified in almost a third of all PCG cases worldwide.

The recognition of a clear P450 genetic marker in glaucoma ushered in new efforts to identify endogenous substrates for the enzyme that might explain the clinical findings. The genetic association was supported by detection of CYP1B1 protein in several eye tissues (Doshi et al. 2006) and generation of a Cyp1b1 null mouse that recapitulated some of the trabecular meshwork abnormalities found in humans (Libby et al. 2003). However, despite expansion of the substrate selectivity of the enzyme to include eicosanoids and retinoids (Choudhary et al. 2004), identification of an endogenous compound(s) whose metabolic pathway(s) are disrupted in carriers of inactivating CYP1B1 mutations has not been achieved to date. One mitigating factor could be the observation that CYP1B1 is but one of several genes that carry mutations in glaucoma patients (Vasiliou and Gonzalez 2008).

An intriguing recent proposal that links all these diverse genes is their potential interference with mitochondrial function (Lascaratos et al. 2012). More work is required to connect the polygenic influence on glaucoma with the metabolic defects in CYP1B1.

CYP4V2 gene mutations are also clearly associated with eye disease. In this case, the disorder is a corneo-retinal dystrophy in which affected individuals exhibit progressive sight loss because of the accumulation of ocular crystalline deposits. The connection between this disease and a P450 gene was made only 10 years ago, when Hejtmancik and colleagues at the National Eye Institute showed that practically all patients diagnosed with Bietti's carried mutations in the coding and intronic regions of the *CYP4V2* gene that is located on the long arm of chromosome 4 (Li et al. 2004).

Biochemical studies with the recombinant protein have demonstrated that CYP4V2 is capable of typical fatty acid ω-hydroxylation reactions, which may provide some clues as to the endogenous substrate for the enzyme (Nakano et al. 2009). Moreover, CYP4V2 protein is expressed in target tissues of the eye (Nakano et al. 2012). However, as is the case for CYP1B1-associated glaucoma, the metabolic pathways that are compromised in carriers of inactivating CYP4V2 polymorphisms remain to be identified. Additional information can be found in our recent review of this 'orphan P450' (Kelly et al. 2011).

21.3 Drug Response: P450 Polymorphisms Affecting Drug Disposition and Toxicity

There is a long and rich history of research that has investigated the effects of genetic variation on drug disposition and response, much of it associated with genes that encode the drug-metabolizing cytochrome P450 enzymes. These enzymes are found on the endoplasmic reticulum in the liver, small intestine, and many other organs of the body. As illustrated next, there are examples in which relatively common variation in the genes encoding the drug-metabolizing P450s confers both reduced and enhanced enzyme activity relative to the wild-type enzyme, and in some instances the complete loss of function. This variation can have a range of effects—small to profound—on the clearance of a drug, depending on the extent to which there are parallel, competing processes of drug elimination. If it is the parent drug that elicits pharmacological or toxicological effects in the body, then the fraction of the drug dose cleared via the polymorphic enzyme pathway will dictate the extent to which systemic drug exposure (e.g., area under the curve, AUC) and response is affected by a genetically determined change in enzyme activity. When it is a drug metabolite that exhibits significant biological effects, the impact of genetic variation on response will depend on the fraction of the metabolite formation clearance that can be attributed to the polymorphic enzyme. In this section, we have focused our attention on those enzymes most commonly implicated in the metabolic clearance of drugs and instances where genetic variation leads to pronounced changes in metabolic activity and pharmacological or toxicological response.

21.3.1 CYP2D6

The gene that encodes the CYP2D6 enzyme is found on the long arm (q13.1) of chromosome 22. It is part of a cluster of homologous, nonfunctional pseudogenes (e.g., *CYP2D7*, *CYP2D8*) that likely arose from gene duplication events in the distant past (Zanger and Klein 2013). CYP2D6 is expressed predominantly in the liver, where it contributes significantly to the clearance of many basic drugs, although much lower levels of mRNA, protein, and associated metabolic activity can be found in other tissues, including the upper small intestine (Madani et al. 1999). As mentioned earlier, some of the earliest investigations of pharmacogenetics involved polymorphic drug metabolism and pharmacological response that was ultimately linked to inherited variation in the *CYP2D6* gene.

During the late 1970s, significant interindividual differences in debrisoquine (Mahgoub et al. 1977) and sparteine (Eichelbaum et al. 1979) metabolism and systemic drug accumulation were first described. This finding resulted in the classification of drug disposition phenotypes into poor, intermediate, and extensive metabolizers, and later ultra-rapid metabolizers, terms that are still in use today. The genetic basis for these CYP2D6 phenotypes was uncovered in subsequent years, with seminal contributions from multiple investigators, as reviewed by (Dahl and Bertilsson 1993). More than 150 different CYP2D6 allelic variants (single SNPs or unique haplotypes) are currently recorded in the P450 Allele database (http://www.cypalleles.ki.se/cyp2d6.htm), suggesting that it is one of the most variable of the human P450 genes; a fact confirmed by results from the NHLBI Exome Sequencing Project (Gordon et al. 2014). However, most of the reduced CYP2D6 activity observed in vivo can be attributed to three relatively common 'null' function alleles: *CYP2D6*3*, **4*, and **5*. Another gene variant, *CYP2D6*10*, confers only a partial loss of function (Yokota et al. 1993), but it still can be of clinical relevance because of its high prevalence in certain world populations (principally eastern Asia) and the existence of sensitive CYP2D6 substrates. One other *CYP2D6* variant of note is the *CYP2D6*2*$\times$*N* allele, where gene duplication has resulted in multiple, functional copies (2 to 13 reported) of the gene on the same chromosome. As expected, this confers an ultra-rapid metabolizer phenotype, with exaggerated drug clearance or bioactivation and a pronounced change in systemic drug/metabolite exposure (Dalen et al. 1998). An important characteristic of *CYP2D6* gene variation (as well as other human P450 genes) is that the prevalence of individual variant alleles (and associated phenotypes) can vary significantly between the different geo-ethnic populations of the world, likely the result of population founder effects, genetic drift, or selective pressure. As a consequence, it is important to understand fully the complement of variation that can influence drug response if the personalization of drug therapy through genetic testing is to achieve a maximal and equitable impact on public health.

There are numerous examples of drugs whose metabolic clearance and pharmacological response is affected by *CYP2D6* gene variation. Some of the most notable include the antidepressant drugs nortripyline and amitripyline, both of which have been highlighted in a recent CPIC guidance for clinicians on how to modify drug treatment in poor and ultra-rapid metabolizers (Hicks et al. 2013). The drug codeine is also an important CYP2D6 substrate, but by virtue of the fact that formation of a minor, active metabolite, morphine, confers much of codeine's analgesic response. Thus, CYP2D6 poor and ultra-rapid metabolizers exhibit marked differences in systemic morphine exposure (Kirchheiner et al. 2007) and codeine efficacy and toxicity (Crews et al. 2012). Interestingly, not all drugs shown to be quantitatively significant substrates of CYP2D6 exhibit polymorphic differences in drug response. In the case of the anti-muscarinic drug tolterodine, poor metabolizers exhibit higher plasma levels of parent drug concentrations than do extensive metabolizers, but because of reduced accumulation of an equipotent active metabolite produced by CYP2D6, 5-hydroxytolterodine, and differences in plasma protein binding between parent and metabolite, there do not appear to be appreciable differences in clinical response (efficacy and safety) between patients with different CYP2D6 metabolic phenotypes (Brynne et al. 1998). Another interesting example is the drug atomoxetine, used for the treatment of attention deficit hyperactivity disorder. It is a very selective substrate for CYPD6 and accordingly poor metabolizers exhibit an approximately tenfold higher AUC of parent drug compared to extensive metabolizers when given equal doses (Sauer et al. 2005). However, there is apparently no clinically meaningful difference in drug response among patients with different CYP2D6 phenotypes because of the high safety margin of the drug. Other notable substrates of CYP2D6 where polymorphic clearance or drug response has been reported include anti-psychotic phenothiazines (chlorpromazine and perphenazine), cardiovascular β-blockers (metoprolol and bufuralol), and tamoxifen (a selective estrogen-receptor modifier) (Teh and Bertilsson 2012).

21.3.2 CYP3A

The functional human CYP3A genes, *CYP3A4*, *CYP3A5*, and *CYP3A7*, are all found on the long arm (q21.1) of chromosome 7 in a gene cluster that includes two pseudogenes and a functional but poorly understood gene, *CYP3A43*. Although there is a high degree of homology between the functional genes and their encoded products, they exhibit clear differences in regulation and substrate specificity. CYP3A7 is expressed primarily in fetal liver (Stevens et al. 2003) and placenta (Schuetz et al. 1993) and can catalyze the 16α-hydroxylation of dehydroepiandrosterone (DHEA), its sulfated metabolite and estrone. Hepatic CYP3A7 expression declines immediately after birth (Lacroix et al. 1997), although it can be detected in some adults, often associated with inheritance of a variant allele, *CYP3A7*1C*, which introduces an ER-6 response element that can bind multiple transcription-enhancing nuclear receptors (Kuehl et al. 2001; Burk et al. 2002). In contrast, CYP3A4 is the principal CYP3A isoform in children and adults and is expressed in liver and the mucosa of the proximal small intestine. CYP3A5 is found in both fetal and adult liver, but its expression is highly polymorphic, as discussed below.

Although CYP3A4 is capable of metabolizing numerous endogenous molecules (e.g., androgens, estrogens, progestins, and corticosteroids, vitamin D, retinoids), it is known principally for the metabolism of a very broad spectrum of xenobiotic molecules, including many drugs (Ince et al. 2013). X-ray crystallography has revealed that the active site of CYP3A4 is large and capable of accommodating multiple ligands simultaneously (Williams et al. 2004). This feature can give rise to unusual enzyme kinetic behavior termed allosterism, where binding of a second ligand can either inhibit or activate the biotransformation of a substrate, influencing both the rate and product specificity of the metabolic process (Atkins 2005). In general, the intrinsic metabolic activity of CYP3A4 toward most drug substrates is much greater than that of CYP3A5 and CYP3A7, with a few notable exceptions as discussed next. Indeed, CYP3A4 is perhaps the most important of the human drug-metabolizing cytochrome P450 enzymes (DMEs) because of its high abundance in key drug clearance organs (liver and intestine) and its very broad substrate specificity. It is also notable for its sensitivity to enzyme inducers, inhibitors, and activators (Houston and Galetin 2005; Zhou et al. 2010).

It has been concluded from the results of twin studies that much of the variability in CYP3A-dependent drug metabolism is heritable (Rahmioglu et al. 2011). Despite this observation, to date, only a small fraction of phenotypic variability can be attributed to specific CYP3A gene variants. Although new discoveries of *CYP3A* gene variation continue to be made (e.g., *CYP3A4*22*), the "missing heritability," if real, is likely to involve variation in CYP3A regulatory genes (Klein and Zanger 2013). Only 43 unique allelic *CYP3A4* variants are reported in the P450 allele database (http://www.cypalleles.ki.se/cyp3a4.htm), and many are relatively rare, particularly nonsynonymous coding variants, such as *CYP3A4*8* (R130Q), **15* (R162Q), **16* (T185S), and **17* (F189S), that are associated with reduced function. Another rare *CYP3A4* variant (*CYP3A4*20*) affects mRNA splicing, resulting in an enzyme inactivating frameshift and a truncated protein (Westlind-Johnsson et al. 2006). Interestingly, data from the NHLBI Exome Sequencing Project suggest that variants in the *CYP3A* genes are fewer in number than that found in most other DME genes (Gordon et al. 2014).

Some of the more common *CYP3A4* variants associated with altered enzyme function in vivo are found in noncoding regions of the gene. For example, a SNP within intron-6 (15389C > T, *CYP3A4*22*) causes reduced mRNA accumulation (Wang et al. 2011) in addition to decreased metabolic clearance of tacrolimus (Elens et al. 2011b) and several of the statin drugs that are metabolically cleared by CYP3A (Elens et al. 2011a; Wang et al. 2011). In addition, a SNP in the 5′-flanking region of *CYP3A4* (–392A > G; *CYP3A4*1B*) has been associated with altered enzyme activity (Rebbeck et al. 1998). It occurs in a putative repressor response element and thus could increase gene transcription and enzyme synthesis

(Amirimani et al. 2003). However, published data on enzyme phenotype associations are conflicting (Garcia-Martin et al. 2002; Westlind et al. 1999), leaving its contribution to variable drug clearance unclear. Nonetheless, *CYP3A4*1B* continues to be associated with altered risk of certain cancers (Zhou et al. 2013), as initially reported. Another notable noncoding SNP in the *CYP3A4* gene is found within intron 10 (20230G > A; *CYP3A4 *1G*) (Fukushima-Uesaka et al. 2004). It has been associated with reduced metabolic activity toward fentanyl (Zhang et al. 2011), tacrolimus (Miura et al. 2011; Li et al. 2014) (Gao et al. 2008), and atorvastatin, most frequently in East Asian populations. In this regard, interpreting the functional impact of the common, noncoding CYP3A variants, such as *CYP3A4*1B* and *CYP3A4*1G*, can be complicated by linkage disequilibrium (LD), which extends across the entire *CYP3A* locus. Multiple haplotypes can exist because of LD between *CYP3A5*1*, *CYP3A4*1B*, *CYP3A4*1G*, and *CYP3A7*1C*. Three of these are putatively gain-of-function alleles, whereas the other, *CYP3A4*1G*, is a loss-of-function allele. Hence, the observed phenotype will depend on the haplotype that is inherited, other factors affecting enzyme expression, and the intrinsic activity of different CYP3A enzymes for the drug. This reality and the fact that the frequency of CYP3A haplotypes can vary between different populations may help explain inconsistencies in study results. For example, the *CYP3A4*1B* allele has not been detected in East Asian populations and thus will be absent from the haplotype containing *CYP3A4*1G*. As mentioned earlier, studies conducted in China and Japan find clear associations between *CYP3A4*1G* and the metabolic clearance of some CYP3A substrates. Interestingly, it was recently reported that there is a relatively high frequency of a *CYP3A* haplotype consisting of the *CYP3A4*1G* allele, wild type at the *CYP3A4*1B* locus, and the loss-of-function *CYP3A5*3* allele (Fohner et al. 2013). With this combination, one might anticipate reduced clearance of CYP3A substrates.

Although LD in the *CYP3A* locus clearly can influence the metabolic clearance of CYP3A substrates, possibly with offsetting effects from the different alleles, there are examples in which the penetrance of one allele is clearly dominant. Perhaps the best known example involves tacrolimus metabolic clearance by CYP3A5. Hepatic CYP3A5 content is highly polymorphic (Wrighton et al. 1989) as a result of gene variants that lead to reduced accumulation of a functional, full-length transcript (Kuehl et al. 2001). The most common loss-of-function allele, *CYP3A5*3*, introduces a cryptic splice site into intron 3 that results in a truncated, nonfunctional enzyme. Inheritance of the wild-type *CYP3A5*1* allele, which is of low frequency in white populations and relatively high frequency in African and East Asian populations, is associated with enhanced tacrolimus clearance and a higher dose requirement to achieve a therapeutic, immunosuppressive blood concentration (Haufroid et al. 2004; Hesselink et al. 2003; Thervet et al. 2003). Not surprisingly, tacrolimus is an excellent substrate for CYP3A5, exhibiting a higher intrinsic clearance for the drug than does CYP3A4 (Dai et al. 2006). A similar genotype–phenotype relationship was seen for vincristine and its associated drug-induced neurotoxicity (Dennison et al. 2007; Egbelakin et al. 2011) as well as the bioactivation of alflatoxin and formation of covalent adducts to albumin (Wojnowski et al. 2004).

Variation in the *CYP3A7* gene (i.e., *CYP3A7*1C*), which confers increased enzyme expression in adults (Kuehl et al. 2001; Burk et al. 2002), has not been linked to interindividual differences in drug clearance, but is associated with reduced plasma DHEA sulfate concentrations (Smit et al. 2005). Inheritance of the *CYP3A7*1C* allele and corresponding reduced plasma DHEA sulfate levels has been associated with reduced bone mineral density (Bacsi et al. 2007). Interestingly, high CYP3A7 activity in the fetal liver generates 16α-hydroxy DHEA sulfate from DHEA sulfate, which in turn can be transferred to the placenta and converted to estriol, thus serving as a major source of estrogen for the placental-maternal unit.

21.3.3 CYP2C9

The CYP2C9 enzyme is a major form of human liver microsomal P450 responsible for the metabolism of a wide variety of principally lipophilic, acidic drugs (Rettie and Jones 2005). Early investigations of its substrate specificity revealed tolbutamide (Relling et al. 1990), phenytoin (Veronese et al. 1991), and (S)-warfarin (Rettie et al. 1992) to be important substrates for the enzyme. Because all three drugs exhibit a low therapeutic index, much of the clinical relevance of polymorphisms in *CYP2C9* stems from exaggerated pharmacological responses to these agents that can result in dangerously lowered blood sugar (oral hypoglycemics), neurotoxicity (phenytoin), and increased risk of bleeding (warfarin) in patients who are prescribed inappropriately high doses for their *CYP2C9* genotype (Ragia et al. 2009; Steward et al. 1997; Holstein et al. 2005; Kidd et al. 2001; Depondt et al. 2011; Aithal et al. 1999). Other important drugs that are metabolized at least in part by CYP2C9 include the antiinflammatory agents flurbiprofen, ibuprofen, and naproxen, other oral hypoglycemics such as glyburide and glipizide, the diuretic and uricosuric torsemide and sulfinpyrazone, respectively, and the angiotensin II blockers losartan, candesartan, and irbesartan (as reviewed by Rettie and Jones 2005).

cDNAs for the *CYP2C9* gene, which is part of the *CYP2C* cluster on the long arm of chromosome 10, were first cloned in the late 1980s (Kimura et al. 1987). More than 50 variants are now listed on the P450 allele website (http://www.cypalleles.ki.se/cyp2c9.htm http://evs.gs.washington.edu/EVS/), but only 4, that is, *CYP2C9*2* (R144C), *CYP2C9*3* (I359L), *CYP2C9*8* (R150H), and *CYP2C9*9* (H251R), are expressed at minor allele frequencies that are greater than 5 % (http://evs.gs.washington.edu/EVS/). Of these 4 relatively common alleles, the **2*, **3*, and **8* alleles encode functionally defective CYP2C9. By far, the greatest pharmacogenomic emphasis has been placed on the **2* and **3* alleles, because they combine the highest population frequencies (at least in white populations) with a substantial degree of impairment of enzyme activity. The first genetic association between a *CYP2C9* polymorphism and altered drug response was reported in a group of warfarin patients carrying the **2* allele who were shown to require a lower maintenance dose of the drug to maintain their INR within the desired therapeutic range (Furuya et al. 1995). These findings have been confirmed and extended to the **3* allele in dozens of studies (Rettie and Tai 2006). Although the U.S. FDA has recognized the importance of these *CYP2C9* polymorphisms in guiding dosing in patients, clinical implementation has been slow. The results of three randomized clinical trials that evaluated the benefit of genotyping coumarin oral anticoagulant patients for both *CYP2C9* as well as the drug target gene *VKORC1* were published in late 2013. It was concluded that pharmacogenetic testing provided no significant gains in terms of the primary endpoint of time-in-therapeutic range of the studies (editorial by Furie 2013). One positive feature was an overall trend toward fewer bleeding events in the pharmacogenetically guided cohort (Zineh et al. 2013), but it seems certain that failure of the trial primary endpoint, along with the introduction of new drugs with different mechanisms of action, will impede progress toward future genetic testing for coumarin anticoagulant therapy.

21.3.4 CYP2C19

The *CYP2C19* gene is another member of the human *CYP2C* locus and historically the most important because defects in the gene underlie poor metabolizer status for (S)-mephenytoin. Mephenytoin is a chiral anticonvulsant agent rarely used today, but in the early 1980s researchers were interested in why the drug caused excessive sedation in some patients. Pharmacokinetic studies revealed a slow metabolizer phenotype resulting from a defect in aromatic hydroxylation in a small proportion of Europeans, and family studies confirmed its inherited nature that was shown to be distinct from the genetically determined debrisoquine polymorphism which was already well recognized (Kupfer and Preisig 1984). This genetic defect affected specifically (S)-mephenytoin hydroxylation (Wedlund et al. 1984) and was more prevalent in Japanese (18 %) than in Caucasians (3 %) (Nakamura et al. 1985).

Purification and cloning of the human liver enzyme responsible for this new P450 drug polymorphism proved to be very challenging, in large part because of the low abundance of the protein in human liver and the complexity of the human *CYP2C* locus. However, purification of the enzyme and cloning were eventually achieved by the Wrighton and Goldstein groups, respectively (Wrighton et al. 1993; Goldstein et al. 1994; de Morais et al. 1994), which enabled its definitive identification as CYP2C19. With the gene in hand, it was found that the principal molecular defect in both Caucasians and Japanese was a single base-pair (G > A) mutation in exon 5 of *CYP2C19* that creates an aberrant splice site (de Morais et al. 1994). This mutation is designated *CYP2C19*2*, and together with a variant termed *CYP2C19*3* that reflects a premature stop codon, these two polymorphisms account for the majority of individuals exhibiting a poor metabolizer phenotype across Caucasian and Asian populations.

CYP2C19 has a preference for generally neutral or basic substrates and contributes to the metabolism of many important drugs including citalopram, omeprazole, diazepam, and nelfinavir; however, only a few can be identified that are metabolized exclusively or almost exclusively by the enzyme (Li-Wan-Po et al. 2010). The anti-convulsants *S*-mephenytoin and *R*-mephobarbital, the anti-malarials proguanil and chlorproguanil, and possibly voriconazole fall into this latter category (Jacqz et al. 1986; Ward et al. 1991), but not all are widely used today. CYP2C19-dependent metabolism of the biguanide anti-malarials to their cyclic active metabolites is an example of the role of the enzyme role in pro-drug activation. Ironically, Southeast Asian populations in which malaria is endemic exhibit the highest incidence of CYP2C19 poor metabolizers, which has implications for drug choice in malaria treatment programs in the area (Kaneko et al. 1997).

The antiplatelet drug clopidogrel also depends on bioactivation by CYP2C19 to form the active *exo*-thiol metabolite. However, resistance to the drug is well documented, and 'non-responders' may constitute as much as 40 % of some patient populations (Angiolillo et al. 2005). *CYP2C19* variants conferring defective catalytic activity are important risk factors for therapeutic resistance to clopidogrel (Mega et al. 2009), but other variables are also important (Shuldiner et al. 2009). The recently discovered *CYP2C19*17* allele is a complicating factor in genotype–phenotype analysis of CYP2C19-dependent drug response because it appears this regulatory polymorphism confers a gain, rather than a loss, of enzyme function resulting from an increase in transcriptional activity (Sim et al. 2006). Research is ongoing to determine whether *CYP2C19*17* homozygotes are truly ultra-metabolizers and what the appropriate phenotype designation is for **2/*17* compound heterozygotes.

21.3.5 CYP2A6

CYP2A6 is a relatively minor drug-metabolizing enzyme in the liver, with narrow substrate specificity. The enzyme prefers small, often bicyclic, substrates typified by coumarin and nicotine, which can be rationalized by the topology and small volume of the CYP2A6 active site (Yano et al. 2005). As a consequence, only a small cadre of drugs, including valproic acid, halothane, letrozole, and disulfiram, are metabolized substantially by CYP2A6. The enzyme also features prominently in the metabolism of several environmental agents and pro-carcinogens such as NNK, 2,6-chlorobenzonitrile, and butadiene (Fernandez-Salguero et al. 1995).

The *CYP2A6* gene is located on the long arm of chromosome 19 within the complex *CYP2ABDG* locus that harbors six different subfamilies, including three members of the *CYP2A* subfamily: *CYP2A6*, *CYP2A7*, and *CYP2A13*. However, only *CYP2A6* encodes a functional hepatic enzyme that significantly catalyzes drug metabolism reactions. Guengerich and coworkers purified CYP2A6 as a by-product of attempts to isolate the mephenytoin hydroxylase (Yun et al. 1991). Gonzalez and coworkers reported the isolation of cDNAs encoding the wild-type enzyme and an

L160H variant, which was named *CYP2A6**2. Expression of the recombinant enzymes suggested that the *2 variant was inactive, perhaps because of failure to incorporate the heme prosthetic group (Yamano et al. 1990). At the time of writing, about 100 CYP2A6 alleles are listed on the P450 alleles website, reflective of the great complexity of the *CYP2A6* locus. A major challenge is accurate gene- and allele-specific genotyping, which is complicated by the presence of the inactive, but structurally highly similar, *CYP2A7* gene. Their physical proximity can result in the generation of chimeras from unequal crossover and gene conversion events between *CYP2A6* and *CYP2A7* (Oscarson et al. 2002; Fukami et al. 2006). Gene deletion is also a prominent cause of the poor metabolizer phenotype, which occurs at relatively higher frequencies in Asian populations (Ku et al. 2010).

Of great public health importance is the role of *CYP2A6* variants in smoking behavior, nicotine dependence, and tobacco-related cancers. A guiding hypothesis has been that an individual will adjust their smoking frequency to maintain a constant level of nicotine exposure, which in turn is influenced by an individual's ability to metabolize the drug (Malaiyandi et al. 2005). It follows then that nicotine-poor metabolizers should need to consume fewer cigarettes daily and thereby have less exposure (among smokers) to health risks of tobacco smoke including chronic obstructive pulmonary disease (COPD) and lung cancer. Indeed, genome-wise association (GWA) studies have found associations in the *CYP2A6* locus with both smoking behavior (Thorgeirsson et al. 2010) and COPD susceptibility (Cho et al. 2012). *CYP2A6* variants also appear to play a role in lung cancer risk, but genetic variation in the *CHRNA5-A3-B4* gene cluster may be a more important factor (Wassenaar et al. 2011).

21.3.6 CYP2B6

The gene encoding CYP2B6 is found in the complex cluster on chromosome 19q13.2 that includes the *CYP2B7* pseudogene, *CYP2A6,* and *CYP2A13*, as well as pseudogenes of *CYP2A* and *CYP2G* (Zanger et al. 2007). Accordingly, genotyping for variants in *CYP2B6* also requires careful attention and consideration of highly homologous genetic regions. In comparison to some of the other drug-metabolizing P450 enzymes, its abundance in the liver is low, approximately 3–6 %, on average, of the total P450 content, but highly variable (Turpeinen and Zanger 2012). Nonetheless, it has been shown to contribute importantly to the metabolic clearance of the drugs efavirenz (Ward et al. 2003), bupropion (Hesse et al. 2000), methadone (Gerber et al. 2004), and ifosfamide (McCune et al. 2005), among other drugs and environmental toxins (Turpeinen and Zanger 2012). The role of CYP2B6 in drug clearance is sometimes erroneously overshadowed by, and difficult to distinguish from, parallel CYP3A-dependent metabolism, as seen for methadone (Totah et al. 2008).

As mentioned, there are considerable interindividual differences in the specific content of CYP2B6 in human liver (Code et al. 1997; Hesse et al. 2004). Some liver specimens contain undetectable amounts, whereas others have a specific content approaching that of CYP3A4. CYP2B6 abundance and catalytic activity can be increased by treatment with agonists of CAR and PXR and reduced by suicide inhibitors such as clopidogrel (Richter et al. 2004). Thus, variable hepatic content as a consequence of prior drug exposure might explain some of the interindividual variability, as could differences in constitutive gene regulation and genetic variation.

More than 60 unique allelic variants have been reported on the P450 Allele website (http://www.cypalleles.ki.se/cyp2b6.htm), and several have been characterized and shown to confer decreased or increased expression and metabolic activity, compared to the wild-type gene. For example, the K262R variant when found alone (*CYP2B6*4*) is associated with increased catalytic activity toward bupropion and efavirenz, whereas when combined as a haplotype with Q172H (*CYP2B6*6*), it exhibits reduced metabolic activity toward some, but not all, CYP2B6 substrates (Ariyoshi et al. 2011). The most consistent finding of clinical importance is increased plasma concentrations of efavirenz (reduced clearance) in individuals homozygous for *CYP2B6*6* and receiving the drug for the treatment of human immunodeficiency virus (HIV) infection (Zanger and Klein 2013).

21.3.7 CYP1A2

The *CYP1A2* gene is found on chromosome 15q24.1, in a cluster that includes *CYP1A1*. The two genes share a common 5′-flanking region and are transcribed in opposite directions. There is evidence that *CYP1A2* can be induced by mechanisms (e.g., AhR and XRE agonism) similar to that of other *CYP1* genes (Zhou et al. 2010). Key drug substrates of CYP1A2 include caffeine, theophylline, ciprofloxacin, and clozapine, and potent drug inhibitors are known that include fluvoxamine and furafylline. For many years, caffeine has been used as an in vitro and in vivo probe of CYP1A2 metabolic activity, particularly for studies of genetic and environmental contributions to variable drug clearance (Perera et al. 2012). Recent GWA studies have identified regions in the *CYP1A2* gene with habitual caffeine consumption (Cornelis et al. 2011). However, most contemporary research focuses on the role of CYP1A2 in bioactivation of pro-toxic arylamine and heterocyclic amines found in our industrialized environment and in genetic susceptibility to cancer (Kim and Guengerich 2005).

There are a number of *CYP1A2* allelic variants reported on the P450 allele website (http://www.cypalleles.ki.se/cyp1a2.htm), including some nonsynonymous coding and splice variants that putatively lead to reduced function. Considerable attention has been given to variants found in the 5′-flanking and promoter regions of the gene (e.g., *CYP1A2*1C*, **1 F*, **1 K*) that may alter basal and inducible gene expression, but the published data to date are conflicting, as reviewed by Zhou et al. (2010).

21.3.8 *CYP2E1*

The *CYP2E1* gene is found on the long arm of chromosome 10 (q26.3) and encodes an enzyme that is relatively abundant in human liver but which displays a restricted substrate repertoire. Classic CYP2E1 substrates are relatively small, polar molecules that include ethanol, acetone, some volatile anesthetics (e.g., enflurane, halothane, isoflurane), nitrosamines, and the enzyme probe drug chlorzoxazone (Lieber 1997). CYP2E1 is inducible by many of its substrates (e.g., ethanol, acetone, isoniazid) through a mechanism that involves protein stabilization with binding of the substrate to the enzyme active site (Roberts et al. 1995) or an enhanced accumulation of mRNA (Takahashi et al. 1993). Thus, chronic exposure to ethanol results in auto-induction of its metabolic clearance (Cederbaum 2012). CYP2E1 can also be induced by aberrant physiological states such as starvation, diabetes, and steatosis (Lieber 1997). Interestingly, the enzyme exhibits a significant degree of uncoupled oxidation, generating reactive oxygen species (ROS), a process that has been linked to nonalcoholic steatohepatitis and alcoholic liver disease (Leung and Nieto 2013).

The extent of *CYP2E1* gene variation reported on the P450 allele website (http://www.cypalleles.ki.se/cyp2e1.htm) is relatively small, in comparison to other P450 genes, perhaps because of more limited clinical interest. Some rare coding variants have been identified, but the most frequently studied variants are found in the 5′-flanking region (tandem repeats, e.g., *CYP2E1*1D*; SNPs, e.g., *CYP2E1*7B*) or other noncoding regions of the gene (Hu et al. 1999; Fairbrother et al. 1998). The *CYP2E1*1D* variant is thought to confer enhanced upregulation by enzyme inducers such as ethanol, whereas the *CYP2E1*7B* variant in the promoter/enhancer region may affect basal gene transcription. However, similar to that seen for *CYP1A2* gene variation, the evidence supporting a functional change in CYP2E1 phenotype from the common *CYP2E1* allelic variants is conflicting, although data associating these variants (Howard et al. 2003) or *CYP2E1* tag SNPs (Peng et al. 2013) with different clinical responses are compelling.

21.4 Summary and Future Directions

Although the P450 field is celebrating its 50th anniversary, the subdiscipline of P450 pharmacogenetics is more nearly 35 years old. The younger sibling has come a long way from the earliest crude phenotypic descriptions of altered drug responses and inborn errors of steroid metabolism to a fairly comprehensive understanding of the molecular biochemistry and pathology underlying these aberrant states. Challenges for the future include improving allele, genotype, and haplotype assignments among the genes that reside in very complex gene loci, as well as improved predictions of phenotype. Improved education for caregivers and clear answers to sensitive questions concerning patient privacy and the ethics of reporting test results to patients are all required for the pharmacogenomics of disease and drug response to advance to the point where molecular diagnostic data can be confidently transmitted as actionable information to clinical databases for routine incorporation into patient health records.

References

Ahn J, Yu K, Stolzenberg-Solomon R, Simon KC, McCullough ML, Gallicchio L, Jacobs EJ, Ascherio A, Helzlsouer K, Jacobs KB, Li Q, Weinstein SJ, Purdue M, Virtamo J, Horst R, Wheeler W, Chanock S, Hunter DJ, Hayes RB, Kraft P, Albanes D (2010) Genome-wide association study of circulating vitamin D levels. Hum Mol Genet 19:2739–2745

Aithal GP, Day CP, Kesteven PJ, Daly AK (1999) Association of polymorphisms in the cytochrome P450 CYP2C9 with warfarin dose requirement and risk of bleeding complications. Lancet 353:717–719

Amirimani B, Ning B, Deitz AC, Weber BL, Kadlubar FF, Rebbeck TR (2003) Increased transcriptional activity of the CYP3A4*1B promoter variant. Environ Mol Mutagen 42:299–305

Anderson PH, Iida S, Tyson JH, Turner AG, Morris HA (2010) Bone CYP27B1 gene expression is increased with high dietary calcium and in mineralising osteoblasts. J Steroid Biochem Mol Biol 121:71–75

Angiolillo DJ, Fernandez-Ortiz A, Bernardo E, Ramirez C, Barrera-Ramirez C, Sabate M, Hernandez R, Moreno R, Escaned J, Alfonso F, Banuelos C, Costa MA, Bass TA, Macaya C (2005) Identification of low responders to a 300-mg clopidogrel loading dose in patients undergoing coronary stenting. Thromb Res 115:101–108

Ariyoshi N, Ohara M, Kaneko M, Afuso S, Kumamoto T, Nakamura H, Ishii I, Ishikawa T, Kitada M (2011) Q172H replacement overcomes effects on the metabolism of cyclophosphamide and efavirenz caused by CYP2B6 variant with Arg262. Drug Metab Dispos 39:2045–2048

Atkins WM (2005) Non-Michaelis–Menten kinetics in cytochrome P450-catalyzed reactions. Annu Rev Pharmacol Toxicol 45:291–310

Bacsi K, Kosa JP, Borgulya G, Balla B, Lazary A, Nagy Z, Horvath C, Speer G, Lakatos P (2007) CYP3A7*1C polymorphism, serum dehydroepiandrosterone sulfate level, and bone mineral density in postmenopausal women. Calcif Tissue Int 80:154–159

Balesaria S, Sangha S, Walters JR (2009) Human duodenum responses to vitamin D metabolites of TRPV6 and other genes involved in calcium absorption. Am J Physiol Gastrointest Liver Physiol 297:G1193–G1197

Barnes HJ, Arlotto MP, Waterman MR (1991) Expression and enzymatic activity of recombinant cytochrome P450 17 alpha-hydroxylase in *Escherichia coli*. Proc Natl Acad Sci USA 88:5597–5601

Blum M, Demierre A, Grant DM, Heim M, Meyer UA (1991) Molecular mechanism of slow acetylation of drugs and carcinogens in humans. Proc Natl Acad Sci USA 88:5237–5241

Brynne N, Dalen P, Alvan G, Bertilsson L, Gabrielsson J (1998) Influence of CYP2D6 polymorphism on the pharmacokinetics and pharmacodynamic of tolterodine. Clin Pharmacol Ther 63:529–539

Bu FX, Armas L, Lappe J, Zhou Y, Gao G, Wang HW, Recker R, Zhao LJ (2010) Comprehensive association analysis of nine candidate genes with serum 25-hydroxy vitamin D levels among healthy Caucasian subjects. Hum Genet 128:549–556

Burk O, Tegude H, Koch I, Hustert E, Wolbold R, Glaeser H, Klein K, Fromm MF, Nuessler AK, Neuhaus P, Zanger UM, Eichelbaum M, Wojnowski L (2002) Molecular mechanisms of polymorphic CYP3A7 expression in adult human liver and intestine. J Biol Chem 277:24280–24288

Cederbaum AI (2012) Alcohol metabolism. Clin Liver Dis 16:667–685
Cheng JB, Motola DL, Mangelsdorf DJ, Russell DW (2003) De-orphanization of cytochrome P450 2R1: a microsomal vitamin D 25-hydroxilase. J Biol Chem 278:38084–38093
Cheng JB, Levine MA, Bell NH, Mangelsdorf DJ, Russell DW (2004) Genetic evidence that the human CYP2R1 enzyme is a key vitamin D 25-hydroxylase. Proc Natl Acad Sci USA 101:7711–7715
Cho MH, Castaldi PJ, Wan ES, Siedlinski M, Hersh CP, Demeo DL, Himes BE, Sylvia JS, Klanderman BJ, Ziniti JP, Lange C, Litonjua AA, Sparrow D, Regan EA, Make BJ, Hokanson JE, Murray T, Hetmanski JB, Pillai SG, Kong X, Anderson WH, Tal-Singer R, Lomas DA, Coxson HO, Edwards LD, Macnee W, Vestbo J, Yates JC, Agusti A, Calverley PM, Celli B, Crim C, Rennard S, Wouters E, Bakke P, Gulsvik A, Crapo JD, Beaty TH, Silverman EK, Investigators I, Investigators E, Investigators CO (2012) A genome-wide association study of COPD identifies a susceptibility locus on chromosome 19q13. Hum Mol Genet 21:947–957
Choudhary D, Jansson I, Stoilov I, Sarfarazi M, Schenkman JB (2004) Metabolism of retinoids and arachidonic acid by human and mouse cytochrome P450 1b1. Drug Metab Dispos 32:840–847
Code EL, Crespi CL, Penman BW, Gonzalez FJ, Chang TK, Waxman DJ (1997) Human cytochrome P4502B6: interindividual hepatic expression, substrate specificity, and role in procarcinogen activation. Drug Metab Dispos 25:985–993
Coeli-Lacchini FB, Turatti W, Elias PC, Elias LL, Martinelli CE Jr, Moreira AC, Antonini SR, De Castro M (2013) A rational, non-radioactive strategy for the molecular diagnosis of congenital adrenal hyperplasia due to 21-hydroxylase deficiency. Gene (Amst) 526:239–245
Cooper DY, Levin S, Narasimhulu S, Rosenthal O (1965) Photochemical action spectrum of the terminal oxidase of mixed function oxidase systems. Science 147:400–402
Cornelis MC, Monda KL, Yu K, Paynter N, Azzato EM, Bennett SN, Berndt SI, Boerwinkle E, Chanock S, Chatterjee N, Couper D, Curhan G, Heiss G, Hu FB, Hunter DJ, Jacobs K, Jensen MK, Kraft P, Landi MT, Nettleton JA, Purdue MP, Rajaraman P, Rimm EB, Rose LM, Rothman N, Silverman D, Stolzenberg-Solomon R, Subar A, Yeager M, Chasman DI, van Dam RM, Caporaso NE (2011) Genome-wide meta-analysis identifies regions on 7p21 (AHR) and 15q24 (CYP1A2) as determinants of habitual caffeine consumption. PLoS Genet 7: e1002033
Crews KR, Gaedigk A, Dunnenberger HM, Klein TE, Shen DD, Callaghan JT, Kharasch ED, Skaar TC, Clinical Pharmacogenetics Implementation Consortium (2012) Clinical Pharmacogenetics Implementation Consortium (CPIC) guidelines for codeine therapy in the context of cytochrome P450 2D6 (CYP2D6) genotype. Clin Pharmacol Ther 91:321–326
Dahl ML, Bertilsson L (1993) Genetically variable metabolism of antidepressants and neuroleptic drugs in man. Pharmacogenetics 3:61–70
Dai Y, Hebert MF, Isoherranen N, Davis CL, Marsh C, Shen DD, Thummel KE (2006) Effect of CYP3A5 polymorphism on tacrolimus metabolic clearance in vitro. Drug Metab Dispos 34:836–847
Dalen P, Dahl ML, Bernal Ruiz ML, Nordin J, Bertilsson L (1998) 10-Hydroxylation of nortriptyline in white persons with 0, 1, 2, 3, and 13 functional CYP2D6 genes. Clin Pharmacol Ther 63:444–452
de Morais SM, Wilkinson GR, Blaisdell J, Nakamura K, Meyer UA, Goldstein JA (1994) The major genetic defect responsible for the polymorphism of *S*-mephenytoin metabolism in humans. J Biol Chem 269:15419–15422
Deluca HF (2008) Evolution of our understanding of vitamin D. Nutr Rev 66:S73–S87
Dennison JB, Jones DR, Renbarger JL, Hall SD (2007) Effect of CYP3A5 expression on vincristine metabolism with human liver microsomes. J Pharmacol Exp Ther 321:553–563
Depondt C, Godard P, Espel RS, Da Cruz AL, Lienard P, Pandolfo M (2011) A candidate gene study of antiepileptic drug tolerability and efficacy identifies an association of CYP2C9 variants with phenytoin toxicity. Eur J Neurol 18:1159–1164
Devlin AH, Thompson P, Robson T, McKeown SR (2010) Cytochrome P450 1B1 mRNA untranslated regions interact to inhibit protein translation. Mol Carcinog 49:190–199

Devore NM, Scott EE (2012) Structures of cytochrome P450 17A1 with prostate cancer drugs abiraterone and TOK-001. Nature (Lond) 482:116–119

Dinour D, Beckerman P, Ganon L, Tordjman K, Eisenstein Z, Holtzman EJ (2013) Loss-of-function mutations of CYP24A1, the vitamin D 24-hydroxylase gene, cause long-standing hypercalciuric nephrolithiasis and nephrocalcinosis. J Urol 190:552–557

Doshi M, Marcus C, Bejjani BA, Edward DP (2006) Immunolocalization of CYP1B1 in normal, human, fetal and adult eyes. Exp Eye Res 82:24–32

Durmaz E, Zou M, Al-Rijjal RA, Bircan I, Akcurin S, Meyer B, Shi Y (2012) Clinical and genetic analysis of patients with vitamin D-dependent rickets type 1A. Clin Endocrinol (Oxf) 77:363–369

Edwards RJ, Adams DA, Watts PS, Davies DS, Boobis AR (1998) Development of a comprehensive panel of antibodies against the major xenobiotic metabolising forms of cytochrome P450 in humans. Biochem Pharmacol 56:377–387

Egbelakin A, Ferguson MJ, Macgill EA, LehmannA S, Topletz AR, Quinney SK, Li L, McCammack KC, Hall SD, Renbarger JL (2011) Increased risk of vincristine neurotoxicity associated with low CYP3A5 expression genotype in children with acute lymphoblastic leukemia. Pediatr Blood Cancer 56:361–367

Eichelbaum M, Spannbrucker N, Steincke B, Dengler HJ (1979) Defective N-oxidation of sparteine in man: a new pharmacogenetic defect. Eur J Clin Pharmacol 16:183–187

Elens L, Becker ML, Haufroid V, Hofman A, Visser LE, Uitterlinden AG, Stricker B, van Schaik RH (2011a) Novel CYP3A4 intron 6 single nucleotide polymorphism is associated with simvastatin-mediated cholesterol reduction in the Rotterdam Study. Pharmacogenet Genomics 21:861–866

Elens L, Bouamar R, Hesselink DA, Haufroid V, van der Heiden IP, van Gelder T, van Schaik RH (2011b) A new functional CYP3A4 intron 6 polymorphism significantly affects tacrolimus pharmacokinetics in kidney transplant recipients. Clin Chem 57:1574–1583

Evans DA, Manley KA, McKusik VA (1960) Genetic control of isoniazid metabolism in man. Br Med J 2:485–491

Fairbrother KS, Grove J, de Waziers I, Steimel DT, Day CP, Crespi CL, Daly AK (1998) Detection and characterization of novel polymorphisms in the CYP2E1 gene. Pharmacogenetics 8:543–552

Fernandez-Salguero P, Hoffman SM, Cholerton S, Mohrenweiser H, Raunio H, Rautio A, Pelkonen O, Huang JD, Evans WE, Idle JR et al (1995) A genetic polymorphism in coumarin 7-hydroxylation: sequence of the human CYP2A genes and identification of variant CYP2A6 alleles. Am J Hum Genet 57:651–660

Fohner A, Muzquiz LI, Austin MA, Gaedigk A, Gordon A, Thornton T, Rieder MJ, Pershouse MA, Putnam EA, Howlett K, Beatty P, Thummel KE, Woodahl EL (2013) Pharmacogenetics in American Indian populations: analysis of CYP2D6, CYP3A4, CYP3A5, and CYP2C9 in the Confederated Salish and Kootenai Tribes. Pharmacogenet Genomics 23:403–414

Fukami T, Nakajima M, Sakai H, McLeod HL, Yokoi T (2006) CYP2A7 polymorphic alleles confound the genotyping of CYP2A6*4A allele. Pharmacogenomics J 6:401–412

Fukushima-Uesaka H, Saito Y, Watanabe H, Shiseki K, Saeki M, Nakamura T, Kurose K, Sai K, Komamura K, Ueno K, Kamakura S, Kitakaze M, Hanai S, Nakajima T, Matsumoto K, Saito H, Goto Y, Kimura H, Katoh M, Sugai K, Minami N, Shirao K, Tamura T, Yamamoto N, Minami H, Ohtsu A, Yoshida T, Saijo N, Kitamura Y, Kamatani N, Ozawa S, Sawada J (2004) Haplotypes of CYP3A4 and their close linkage with CYP3A5 haplotypes in a Japanese population. Hum Mutat 23:100

Furie B (2013) Do pharmacogenetics have a role in the dosing of vitamin K antagonists? N Engl J Med 369:2345–2346

Furuya H, Fernandez-Salguero P, Gregory W, Taber H, Steward A, Gonzalez FJ, Idle JR (1995) Genetic polymorphism of CYP2C9 and its effect on warfarin maintenance dose requirement in patients undergoing anticoagulation therapy. Pharmacogenetics 5:389–392

Gao Y, Zhang LR, Fu Q (2008) CYP3A4*1G polymorphism is associated with lipid-lowering efficacy of atorvastatin but not of simvastatin. Eur J Clin Pharmacol 64:877–882

Garcia-Martin E, Martinez C, Pizarro RM, Garcia-Gamito FJ, Gullsten H, Raunio H, Agundez JA (2002) CYP3A4 variant alleles in white individuals with low CYP3A4 enzyme activity. Clin Pharmacol Ther 71:196–204

Geller DH, Auchus RJ, Mendonca BB, Miller WL (1997) The genetic and functional basis of isolated 17,20-lyase deficiency. Nat Genet 17:201–205

Gerber JG, Rhodes RJ, Gal J (2004) Stereoselective metabolism of methadone N-demethylation by cytochrome P4502B6 and 2C19. Chirality 16:36–44

Gilep AA, Sushko TA, Usanov SA (2011) At the crossroads of steroid hormone biosynthesis: the role, substrate specificity and evolutionary development of CYP17. Biochim Biophys Acta 1814:200–209

Goldstein JA, Faletto MB, Romkes-Sparks M, Sullivan T, Kitareewan S, Raucy JL, Lasker JM, Ghanayem BI (1994) Evidence that CYP2C19 is the major (S)-mephenytoin 4′-hydroxylase in humans. Biochemistry 33:1743–1752

Gordon AS, Tabor HK, Johnson AD, Snively BM, Assimes TL, Auer PL, Ioannidis JP, Peters U, Robinson JG, Sucheston LE, Wang D, Sotoodehnia N, Rotter JI, Psaty BM, Jackson RD, Herrington DM, O'Donnell CJ, Reiner AP, Rich SS, Rieder MJ, Bamshad MJ, Nickerson DA, on behalf of the N. G. O. E. S. P. (2014) Quantifying rare, deleterious variation in 12 human cytochrome P450 drug-metabolism genes in a large-scale exome dataset. Hum Mol Genet 23:1957–1963

Haufroid V, Mourad M, van Kerckhove V, Wawrzyniak J, de Meyer M, Eddour DC, Malaise J, Lison D, Squifflet JP, Wallemacq P (2004) The effect of CYP3A5 and MDR1 (ABCB1) polymorphisms on cyclosporine and tacrolimus dose requirements and trough blood levels in stable renal transplant patients. Pharmacogenetics 14:147–154

Haussler MR, Whitfield GK, Kaneko I, Haussler CA, Hsieh D, Hsieh JC, Jurutka PW (2013) Molecular mechanisms of vitamin D action. Calcif Tissue Int 92:77–98

Hesse LM, Venkatakrishnan K, Court MH, von Moltke LL, Duan SX, Shader RI, Greenblatt DJ (2000) CYP2B6 mediates the in vitro hydroxylation of bupropion: potential drug interactions with other antidepressants. Drug Metab Dispos 28:1176–1183

Hesse LM, He P, Krishnaswamy S, Hao Q, Hogan K, Von Moltke LL, Greenblatt DJ, Court MH (2004) Pharmacogenetic determinants of interindividual variability in bupropion hydroxylation by cytochrome P450 2B6 in human liver microsomes. Pharmacogenetics 14:225–238

Hesselink DA, van Schaik RH, van der Heiden IP, van der Werf M, Gregoor PJ, Lindemans J, Weimar W, van Gelder T (2003) Genetic polymorphisms of the CYP3A4, CYP3A5, and MDR-1 genes and pharmacokinetics of the calcineurin inhibitors cyclosporine and tacrolimus. Clin Pharmacol Ther 74:245–254

Hicks JK, Swen JJ, Thorn CF, Sangkuhl K, Kharasch ED, Ellingrod VL, Skaar TC, Muller DJ, Gaedigk A, Stingl JC, Clinical Pharmacogenetics Implementation Consortium (2013) Clinical Pharmacogenetics Implementation Consortium guideline for CYP2D6 and CYP2C19 genotypes and dosing of tricyclic antidepressants. Clin Pharmacol Ther 93:402–408

Holstein A, Plaschke A, Ptak M, Egberts EH, El-Din J, Brockmoller J, Kirchheiner J (2005) Association between CYP2C9 slow metabolizer genotypes and severe hypoglycaemia on medication with sulphonylurea hypoglycaemic agents. Br J Clin Pharmacol 60:103–106

Houston JB, Galetin A (2005) Modelling atypical CYP3A4 kinetics: principles and pragmatism. Arch Biochem Biophys 433:351–360

Howard LA, Ahluwalia JS, Lin SK, Sellers EM, Tyndale RF (2003) CYP2E1*1D regulatory polymorphism: association with alcohol and nicotine dependence. Pharmacogenetics 13:321–328

Hu Y, Hakkola J, Oscarson M, Ingelman-Sundberg M (1999) Structural and functional characterization of the 5′-flanking region of the rat and human cytochrome P450 2E1 genes: identification of a polymorphic repeat in the human gene. Biochem Biophys Res Commun 263:286–293

Ince I, Knibbe CA, Danhof M, de Wildt SN (2013) Developmental changes in the expression and function of cytochrome P450 3A isoforms: evidence from in vitro and in vivo investigations. Clin Pharmacokinet 52:333–345

Jacqz E, Hall SD, Branch RA, Wilkinson GR (1986) Polymorphic metabolism of mephenytoin in man: pharmacokinetic interaction with a co-regulated substrate, mephobarbital. Clin Pharmacol Ther 39:646–653

Jones G, Prosser DE, Kaufmann M (2012) 25-Hydroxyvitamin D-24-hydroxylase (CYP24A1): its important role in the degradation of vitamin D. Arch Biochem Biophys 523:9–18

Jones G, Prosser DE, Kaufmann M (2014) Cytochrome P450-mediated metabolism of vitamin D. J Lipid Res 55:13–31

Kaneko A, Kaneko O, Taleo G, Bjorkman A, Kobayakawa T (1997) High frequencies of CYP2C19 mutations and poor metabolism of proguanil in Vanuatu. Lancet 349:921–922

Katagiri M, Takemori S, Itagaki E, Suhara K, Gomi T (1976) Characterization of purified cytochrome P-450scc and P-450(11)beta from bovine adrenocortical mitochondria. Adv Exp Med Biol 74:281–289

Kelly EJ, Nakano M, Rohatgi P, Yarov-Yarovoy V, Rettie AE (2011) Finding homes for orphan cytochrome P450s: CYP4V2 and CYP4F22 in disease states. Mol Interv 11:124–132

Kidd RS, Curry TB, Gallagher S, Edeki T, Blaisdell J, Goldstein JA (2001) Identification of a null allele of CYP2C9 in an African-American exhibiting toxicity to phenytoin. Pharmacogenetics 11:803–808

Kim D, Guengerich FP (2005) Cytochrome P450 activation of arylamines and heterocyclic amines. Annu Rev Pharmacol Toxicol 45:27–49

Kim JH, Sherman ME, Curriero FC, Guengerich FP, Strickland PT, Sutter TR (2004) Expression of cytochromes P450 1A1 and 1B1 in human lung from smokers, non-smokers, and ex-smokers. Toxicol Appl Pharmacol 199:210–219

Kimura S, Pastewka J, Gelboin HV, Gonzalez FJ (1987) cDNA and amino acid sequences of two members of the human P450IIC gene subfamily. Nucleic Acids Res 15:10053–10054

Kirchheiner J, Schmidt H, Tzvetkov M, Keulen JT, Lotsch J, Roots I, Brockmoller J (2007) Pharmacokinetics of codeine and its metabolite morphine in ultra-rapid metabolizers due to CYP2D6 duplication. Pharmacogenomics J 7:257–265

Klein K, Zanger UM (2013) Pharmacogenomics of cytochrome P450 3A4: recent progress toward the "Missing Heritability" problem. Front Genet 4:12

Ku CS, Pawitan Y, Sim X, Ong RT, Seielstad M, Lee EJ, Teo YY, Chia KS, Salim A (2010) Genomic copy number variations in three Southeast Asian populations. Hum Mutat 31:851–857

Kuehl P, Zhang J, Lin Y, Lamba J, Assem M, Schuetz J, Watkins PB, Daly A, Wrighton SA, Hall SD, Maurel P, Relling M, Brimer C, Yasuda K, Venkataramanan R, Strom S, Thummel K, Boguski MS, Schuetz E (2001) Sequence diversity in CYP3A promoters and characterization of the genetic basis of polymorphic CYP3A5 expression. Nat Genet 27:383–391

Kupfer A, Preisig R (1984) Pharmacogenetics of mephenytoin: a new drug hydroxylation polymorphism in man. Eur J Clin Pharmacol 26:753–759

Lacroix D, Sonnier M, Moncion A, Cheron G, Cresteil T (1997) Expression of CYP3A in the human liver–evidence that the shift between CYP3A7 and CYP3A4 occurs immediately after birth. Eur J Biochem 247:625–634

Lascaratos G, Garway-Heath DF, WilloughbyC E, Chau KY, Schapira AH (2012) Mitochondrial dysfunction in glaucoma: understanding genetic influences. Mitochondrion 12:202–212

Lasky-Su J, Lange N, Brehm JM, Damask A, Soto-Quiros M, Avila L, Celedon JC, Canino G, Cloutier MM, Hollis BW, Weiss ST, Litonjua AA (2012) Genome-wide association analysis of circulating vitamin D levels in children with asthma. Hum Genet 131:1495–1505

Leung TM, Nieto N (2013) CYP2E1 and oxidant stress in alcoholic and non-alcoholic fatty liver disease. J Hepatol 58:395–398

Li A, Jiao X, Munier FL, Schorderet DF, Yao W, Iwata F, Hayakawa M, Kanai A, Shy Chen M, Alan Lewis R, Heckenlively J, Weleber RG, Traboulsi EI, Zhang Q, Xiao X, Kaiser-Kupfer M, Sergeev YV, Hejtmancik JF (2004) Bietti crystalline corneoretinal dystrophy is caused by mutations in the novel gene CYP4V2. Am J Hum Genet 74:817–826

Li N, Zhou Y, Du L, Wei M, Chen X (2011) Overview of cytochrome P450 1B1 gene mutations in patients with primary congenital glaucoma. Exp Eye Res 93:572–579

Li CJ, Li L, Lin L, Jiang HX, Zhong ZY, Li WM, Zhang YJ, Zheng P, Tan XH, Zhou L (2014) Impact of the CYP3A5, CYP3A4, COMT, IL-10 and POR genetic polymorphisms on tacrolimus metabolism in Chinese renal transplant recipients. PLoS One 9:e86206

Libby RT, Smith RS, Savinova OV, Zabaleta A, Martin JE, Gonzalez FJ, John SW (2003) Modification of ocular defects in mouse developmental glaucoma models by tyrosinase. Science 299:1578–1581

Lieber CS (1997) Cytochrome P-4502E1: its physiological and pathological role. Physiol Rev 77:517–544

Lifton RP, Dluhy RG, Powers M, Rich GM, Cook S, Ulick S, Lalouel JM (1992) A chimaeric 11 beta-hydroxylase/aldosterone synthase gene causes glucocorticoid-remediable aldosteronism and human hypertension. Nature (Lond) 355:262–265

Li-Wan-Po A, Girard T, Farndon P, Cooley C, Lithgow J (2010) Pharmacogenetics of CYP2C19: functional and clinical implications of a new variant CYP2C19*17. Br J Clin Pharmacol 69:222–230

Madani S, Paine MF, Lewis L, Thummel KE, Shen DD (1999) Comparison of CYP2D6 content and metoprolol oxidation between microsomes isolated from human livers and small intestines. Pharm Res 16:1199–1205

Mahgoub A, Idle JR, Dring LG, Lancaster R, Smith RL (1977) Polymorphic hydroxylation of debrisoquine in man. Lancet 2:584–586

Mak PJ, Gregory MC, Sligar SG, Kincaid JR (2014) Resonance Raman spectroscopy reveals that substrate structure selectively impacts the heme-bound diatomic ligands of CYP17. Biochemistry 53:90–100

Malaiyandi V, Sellers EM, Tyndale RF (2005) Implications of CYP2A6 genetic variation for smoking behaviors and nicotine dependence. Clin Pharmacol Ther 77:145–158

McCune JS, Risler LJ, Phillips BR, Thummel KE, Blough D, Shen DD (2005) Contribution of CYP3A5 to hepatic and renal ifosfamide N-dechloroethylation. Drug Metab Dispos 33:1074–1081

McGrath JJ, Saha S, Burne TH, Eyles DW (2010) A systematic review of the association between common single nucleotide polymorphisms and 25-hydroxyvitamin D concentrations. J Steroid Biochem Mol Biol 121:471–477

Mega JL, Close SL, Wiviott SD, Shen L, Hockett RD, Brandt JT, Walker JR, Antman EM, Macias W, Braunwald E, Sabatine MS (2009) Cytochrome p-450 polymorphisms and response to clopidogrel. N Engl J Med 360:354–362

Meyer UA (2004) Pharmacogenetics: five decades of therapeutic lessons from genetic diversity. Nat Rev Genet 5:669–676

Miura M, Satoh S, Kagaya H, Saito M, Numakura K, Tsuchiya N, Habuchi T (2011) Impact of the CYP3A4*1G polymorphism and its combination with CYP3A5 genotypes on tacrolimus pharmacokinetics in renal transplant patients. Pharmacogenomics 12:977–984

Mizuochi T, Kimura A, Suzuki M, Ueki I, Takei H, Nittono H, Kakiuchi T, Shigeta T, Sakamoto S, Fukuda A, Nakazawa A, Shimizu T, Kurosawa T, Kasahara M (2011) Successful heterozygous living donor liver transplantation for an oxysterol 7alpha-hydroxylase deficiency in a Japanese patient. Liver Transpl 17:1059–1065

Monte M-J (2009) Bile acids: chemistry, physiology, and pathophysiology. World J Gastroenterol 15:804

Murray GI, Taylor MC, McFadyen MC, McKay JA, Greenlee WF, Burke MD, Melvin WT (1997) Tumor-specific expression of cytochrome P450 CYP1B1. Cancer Res 57:3026–3031

Murray GI, Melvin WT, Greenlee WF, Burke MD (2001) Regulation, function, and tissue-specific expression of cytochrome P450 CYP1B1. Annu Rev Pharmacol Toxicol 41:297–316

Nakamura K, Goto F, Ray WA, McAllister CB, Jacqz E, Wilkinson GR, Branch RA (1985) Interethnic differences in genetic polymorphism of debrisoquin and mephenytoin hydroxylation between Japanese and Caucasian populations. Clin Pharmacol Ther 38:402–408

Nakano M, Kelly EJ, Rettie AE (2009) Expression and characterization of CYP4V2 as a fatty acid omega-hydroxylase. Drug Metab Dispos 37:2119–2122

Nakano M, Kelly EJ, Wiek C, Hanenberg H, Rettie AE (2012) CYP4V2 in Bietti's crystalline dystrophy: ocular localization, metabolism of omega-3-polyunsaturated fatty acids, and functional deficit of the p.H331P variant. Mol Pharmacol 82:679–686

Nelson DR (2003) Comparison of P450s from human and fugu: 420 million years of vertebrate P450 evolution. Arch Biochem Biophys 409:18–24

Nesterova G, Malicdan MC, Yasuda K, Sakaki T, Vilboux T, Ciccone C, Horst R, Huang Y, Golas G, Introne W, Huizing M, Adams D, Boerkoel CF, Collins MT, Gahl WA (2013) 1,25-(OH)2D-24 hydroxylase (CYP24A1) deficiency as a cause of nephrolithiasis. Clin J Am Soc Nephrol 8:649–657

Nimkarn S, New MI (2010) Congenital adrenal hyperplasia due to 21-hydroxylase deficiency: a paradigm for prenatal diagnosis and treatment. Ann N Y Acad Sci 1192:5–11

Okamoto M, Nonaka Y (1992) Molecular biology of rat steroid 11 beta-hydroxylase [P450 (11 beta)] and aldosterone synthase [P450(11 beta, aldo)]. J Steroid Biochem Mol Biol 41:415–419

Omura T, Sato R (1964) The carbon monoxide-binding pigment of liver microsomes. II. Solubilization, purification, and properties. J Biol Chem 239:2379–2385

Oscarson M, McLellan RA, Asp V, Ledesma M, Bernal Ruiz ML, Sinues B, Rautio A, Ingelman-Sundberg M (2002) Characterization of a novel CYP2A7/CYP2A6 hybrid allele (CYP2A6*12) that causes reduced CYP2A6 activity. Hum Mutat 20:275–283

Otto S, Marcus C, Pidgeon C, Jefcoate C (1991) A novel adrenocorticotropin-inducible cytochrome P450 from rat adrenal microsomes catalyzes polycyclic aromatic hydrocarbon metabolism. Endocrinology 129:970–982

Pang S, Murphey W, Levine LS, Spence DA, Leon A, Lafranchi S, Surve AS, New MI (1982) A pilot newborn screening for congenital adrenal hyperplasia in Alaska. J Clin Endocrinol Metab 55:413–420

Pascoe L, Curnow KM, Slutsker L, Connell JM, Speiser PW, New MI, White PC (1992) Glucocorticoid-suppressible hyperaldosteronism results from hybrid genes created by unequal crossovers between CYP11B1 and CYP11B2. Proc Natl Acad Sci U S A 89:8327–8331

Peng H, Xie SK, Huang MJ, Ren DL (2013) Associations of CYP2E1 rs2031920 and rs3813867 polymorphisms with colorectal cancer risk: a systemic review and meta-analysis. Tumour Biol 34:2389–2395

Perera V, Gross AS, McLachlan AJ (2012) Measurement of CYP1A2 activity: a focus on caffeine as a probe. Curr Drug Metab 13:667–678

Pikuleva IA, Waterman MR (2013) Cytochromes p450: roles in diseases. J Biol Chem 288:17091–17098

Pottenger LH, Christou M, Jefcoate CR (1991) Purification and immunological characterization of a novel cytochrome P450 from C3H/10T1/2 cells. Arch Biochem Biophys 286:488–497

Pullinger CR, Eng C, Salen G, Shefer S, Batta AK, Erickson SK, Verhagen A, Rivera CR, Mulvihill SJ, Malloy MJ, Kane JP (2002) Human cholesterol 7α-hydroxylase (CYP7A1) deficiency has a hypercholesterolemic phenotype. J Clin Invest 110:109–117

Ragia G, Petridis I, Tavridou A, Christakidis D, Manolopoulos VG (2009) Presence of CYP2C9*3 allele increases risk for hypoglycemia in type 2 diabetic patients treated with sulfonylureas. Pharmacogenomics 10:1781–1787

Rahmioglu N, Heaton J, Clement G, Gill R, Surdulescu G, Zlobecka K, Hodgkiss D, Ma Y, Hider RC, Smith NW, Ahmadi KR (2011) Genetic epidemiology of induced CYP3A4 activity. Pharmacogenet Genomics 21:642–651

Rebbeck TR, Jaffe JM, Walker AH, Wein AJ, Malkowicz SB (1998) Modification of clinical presentation of prostate tumors by a novel genetic variant in CYP3A4. J Natl Cancer Inst 90:1225–1229

Relling MV, Aoyama T, Gonzalez FJ, Meyer UA (1990) Tolbutamide and mephenytoin hydroxylation by human cytochrome P450s in the CYP2C subfamily. J Pharmacol Exp Ther 252:442–447

Rettie AE, Jones JP (2005) Clinical and toxicological relevance of CYP2C9: drug–drug interactions and pharmacogenetics. Annu Rev Pharmacol Toxicol 45:477–494

Rettie AE, Tai G (2006) The pharmocogenomics of warfarin: closing in on personalized medicine. Mol Interv 6:223–227

Rettie AE, Korzekwa KR, Kunze KL, Lawrence RF, Eddy AC, Aoyama T, Gelboin HV, Gonzalez FJ, Trager WF (1992) Hydroxylation of warfarin by human cDNA-expressed cytochrome P-450: a role for P-4502C9 in the etiology of (S)-warfarin-drug interactions. Chem Res Toxicol 5:54–59

Richter T, Murdter TE, Heinkele G, Pleiss J, Tatzel S, Schwab M, Eichelbaum M, Zanger UM (2004) Potent mechanism-based inhibition of human CYP2B6 by clopidogrel and ticlopidine. J Pharmacol Exp Ther 308:189–197

Roberts BJ, Song BJ, Soh Y, Park SS, Shoaf SE (1995) Ethanol induces CYP2E1 by protein stabilization. Role of ubiquitin conjugation in the rapid degradation of CYP2E1. J Biol Chem 270:29632–29635

Rose K, Allan A, Gauldie S, Stapleton G, Dobbie L, Dott K, Martin C, Wang L, Hedlund E, Seckl JR, Gustafsson JA, Lathe R (2001) Neurosteroid hydroxylase CYP7B: vivid reporter activity in dentate gyrus of gene-targeted mice and abolition of a widespread pathway of steroid and oxysterol hydroxylation. J Biol Chem 276:23937–23944

Ryan KJ, Engel LL (1957) Hydroxylation of steroids at carbon 21. J Biol Chem 225:103–114

Sauer JM, Ring BJ, Witcher JW (2005) Clinical pharmacokinetics of atomoxetine. Clin Pharmacokinet 44:571–590

Schlingmann KP, Kaufmann M, Weber S, Irwin A, Goos C, John U, Misselwitz J, Klaus G, Kuwertz-Broking E, Fehrenbach H, Wingen AM, Guran T, Hoenderop JG, Bindels RJ, Prosser DE, Jones G, Konrad M (2011) Mutations in CYP24A1 and idiopathic infantile hypercalcemia. N Engl J Med 365:410–421

Schuetz JD, Kauma S, Guzelian PS (1993) Identification of the fetal liver cytochrome CYP3A7 in human endometrium and placenta. J Clin Invest 92:1018–1024

Schwarz M, Lund EG, Setchell KD, Kayden HJ, Zerwekh JE, Bjorkhem I, Herz J, Russell DW (1996) Disruption of cholesterol 7alpha-hydroxylase gene in mice. II. Bile acid deficiency is overcome by induction of oxysterol 7alpha-hydroxylase. J Biol Chem 271:18024–18031

Setchell KD, Schwarz M, O'Connell NC, Lund EG, Davis DL, Lathe R, Thompson HR, Tyson WR, Sokol RJ, Russell DW (1998) Identification of a new inborn error in bile acid synthesis: mutation of the oxysterol 7alpha-hydroxylase gene causes severe neonatal liver disease. J Clin Invest 102:1690–1703

Shear NH, Spielberg SP, Grant DM, Tang BK, Kalow W (1986) Differences in metabolism of sulfonamides predisposing to idiosyncratic toxicity. Ann Intern Med 105:179–184

Shinkyo R, Sakaki T, Kamakura M, Ohta M, Inouye K (2004) Metabolism of vitamin D by human microsomal CYP2R1. Biochem Biophys Res Commun 324:451–457

Shuldiner AR, O'Connell JR, Bliden KP, Gandhi A, Ryan K, Horenstein RB, Damcott CM, Pakyz R, Tantry US, Gibson Q, Pollin TI, Post W, Parsa A, Mitchell BD, Faraday N, Herzog W, Gurbel PA (2009) Association of cytochrome P450 2C19 genotype with the antiplatelet effect and clinical efficacy of clopidogrel therapy. JAMA 302:849–857

Signorello LB, Shi J, Cai Q, Zheng W, Williams SM, Long J, Cohen SS, Li G, Hollis BW, Smith JR, Blot WJ (2011) Common variation in vitamin D pathway genes predicts circulating 25-hydroxyvitamin D levels among African Americans. PLoS One 6:e28623

Sim SC, Risinger C, Dahl ML, Aklillu E, Christensen M, Bertilsson L, Ingelman-Sundberg M (2006) A common novel CYP2C19 gene variant causes ultrarapid drug metabolism relevant for the drug response to proton pump inhibitors and antidepressants. Clin Pharmacol Ther 79:103–113

Smit P, van Schaik RH, van der Werf M, van den Beld AW, Koper JW, Lindemans J, Pols HA, Brinkmann AO, de Jong FH, Lamberts SW (2005) A common polymorphism in the CYP3A7 gene is associated with a nearly 50% reduction in serum dehydroepiandrosterone sulfate levels. J Clin Endocrinol Metab 90:5313–5316

Speiser PW, White PC (2003) Congenital adrenal hyperplasia. N Engl J Med 349:776–788
Stevens JC, Hines RN, Gu C, Koukouritaki SB, Manro JR, Tandler PJ, Zaya MJ (2003) Developmental expression of the major human hepatic CYP3A enzymes. J Pharmacol Exp Ther 307:573–582
Steward DJ, Haining RL, Henne KR, Davis G, Rushmore TH, Trager WF, Rettie AE (1997) Genetic association between sensitivity to warfarin and expression of CYP2C9*3. Pharmacogenetics 7:361–367
Stiles AR, McDonald JG, Bauman DR, Russell DW (2009) CYP7B1: one cytochrome P450, two human genetic diseases, and multiple physiological functions. J Biol Chem 284:28485–28489
Stoilov I, Akarsu AN, Sarfarazi M (1997) Identification of three different truncating mutations in cytochrome P4501B1 (CYP1B1) as the principal cause of primary congenital glaucoma (buphthalmos) in families linked to the GLC3A locus on chromosome 2p21. Hum Mol Genet 6:641–647
Strushkevich N, Usanov SA, Plotnikov AN, Jones G, Park HW (2008) Structural analysis of CYP2R1 in complex with vitamin D3. J Mol Biol 380:95–106
Sutter TR, Tang YM, Hayes CL, Wo YY, Jabs EW, Li X, Yin H, Cody CW, Greenlee WF (1994) Complete cDNA sequence of a human dioxin-inducible mRNA identifies a new gene subfamily of cytochrome P450 that maps to chromosome 2. J Biol Chem 269:13092–13099
Takahashi T, Lasker JM, Rosman AS, Lieber CS (1993) Induction of cytochrome P-4502E1 in the human liver by ethanol is caused by a corresponding increase in encoding messenger RNA. Hepatology 17:236–245
Teh LK, Bertilsson L (2012) Pharmacogenomics of CYP2D6: molecular genetics, interethnic differences and clinical importance. Drug Metab Pharmacokinet 27:55–67
Thervet E, Anglicheau D, King B, Schlageter MH, Cassinat B, Beaune P, Legendre C, Daly AK (2003) Impact of cytochrome p450 3A5 genetic polymorphism on tacrolimus doses and concentration-to-dose ratio in renal transplant recipients. Transplantation 76:1233–1235
Thorgeirsson TE, Gudbjartsson DF, Surakka I, Vink JM, Amin N, Geller F, Sulem P, Rafnar T, Esko T, Walter S, Gieger C, Rawal R, Mangino M, Prokopenko I, Magi R, Keskitalo K, Gudjonsdottir IH, Gretarsdottir S, Stefansson H, Thompson JR, Aulchenko YS, Nelis M, Aben KK, Den Heijer M, Dirksen A, Ashraf H, Soranzo N, Valdes AM, Steves C, Uitterlinden AG, Hofman A, Tonjes A, Kovacs P, Hottenga JJ, Willemsen G, Vogelzangs N, Doring A, Dahmen N, Nitz B, Pergadia ML, Saez B, de Diego V, Lezcano V, Garcia-Prats MD, Ripatti S, Perola M, Kettunen J, Hartikainen AL, Pouta A, Laitinen J, Isohanni M, Huei-Yi S, Allen M, Krestyaninova M, Hall AS, Jones GT, van Rij AM, Mueller T, Dieplinger B, Haltmayer M, Jonsson S, Matthiasson SE, Oskarsson H, Tyrfingsson T, Kiemeney LA, Mayordomo JI, Lindholt JS, Pedersen JH, Franklin WA, Wolf H, Montgomery GW, Heath AC, Martin NG, Madden PA, Giegling I, Rujescu D, Jarvelin MR, Salomaa V, Stumvoll M, Spector TD, Wichmann HE, Metspalu A, Samani NJ, Penninx BW, Oostra BA, Boomsma DI, Tiemeier H, van Duijn CM, Kaprio J, Gulcher JR, Consortium E, McCarthy MI, Peltonen L, Thorsteinsdottir U, Stefansson K (2010) Sequence variants at CHRNB3-CHRNA6 and CYP2A6 affect smoking behavior. Nat Genet 42:448–453
Totah RA, Sheffels P, Roberts T, Whittington D, Thummel K, Kharasch ED (2008) Role of CYP2B6 in stereoselective human methadone metabolism. Anesthesiology 108:363–374
Turpeinen M, Zanger UM (2012) Cytochrome P450 2B6: function, genetics, and clinical relevance. Drug Metabol Drug Interact 27:185–197
Ueki I, Kimura A, Nishiyori A, Chen HL, Takei H, Nittono H, Kurosawa T (2008) Neonatal cholestatic liver disease in an Asian patient with a homozygous mutation in the oxysterol 7alpha-hydroxylase gene. J Pediatr Gastroenterol Nutr 46:465–469
Vasiliou V, Gonzalez FJ (2008) Role of CYP1B1 in glaucoma. Annu Rev Pharmacol Toxicol 48:333–358
Vatsis KP, Martell KJ, Weber WW (1991) Diverse point mutations in the human gene for polymorphic *N*-acetyltransferase. Proc Natl Acad Sci USA 88:6333–6337

Veronese ME, Mackenzie PI, Doecke CJ, McManus ME, Miners JO, Birkett DJ (1991) Tolbutamide and phenytoin hydroxylations by cDNA-expressed human liver cytochrome P4502C9. Biochem Biophys Res Commun 175:1112–1118

Wang JT, Lin CJ, Burridge SM, Fu GK, Labuda M, Portale AA, Miller WL (1998) Genetics of vitamin D 1-alpha-hydroxylase deficiency in 17 families. Am J Hum Genet 63:1694–1702

Wang TJ, Zhang F, Richards JB, Kestenbaum B, van Meurs JB, Berry D, Kiel DP, Streeten EA, Ohlsson C, Koller DL, Peltonen L, Cooper JD, O'Reilly PF, Houston DK, Glazer NL, Vandenput L, Peacock M, Shi J, Rivadeneira F, McCarthy MI, Anneli P, de Boer IH, Mangino M, Kato B, Smyth DJ, Booth SL, Jacques PF, Burke GL, Goodarzi M, Cheung CL, Wolf M, Rice K, Goltzman D, Hidiroglou N, Ladouceur M, Wareham NJ, Hocking LJ, Hart D, Arden NK, Cooper C, Malik S, Fraser WD, Hartikainen AL, Zhai G, Macdonald HM, Forouhi NG, Loos RJ, Reid DM, Hakim A, Dennison E, Liu Y, Power C, Stevens HE, Jaana L, Vasan RS, Soranzo N, Bojunga J, Psaty BM, Lorentzon M, Foroud T, Harris TB, Hofman A, Jansson JO, Cauley JA, Uitterlinden AG, Gibson Q, Jarvelin MR, Karasik D, Siscovick DS, Econs MJ, Kritchevsky SB, Florez JC, Todd JA, Dupuis J, Hypponen E, Spector TD (2010) Common genetic determinants of vitamin D insufficiency: a genome-wide association study. Lancet 376:180–188

Wang D, Guo Y, Wrighton SA, Cooke GE, Sadee W (2011) Intronic polymorphism in CYP3A4 affects hepatic expression and response to statin drugs. Pharmacogenomics J 11:274–286

Ward SA, Helsby NA, Skjelbo E, Brosen K, Gram LF, Breckenridge AM (1991) The activation of the biguanide antimalarial proguanil co-segregates with the mephenytoin oxidation polymorphism–a panel study. Br J Clin Pharmacol 31:689–692

Ward BA, Gorski JC, Jones DR, Hall SD, Flockhart DA, Desta Z (2003) The cytochrome P450 2B6 (CYP2B6) is the main catalyst of efavirenz primary and secondary metabolism: implication for HIV/AIDS therapy and utility of efavirenz as a substrate marker of CYP2B6 catalytic activity. J Pharmacol Exp Ther 306:287–300

Wassenaar CA, Dong Q, Wei Q, Amos CI, Spitz MR, Tyndale RF (2011) Relationship between CYP2A6 and CHRNA5-CHRNA3-CHRNB4 variation and smoking behaviors and lung cancer risk. J Natl Cancer Inst 103:1342–1346

Wedlund PJ, Aslanian WS, McAllister CB, Wilkinson GR, Branch RA (1984) Mephenytoin hydroxylation deficiency in Caucasians: frequency of a new oxidative drug metabolism polymorphism. Clin Pharmacol Ther 36:773–780

Westlind A, Lofberg L, Tindberg N, Andersson TB, Ingelman-Sundberg M (1999) Interindividual differences in hepatic expression of CYP3A4: relationship to genetic polymorphism in the 5′-upstream regulatory region. Biochem Biophys Res Commun 259:201–205

Westlind-Johnsson A, Hermann R, Huennemeyer A, Hauns B, Lahu G, Nassr N, Zech K, Ingelman-Sundberg M, Von Richter O (2006) Identification and characterization of CYP3A4*20, a novel rare CYP3A4 allele without functional activity. Clin Pharmacol Ther 79:339–349

White PC, Obeid J, Agarwal AK, Tannin GM, Nikkila H (1994) Genetic analysis of 11 beta-hydroxysteroid dehydrogenase. Steroids 59:111–115

Williams PA, Cosme J, Vinkovic DM, Ward A, Angove HC, Day PJ, Vonrhein C, Tickle IJ, Jhoti H (2004) Crystal structures of human cytochrome P450 3A4 bound to metyrapone and progesterone. Science 305:683–686

Wilson RC, Nimkarn S, Dumic M, Obeid J, Azar MR, Najmabadi H, Saffari F, New MI (2007) Ethnic-specific distribution of mutations in 716 patients with congenital adrenal hyperplasia owing to 21-hydroxylase deficiency. Mol Genet Metab 90:414–421

Wojnowski L, Turner PC, Pedersen B, Hustert E, Brockmoller J, Mendy M, Whittle HC, Kirk G, Wild CP (2004) Increased levels of aflatoxin-albumin adducts are associated with CYP3A5 polymorphisms in The Gambia, West Africa. Pharmacogenetics 14:691–700

Wrighton SA, Ring BJ, Watkins PB, Vandenbranden M (1989) Identification of a polymorphically expressed member of the human cytochrome P-450III family. Mol Pharmacol 36:97–105

Wrighton SA, Stevens JC, Becker GW, Vandenbranden M (1993) Isolation and characterization of human liver cytochrome P450 2C19: correlation between 2C19 and *S*-mephenytoin 4′-hydroxylation. Arch Biochem Biophys 306:240–245

Xu Z, Chen W, Merke DP, McDonnell NB (2013) Comprehensive mutation analysis of the CYP21A2 gene: an efficient multistep approach to the molecular diagnosis of congenital adrenal hyperplasia. J Mol Diagn 15:745–753

Yamano S, Tatsuno J, Gonzalez FJ (1990) The CYP2A3 gene product catalyzes coumarin 7-hydroxylation in human liver microsomes. Biochemistry 29:1322–1329

Yano JK, Hsu MH, Griffin KJ, Stout CD, Johnson EF (2005) Structures of human microsomal cytochrome P450 2A6 complexed with coumarin and methoxsalen. Nat Struct Mol Biol 12:822–823

Yokota H, Tamura S, Furuya H, Kimura S, Watanabe M, Kanazawa I, Kondo I, Gonzalez FJ (1993) Evidence for a new variant CYP2D6 allele CYP2D6J in a Japanese population associated with lower in vivo rates of sparteine metabolism. Pharmacogenetics 3:256–263

Yun CH, Shimada T, Guengerich FP (1991) Purification and characterization of human liver microsomal cytochrome P-450 2A6. Mol Pharmacol 40:679–685

Zanger UM, Klein K (2013) Pharmacogenetics of cytochrome P450 2B6 (CYP2B6): advances on polymorphisms, mechanisms, and clinical relevance. Front Genet 4:24

Zanger UM, Klein K, Saussele T, Blievernicht J, Hofmann MH, Schwab M (2007) Polymorphic CYP2B6: molecular mechanisms and emerging clinical significance. Pharmacogenomics 8:743–759

Zhang W, Yuan JJ, Kan QC, Zhang LR, Chang YZ, Wang ZY, Li ZS (2011) Influence of CYP3A5*3 polymorphism and interaction between CYP3A5*3 and CYP3A4*1G polymorphisms on post-operative fentanyl analgesia in Chinese patients undergoing gynaecological surgery. Eur J Anaesthesiol 28:245–250

Zhao B, Lei L, Kagawa N, Sundaramoorthy M, Banerjee S, Nagy LD, Guengerich FP, Waterman MR (2012) Three-dimensional structure of steroid 21-hydroxylase (cytochrome P450 21A2) with two substrates reveals locations of disease-associated variants. J Biol Chem 287:10613–10622

Zhou SF, Wang B, Yang LP, Liu JP (2010) Structure, function, regulation and polymorphism and the clinical significance of human cytochrome P450 1A2. Drug Metab Rev 42:268–354

Zhou LP, Yao F, Luan H, Wang YL, Dong XH, Zhou WW, Wang QH (2013) CYP3A4*1B polymorphism and cancer risk: a HuGE review and meta-analysis. Tumour Biol 34:649–660

Zhu JG, Ochalek JT, Kaufmann M, Jones G, Deluca HF (2013) CYP2R1 is a major, but not exclusive, contributor to 25-hydroxyvitamin D production in vivo. Proc Natl Acad Sci USA 110:15650–15655

Zineh I, Pacanowski M, Woodcock J (2013) Pharmacogenetics and coumarin dosing–recalibrating expectations. N Engl J Med 369:2273–2275

Epilogue

This is a book that deals with all aspects of cytochrome P450s in a single volume. There are descriptions of almost all that is known about P450s in fields such as medicine, agriculture, and biotechnology. I believe that such an understanding of the close relationship between P450s and humans will increase awareness of the importance of P450s. The field of P450 research has its roots in studies of metabolism, but the true beginning was in 1962 with the publication of the communication with Drs. Tsuneo Omura and Ryo Sato in *The Journal of Biological Chemistry* (J. Biol. Chem. 237, PC1375–PC1376). The contributions of the pioneers who made highly important contributions to the discovery of P450 and the elucidation of its molecular and functional properties in the early years are summarized by Dr. Omura, for about two decades from the beginning of the research on this unique hemoprotein, focusing on biochemical aspects of P450 research.

This book is a kind of fruit of the 50th Anniversary Symposium on Cytochrome P450 in Fukuoka, chaired by Dr. Yoshitsugu Shiro, RIKEN SPring-8 Center, Japan (see Fig. 1 in the Preface). I was involved in the organizing committee as a co-chair along with Dr. Masahiko Negishi, NIEHS, USA, and Dr. Toshiyuki Sakaki, Toyama Prefectural University, Japan, my co-chairs. Drs. Yuri Aoyama, Soka University; Hirofumi Ichinose, Kyushu University; and Shingo Nagano, Tottori University, Japan, worked very hard as important committee members to make this symposium successful. I am greatly pleased to introduce the special letter from the editorial office of *The Journal of Biological Chemistry* to Dr. Omura, as shown here (Fig. 1), which was kindly provided by Dr. Fred Peter Guengerich, a current associate editor of the journal and the author of the second chapter of this book. Issues addressed in the book include biochemistry, biophysics, and gene regulation as well as new applications of P450 science to today's practical problems, as P450 has become a more mature field. I appreciated very much all their great efforts for this special occasion celebrating the 50th anniversary of the first publication on P450.

After editing this book, I thank the contributors of all the chapters for introducing valuable history and cutting-edge knowledge on P450s from various fields. In addition, I hope that our readers will discover a new side to P450s and broaden

H. Yamazaki (ed.), *Fifty Years of Cytochrome P450 Research*,
DOI 10.1007/978-4-431-54992-5, © Springer Japan 2014

ASBMB

The Journal of Biological Chemistry
11200 Rockville Pike
Rockville, MD 20852-3110
USA

Published by American Society for Biochemistry and Molecular Biology

MARTHA J. FEDOR
Editor-in-Chief

November 23, 2012

Dr. Tsuneo Omura
Professor Emeritus
Department of Molecular Biology
Kyushu University Medical School
Higashi-ku, Fukuoka 812-82
JAPAN

Dear Dr. Omura:

The field of cytochrome P450 research had its roots in studies of the metabolism of steroids, drugs, and carcinogens but the true beginning was in 1962 with the publication of your Communication with Professor Ryo Sato in *The Journal of Biological Chemistry* (*J. Biol. Chem.* **237**, PC1375-PC1376). We congratulate you and other researchers in the cytochrome P450 field on the 50th anniversary of this seminal work.

Your 1962 Communication was followed by the two classic Omura and Sato papers in 1964 (*J. Biol. Chem.* **239**, 2370-2378 and 2379-2385), which were featured in a Classic many years later (*J. Biol. Chem.* **281**, e15). More than one hundred prominent cytochrome P450 papers were published in a 2-issue electronic and print set in 2009 (http://www.jbc.org/site/collections/p450/).

While congratulating you and your fellow cytochrome P450 scientists on this important anniversary, we also thank you for publishing so much of the best work in *The Journal of Biological Chemistry*. We are grateful for your interest in our journal, and we consider it a privilege to be able to advance the field of cytochrome P450 research in our journal.

Of course, cytochrome P450 research has developed tremendously since your early studies with rat liver microsomes. The success of the research has had tremendous implications in fields such as medicine, agricultures, and biotechnology. We wish all of you in the field continued success, and *The Journal of Biological Chemistry* continues to publish original papers and reviews in cytochrome P450. We have truly enjoyed this symbiotic relationship.

Again, congratulations and our best wishes for continued success.

Sincerely yours,

Marty

Martha J. Fedor, Editor-in-Chief

Herbert Tabor

Herbert Tabor, Co-Editor (Editor 1969-2010)

Tel: 240-283-6620 • Fax: 301-881-2573 • E-mail: mfedor@asbmb.org

Fig. 1 A letter with congratulatory remarks from the editorial office of *The Journal of Biological Chemistry* to Dr. Tsuneo Omura and P450 researchers

their outlook on this P450 research area that has such tremendous implications. Finally, I fervently hope, with all sincerity, that this P450 research will continue and contribute to our health and lives.

Hiroshi Yamazaki

Index

H. Yamazaki (ed.), *Fifty Years of Cytochrome P450 Research*,
DOI 10.1007/978-4-431-54992-5, © Springer Japan 2014

CPSIA information can be obtained at www.ICGtesting.com
Printed in the USA
LVOW01*0426200814

399909LV00001B/47/P

9 784431 549918